Jan Schwarzbauer

**Organic Contaminants in Riverine and Groundwater Systems**

Jan Schwarzbauer

# Organic Contaminants in Riverine and Groundwater Systems

## Aspects of the Anthropogenic Contribution

With 100 Figures

Springer

AUTHOR:

DR. JAN SCHWARZBAUER
LEHRSTUHL FÜR GEOLOGIE,
GEOCHEMIE UND LAGERSTÄTTEN
DES ERDÖLS UND DER KOHLE
RWTH AACHEN
LOCHNERSTR. 4-20
52056 AACHEN
GERMANY

E-mail: schwarzbauer@lek.rwth-aachen.de

ISBN 10 3-540-31169-6 **Springer Berlin Heidelberg New York**
ISBN 13 978-3-540-31169-0 **Springer Berlin Heidelberg New York**

Library of Congress Control Number: 2006922188



**Springer is a part of Springer Science+Business Media**
springeronline.com

Printed in The Netherlands



Cover design: E. Kirchner, Heidelberg
Production: A. Oelschläger
Typesetting: Camera-ready by the Author

Printed on acid-free paper 30/2132/AO 543210

# Preface

It is known for milleniums, that water is forming the principle base for the total biosphere including human beings (*'Thales [of Milet] says that it [the nature of things] is water' Aristoteles, Metaph. 983 b20*), Nevertheless, basically the anthropogenic activities and human requirements, in particular our enormous consume of chemicals, are affecting worldwide the aquatic environment on a rapidly increasing level. Consequently, huge scientific efforts, worldwide, focussed intensively on increasing the knowlegde about the state of pollution and on the most important processes harming the aquatic systems. The overall aim is the rehabilitation and protection of this unique and essential compartment.

Environmental studies on riverine and groundwater systems are base research activities at the Institute of Geology and Geochemistry of Petroleum and Coal, RWTH Aachen University. This book summarizes the analytical and organic-geochemical work performed between 1998 and 2003 at this institute. All the individual studies and the corresponding results were separately published in various scientific journals. However, I felt the substantial need to combine these individual parts in order to reveal a comprehensive overview on all the different and novel aspects concerning the organic contamination of rivers and groundwater.

I wish here to thank all my coworkers (senior scientists, PhD-students, diploma students and the technical staff of the *Laboratory for organic-geochemical Analysis*, RWTH Aachen University) which were deeply involved in all the environmental investigations presented in this book. In detail these are especially (in alphabetical order): Sabine Brincker, Larissa Dsikowitzky, Yvonne Esser, Stephan Franke, Sabine Heim, Alexander Kronimus, Mathias Ricking and Gero Vinzelberg. In particular, I appreciate the support and contributions of Ralf Littke, who enabled me the realisation of the environmental projects at his institute.

Jan Schwarzbauer Aachen, November 2005

# Contents

## List of abbreviations

| | |
|---|---|
| AAS | Atomic absorption spectroscopy |
| ABS | Tetrapropylene-based alkylbenzenesulfonates |
| ADBI | Celestolide |
| AHMI | Pantolide |
| AHTN | Tonalide |
| AOX | Adsorbable organic halogens |
| ASPE | Alkylsulfonic acid esters |
| ATTI | Traseolide |
| BHA | *tert*-Butylhydroxyanisole |
| BHT | Ionol, 2,6-Di-*tert*-butyl-4-methylphenol |
| BOD | Biological oxygen demand |
| CN | Chloronaphthalene |
| COD | Chemical oxygen demand |
| DBP | Dichlorobenzophenone |
| DBT | Dibutyl tin |
| DCM | Dichloromethane |
| DCN | Dichloronaphthalenes |
| DDA | Bis(chlorophenyl)acetic acid |
| DDCN | 2,2-bis(chlorophenyl)acetonitril |
| DDE | 2,2-Bis(chlorophenyl)-1,1-dichloroethene |
| DDEt | 1,1-Bis(chlorophenyl)ethane |
| DDM | Bis(chlorophenyl)methane |
| DDMU | 2,2-Bis(chlorophenyl)-1-chloroethene |
| DDMS | 2,2-Bis(chlorophenyl)-1-chloroethane |
| DDNU | 1,1-Bis(chlorophenyl)ethene |
| DDOH | 2,2-Bis(chlorophenyl)-ethanol |
| DDS | Bis(chlorophenyl)sulfide |
| DDT | 2,2-Bis(chlorophenyl)-1,1,1-trichloroethane |
| DEET | N,N-Diethyltoluamide |
| DEHP | Bis(2-ethylhexyl)phthalate |

| | |
|---|---|
| DiBP | Di-*iso*-butylphthalate |
| DnBP | Di-*n*-butylphthalate |
| DOC | Dissolved organic carbon |
| DPMI | Cashmeran |
| | |
| EB | Eckman-Birge-grab sampler |
| ECD | Electron capture detector |
| EDTA | Ethylenediaminetetraacetic acid |
| $EI^+$ | Postive electron impact ionisation |
| EPA | Environmental protection agency (USA) |
| EPTC | S-Ethyldipropylcarbamate |
| ESR | Electron spin resonance (spectroscopy) |
| EU | European union |
| | |
| FCKW | Fluorchlorkohlenwasserstoffe, Fluorochlorohydrocarbons |
| FID | Flame ionization detector |
| | |
| GC | Gas chromatograph, gas chromatography |
| GC/irmMS | Gas chromatography / isotope ratio monitoring mass spectrometry |
| GC/MS | Gas chromatography / mass spectrometry |
| GDR | German Democratic republic |
| | |
| HCB | Hexachlorobenzene |
| HCBD, HCBu | Hexachloro-1,3-butadiene |
| HCH | Hexachlorocyclohexanes |
| HGA | Heated graphite furnace |
| HHCB | Galaxolide |
| HMN | 2,2,4,4,6,8,8-Heptamethylnonane |
| | |
| ICP | Inductively coupled plasma |
| IKSE | Internationale Kommission zum Schutz der Elbe, International Elbe river protection agency |
| IKSR | Internationale Kommission zum Schutz des Rheins, International Rhine river protection agency |
| IR | Infrared (spectroscopy) |
| irmMS | Isotope ratio monitoring mass spectrometer, isotope ratio monitoring mass spectrometry |

| | |
|---|---|
| $L_{org}$ | organic load |
| LAB | Linear alkylbenzenes |
| LAS | Linear alkylbenzenesulfonates |
| LC | Liquid chromatography |
| LOD | Limit of detection |
| LOQ | Limit of quantitation |
| | |
| MBT | Monobutyl tin |
| MCPP | Mecoprop |
| MDA | Bis(methoxyphenyl)acetic acid |
| MDB | Dimethoxybenzophenone |
| MDD | 2,2-Bis(methoxyphenyl)-1,1-dichloroethane |
| MDE | 2,2-Bis(methoxyphenyl)-1,1-dichloroethene |
| MDT | 2,2-bis(methoxyphenyl)-1,1,1-trichloroethane |
| MK | Musk ketone |
| MOM | Macromolecular organic matter |
| MQ | Median river runoff |
| MS | Mass spectrometer, mass spectrometry |
| MSTFA | N-Methyl-N-(trimethylsilyl)trifluoroacetamide |
| MTB | 2-Methylthiobenzothiazole |
| MX | Musk xylene |
| | |
| NBBS | N-Butylbenzenesulfonamide |
| NF | Liquid nitrogen deep freeze method |
| NFP | N-Formylpiperidine |
| NIST | National institute of standards (USA) |
| NIH | National institute of health (USA) |
| NMR | Nuclear magnetic resonance (spectroscopy) |
| NP | Nonylphenols |
| NTA | Nitrilotriacetic acid |
| | |
| OCS | Octachlorostyrene |
| OES | Optical emission spectroscopy |
| | |
| PA | Polyamide |
| PAC | Polycyclic aromatic compounds |
| PAH | Polycyclic aromatic hydrocarbons |
| PBDE | Polybrominated diphenylethers |
| PBN | Polybrominated naphthalenes |
| PCB | Polychlorinated biphenyls |
| PCDBT | Polychlorinated dibenzothiophens |
| PCDD | Polychlorinated dibenzo-*p*-dioxines |

| | |
|---|---|
| PCDE | Polychlorinated diphenylethers |
| PCDF | Polychlorinated dibenzofuranes |
| PCN | Polychlorinated naphthalenes |
| PCT | Polychlorinated terphenyls |
| PCTPM | Polychlorinated triphenylmethanes |
| PER | Tetrachloroethylene |
| PMF | Polycyclic synthetic musk substitutes |
| POM | Particulate organic matter |
| POP | Persistent organic pollutant |
| PVC | Polyvinyl chloride |
| RIC | Reconstructed ion chromatogram |
| SPE | Solid phase extraction |
| STP | Sewage treatment plant |
| $T_m$ | Trisnorhopane |
| $T_s$ | Trisnorneohopane |
| TAB | Tetrapropylene-based alkylbenzenes |
| TAED | N,N,N`,N`-Tetraacetyl ethylenediamine |
| TBP | Tributyl phosphates |
| TBT | Tributyl tin |
| TCBT | Tetrachlorobenzyltoluenes |
| TCEP | Tris(2-chloroethyl)phosphate |
| TeBT | Tetrabutyl tin |
| TIC | Total ion chromatogram |
| TMDD | 2,4,7,9-Tetramethyl-5-decyne-4,7-diol |
| TMSH | Trimethylsulfoniumhydroxide |
| TOC | Total organic carbon |
| TOM | Terrestrial organic matter |
| TPDB, TXIB | 2,4,4-Trimethylpentane-1,3-dioldi-*iso*-butyrate |
| TTP | Tritolyl phosphates |
| USA | United States of America |
| UV/VIS | Ultraviolet / visible (spectroscopy) |
| VOC | Volatile organic compounds |
| VPDB | Vienna Pee Dee Belemnite |
| WFD | Water Framework Directive |

# 1 Introduction

## 1.1 Riverine and groundwater systems as priority environmental compartments

Water is the most fundamental substance for life. In particular available fresh water (groundwater, lake and river water) usable as drinking water is the most important aquatic resource with respect to the sustainment of terrestrial organism. Consequently, human beings are highly interested in exploration, conditioning and protection of their drinking water resources. In addition, the usage as nutrient water is also of great important in agriculture and, especially in the modern world, process water is necessary mainly for industrial purposes. Although water is essential for human beings anthropogenic activities resulted in a significant impairment of the aquatic environment. Especially rivers are used over a longe periods of time as 'solvent' and means of transportation for liquid and solid human waste due to the high dynamics of riverine systems. Apart from chemical waste discharges also physical interferences e.g. radiation, noise, thermal and mechanical stress as well as artificial structural modifications affected strongly the natural systems.

In the 20$^{th}$ century the water resources became more and more polluted by increasing anthropogenic emissions as a result of the intensive industrialisation. Nevertheless, it took a long period of time before the public interest turned to a detailed view on the chemical and microbial quality of the aquatic environment (partially initiated by the publication of 'Silent Spring' by R. Carson 1962). Apart from the introduction of hazardous inorganic pollutants (heavy metals, cyanides etc.) a huge proportion of the pollution has to be attributed to the organic substances. As a new aspect, since the industrialisation not only the amount of compounds discharged to the aquatic environment accumulated rapidly, but also the structural variety of organic contaminants increased. Of particular importance was the introduction of a new class of chemicals, the xenobiotical substances. These artificial compounds have no counterpart in the bio- and geosphere and alterated the natural environment significantly.

Most of the 'notorious' pollutants, which are objects of intense public discussions, belong to the group of xenobiotics (e.g DDT, PCB and PVC). However, comparing the introduction of organic xenobiotics to the technosphere with their point in time of identification in the natural environment and the subsequent monitoring a significant time lag can be stated as illustrated in fig. 1.

With respect to drinking water supply particularly groundwater is a major source, because it combines frequently a high chemical as well as microbial purity as the result of physico-chemical cleaning processes during the soil passage. But an over-intensive usage or the absence of appropriate groundwater resources led to an increasing necessity of surface water for the drinking water supply within the last 20 years. Even in Germany a significant proportion of drinking water derives from water supply dam, river water or bank filtrate.

However, drinking water quality in terms of chemical purity is linked closely to the contamination level of the receiving systems. Hence, many scientific studies were performed since the early 70'ies in order to document the state of pollution of rivers, lakes and aquifers. Subsequently, huge technical efforts were conducted to improve the quality of surface water and groundwater systems.

As a result of an increasing public attention to the state of riverine contamination the research activities on the description and protection of river systems as well as first political ambitions led to the foundation of international protection organisations. For instance the Rhine river system, highly polluted in the 70'ies, is the objective of the IKSR (Internationale Kommission zum Schutz des Rheins) founded in 1963. As a consequence of their activities the quality of water and sediments of the Rhine river increased significantly up to now. After the German reunification a similar approach was introduced for the Elbe river (IKSE, Internationale Kommission zum Schutz der Elbe) in 1990. Although the number of environmental investigations regarding fresh water resources increases permanently, the knowledge about riverine and groundwater pollution still remains limited.

Further on, the protection of the fresh water systems is a big challenge also for the politics not only of the individual nations but also of the European Union. As a first result of EU-activities the *Water Framework Directive (WFD)* was introduced in 2002 in order to organise and optimise the national activities in the field of water science and technology (s.a. Irmer 2000).

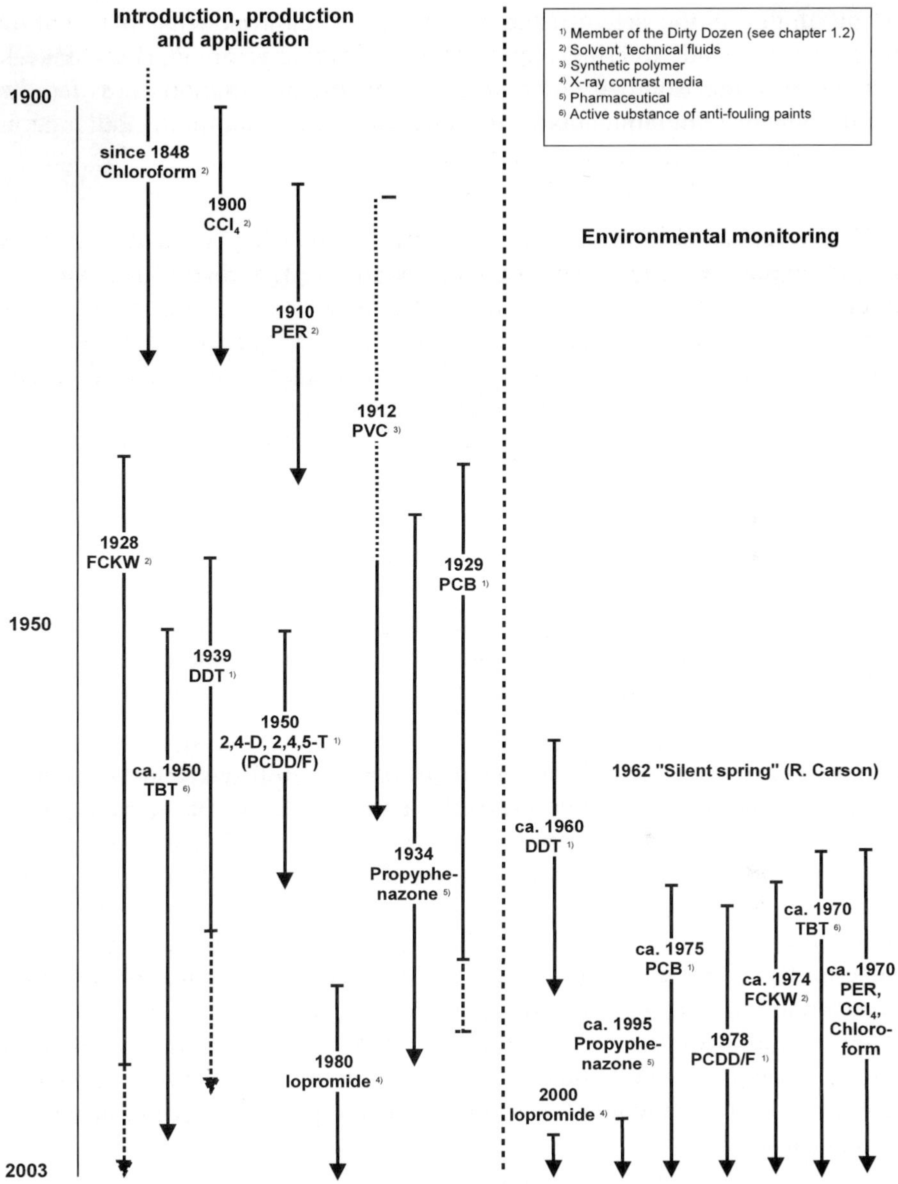


**Fig. 1:** Time period of application of selected xenobiotics in the technosphere contrasted to the time period of environmental monitoring. Data are adapted from Senthilkumar et al. (1999), Roempp (1995), Ternes and Hirsch (2000) and Scheringer (1999).

In summary, the scientific research on the anthropogenic pollution of groundwater and riverine systems, especially by organic substances, comprising the sources, emission pathways, distribution and fate of anthropogenic contaminants still remains as an important challenge in environmental sciences.

## 1.2 Organic contaminants in groundwater and riverine systems

Organic contaminants discharged to the aquatic environment are characterised by a high diversity with respect to their molecular structures and the resulting physico-chemical properties. A similar complexity with respect to physical and chemical properties exists also in the natural systems of ground- and surface water. Since the environmental transformation and transport depends on both the substance specific as well as the compartment specific properties, a unique fate and distribution for each individual contaminant in the respective natural systems has to be stated. Therefore, only limited information on the distribution and the transformation of organic contaminants in riverine and groundwater systems can be given on a principal base.

However, in the aquatic environment two quite different compartments exist, the particulate matter as well as the water phase, representing a hydrophilic and a more lipophilic phase, respectively. Consequently, most of the organic substances accumulate either in the water phase or in the sediments depending on their polarity and the resulting hydro- or liphophilicity as espressed by their $K_{OC}$-values. This partition also determines the principal transport processes and, consequently, the distribution of the organic contaminants. Particle associated substances in riverine systems are transported either by the horizontal flux as the result of sedimentation and resuspension. Superimposed a vertical transport of the sedimentary fraction by saltation and reptation as well as of the suspended fraction within the water phase according to the general flow direction has to be considered. This latter transport is responsible exclusively for the distribution of dissolved organic components within river systems.

Additionally the water phase including the suspended particulate matter represents an aerobic environment, whereas the sedimentary compartment is mainly more anaerobic. This difference results in quite different degradation pathways for most of the organic compounds as the result of varying microbial communities.

In groundwater a vertical flux in the water unsaturated zone as well as a horizontal flux in the water saturated zone has to be stated and, consequently, an associated transport of dissolved and particle bound contaminants can be observed. Concurrently, corresponding aerobic and anaerobic zones have to be differentiated with respect to the microbial degradation processes.

A comprehensive review on the environmental behaviour of anthropogenic contaminants in the sedimentary matter of freshwater systems was published very recently (Warren et al. 2003)

As a result of the high diversity of organic contaminants and the complex processes affecting their distribution within riverine and groundwater systems the description of anthropogenic contribution to the pollution is frequently restricted to analysis of chemical and ecotoxicological bulk parameters such as AOX, DOC, COD or BOD. However, whereas these data are appropriate to characterise the level of pollution of groundwater or river systems, a fundamental knowledge about the anthropogenic contamination needs more detailed information on the molecular level.

### 1.2.1 Natural compounds

Natural compounds in groundwater, river water and corresponding sediments reflect on the one hand the biological activity with its autochthonous emissions, on the other hand the surrounding biosphere as characterised by the allochtonous contributions. Hence, many biogenic compounds of low molecular weight (e.g. fatty acids, fatty alcohols, long chain *n*-aldehydes, functionalized terpenoids, sterols, phytol, squalene, di-*iso*-propyldisulfide) but also of high molecular weight (e.g. carbohydrates, peptides, humic acids) can be detected and attributed to different biogenic sources. Generally, in contrast to the more unaffected biogenic compounds appearing in the water phase numerous low molecular weight substances in the sediments are degradation products of biogenic precursors. Known examples are loliolide, ionenes, different isomers of phytenes and phytols, steradienes and 4,8,12,16-tetramethylheptadecan-4-olide. They correspond to the biogenic input of chlorophyl, carotenoids, sterols and tocopherols (see also Cranwell 1981a; Eganhouse and Kaplan 1985; Cranwell et al.

1987; Prahl and Pinto 1987; Ittekkot 1988; Stephanou 1989; Riley et al. 1991; Hedges et al. 1994; Franke et al. 1995; Schwarzbauer et al. 2000).

For instance several detailed biogeochemical investigations were applied to riverine systems of South America. The partition of amino acids and carbohydrates between the liquid and the solid phase of the Amazonas river was studied in order to differentiate biogenic sources and to determine degradation processes (Hedges et al. 1994). A comprehensive view on the biogeochemistry of the Orinoco river analysing *n*-alcohols, fatty acids, terpenoic hydrocarbons and sterols has been given by Jaffe et al. (1995, 1996).

However, a major part of biogeochemical investigations on river systems focussed on the terrestrial contribution to the marine environment. As initial steps river specific compounds of both low molecular as well as high molecular weight were characterised and quantified and, consequently, monitored in marine water and sediments (Hedges et al. 1997). This approach was used in order to document the riverine contributions for example of the Lena, Indus, McKenzie, Loire, York, Godavaria and Mexico rivers to the corresponding coastal areas (Laptev Sea, Arabian Sea, Beaufort Sea, Gulf of Biscaya, Chesapeake Bay, Bay of Bengal and Gulf of Mexico). The organic substances analysed included carbohydrates, amino carbohydrates, amino acids, carboxylic acids, aliphatic and aromatic hydrocarbons. Isotopic data were partially used to differentiate riverine and marine contributions to the organic matter of coastal areas (Hedges and Parker 1976; Saliot et al. 1988, Yunker et al. 1991 and 1993; Zegouagh et al. 1996 and 1998, Countway et al. 2003).

In addition to the individual biogenic compounds introduced above a major proportion of the organic matter in riverine and groundwater systems can be attributed to the humic substances. These natural compounds with unordered molecular structures represent an organic pool, that is objective of numerous investigations especially with respect to structure analysis (e.g. Thurman and Malcolm 1983; Sonnenberg et al. 1989; Schulten and Leineweber 1996; Göbbels and Püttmann 1997; Kumke et al. 1999; Esteves and Duarte 2000) or the interaction with natural occurring or anthropogenic low molecular weight substances (e.g Huber et al. 1992; Jensen-Korte et al. 1987; Klaus et al. 1998a/b; Zwiener et al. 1999, Northcot and Jones 2000).

### 1.2.2 Anthropogenic pollutants

In nearly all industrialized, densely populated or agricultural stressed regions and countries (especially in North America and Europe) the riverine organic matter is superimposed by huge anthropogenic emissions as described above.

An excellent screening study on anthropogenic contaminants in a river system dated back to the middle 70'ies. Hites et al. (1974) published detailed GC-MS based analyses applied to water samples of the Delaware river. This study reveals more than 100 organic substances as constituents of the river water. A similar approach was used by Franke et al. (1995) in order to document the state of pollution of the Elbe river. Especially the contributions of individual tributaries were pointed out by characteristic organic compounds derived from industrial, municipal or agricultural sources. More than 250 organic contaminants were described. Additionally similar comprehensive and detailed analyses were applied to sewage effluents characterising a major anthropogenic contribution to the riverine environment (Paxeus 1996). All these approaches revealed a high number of organic contaminants and lead to the structure elucidation and identification of a wide variety of still unknown or unreported organic contaminants. Hence, such comprehensive and detailed studies build up the base for advanced investigations on the complex mixture of anthropogenic organic matter in groundwater and river systems on a molecular level.

However, screening analyses has been rarely performed and most of the published environmental investigations on anthropogenic emissions in riverine or groundwater systems focused on preselected contaminants. The state of knowledge with respect to the individual contaminants can be divided roughly into two pools of studied groups of organic pollutants: the 'priority pollutants' and a 'new generation' of contaminants.

#### *1.2.2.1. Priority pollutants*

Earlier environmental investigations dealing with anthropogenic organic contaminants in riverine and groundwater systems focused mainly on the so-called 'priority pollutants' (see also Keith and Telliard 1979). These compounds are characterised by elevated geo- and bioaccumulation rates as the result of a high environmental stability and lipophilicity. These physico-chemical properties led to a widespread, frequently ubiquitous distribution in the natural compartments. Additionally, they represent also toxicological relevant compounds. Generally, most of the priority pollutants are integrated in environmental monitoring programs. E.g. the EU WFD defined 32 priority pollutants, that should be monitored in order to asses the chemical status of surface water systems. On the contrary the US Environmental Protection Agency (US EPA) defined more than 129 priority pollutants (Keith and Telliard 1979). It can be stated that a comprehensive knowledge about these widespread distributed organic pollutants exist.

A major proportion of the priority pollutants are covered by the so-called 'persistent organic pollutants' (POPs) comprising e.g. PCBs, DDT-related compounds as well as further chlorinated aromatics.

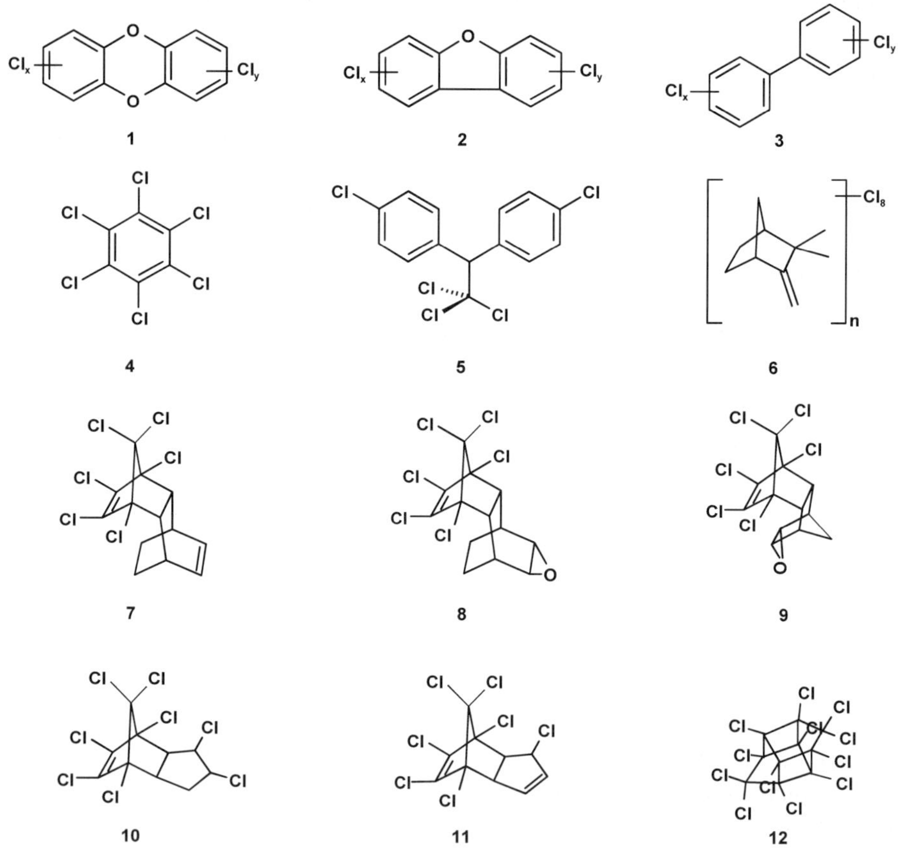


**Fig. 2:** Molecular structures of the 'Dirty Dozen'. 1 - Polychlorinated dibenzo-*p*-dioxines, PCDD ; 2 - Polychlorinated dibenzofuranes, PCDF; 3 - Polychlorinated biphenyls, PCB; 4 - Hexachlorobenzene, HCB ; 5 - 2,2-Bis(4-chlorophenyl)-1,1,1-trichloroethane, DDT ; 6 - Toxaphene ; 7 - Aldrin ; 8 - Dieldrin ; 9 - Endrin ; 10 - Chlordane ; 11 - Heptachlor ; 12 - Mirex.

The environmental relevance of the POPs leads to the ban of the most important representatives, the 'Dirty Dozen', by the Stockholm Convention in May 2001 (see also http:\\www.pops.int ). Prior to the prohibition of production and emission of these 'Dirty Dozen' numerous scientific investigations considering the occurrence and fate of these compounds especially in the aquatic environment were performed (see also Jones and Voogt 1999).

A major proportion of persistent contaminants belongs to the group of polyhalogenated aromatics. Among the POP's the polychlorinated dibenzo-*p*-dioxines and dibenzofuranes (PCDD, PCDF) as well as associated coplanar PCB's represent the most toxic category. PCDD/F's appear in the aquatic environment on an ultra trace concentration level. Numerous congeners are characterized not only by a high acute toxicity, but they are also classified as carcinogenic and as endocrine disrupters (Cikryt 1991; Focant et al. 2001). Therefore, especially their high bioaccumulation tendency caused for concern. Due to their lipophilicity they accumulate also in the particulate matter within groundwater or riverine systems. Thus, numerous studies identified and quantified the dioxines in sediments and suspended particulate matter of German riverine systems, for example the Elbe, Odra, Lippe and Rhine rivers or in minor riverine systems in Mecklenburg-West Pomerania (Evers et al. 1988; Klös and Schoch 1992; Götz et al. 1993; Dannenberger and Lerz 1998;Evers et al. 1998; Dannenberger and Lerz 1999). The anaerobic dechlorination as the most important degradation process in the sedimentary aquatic environment has been particularly investigated for the most toxic tetrachlorinated isomers (e.g. Bunge et al. 2001).

A very similar environmental behaviour was stated for a further congeneric group of chlorinated aromatics, the PCB as reviewed by Lang (1992). But in contrast to the PCDD/F's these compounds are xenobiotics, technically synthesised by direct chlorination of biphenyl using iron based catalysts. Until a strict regulation of production and technical application in the late 70'ies PCBs were widely used e.g as hydraulic fluids, refrigerants, insulator fluids or plasticizers in lacquers. Due to their ubiquitous distribution as well as their toxicological effects (e.g. Lang 1992; Jacobson and Jacobson 1996) PCBs are frequently analysed in groundwater and river systems. Data are presented e.g. for sediments of the Weser, Rhine, Main, Odra, Spree, Elbe, Saar and Lippe rivers (Evers et al. 1988; Bohlen et al. 1989; Klös and Schoch 1992; Trapp et al. 1992; Breitung 1992; Götz et al. 1994; Terytze 1995; Müller et al. 2002) but also e.g. of Swedish, Finnish, Czech, Chinese and American rivers (Järnberg et al. 1993; Holoubek et al. 1994; Koistinen et al. 1995; Loganathan et al.

1995; Rostad 1997; Foster et al. 2000; Mai et al. 2002). Additionally a variety of investigations dealt with the environmental fate of PCBs including transport, partition, transformation and degradation in groundwater and river systems. Determining relevant physico-chemical properties such as $K_{OH}$-values the distribution and transport of these compounds in the aquatic environment has been pointed out (e.g. Bopp et al. 1981; Brownawell and Farrington 1986; Bruijn et al. 1989; McGroddy et al. 1996; Eisele and Schorer 1997; Connolly et al. 2000). Studies on the degradation and transformation processes of PCB in the environment focussed mainly on the methylsulfonated and hydroxylated metabolites (e.g. Buser et al. 1992; deVoogt and Häggberg 1993; Adriaens and Grbic-Galic 1994) but also on the reductive dechlorination processes in anaerobic sediments (Alder et al. 1993; Abramowicz et al. 1993). Further on, comparison of isomeric pattern is an important analytical tool to differentiate emission sources or technical formulations and was also applied for the environmental monitoring of PCB in surface water systems (e.g. Dunn et al. 1984; Schulte and Malisch 1983; Buchert and Ballschmiter 1985). But due to the fact that in accordance with the complex composite of PCDD/F's also PCB's appear as congeneric mixtures with up to 209 individual congeners, the complexicity hinders frequently a successful isomer specific determination. Hence, partially hyphenated analytical approaches as multidimensional chromatography or high resolution mass spectrometry coupled with gas chromatography are requested (e.g. Pellizzari et al. 1985; Duinker et al. 1988; Fuoco et al. 1993). Nevertheless, all congener specific analyses of the PCB are based mainly on the work of Ballschmiter dated back to the early 80'ies (Ballschmiter and Zell 1980). On the contrary very recently PCBs became objectives of isotopic studies in order to discriminate different technical sources or to monitor microbial degradation processes (e.g. Jarman et al. 1998; Yanik et al. 2003a).

A second group of important POP's is represented by chlorinated pesticides. Major attention was attributed to the environmental occurrence and fate of DDT and its numerous metabolites. As a result a comprehensive knowledge about the major properties with respect to environmental considerations (environmental stability, toxicological effects, transport processes including its global distribution etc.) exists. For instance the information on the degradation pathway under aerobic as well as under anaerobic conditions is nearly complete as illustrated in Fig. 3.

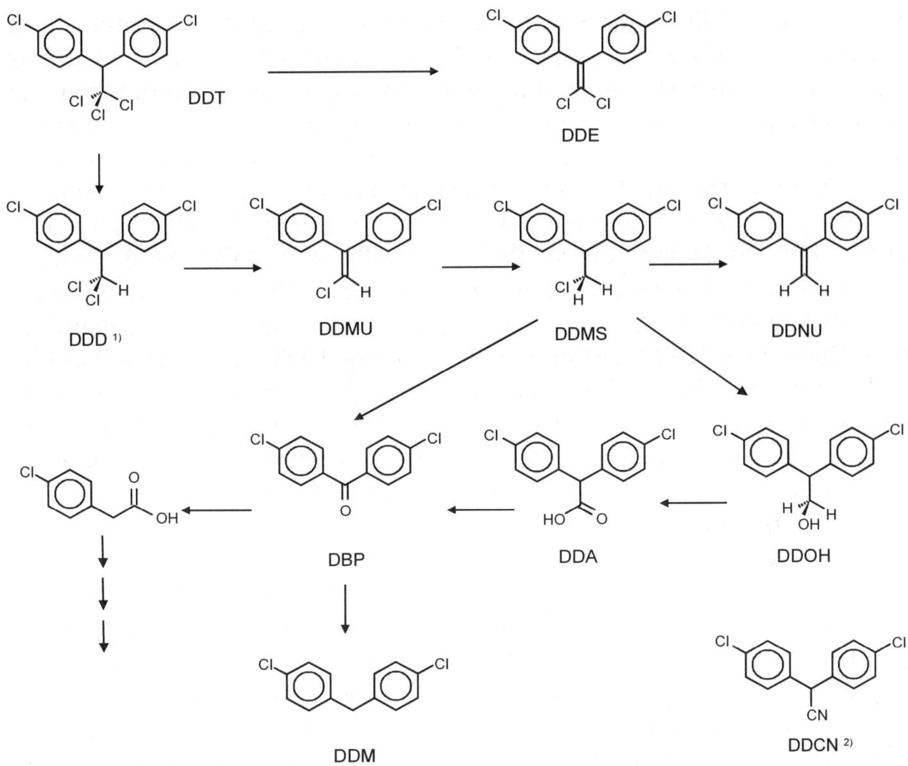


**Fig. 3:** Anaerobic degradation pathway of DDT in the environment adapted from *http://umbbd.ahc.umn.edu/ddt2/ddt2_map.html* .
[1)] = main metabolite; [2)] metabolite is not integrated so far

With respect to the German situation the prolonged usage of DDT until the mid 80'ies in the formerly German Democratic Republik (DDR) caused still significant contamination levels in selected riverine systems as the Mulde, Elbe, Odra or Spree rivers. Consequently, since the reunification many investigations were performed on DDT-residues in these riverine systems (e.g. Götz et al. 1994; Pietsch 1995, Terytze 1995, Dannenberger and Lerz 1998; Heberer and Dünnbier 1999). In addition, other German river systems (e.g. Rhine and Weser rivers) were investigated with respect to their DDT-contamination level (Bohlen et al. 1989; Trapp et al. 1992, Brauch 1993). It has to be noted that mainly DDT and its main metabolites of the anaerobic and aerobic degradation pathway, DDD and DDE, respectively, were analysed in most of the studies performed on DDT contamination. Although the ecotoxicological risk is not restricted to the main metabolites (Schulze et al. 2003), only very few studies considered the wide variety of further metabolites and their contribution to the amount and quality of DDT-emissions (e.g Schwarzbauer et al. 2001). Hence, a significant proportion of the real toxic potential often remains underestimated.

Further on, the pesticides γ-hexachlorocyclohexanes (γ-HCH, lindane) and its aromatic analogue, hexachlorobenzene (HCB), attract attention with respect to riverine pollution. The environmental appearance of HCH is not restricted to the γ-isomer lindane, because technical HCH mixtures include mainly the α-, β-, γ- and δ-enantiomers. Furthermore beside the industrial discharge of purification residues derived from the lindane production, also the technical mixture itself was formerly used as pesticide formulation. Nevertheless, the most harmful enantiomers are the synthetically desired ones and, accordingly, nearly all environmental studies included the prominant isomers. In aquatic ecosystems HCHs appear associated with the particulate matter as well as dissolved in the water body as a result of an elevated solubility as compared to the compounds discussed so far. Examples include the determination of HCH in sediments or water of the Gallego, Weser, Spree, Elbe rivers as well as of riverine systems in Taiwan, China, UK and USA (Nerin et al. 1991; Oxynos et al. 1995; Terytze 1995; Zhou et al. 1998; Foster 2000; Schwarzbauer et al. 2001; Doong et al. 2002; Sun et al. 2002). Due to its specific stereochemical properties HCHs are model compounds for the enantioselective monitoring of environmental contaminants. Both biotic transformation and physiological partition processes affect the composition of the individual enantiomers. Thus, an enantioselective analysis of HCHs can reveal information on these environmental processes (e.g. Hühnerfuss et al. 1993; Möller et al. 1994; Kallenborn and Hühnerfuss 2001; Bethan et al. 2001).

Its physico-chemical properties (lipophilicity, vapor pressure, persistence) and a widespread emission combine to facilitate the global distribution of hexachlorobenzene (HCB). The tendency to bio- and geoaccumulate as well as a potential toxicity characterizes HCB as an priority pollutant. In contrast to nearly all other pesticides the usage and subsequent emission of HCB is not restricted to agricultural application. The manifold emission sources have been reviewed recently by Bailey (2001). In summary the discharge of HCB to the aquatic environment can be attributed to (i) agricultural usage of HCB containing fungicides, (ii) industrial discharge from chemical plants, (iii) pyro-synthesis during combustion processes, and (iv) generation in metals industry during carbon anode treatment or aluminium fluxing and degassing. As a consequence of its environmental interest HCB was analysed prevalently in groundwater and river systems including its environmental behaviour (e.g. Wilken and Wirth 1986; Evers et al. 1988; Bohlen et al. 1989; Götz et al. 1994; Oxynos et al. 1995; Amaral et al. 1996; Rostad 1997; Dannenberger and Lerz 1998; Sun et al. 2002).

Due to their environmental persistence and risk most of the pesticides described above have been substituted within the last three decades. The new generation of modern pesticides are charactericed by higher microbial degradation rates, lower lipophilicity and less toxicity (for comprehensive information on the toxicity of pesticides to aquatic organism see DeLorenzo et al. 2001). On the other hand, their higher hydrophilicity represents the possibility to enter aquifers after agricultural application. Therefore, numerous studies focused on the groundwater contamination by pesticides (e.g. Reupert and Plöger 1989; Pereira et al. 1990; Abke et al. 1993; Skark and Zullei-Seibert 1994; Suzuki et al. 1998, Kolpin et al. 2001).

Representatives of more modern pesticides based structurally on carbamates and thiocarbamates (e.g. carbendazim, EPTC), phosphates (e.g. malathion, dichlorvos), sulfonylureas (e.g. cinosulforon), triazines (e.g. simazine, atrazine) and further nitrogen, sulfur and phosphorous containing moieties. They are objectives of numerous studies on degradation and transformation processes in surface water areas e.g. by photooxidation or hydrolysis (e.g. Lartiges and Garigues 1995; Liu et al. 2001; Abu-Qare and Duncan 2002; Sakkas et al. 2002). Comprehensive investigations on the occurrence and distribution of pesticides in surface water, including partially an attribution to application times or areas, have been performed world wide on many river systems e.g. in Spain, Malaysia, Greece, and USA (Tan and Vijayaletchumy 1994; Miliades and Malatou 1997; Planas et al. 1997; Barber and Writer 1998; Buser et al. 1998; Battaglin and Goolsby 1999; de Almeida Azevedo et al. 2000; Müller et al. 2003). German rivers analysed by their pesticide content comprise for example the Stör, Main, Elbe und Rhine rivers (Trapp et al. 1992; Brauch 1993; Götz et al. 1994; Pietsch et al. 1995; Bethan et al. 1997)

Next to persistent halogenated aromatics or pesticides further contaminants can be attributed to the group of priority pollutants. From the beginning of the $20^{th}$ century the environmental concentrations of polycyclic aromatic hydrocarbons (PAHs) was steadily increasing as the result of anthropogenic emission superimposing the naturally derived content. Accordingly, these compounds are ubiquitously present in river water and sediments as well as in aquifers. Although they are preferentially bound to the particulate matter, a few representatives were regularly detected in the water phase comprising mainly naphthalene, fluorene, phenanthrene and anthracene (e.g. Trapp et al 1992; Franke et al 1995). However, principal transport processes distributing PAH within riverine and groundwater systems are particle associated. Hence, numerous studies were performed on the particle bound PAH transport including adsorption and desorption phenomona (e.g. Murphy et al. 1994; Johnson and Amy 1995; McCrody et al. 1996; Furlong et al. 1997; Leppard et al. 1998)

PAHs were frequently analysed in German rivers (e.g. Buchert et al. 1982; Trapp et al. 1992; Francke et al. 1995; Franke et al 1995; Franke et al 1996; Dannenberger and Lerz 1998; Müller et al. 2002) and in international riverine systems (e.g. Holoubek et al. 1994; Picer et al. 2001; Mai et al. 2002). The origin of anthropogenic PAH and related substances (e.g. nitrogen-, oxygen- or sulphur-containing polycyclic aromatic compounds, PACs) can be generally attributed to pyrogenic or petrogenic sources. Specific pyrosynthesized PAHs, characterized by higher condensed aromatic ring systems and minor alkylsubstitution, enter the aquatic environment mainly in form of a particle associated atmospheric deposition or via sewage effluents. Petrogenic contributions, dominated mainly by two- and three-member ring systems with more alkysubstituents, are released as the result of navigation, industrial discharge or municipial effluents. As a consequence, ratios of source specific PAHs allow the differentiation of petrogenic or pyrolytic origins (e.g. Venkatesan et al. 1987; Colombo et al. 1989; Schulz et al. 1991; Bayona et al. 1994; Benner et al.1995; Burns et al. 1997; Volkman et al. 1997). The environmental interest on PAHs is mainly based on their toxicological and ecotoxicological impact. Especially molecules with higher condensed aromatic systems (e.g. benz[a]anthracene, benz[a]pyrene, chrysene) exhibit a high carcinogenic and mutagenic potential (e.g. Schuetzle et al. 1981; West et al. 1986; Wenzel-Hartung et al. 1990; Menzie et al. 1992). Consequently also the microbial or abiotic degradation of PAHs and PACs in surface and groundwater systems are of common environmental interest (e.g. Goßel and Püttmann 1991; McNally et al. 1998; Yuan et al. 2001). Very recently the stable carbon isotopic fractionation during degradation processes of PAHs as well as an isotopic fingerprint approach have been discussed (McRae et al. 1999; Mazeas et al. 2002; Mazeas and Budzinski 2002, Yanik et al. 2003b).

In addition to the pollutants discussed in detail there remain only few further contaminants comprising halogenated and nitro-substituted phenols and anilines as well as volatile organic compounds (VOCs; e.g.

chloroform, PER), which belong also to the group of priority pollutants (see also Keith and Telliard 1979). These compounds were also integrated in comprehensive monitoring analyses of German riverine water and sediments periodically performed by surveillance authorities especially the Federal State Environmental Protection Agencies.

### *1.2.2.2 'New' pollutants*

Apart from the well-recognised more conventional contaminants comprising mainly the priority pollutants further substances were identified and monitored in riverine systems and groundwater within the last decade, which represent a 'new generation' of anthropogenic contaminants (see also Alcock and Jones 1999, Paasivirta 2000). These 'new' pollutants will be shortly introduced in the following section.

A first group of these contaminants includes halogenated aromatics, which are structurally related to PCDD/Fs and PCBs. Interestingly, apart from the chlorinated aromatics the quality and quantity of brominated contaminants analysed in the aquatic environment increase as a result of rising usage of polybrominated aromatics as flame retardants (deWit 2002).

In detail polychlorinated and polybrominated naphthalenes (PCN, PBN), tetrachlorobenzyltoluenes (TCBT), polychlorinated and polybrominated diphenylethers (PCDE, PBDE), polychlorinated triphenylmethanes (PCTPM), polychlorinated terphenyls (PCT), tetrabromobisphenol A and polychlorinated dibenzothiophenes (PCDBT) were analysed in river water and sediments ( e.g. Poppe et al. 1991; Breitung 1992; Järnberg et al. 1993; Claus et al. 1998, Sellström et al. 1998; Schwarzbauer et al. 2001; Ikonomou et al. 2002). The molecular structures of selected halogenated aromatics are illustrated in Fig. 4. The state of knowledge concerning the environmental behaviour and effects of the individual contaminants varies strongly depending among other things on the time period since their identification. On the one hand information on ecotoxicological effects, emission sources and environmental transport of PCNs, PBDEs and PCDEs are available (Akerblom et al. 2000; Pesonen et al. 2000; Imagawa and Lee 2001; McDonald 2002; Hardy 2002; Palm et al. 2002), on the other hand the knowledge about the environmental behaviour of PCDBT or PCTPMs is very limited or absent.

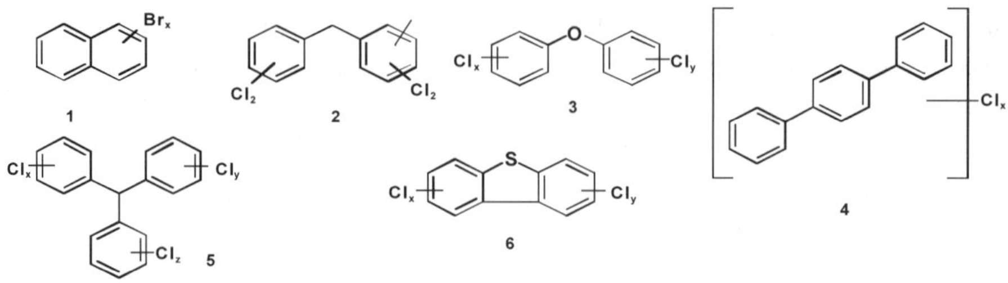


**Fig. 4:** Molecular structures of **1** - Polybrominated naphthalenes, PBNs, **2** - Tetrachlorobenzyltoluenes, TCBTs, **3** - Polybrominated diphenylethers, PBDEs, **4** - Polychlorinated terphenyls, PCTs, **5** - Polychlorinated triphenylmethanes, PCTPM, **6** - Polychlorinated dibenzothiophenes, PCDBT.

A second group of 'new' pollutants is represented by organo-metallic compounds. Several metals are able to bind to organic moieties and form substances that are not only of technical interest but are also characterized by an elevated environmental stability. The most important organo-metallic contaminants of aquatic systems in terms of contents are butyl tin compounds. Based on various physiological and physical effects ionic butylated tin compounds are widely applied and, consequently, discharged to the aquatic environment. Since mono- and dibutylated substances are used dominantly as stabilizer in PVC, desinfectants or industrial biocides, they are introduced to river systems mainly via sewage effluents and waste discharge. Tributyl tin (TBT) was the active ingredient in antifouling paint formulations. As a consequence of its desired biocidal effects it was found to be one of the most acute toxic substances for aquatic organisms especially molluscs (Maguire 1987). Hence, the use of TBT has been regulated in Europe, North America and Japan in the late 80'ies, restricting the amount and release rate of TBT based paints, establishing an environmental quality target in water (0.8 ng/L Sn) and controlling sale as well as commercial handling of technical formulations (Gomez-Ariza et al. 2001a). Nevertheless TBT and other tin organic contaminants have been detected in numerous river systems until now (e.g. Jantzen et al. 1993; Wilken et al. 1994; Francke et al. 1995; Steffen et al. 1995; Ansari et al. 1998; Shawky and Emons 1998; Senthilkumar et al. 1999; Gomez-Ariza et al. 2001a).

In addition organo-mercury compounds are of environmental interest in aquatic systems not only since the Minamata disaster (Takizawa 2000). Especially methylated mercury species were detected and monitored in riverine systems as the result of biotic transformation of inorganic mercury ions (Hintelmann and Wilken 1994; Jantzen and Prange 1995). To a minor extend also organo-substituted lead, selenium and arsenic are subject of environmental studies (e.g. Jantzen and Prange 1995; Paneli et al. 1997; Szostek and Aldstadt 1998; Killelea and Aldstadt 2002).

Generally, the ionic character of a lot of organo-metallic compounds and the significant changes of ecotoxicologial properties depending on the degree of substition and the quality of the alkyl moieties leads to the introduction of metal speciation as a new field in environmental analysis. Concurrently, numerous new analytical methods were developed in order to isolate and detect the different metal species (e.g. Harino and Fukushima 1992; Rapsomanikis et al 1986; Chau and Wong 1991; Reader 1992; Cai et al. 1993; Abalos et al. 1997; Gomez-Ariza et al. 2001b).

Further 'new' pollutants are not defined by their molecular structures but by their field of application as well as their mode of ecotoxicogical or physiological impact. In the mid 90'ies the analytical interest turned partially to a group of compounds with estrogenic effects, the endocrine disrupters. As a major ecotoxicological effect in river and groundwater systems these compounds can affect the fertility of aquatic species and are suspected to induce similar disturbances in human organism, for instance as a consequence of contaminated drinking water consumption (Hock et al. 2000). Examples of xenobiotical endocrine disrupters (the so-called 'xenoestrogens') include on the one hand well recognized pollutants as DDT and PCBs, on the other hand further so far not accounted contaminants widely used in the anthroposphere comprising phthalates (plasticizers, see below), nonylphenols (NPs, degradation products of polyetholxylated parent compounds, widely used as non-ionic surfactants), bisphenol A (antioxidant) and synthetic estrogens (e.g. ethynylestradiol, active ingredient in contraceptives). In addition, natural occurring estrogens, comprising for example 17β-estradiol and dihydrotestosterone, account to the group of endocrine disrupters. All these substances are introduced to rivers by sewage effluents (Stumpf et al. 1996a) and, consequently, were detected in the corresponding water and sediments (e.g. Tan 1995; Ding and Tzing 1998; Pedersen and Lindholst 1999; Ahel et al. 2000; Heemken et al. 2001; Bolz et al. 2001; Adler et al. 2001; Penalver et al. 2001).

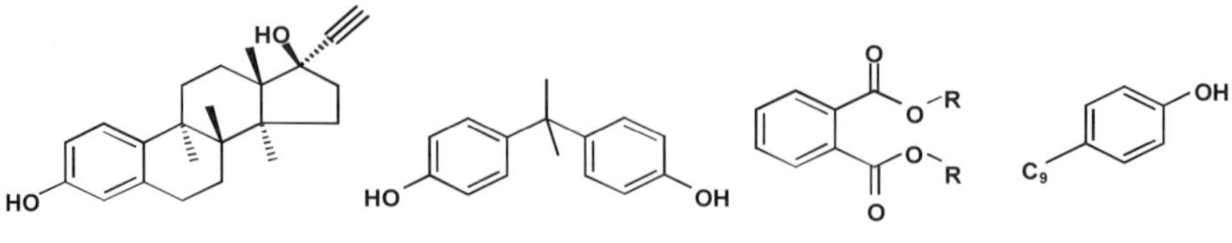


1 2 3 4

**Fig. 5:** Molecular structures of selected xenoestrogens: **1** - Ethynylestradiol, **2** - Bisphenol A, **3** - Phthalates, **4** - Nonylphenols

A second set of compounds with obviously physiological effects comprises the pharmaceuticals. Since the first publication concerning the appearance of drug residues in the environment (Garrison et al. 1979) many investigations confirmed the presence of numerous pharmaceuticals in the aquatic systems. In particular clofibric acid (metabolite of ethyl clofibrate, a blood lipid regulator), phenazone, diclofenac, ibuprofen, carbamazepine and selected antimicrobials (molecular structures see Fig. 6) are common contaminants of river and groundwater water (e.g. Waggot 1981; Richardson and Bowron 1985; Stan and Linkerhäger 1994; Holm et al. 1995; Heberer et al. 1998; Öllers et al. 2001; Ternes et al. 2001; Farre et al. 2001, Sacher et al. 2001; Lindsay et al. 2001; Reddersen et al. 2002). As the main emission, source sewage effluents were pointed out (Stumpf et al. 1996b; Kümmerer 2001). Although the acute physiological properties of pharmaceuticals are certainly well known, the long term effects of an continuous uptake by aquatic organisms or mammals are still unknown.

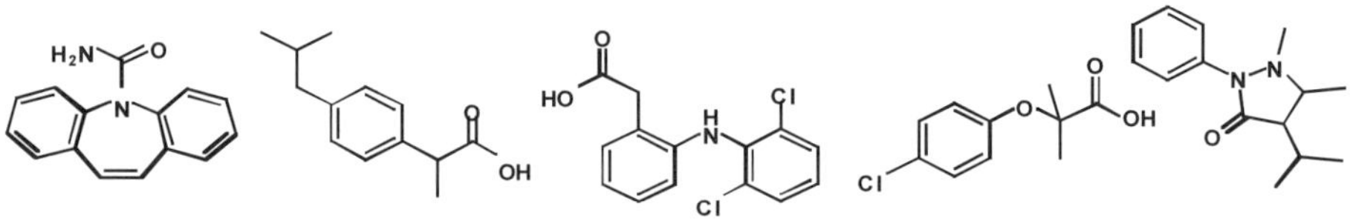


1 2 3 4 5

**Fig. 6:** Molecular structures of selected pharmaceuticals known as aquatic contaminants:

**1** - Carbamazepine, **2** - Ibuprofene, **3** - Diclofenac, **4** - Clofibric acid, **5** - Propyphenazone

Further on, an increasing number of active constituents of personal care products and washing agents as well as polymers has been identified in the aquatic environment within the last 10 years (see also Daughton and Ternes 1999). Representatives are fragrances, surfactants, desinfectants and plastizicers. Individual substances are presented in Fig. 7. Polycyclic musk fragrances are frequently detected in the aquatic environment, especially galaxolide and tonalide (e.g. Eschke 1996; Lach and Steffen 1997; Winkler et al. 1998; Heberer and Dünnbier 1999; Verbruggen et al. 2000; Standley et al. 2000). These compounds are used in almost all scented consumer products, such as perfumes, cosmetics or laundry detergents. With respect to their molecular structure they are functionalized indane and tetraline derivatives highly substituted mainly by methylgroups. A comprehensive review on the environmental occurrence and fate of polycyclic musk fragrances is recently published (Rimkus 1999). Further known constituents of personal care products that are analysed in surface water systems, are the insecticide DEET and the bactericide triclosan (e.g. Franke et al. 1995; Lindström et al. 2002; Adolfsson-Erici et al. 2002). Further on, several ingredients of laundry formulations are prominant contaminants of rivers or aquifers. In particular the complexing agents ethylenediaminetetraacetic acid (EDTA) and to a minor extend nitrilotriacetic acid (NTA), both used as softener, were frequently detected especially in German rivers (e.g. Hansen 1986; Frimmel et al. 1989; Trapp et al. 1992; Sacher et al. 1998).

Plasticizers, important polymer additives, are used in high amounts for the optimisation of physico-chemical properties of polymers such as PVC or PA. Important representatives in terms of environmental contamination are phthalates and phosphates. Due to permant leaching processes these compounds migrate to environmental compartments and, consequently, were detected in river and groundwater systems (Saeger et al. 1979; Bohlen et al. 1989; Bohlen et al. 1990; Lenhart and Lemm 1993; Tan 1995, Fries and Püttmann 2001). Both groups also exhibit partially toxicological, estrogenic and ecotoxicological properties.

Generally, all active constituents of personal care products and washing agents presented as well as the major amount of polymer additives are discharged to the aquatic environment by sewage effluents.

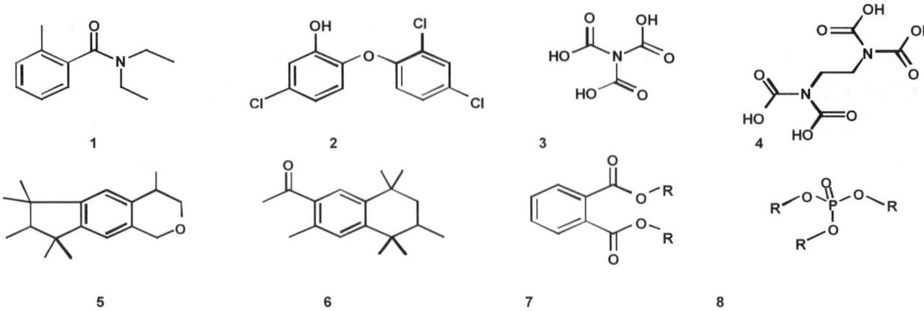

**Fig. 7:** Molecular structures of selected active constituents of personal care products, washing agents or polymers: **1** - N,N-Diethyltoluamide, DEET, **2** - Triclosan, **3** - Nitrilo triacetate, NTA, **4** - Ethylenediaminetetraacetic acid, EDTA, **5** - Galaxolide, **6** - Tonalide, **7** - Phthalates, **8** - Phosphates

Beside all pollutants described above, several organic contaminants identified and monitored in rivers and groundwater remain undiscussed, because they were either only rarely detected so far or of minor environmental interest in terms of ecotoxicological aspects. Examples include e.g. N-methylpyrrolidinone, dinapthylsulfones, polychlorinated *n*-alkanes, alkylsulfonic acid aryl esters, hexabromobenzene and X-ray contrast media (Watanabe et al. 1986; Franke et al. 1998a; Ternes and Hirsch 2000; Nicholls et al. 2001; Headley et al. 2001; Schwarzbauer and Franke 2003b).

### 1.2.3 Anthropogenic marker compounds

In addition to the above mentioned pollutants, which are characterized by harmful ecotoxicological or toxicological effects, further anthropogenic contaminants were analysed frequently, the so-called 'anthropogenic markers'. An impressive review on the evaluation and use of anthropogenic marker compounds has been published recently by Takada and Eganhouse (1998).

These substances are appropriate to reflect anthropogenic emissions to natural systems especially in terms of source specifity. On the contrary all contaminants discussed so far are only general indicators of human activities, because they derived from multiple sources.

Anthropogenic marker compounds are both natural substances derived from human feces or urine and xenobiotics. Apart from information on the emission sources they also provide information on transport pathways and fate of anthropogenic contaminants. In case of undisturbed sedimentation the analysis of marker compounds also reveals geochronological information. The marker approach is based generally on the unique molecular structures that can be unambiguously linked to specific origins. An imposing example is presented in Fig. 8. Urobilin has been recently proposed to be an anthropogenic marker based on its closed linkage with human urine and feces.

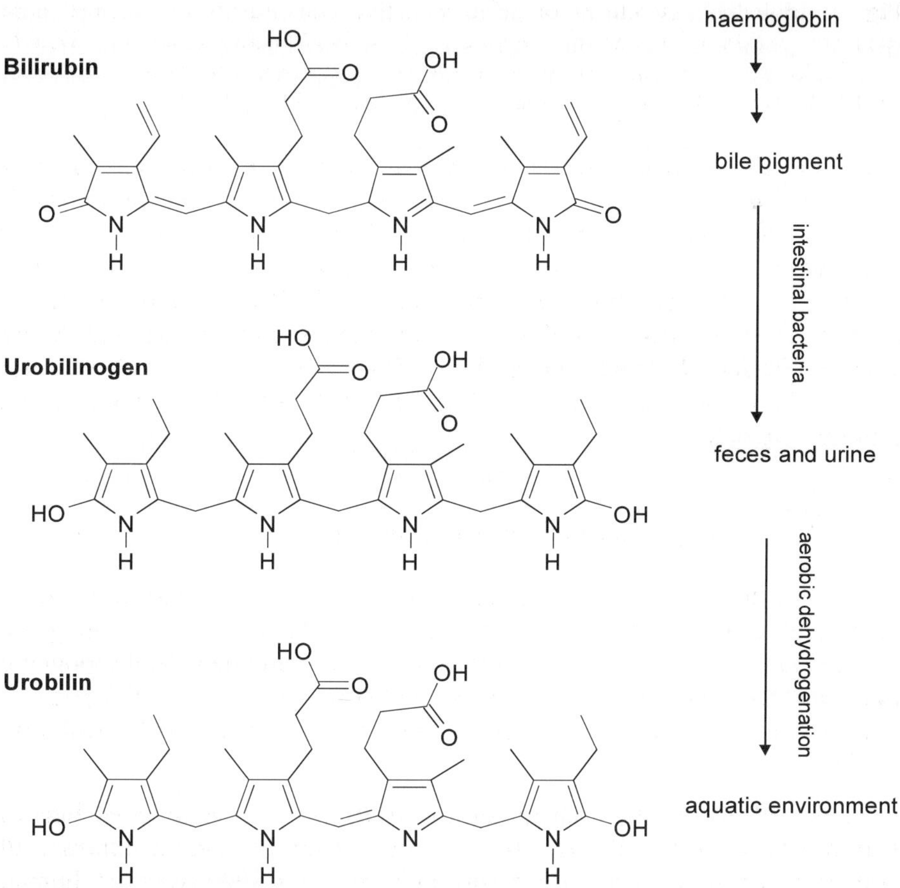


**Fig. 8:** Origin and transformation pathway of the anthropogenic marker compound urobilin, partially adapted from Takada and Eganhouse (1998).

Urobilin is an oxidation product of urobilinogen that is formed from bilirubin, a bile pigment, by human intestine. Eventually, all these compounds are natural degradation products of haemoglobin, the blood pigment. Tracing urobilin it was able to monitor municipal waste water effluents in a Japanese river system (Takada and Eganhouse 1998).

However, anthropogenic marker compounds has to fulfill the following criteria: (i) source specifity, (ii) widespread and massive usage and (iii) environmental persistence (Takada and Eganhouse 1998). Based on their physico-chemical properties and the resulting environmental behaviour they can be divided into hydrophilic and hydrophobic marker compounds that are associated with the water or the particulate phase, respectively. Hydrophobic marker compounds are widely used to trace the transport and mobilization of particle-bound contaminations, whereas hydrophilic substances are applied to monitor groundwater plumes. A classification of selected marker compounds appropriate for tracing sewage input and street runoff is presented in Table 1.

**Tab. 1:** Well known anthropogenic marker compounds, their sources and occurrence (adapted from Takada and Eganhouse 1998)

| Source | Hydropohobic marker | Hydrophilic marker |
|---|---|---|
| | Sediments, SPM | Water |
| Sewage | | |
| Natural products | Coprostanol/Coprostanon<br>α-Tocopherolacetate | Urobilin<br>Aminopropane<br>Coffeine |
| Synthetic detergents | Linear alkylbenzenes (LABs) | Linear alkylbenzenesulfonates (LAS) |
| | Tetrapropylen-based alkylbenzenes (TABs) | Tetrapropylen-based alkylbenzenesulfonates (TABs) |
| | Trialkylamines<br>Long chain alkylnitriles | Dialkyltetralinsulfonates<br>Fluorescent whitening agents |
| Street runoff | | |
| | Benzothiazole derivatives | Benzothiazole derivatives |

The best known anthropogenic marker compounds comprise the group of alkylated benzenes and benzenesulfonates. Tetrapropylene-based alkylbenzenes (TABs) are byproducts of the corresponding sulfonates (ABS), which are synthetic cationic detergents widely used until the mid 70'ies. Due to a poor biotic degradation potential they were substituted by linear alkylbenzenesulfonates (LAS), but nevertheless the non-sulfonated benzenes (LABs) remained as byproducts. Hence the occurrence of LABs and TABs in the environment can be clearly attributed to a time depending usage of synthetic detergents and, consequently, they have been widely used as anthropogenic marker of domestic sewage effluents. Based on their lipophilicity TABs and LABs are generally analysed in sediments and suspended particulate matter within aquatic systems (e.g. Takada and Ishiwatari 1990; Chaulax et al. 1995; Barber and Writer 1998, Takada and Eganhouse 1998 and citation therein). Another feature of the LABs is the systematic change in isomeric composition which provides information on the extent of microbial degradation processes. This approach is based on the observation of a relative depletion of external substituted isomers during incubation experiments indicating an isomer-selective degradation (Takada and Ishiwatari 1990).

Apart from LABs and TABs also steroids are frequently used to trace municipal sewage emissions. The ratio of naturally widespread cholesterol and coprostanol, one of the major fecal steroids, is an appropriate indicator to assess sewage contributions to the aquatic environment (e.g. Chalaux et al. 1995; Takada et al. 1997; Chan et al. 1998; Eganhouse and Sherblom 2001). This approach is also applicable on the oxidation products cholestanone and coprostanone in a similar way.

Marker compounds were also used to evaluate petrogenic emissions. Characteristic fossil biomakers were used to identifiy sources of petroleum contamination in the aquatic environment. Appropriate compounds are *n*-alkanes, steranes, hopanes and acyclic isoprenoids (e.g. Jovancicevic et al. 1997; Jovancicevic et al. 1998, Kaplan et al. 1997; Tran et al. 1997; Volkmann et al. 1997, Wehner et al. 1998: Faure et al. 2000; Jovancicevic et al. 2003 ).

## 1.3 Overview on the presented studies

Although the chemical analysis of anthropogenic contamination in groundwater and river systems is still an important feature, especially with respect to the identification and structure elucidation of novel contaminants, the focus of environmental studies has changed. Firstly, in the last 20 years the research on the environmental behaviour of organic pollutants became more and more important in order to evaluate and assess their risk as a result of their discharge to natural systems. Consequently, the knowledge not only about the occurrence and amount of anthropogenic contaminants but also about their state of bioavailability, their exchange within different compartments, as well as on molecular aspects of degradation or metabolism processes increases significantly. Secondly, environmental studies on the anthropogenic pollution of aquatic systems do not remain on the level of the analysis of pre-selected substances. Considering the currently use of over 1000.000 man-made chemicals in the EU (from which over 2000 are 'high volume' chemicals with production rates of more than 1000 t per year, see Alcock and Jones 1999) it has to be assumed, that analytical investigations on environmental systems solely focussed on a few pollutants still do not depict the 'real' pollution level. Therefore, the number of more extensive and detailed investigations considering a wide spectra of relevant organic pollutants increases slowly. Such investigations reveal a more comprehensive view on the state of pollution of a natural system.

The following work (chapters 2 to 6) represents a selection of organic-geochemical investigations performed between 1998 and 2003 at the Institute of Geology and Geochemistry of Petroleum and Coal, Aachen University. These studies consider several quite different aspects on the anthropogenic contribution to the organic contamination of riverine and groundwater systems. Nevertheless, all presented investigations are based on various common elements:

The areas of investigation are spatially restricted. Studies on river water or sediments focussed on selected areas of the Elbe and Rhine river systems comprising the Havel and Spree river (exclusively in the urban area of Berlin), the Elbe estuary (German Bight), the Teltow Canal, as well as the Lippe river and the middle Rhine river. Investigations on groundwater contamination were exclusively restricted to two small areas in North-Rhine/Westfalia (Fig. 9).

Further on, a methodological concordance exist. All investigation include comprehensive non-target screening analyses in order to identify and quantify a wide variety of individual organic contaminants. The spectrum of substances is only restricted by the analytical techniques applied. Since GC/MS is used, lipophilic and volatile to semi-volatile compounds were detected, whereas very polar or non-volatile substances were neglected. However, information on macromolecular organic matter and associated low molecular substances were also obtained after application of pyrolytic or chemical degradation procedures. All investigations performed aimed on a comprehensive detection of the anthropogenic impact of anthropogenic emissions on the natural systems studied.

A huge proportion of the presented investigations used organic marker compounds in order to differentiate the various anthropogenic emissions to the river and groundwater systems. The studies are not restricted to the application of established marker compounds, but numerous new marker substances were introduced and evaluated.

For a more detailed overview on the areas of investigations as well as on the compartments studied (with exception of study chapter 4.1.1) see Fig. 9. In the following all individual studies are briefly introduced.

*Chapter 2* contains two studies related to groundwater contamination as a result of two different anthropogenic activities. The first study (chapter 2.1) focuses on emissions derived from a leaking waste deposit landfill. The chemical characterisation as well as the identification of specific contaminants are main objectives. Further on, using a set of specific waste deposit derived marker compounds the spatial distribution as well as a time trend was characterised in order to assess the risk of the waste deposit derived effluents.

On the contrary the second investigation (chapter 2.2) deals with a very immobile contamination as the result of the activities of a former military base since the late 1950s. For impregnation of woody material coal tar based additives and technical formulations of mono- and dichlorinated naphthalenes were applied and due to a careless usage emitted continuously to the soil and the corresponding groundwater. Innovative remediation strategies e.g. by Natural Attenuation or Enhanced Natural Attenuation have been discussed intensively for this contamination. However, with respect to these remediation activities the information on occurrence and rate of microbial degradation activities is of fundamental importance. Hence the presented study was performed in order to evaluate the state of pollution as well as to reveal molecular markers for a microbial activity with respect to the main organic pollutants.

In *chapter 3* the focus of the presented studies turns to riverine systems. Firstly, the anthropogenic contamination of the Lippe river system is characterized and discussed by analyses of water and sediments (chapter 3.1) in a longitudinal profile. Beside the attribution of the individual substances to different anthopogenic input pathways, quantitative analyses provide information on the environmental distribution and fate of selected characteristic contaminants. Further on, four sampling campaigns performed within two years allow a first insight into the time depending behaviour of the organic load not only of the Lippe river water but also of the corresponding sediments.

The same approach was applied to water samples of the Rhine river between Koblenz and the border to the Netherlands (chapter 3.2). Similarly, a comprehensive qualitative analysis of the anthropogenic organic contaminants and quantitative measurements of selected substances are the main objectives. The results are discussed with respect to emission sources, environmental behaviour of the pollutants as well as their marker properties.

The last sub chapter (chapter 3.3.) deals with organic-geochemical analyses applied to sediment samples of the Havel and Spree rivers. Both qualitative and quantitative results are discussed exclusively by the assessment of new anthropogenic marker compounds. Appropriate candidates revealed by intense screening analysis are discussed considering their source specifity, environmental stability and distribution within the technosphere. The region of investigation was selected due to special conditions in the aquatic environment of the Havel and Spree Rivers, which are characterized by high inputs of anthropogenic contaminants into an eutrophic to hypertrophic riverine system with very slow flowing conditions. These conditions are ideal for searching novel anthopogenic markers. It has to be noted that the experimental part of this study was performed between 1996 to 1997 at the University of Hamburg.

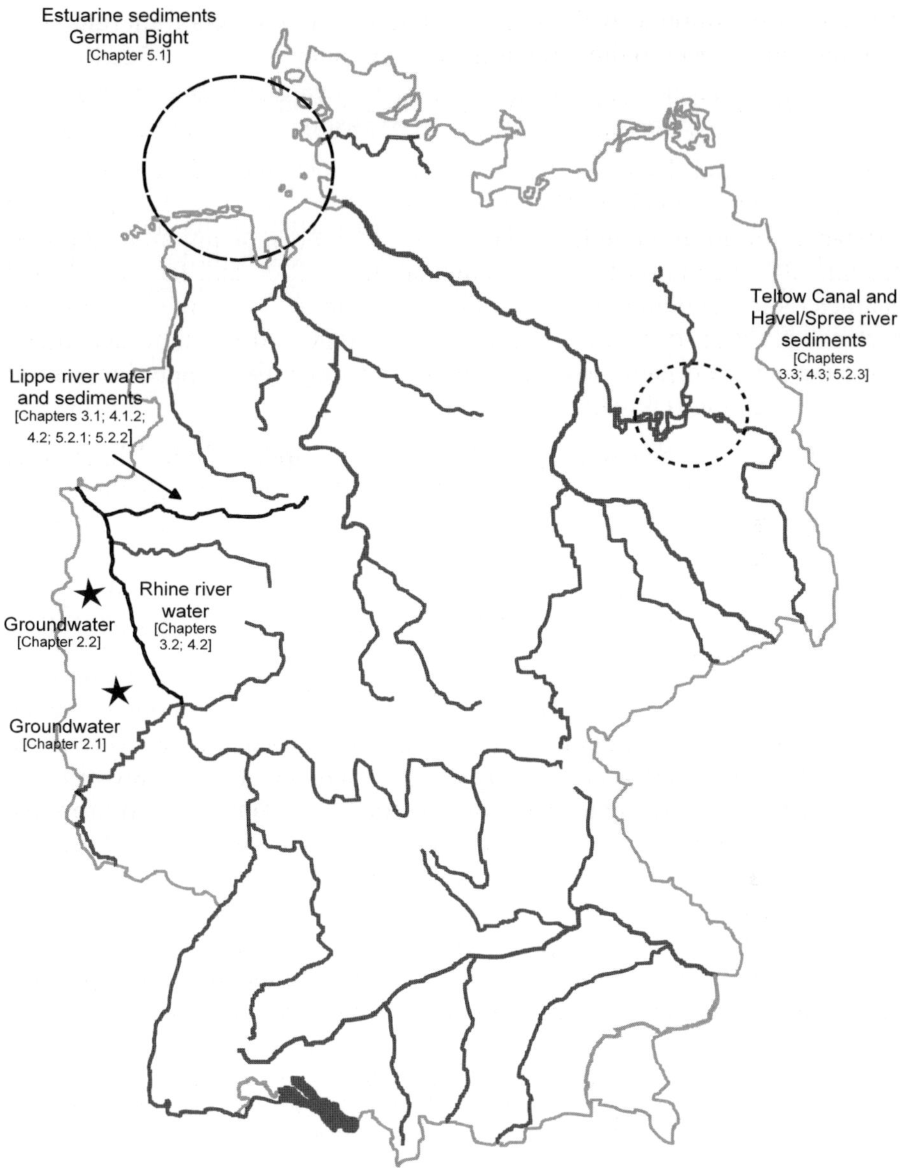


**Fig. 9:** Sampling locations and compartments investigated of nearly all studies presented

*Chapter 4* covers studies which deal intensively with the fate of anthropogenic substances within riverine systems. In a first section (chapter 4.1.) the environmental behaviour of known contaminants, the polycyclic musk fragrances, are examined. At first different emission profiles of selected prominent representatives in sewage effluents from Sweden and Canada are discussed and significant differences evaluated (Swedish data were provided from Dr. A Svenson, Swedish Environmental Research Institute, Stockholm). Following, the spatial and time depending distribution of these compounds within the Lippe river system comprising water and particulate matter is presented. In particular the monitoring of galaxolide/tonalide ratios, both the dominant musk substitutes (see chapter 1.2), provides information on emission sources and environmental partition processes.

Chapter 4.2 introduces a novel analytical tool, the compound specific isotope analysis (CSIA), to evaluate the origin and fate of organic substances in riverine systems. Carbon isotope ratios are useful data not only for evaluating transport and transformation processes (e.g. vaporization, adsorption, abiotic and biotic degradation) but also for distinguishing between different emission sources discharging the same contaminants. But common environmental applications are performed on systems with elevated contamination levels as well as on simple compound mixtures and, consequently, avoid the two major limitations of compound-specific isotope analyses. These limitations lie in the lower sensitivity as compared to traditional GC/MS analyses and in the requirement of complete gas chromatographic separation of the contaminants. However, natural aquatic systems, especially riverine or groundwater systems, are mostly contaminated at a low or medium level. Therefore, the present investigation focussed on the evaluation of an appropriate analytical procedure for compound-specific stable carbon isotope analyses of riverine contaminants. In a second part this analytical method was applied to water samples of the Rhine and Lippe rivers, already introduced in chapters 3.1, 3.2 and 4.1. $\delta^{13}C$-values of selected anthropogenic contaminants are presented and discussed with respect to information on their sources and environmental fate.

The last aspect presented regarding the environmental behaviour of organic contaminants considers the incorporation of low molecular weight compounds into the nonextractable particulate fraction, preferentially into the macromolecular organic matter (MOM). These phenomena, well known as the so-called 'bound residues', are well investigated for organic pollutants, mainly pesticides, introduced to the pedosphere. However, detailed studies of bound residues in the aquatic environment, especially comprehensive analyses on a molecular base, are rarely published. Nevertheless, anthropogenic activities induce significant alterations of the macromolecular organic matter in riverine systems mainly by emission of pollutants and their subsequent incorporation into geopolymers. Hence, the studies presented in *chapter 4.3* characterize the non-extractable residues of highly polluted riverine sediments (Spree river, Teltow Canal) in order to investigate the occurrence, alteration and distribution of several organic xenobiotics *in situ*. Therefore the investigations intend a comprehensive characterization of riverine MOM and associated bound residues combining different analytical techniques (pyrolytic analyses and chemical degradation procedures), in order to provide information concerning the incorporation mechanism and the mode of binding of organic pollutants. A comprehensive qualitative characterization of both the macromolecular and the low molecular weight fraction is given in the first section, whereas the following sections focus on quantitative determinations of selected contaminants as revealed by detailed screening analyses. The contents of bound DDT-related compounds as well as of specific marker compounds or still unnoted riverine contaminants are analysed and discussed.

Apart from the above introduced studies considering emission sources, distribution and fate of organic contaminants within riverine systems (as presented in the chapters 2,3 and 4), the output of organic contaminants from riverine systems is treated in *chapter 5* –. As mentioned in chapter 1.2 an important transport pathway especially for lipophilic contaminants is the particle associated displacement. Generally two different target areas exist for the sedimentation of riverine particulate matter outside the river system. Both estuarine areas as well as terrestrial flood plains (e.g. riparian wetlands) represent an important sink of fluviatile particles and associated contaminants. These regions are objectives of two different studies. Sediments of the German Bight are investigated not only in order to document the state of organic contamination but also to identify specific molecular markers that are appropriate to estimate the discharge of anthropogenic compounds derived from the Elbe river (chapter 5.1). Following, selected contaminants were quantified representing proposed Elbe specific marker compounds. For the evaluation the quantitative and spatial distribution was investigated and successfully discussed as the most important criterion to define their source specifity.

A second set of studies deals with the riverine contribution of Lippe river derived substances to the organic contamination of a corresponding periodically flooded riparian wetland (chapter 5.2.1 and 5.2.2). Dated core samples allow to estimate the geochronological occurrence of numerous persistent contaminants reflecting the pollution history of the Lippe river system since the beginning of the 20$^{th}$ century. The radiological dating was performed by Clemens Woda and Prof. Dr. Mangini, University of Heidelberg.

Furthermore, a long term storage in flow restricted subaquatic areas (e.g. harbours or bayous) represent a metastable output of sedimentary riverine matter. This phenomena is considered in an investigation performed on a river sediment core from the Teltow Canal (chapter 5.2.3). In particular DDT-related substances are analysed in order to obtain information on the long time emission of an industrial point source. Also an undisturbed sedimentation allows the correlation of the quantitative data obtained from the core samples with a geochronological determination of the pollutants.

The results and the most important conclusions of all investigations described are summarized in *chapter 6*. A major proportion of the presented results are separately published in more detail in scientific journals:

Chapter 2.1.1 : Schwarzbauer J, Heim S, Brinker S, Littke R (2002) Wat Res 36: 2275-2287.

Chapter 2.1.2[*]: Heim S, Schwarzbauer J (2003) Environ Chem Lett 2: 21-25.

Chapter 2.2.1[*]: Vinzelberg G, Schwarzbauer J, Littke R (2005) Chemosphere 61: 770-782.

Chapter 3.1.1 and 3.1.2[*]: Dsikowitzky L, Schwarzbauer J, Kronimus A, Littke R (2003) Chemosphere 57: 1275-1287 and 1288-1293

Chapter 3.1.3[*]: Kronimus A, Schwarzbauer J, Heim S, Littke R (2004) Wat Res 38: 3473-3484

Chapter 3.2.1[*]: Schwarzbauer J, Heim S (2005) Wat Res 39: 4735-4748

Chapter 3.3.1[*]: Ricking, M., Schwarzbauer, J., Franke, S. (2003) Wat Res 37: 2607-2617.

Chapter 4.1.1[*]: Ricking, M., Schwarzbauer, J., Hellou, J., Svenson, J., Zitko, A. (2003) Marine Poll Bull 46: 410-417.

Chapter 4.1.2 : Dsikowitzky L, Schwarzbauer J, Littke R (2002) Org Geochem 33: 1747-1758.

Chapter 4.2.1[*]: Schwarzbauer J, Dsikowitzky L, Heim S, Littke R (2005) Int J Environ Anal Chem 85: 349-364

Chapter 4.3.1*: Schwarzbauer J, Ricking M, Gieren B, Keller R, Littke R (2005) In: Environmental Chemistry, Lichtfouse, Schwarzbauer, Didier (Eds), Springer Verlag, Weinheim: 329-352.

Chapter 4.3.2*: Schwarzbauer J, Ricking M, Littke R (2002) Environ Sci Technol 37: 488-495.

Chapter 4.3.3*: Schwarzbauer J, Ricking M, Littke R (2003) Acta Hydrochim Hydrobiol 31: 469-481

Chapter 5.1.1 : Schwarzbauer J, Littke R, Weigelt V (2000) Org Geochem 31: 1713-1731.

Chapter 5.1.2*: Schwarzbauer J, Littke R (2003) J Soils Sed 4: 177-183

Chapter 5.2.1*: Heim S, Schwarzbauer J, Kronimus A, Littke R, Hembrock-Heger A (2003) Env Chem Lett 1: 169-173

Chapter 5.2.2*: Heim S, Schwarzbauer J, Kronimus A, Littke R, Woda C, Mangini A (2003) Org Geochem 35: 1409-1425

Chapter 5.2.3 : Ricking M, Heim S, Schwarzbauer J (2004) Organohalogen Comp 66: 1455-1462.

* Reprinted with permission from the following publishers: Elsevier, ACS, Springer, Taylor and Francis, Wiley-VCH and Ecomed.

# 2 Low molecular weight anthropogenic compounds in groundwater

## 2.1 Waste deposit derived contamination of groundwater

### 2.1.1 Occurrence and alteration of organic contaminants in seepage and leakage water from a waste deposit landfill ‡

#### Introduction

Seepage water of a waste deposit landfill is characterized by high values of chemical as well as biological oxygen demand (COD, BOD) and is contaminated by a complex mixture of waste-derived organic compounds. The pattern of the lipophilic and low molecular-weight organic substances may vary considerably depending on the composition of the deposited waste, the method of disposal, the chemical behavior of the contaminants, the physico-chemical parameters existing within the deposit and different transport mechanisms (Lema et al., 1988). It has to be noted that several studies confirmed the potential toxicological risk of landfill leachates (Cameron et al., 1980; Atwater et al., 1983).

Besides structural investigations of the macromolecular organic matter in landfill leachates (Göbbels and Püttmann, 1997) several studies have focused on the occurrence and quantitation of selected low molecular-weight pollutants in seepage water due to their potential health risk or physico-chemical behavior. Examples include the determination of PCDD/PCDF and chlorinated benzenes in oil and water leachate of a sanitary landfill in Hamburg, Germany, (Götz, 1985), the detection of low volatile compounds in different landfill leachates (e.g. Först et al. 1989) as well as an investigation concerning the distribution of alkylated benzenes and naphthalenes in seepage water of a municipal solid waste deposit in Denmark (Rügge et al., 1995). Only a few studies considered the composition of organic components in leachate samples in a more

‡ Mainly adapted/reprinted from Schwarzbauer et al. 2002

comprehensive way (Schultz and Kjeldsen, 1986; Harmsen, 1993; Öman and Hynning, 1993). Investigations concerning transformation or degradation processes of organic pollutants within the deposit are reported rarely (Paxeus, 2001).

Whereas controlled decontamination of seepage water by biological or chemical waste water treatment is usually performed leachate can also infiltrate the groundwater due to an insufficient bottom sealing (Murray et al., 1981; Albaiges et al., 1986; Rügge et al., 1995; Looser et al., 1999). In this case organic pollutants of landfill leachates can seriously affect groundwater quality. Information about the fate of organic In order to assess their impact on the groundwater identification and quantification of landfill-derived substances has to be undertaken in a systematic way. Furthermore information about the fate of organic contaminants after release of the deposit are very limited although the change from anaerobic to more aerobic conditions outside the deposit should cause significant alterations to the organic matter.

This study presents detailed information about the organic constituents in groundwater-contaminating seepage water of a waste deposit landfill. Based on detailed gas chromatographic-mass spectrometric analyses a comprehensive view on the composition of organic contaminants will be given.

The investigated seepage water samples were taken from a sanitary landfill in Germany where a leak in the bottom sealing was discovered. Although the wastewater was found to seep down into a former mining system situated below the deposit, the subsequent route of the contaminants is still unknown. Because leakage water could be sampled directly within the mining system, this study also presents results on the alteration of seepage water as a consequence of a short soil passage.

## Samples

In 1998 two seepage wastewater samples were taken from a waste deposit landfill in Germany (see Tab. 1 and Fig.1). Sampling was performed at two different haulings directly at the top of the landfill. In addition a water sample of leakage water was taken in the underlying mining system (see Tab. 1). All water samples had a pH-value of 7.9.

Details of quantitative analyses are given in Table 2. The detection limit was in the range of 50 ng/L, no attempts were made to quantify components at concentrations of less than 1 μg/L.

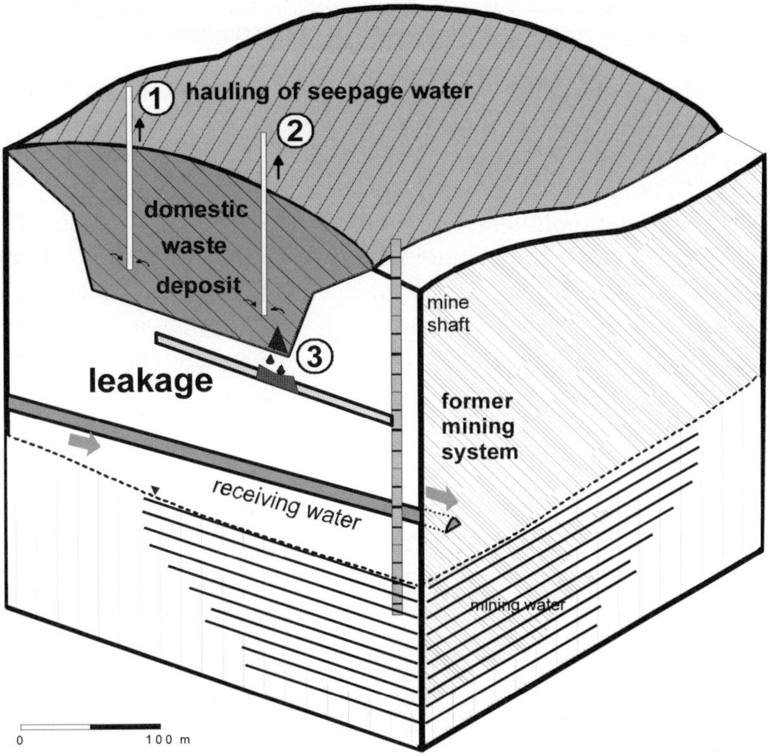


**Fig 1:** Sampling locations of seepage water (1,2) and leakage water (3) of a waste deposit landfill, Germany

**Tab. 1:** Samples of seepage water and leakage water.

| Abbreviation | Sampling location | Date of sampling | pH |
|---|---|---|---|
| SW1 | Seepage water of hauling 1 | June 1998 | 7.9 |
| SW2 | Seepage water of hauling 2 | June 1998 | 7.9 |
| LW | Leakage water | June 1998 | 7.9 |

**Tab. 2:** Selected contaminants and affiliated ion fragments used for quantification as well as determined recovery rates.

| Compounds | Ion fragments | Recovery |
|---|---|---|
| *Terpenoids* | | |
| Limonene | 68,93,136 | 16 |
| 1,8-Cineole | 81,108,154 | 16 |
| Fenchone | 69,81 | 25 |
| Camphor | 95,108,152 | 20 |
| Borneol | 110,139 | 23 |
| Thymol | 135,150 | 22 |
| α-Terpineol | 93,136 | 27 |
| *Phenols* | | |
| *p*-Cresol | 90,107 | 36 |
| Bisphenol A | 119,213,228 | 41 |
| *N-containing substances* | | |
| N,N-Diethyltoluamide (DEET) | 91,190 | 60 |
| Nicotine | 84,133,162 | 60 |
| Cotinine | 98,176 | 60 |
| 1H-Indol | 90,117 | 58 |
| 3-Methylindol, scatol | 93,130 | 58 |
| Benzthiazol | 108,135 | 37 |
| Propyphenazone | 215,230 | 60 |
| *S-containing substances* | | |
| Diphenylsulfone | 77,125 | 55 |
| N-Butylbenzenesulfonamide (NBBS) | 141,170 | 60 |
| *P-containing substances* | | |
| Tri-*iso*-butylphosphate | 99,155,211 | 45 |
| Tri-*n*-butylphosphate | 99,155,211 | 45 |
| Tris-(2-chloroethyl)phosphate | 205,249,251 | 44 |
| Triethylphosphate | 99,127 | 44 |

## Results and discussion

GC/MS screening analyses of the seepage water and leakage water samples revealed a wide variety of lipophilic organic substances. The more than 180 individual organic contaminants are summarized in Tab. 3 and quantitative data of selected compounds are given in Tab. 4.

The sequential extraction of the seepage water and leakage water samples yielded three extract fractions with low polar, moderately polar and acidic compounds, respectively. Occurrence and distribution of individual organic compounds in each extract are illustrated in Fig. 2 which comprises the gas chromatograms of the two non derivatized extracts and the methylated as well as silylated third extract of sample SW2. The composition of the organic contaminants in seepage water samples (SW1, SW2) differed significantly as compared to the leakage water sample (LW). These differences in the organic-geochemical signature are depicted in Fig.4.

**Tab. 3:** Organic compounds identified in seepage and leakage water of a German waste deposit landfill.

| | **Compounds** | **SW1** | **SW2** | **LW** |
|---|---|---|---|---|
| | *Aliphatics* | | | |
| 1 | *n*-Tricosane [a] | + | + | |
| 2 | *n*-Tetracosane [a] | + | + | |
| 3 | *n*-Pentacosane [a] | + | + | |
| 4 | *n*-Hexacosane [a] | + | + | |
| 5 | *n*-Heptacosane [a] | + | + | |
| 6 | *n*-Octacosane [a] | + | + | |
| 7 | *n*-Nonacosane [a] | | + | |
| 8 | *n*-Triacontane [a] | | + | |
| | *Terpenoids* | | | |
| 9 | Menthone [b] | + | + | |
| 10 | Limonene [a] | | + | |
| 11 | Thymol [a] | + | + | + |
| 12 | Terpineol (2 isomers) [a] | + | + | + |
| 13 | 1,8-Cineole [a] | + | + | |
| 14 | Carvomenthone [b] | + | + | |
| 15 | Fenchone [a] | + | + | |
| 16 | Borneol [a] | | + | |
| 17 | Camphor [a] | + | + | + |

| | Compounds | SW1 | SW2 | LW |
|---|---|---|---|---|
| 18 | Camphorchinone [e] | | | + |
| 19 | α-Campholenic acid [# b] | + | + | |
| 20 | 3,3-Dimethylnorbornane-2-carboxclic acid [§ e] | + | | |
| 21 | Bicyclo[2.2.1]heptenedicarboxylic acid [# d e] | + | + | + |
| 22 | Norborn-5-ene-2,3-dicarboxylic acid [§ e] | + | + | |
| 23 | Nopinone [b] | + | + | |
| 24 | Methylnopinone [b] | + | + | + |
| | | | | |
| 25 | Dehydroabietinic acid [§ # a] | + | + | + |
| 26 | Levopimaric acid [# c] | + | + | + |
| 27 | Podocarbic acid [§ # e] | | + | |
| 28 | Tetrahydroretene [a] | + | + | |
| | | | | |
| 29 | 7-Acetyl-2-hydroxy-2-methyl-5-iso-propyl-bicyclo[4.3.0.]nonane [e] | + | + | + |
| 30 | Squalene [a] | + | | + |
| | | | | |
| | *Benzenes* | | | |
| 31 | p-Xylene [a] | + | + | |
| 32 | $C_3$-Benzene (2 isomers) [c] | + | + | |
| 33 | $C_4$-Benzene [c] | | + | |
| | | | | |
| | Aromatics | | | |
| 34 | Indane [a] | + | + | |
| 35 | Fluorene [a] | + | + | |
| 36 | Naphthalene [a] | + | + | |
| 37 | Trimethyltetralin [d e] | + | | |
| 38 | Tetrahydrohydroxynaphthoic acid [§ e] | + | | |
| 39 | $C_4$-Phenanthrene (1 isomer) [d] | | | + |
| | | | | |
| | *Alcohols and ethers* | | | |
| 40 | Heptanol [b] | + | | |
| 41 | 1-(1-Ethoxyisopropoxy)-2-propanol [e] | | + | |
| 42 | 2-Phenylpropan-2-ol [b] | | | + |
| 43 | Glycol [a] | + | + | |
| 44 | Diphenylether [a] | + | + | |
| | | | | |
| | *Phenols and thiophenols* | | | |
| 45 | Phenol [a] | + | + | + |
| 46 | p-Cresol [a] | + | + | + |
| 47 | $C_2$-Phenol (5 isomers) [c] | (+) | + | (+) |
| 48 | $C_3$-Phenol (5 isomers) [c] | (+) | + | (+) |

| | Compounds | SW1 | SW2 | LW |
|---|---|---|---|---|
| 49 | $C_4$-Phenol (2 isomers) [c] | + | + | (+) |
| 50 | Ionol [b] | | + | |
| 51 | 2-(2-Hydroxyphenyl)-2-(4-hydroxyphenyl)-propane [e] | | + | |
| 52 | Bisphenol A [a] | + | + | + |
| 53 | Thiophenol [b] | | | + |
| 54 | Thiocresol [b] | | | + |
| | *Aldehydes and ketones* | | | |
| 55 | iso-Propylbenzaldehyde [b] | | + | |
| 56 | Methylcyclohexanone [d] | | | + |
| 57 | 3,3,5-Trimethylcyclohexanone [b] | + | + | |
| 58 | Acetophenone [a] | + | + | + |
| 59 | *iso*-Propylacetophenone [d] | | | + |
| 60 | (Hydroxy-iso-propyl)acetophenone [d e] | | + | + |
| 61 | Propiophenone [a] | | + | |
| 62 | 2,6-Di-tert-butylchinone [a] | + | + | + |
| 63 | Coumaranone [# b] | | | + |
| 64 | Benzophenone [a] | + | + | |
| | *Aliphatic acids* | | | |
| 65 | Butyric acid [# a] | + | + | |
| 66 | Valeric acid [§ # a] | | + | + |
| 67 | Hexanoic acid [§ # a] | + | + | |
| 68 | Octanoic acid [§ # a] | + | + | + |
| 69 | Pelargonic acid [§ # a] | + | + | + |
| 70 | Decanoic acid [# a] | | + | |
| 71 | Lauric acid [§ # a] | + | + | |
| 72 | Myristic acid [§ # a] | + | + | |
| 73 | Palmitic acid [§ # a] | | | + |
| 74 | Stearic acid [# a] | + | + | |
| 75 | Oleic acid [# a] | + | + | |
| 76 | Linoleic acid [# a] | + | + | + |
| 77 | Malonic acid [§ # a] | + | | + |
| 78 | Fumaric acid [§ # a] | + | + | + |
| 79 | Succinic acid [§ # a] | + | + | + |
| 80 | Glutaric acid [§ # a] | + | + | |
| 81 | Adipic acid [§ # a] | + | + | + |

| | Compounds | SW1 | SW2 | LW |
|---|---|---|---|---|
| 82 | Pimelic acid [§ # a] | + | | + |
| 83 | Azelaic acid [§ # a] | + | + | + |
| 84 | Sebacic acid [§ # a] | | + | + |
| 85 | Branched butyric acid [# d] | | + | + |
| 86 | Branched pentanoic acid (3 isomers) [# d] | + | + | + |
| 87 | Branched hexanoic acid (2 isomers) [# d] | + | | + |
| 88 | Branched heptanoic acid (2 isomers) [# d] | | + | |
| 89 | Branched octanoic acid (2 isomers) [# d] | (+) | (+) | (+) |
| 90 | Branched nonanoic acid [# d] | | + | |
| 91 | Glycolic acid [§ # e] | + | + | + |
| 92 | Hydroxybutyric acid [§ # e] | + | | |
| 93 | $C_2$-Hydroxybutyric acid [# d] | | | + |
| 94 | Dihydroxybutyric acid [# e] | + | + | |
| 95 | Hydroxypivalinic acid [§ # d] | + | + | + |
| 96 | Hydroxypentanoic acid [§ d] | + | + | |
| 97 | Hydroxyheptanoic acid [§ d] | + | + | + |
| | *Cyclic and aromatic carboxylic acids* | | | |
| 98 | Cyclopentane carboxylic acid [§ e] | | | + |
| 98 | Cyclohexane carboxylic acid [# § e] | | | + |
| 100 | $C_3$-Cyclohexane carboxylic acid [# d] | + | + | |
| 101 | Cyclohexylacetic acid [§ # d] | + | | |
| 102 | Hydroxycyclohexane carboxylic acid [§ # d] | + | | + |
| 103 | Benzoic acid [# § a] | + | | + |
| 104 | Phenylacetic acid [# § a] | | | + |
| 105 | Phenylpropionic acid [# e] | + | + | + |
| 106 | Phenoxypropionic acid [§ # e] | | | + |
| 107 | Phenylbutyric acid (2 isomers) [§ # e] | | (+) | (+) |
| 108 | Methylphenylbutyric acid [§ # d] | + | + | |
| 109 | Phenylsuccinic acid [§ # e] | + | + | + |
| 110 | $C_1$-Benzoic acid (2 Isomers) [§ # d] | (+) | + | + |
| 111 | $C_2$-Benzoic acid (7 Isomers) [§ # d] | + | + | |
| 112 | $C_3$-Benzoic acid [§ # d] | + | + | |
| 113 | $C_4$-Benzoic acid [§ # d] | | | |
| 114 | $C_2$-Phenylacetic acid [# d] | + | + | + |

| | Compounds | SW1 | SW2 | LW |
|---|---|---|---|---|
| 115 | Thiophencarboxylic acid [# § e] | + | + | + |
| | | | | |
| 116 | Salicylic acid [# § a] | + | | + |
| 117 | Methylhydroxybenzoic acid [§ # d] | + | | |
| 118 | Hydroxyphenylacetic acid [§ e] | | | + |
| 119 | Hydroxyphenylpropionic acid [§ e] | | | + |
| 120 | Hydroxyphenylbutyric acid [§ e] | | | + |
| 121 | Hydroxycinnamic acid [§ e] | + | + | |
| 122 | Dihydroxydihydrocinnamic acid [§ # e] | + | + | + |
| | | | | |
| 123 | Biphenyl-2,2-dicarbonic acid [# e] | + | + | |
| 124 | 1,2,4-Benzenetricaboxylic acid [# e] | | + | |
| | *Esters* | | | |
| 125 | Phenylbenzoate [a] | | + | |
| 126 | Bis(2-ethylhexyl)adipate [e] | | | + |
| | *Phthalates und metabolites* | | | |
| 127 | Cyclohexane dicarboxylic acid (3 isomers)[§ # c] | + | + | (+) |
| 128 | Methyl-1,2-cyclohexane dicarboxylic acid [§ # c] | + | | |
| | | | | |
| 129 | Cyclohexene-1,2-dicarboxylic acid (2 isomers) [§ # c] | + | + | |
| 130 | Methylcyclohexene-1,2-dicarboxylic acid (2 isomers) [§ # c] | (+) | (+) | + |
| | | | | |
| 131 | Phthalic acid [# § a] | + | + | + |
| 132 | Isophthalic acid [§ # b] | + | + | + |
| 133 | Therephthalic acid [§ # b] | + | + | + |
| 134 | Methylisophthalic acid [# d] | + | + | |
| 135 | Diethylphthalate [a] | + | + | |
| 136 | Di-*iso*-butylphthalate [a] | + | + | + |
| 137 | Di-*n*-butylphthalate [a] | + | + | + |
| 138 | Bis(2-ethylhexyl)phthalate (DEHP) [a] | + | + | + |
| | *Pharmaceuticals* | | | |
| 139 | Ibuprofene [# a] | + | + | + |
| 140 | Propyphenazone [a] | + | + | + |
| 141 | Clofibric acid [# a] | + | + | + |
| | *Pesticides* | | | |
| 142 | N,N-Diethyltoluamide (DEET) [a] | + | + | + |

| | Compounds | SW1 | SW2 | LW |
|---|---|---|---|---|
| 143 | Mecoprop (MCPP) [# a] | + | + | + |
| | *Anilines* | | | |
| 144 | Aniline [a] | + | + | |
| | *Pyridines* | | | |
| 145 | Methylpyridine [d] | | + | |
| 146 | $C_2$-Pyridine [d] | | + | |
| 147 | Nicotine [a] | + | + | + |
| 148 | Cotinine [a] | + | + | + |
| | *Pyrazines* | | | |
| 149 | $C_2$-Pyrazine [d] | + | + | + |
| 150 | $C_3$-Pyrazine (2 isomers) [d] | + | + | + |
| 151 | $C_4$-Pyrazine (2 isomers) [d] | (+) | + | (+) |
| 152 | $C_5$-Pyrazine (2 isomers) [d] | | + | + |
| 153 | Methyl-6,7-dihydro-5H-cyclopentapyrazine (1 isomer) [c] | | + | |
| 154 | Dimethyl-6,7-dihydro-5H-cyclopenta-pyrazine (2 isomers) [c] | + | + | |
| 155 | Trimethyl-6,7-dihydro-5H-cyclopenta-pyrazine (1 isomer) [c] | + | + | |
| | *Indoles* | | | |
| 156 | 1H-Indole [a] | | + | + |
| 157 | 3-Methylindole, skatole [a] | + | + | + |
| 158 | 1H-Indoleacetic acid [§ # b] | | | + |
| 159 | Hydroxyindolecarboxylic acid [§ e] | | | + |
| | *Benzthiazoles* | | | |
| 160 | Benzthiazol [a] | + | | |
| 161 | Benzthiazolone (2 isomers) [e] | (+) | + | (+) |
| | *Sulfones and sulfonamides* | | | |
| 162 | Diphenylsulfone [a] | + | + | + |
| 163 | Toluenesulfonamide [c] | | | + |
| 164 | N,N,4-Trimethylbenzenesulfonamide [e] | | | |
| 165 | N-Phenylbenzenesulfonamide [c] | + | | + |
| 166 | N-Butylbenzenesulfonamide [a] | + | + | + |
| 167 | N- Ethyl-*p*-toluenesulfonamide [e] | + | + | + |
| 168 | N-Ethyl-*o*-toluenesulfonamide [e] | | | + |

| | Compounds | SW1 | SW2 | LW |
|---|---|---|---|---|
| | *Phosphates* | | | |
| 169 | Tri-*iso*-butylphosphate [a] | + | + | + |
| 170 | Tri-*n*-butylphosphate [a] | + | + | + |
| 171 | Triethylphosphate [a] | + | + | + |
| 172 | Tris(2-chloroethyl)phosphate [a] | | + | |
| 173 | Tris(butoxyethyl)phosphate [a] | + | + | |
| | *Chlorinated aromatics* | | | |
| 174 | Chloromethylpyridine [d] | | + | |
| 175 | 2,4-Dichlorobenzene [a] | | + | |
| 176 | Chloromethylphenol [d] | | + | |
| 177 | Dichloroaniline (1 isomer) [c] | | + | |
| 178 | 4-Chlorobenzoic acid [# § a] | + | + | |
| 179 | 2,4-Dichlorobenzoic acid [# a] | + | + | |
| | *Siloxanes* | | | |
| 180 | Octamethylcyclotetrasiloxane [a] | + | + | + |
| 181 | Decamethylcyclopentasiloxane [a] | + | | + |
| 182 | Dodecamethylcyclohexasiloxane [a] | + | + | |
| | *Inorganic acids* | | | |
| 183 | Boronic acid [$ e] | + | + | + |
| 184 | Phosphoric acid [§ # e] | + | + | |

(+) = not all isomers were detected
§ = detected as trimethylsilyl ester
# = detected as methyl ester
a) = identified by comparision of gas chromatographic and mass spectral data with those of reference compounds
b) = identified by comparison of gas chromatographic and mass spectral data with those of mass spectral data bases and published retention indices.
c) = identified by comparison of gas chromatographic and mass spectral data with those of mass spectral data bases and published gas chromatographic elution patterns.
d) = molecular structure is not more specified.
e) = identified by comparison of mass spectral data with those of mass spectral data bases

The organic components of the low-polar extract of the seepage water samples were dominated by terpenoic alcohols and ketones including camphor, fenchone, α-terpineol, methylnopinone and 1,8-cineole (see Fig. 2A). Concentrations of the most abundant terpenoics were in the range between 400 and 8000 μg/L. Apart from thymol the amounts of the terpenoics were lower in the leakage water sample indicating a rapid degradation of these compounds during the short soil passage (see Tab. 4). Terpenoic compounds are major components in etheric oils extractable from wood, leaves, blossoms, roots and cortices. Hence their occurrence in seepage water can be mainly attributed to emissions of deposited plant material. A significant contribution of technical terpenes mixtures (e.g. turpentine) to the contamination of the seepage water was excluded, because of the absence of their main components, the pinenes. Diterpenoic acids including dehydroabietinic, podocarbic and laevipomaric acid were detected in the acidic fraction accompanied by the degradation product tetrahydroretene. These resin acids indicated also plant-derived contributions to the organic contamination of the seepage water. A major contribution of terpenoic compounds to the organic matter of leachate water has been reported earlier (Reinhard et al., 1984).

Apart from terpenoic compounds a significant proportion of 3-methylindole (skatole) with up to 1400 μg/L was detected mainly in the extracts with low to moderately polar substances (see Fig. 2a). Skatole is a well known degradation product of the nitrogen-containing amino acid tryptophane. Generally the occurence of skatole accompanied by phenol and cresol can be regarded as an indicator for putrefaction processes.

**Fig. 2:** Total ion chromatograms of *n*-pentane (A), dichloromethane (B) and dichloromethane extracts after acidication and derivatization with TMSH (C) or MSTFA (D). Peaks are numbered according to Table 3 (A = artefacts).

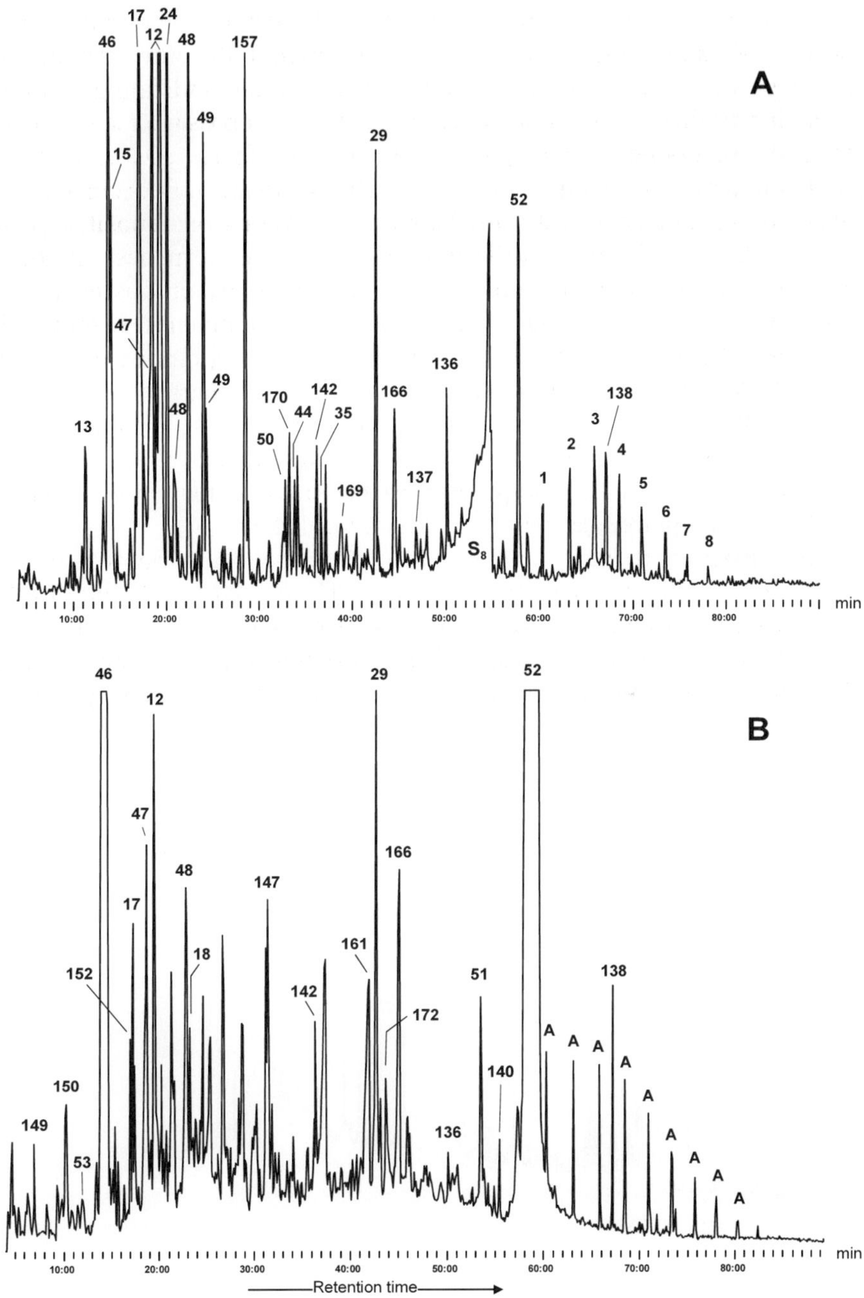

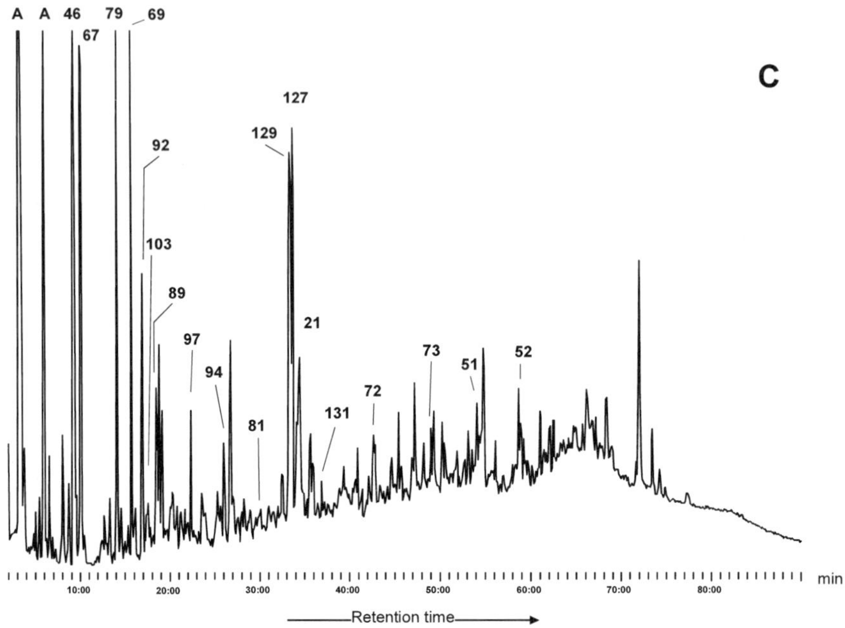
A
A
46
67
79
69
C
127
129
92
103
89
21
97
94
73
51
52
72
131
81
10:00
20:00
30:00
40:00
50:00
60:00
70:00
80:00
min
Retention time

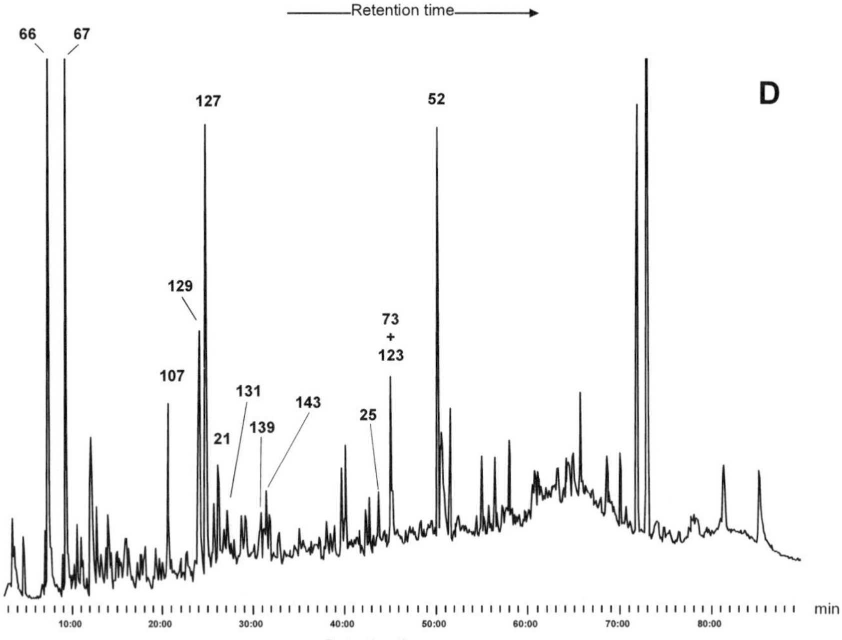
66
67
D
127
52
129
73
+
123
107
131
143
139
21
25
10:00
20:00
30:00
40:00
50:00
60:00
70:00
80:00
min
Retention time

**Tab. 4:** Quantitative data of selected contaminats in seepage and leakage water samples (µg/L).

| Compounds | SW1 | SW2 | LW |
|---|---|---|---|
| *Terpenoids* | | | |
| Limonene | <1 | 60 | <1 |
| 1,8-Cineol | 540 | 380 | <1 |
| Fenchone | 570 | 720 | <1 |
| Camphor | 7100 | 7400 | 370 |
| Borneol | <1 | 2300 | <1 |
| Thymol | 300 | 1300 | 380 |
| α-Terpineol | 1000 | 4300 | 740 |
| *Phenols* | | | |
| p-Cresol | 260 | 12000 | 8700 |
| Bisphenol A | 13000 | 25000 | 4200 |
| *N-containing substances* | | | |
| N,N-Diethyltoluamide (DEET) | 200 | 320 | 310 |
| Nicotine | 710 | 550 | 960 |
| Cotinine | 88 | 140 | 180 |
| 1H-Indol | 130 | 760 | 1400 |
| 3-Methylindol, scatol | 170 | 1200 | 1400 |
| Benzthiazol | 130 | <1 | <1 |
| Propyphenazone | 110 | 120 | 140 |
| *S-containing substances* | | | |
| Diphenylsulfone | 8 | 73 | 32 |
| N-Butylbenzenesulfonamide (NBBS) | 420 | 710 | 590 |
| *P-containing substances* | | | |
| Tri-*iso*-butylphosphate | 110 | 350 | 250 |
| Tri-*n*-butylphosphate | 18 | 50 | 110 |
| Tris(2-chloroethyl)phosphate | <1 | 240 | <1 |
| Triethylphosphate | 110 | 53 | 40 |

Within the moderately polar substances of the second extracts phenolic compounds were most abundant (see Fig. 2B). Next to *p*-cresol (260–12000 µg/L), a further product of amino acid degradation and putrefaction processes, the xenobiotic compound 2,2-bis(4-hydroxyphenyl)propane (bisphenol A) was detected with concentrations up to 25000 µg/L in seepage water samples. The amount in the sample LW was significantly lower (4200 µg/L) indicating a rapid degradation of these phenols during the short soil passage. Bisphenol A is widely used as plasticizer, fungicide and intermediate for polymer syntheses and exhibits estrogenic activity (Safe and Gaido, 1998). The occurence of bisphenol A in leachate water of hazardous waste deposit landfills has been reported recently (Yamamoto et al., 2001). In addition, numerous other plasticizers were identified including phthalates, phosphates and sulfonamides. Phthalates are well known and ubiquitous pollutants due to their persistence. Besides the 2-ethylhexyl, butyl, ethyl and methyl esters as well as the mixed methyl-2-ethylhexylester, phthalic acid and minor concentrations of terephthalic acid, isoterephthalic acid as well as a methylated phthalic acid were identified in the third extract. It has to be noted that cyclohexanedicarbonic acids and cyclohexenedicarbonic acids, hydrogenation products of the corresponding aromatic compounds, also appeared in the seepage water samples. The coincident occurrence of phthalic acids and its hydrogenation products indicates transformation processes of phthalates in waste deposits via hydrogenation and hydrolysis as illustrated in Fig. 3.

Within the group of sulfonamide plasticizers the N-butylbenzenesulfonamide (NBBS) was most abundant (420-710 µg/L). The concentrations of this compound in the leakage water sample were not reduced with respect to the seepage water. NBBS is a known neurotoxic pollutant e.g. of surface water (e.g. Sheldon and Hites, 1979; Duffield et al., 1994), groundwater (e.g. Guardiola et al., 1989; Duffield et al, 1994), waste water (e.g. Huppert et al., 1998) and also landfill leachates (Albaiges et al., 1986; Öman and Hynning, 1993; Duffield et al., 1994). Benzenesulfonamides with methyl, N-methyl, N-phenyl and N-ethyl substitution were identified in minor concentrations.

Phosphate plasticizers are also common xenobiotics. In the seepage water samples the butyl and ethyl esters were dominant with concentrations between 20 and 250 µg/L. The chlorinated substance tris(2-chloroethyl)phosphate, mainly used as a flame retardant, appeared only in sample SW2 with a concentration of 240 µg/L.

Several compounds were identified both in the seepage water samples and the leakage water sample, that were attributed to the group of pharmaceuticals. Propyphenazone, a widely used analgesic and antipyretic, was detected with concentrations of 110 to 140 µg/L. Ibuprofen (analgesic, antiphlogistic) detected in samples SW1, SW2 and LW has been formerly reported as a waste water contaminant (Huppert et al., 1998). The third pharmceutical compound, clofibric acid, is an environmental metabolite of the corresponding ethylester used as a blood lipid regulator (Buser et al., 1998).

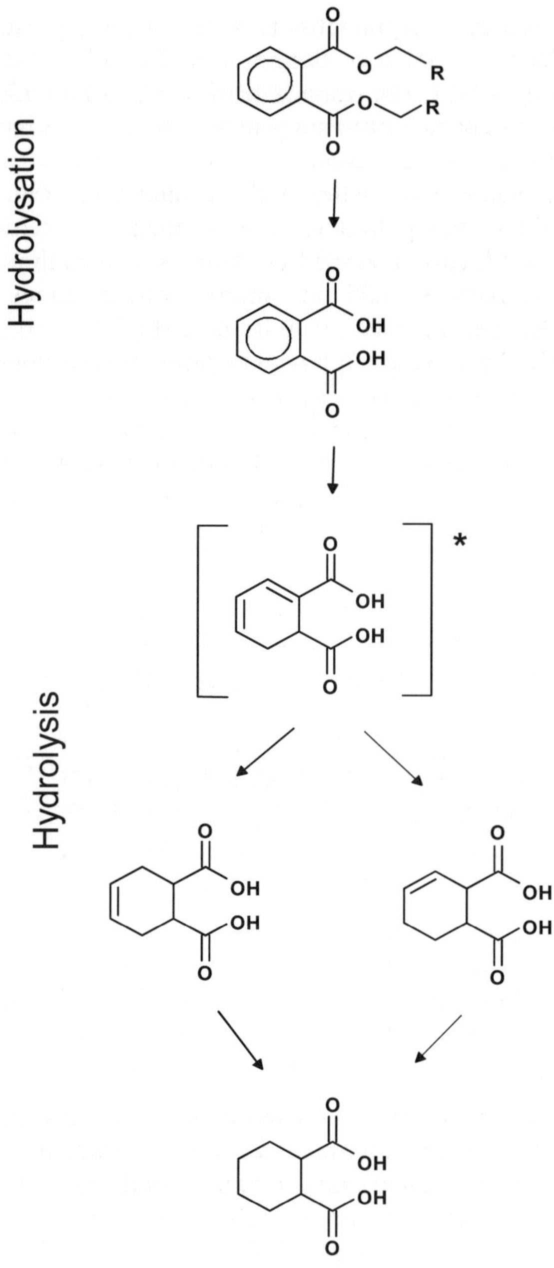


**Fig. 3:** Postulated degradation pathway of phthalates in waste deposit landfills. *= not identified in the examined seepage water samples.

The insect repellent N,N-diethyltoluamide (DEET) and the herbicide mecoprop (MCPP) were also identified in the samples SW1, SW2 and LW. The amounts of DEET ranged between 200 and 320 µg/L. Similar concentrations in all samples suggest that DEET is moderately stable under anaerobic and aerobic conditions. DEET and MCPP are widely used and the resulting contamination of surface water has been reported (e.g. Franke et al., 1995; Buser et al., 1998). The herbicide has also been reported as contaminant in landfill leachates with a high persistence under the anaerobic conditions existing within the deposits (e.g. Schultz und Kjeldsen, 1986; Gintautas et al., 1992).

Apart from 2,4-dichlorobenzoic acid several chlorinated aromatics including chlorinated methylpyridine, benzene, methylphenol, aniline and benzoic acid occurred in the seepage water samples only in minor amounts. 2,4-dichlorobenzoic acid was also detected in the leakage water sample due to its higher abundance.

Furthermore, nicotine and benzothiazol were specific contaminants of the seepage water, accompanied by their oxidation products cotinine and benzothiazolones. The tobacco alkaloid cotinine, which is also a mammal metabolite of nicotine, appeared in minor concentrations. Concentrations of nicotine and cotinine ranged from 550 to 710 µg/L and from 90 to 180 µg/L, respectively. As compared to the seepage water samples no signficant reduction of the concentrations in the leakage water was observed. The origin of benzothiazolone can be attributed either to an oxidation of benzothiazol or to the hydrolysis of substituted benzothiazols like 2-methylthiobenzothiazol, a common rubber additive.

A contamination with several $C_1$-$C_4$ alkylated pyrazines, cyclopentapyrazines and pyridines was discovered in the second extracts. These compounds are described in detail as primary products of the Maillard reaction, the non-enzymic browning reaction between carbohydrates and amino acids (e.g. Mulders, 1973; Chen et al., 1998). Therefore, the occurence of these specific pyrazines and pyridines suggested either contemporary degradation processes of carbohydrates with amino acids or ammonia within the deposit landfill or the emission of waste derived Maillard products generated already before deposition.

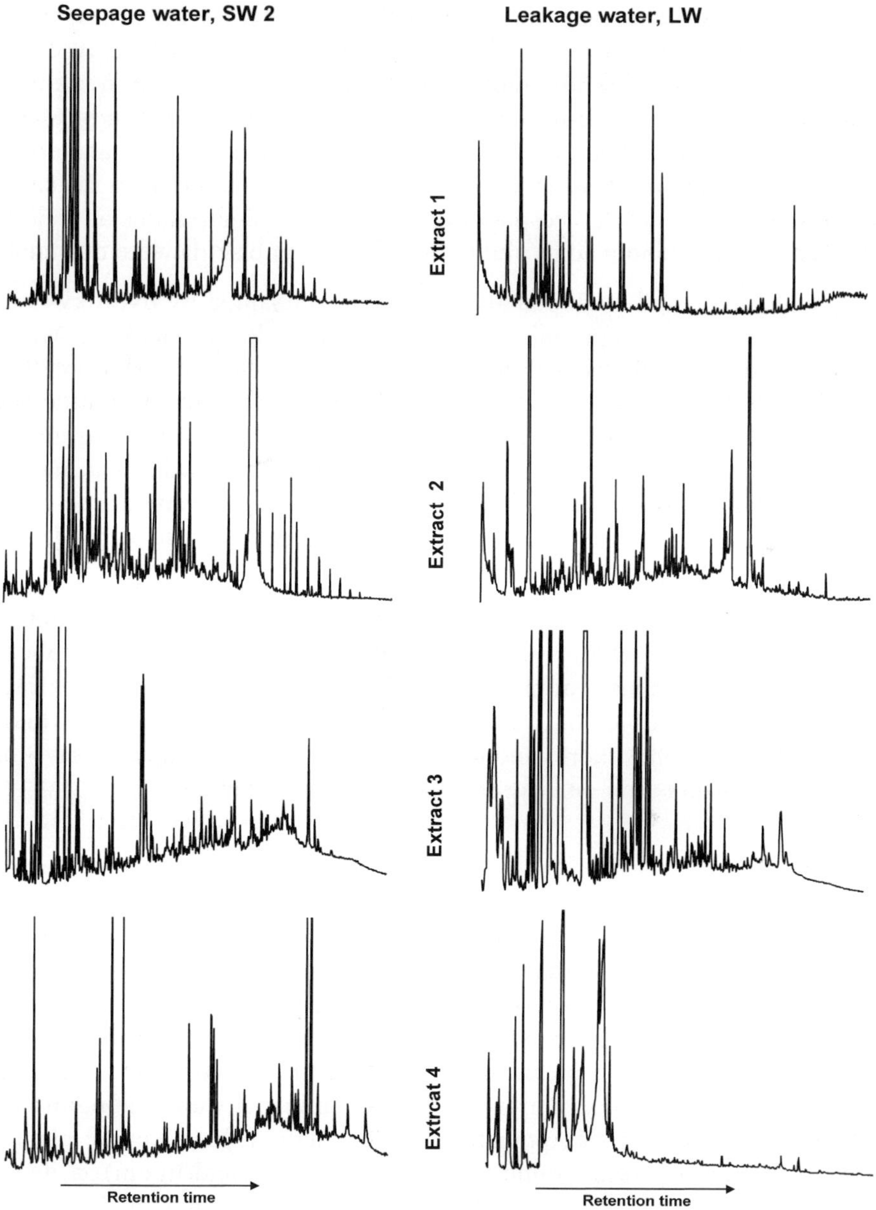


**Fig. 4:** Comparision of organic-geochemical signature of the seepage water sample SW2 and the leakage water sample LW. *Extract 1:* pentane extract. *Extract 2:* dichloromethane extract. *Extract 3:* dichloromethane extract, pH 2, derivatized with MSTFA. *Extract 4:* dichloromethane extract, pH 2, derivatized with TMSH.

According to Chian and deWalle (1977) as well as Albaiges et al. (1986) the acidic fraction has been observed previously to represent a major contribution to the low molecular-weight organic matter in seepage water samples (see Fig. 2c-d). Especially the abundant occurence of $C_2$-$C_7$ alkanoic acids has formerly been described as a result of the anaerobic fermentation (Yan and Jen, 1992; Harmsen, 1993) In addition to the short chain alkanoic acids the corresponding hydroxylated compounds as well as branched $C_4$-$C_9$ alkanoic acids and $C_4$-$C_9$ dicarboxylic acids were detected.

Cyclohexanoic and cyclopentanoic acids accompanied by hydroxy- and methyl-substituted compounds, mainly identified in the leakage water sample, were propably generated by hydrogenation of the corresponding benzenecarboxylic acids according to the transformation processes previously described for the phthalates. Cycloalkanoic as well as aromatic acids including the hydroxylated and methylated isomers derived mainly from the microbial degradation of lignin. In addition, phenylacetic acid, phenylpropionic acid and phenylbutanoic acids represented degradation products of proteins (Albaiges et al., 1986).

It is noteworthy that, apart from the organic compounds described above, a few extractable inorganic compounds were identified. A contribution of elemental sulphur ($S_8$) was observed in the pentane extract indicating the anaerobic conditions within the deposit (see Fig.3A). Minor concentrations of phosphoric and boronic acid were detected as methyl- or trimethylsilylesters after derivatization of the acidic extract.

## Conclusions

In order to provide a comprehensive view on the organic-geochemical composition of seepage water and leakage water derived from a waste deposit landfill, detailed GC/MS screening analyses were applied to two leachate samples and one leakage water sample. With respect to the molecular structures, the quantitative data and their occurence in seepage or leakage water samples two groups of contaminants were distingiushed. Plant material-derived compounds and degradation products of natural substances (e.g. peptides, amino acids, carbohydrates and lignin) made up a major proportion of the low molecular-weight organic matter. Examples include terpenoic substances, skatole, *p*-cresol, substituted pyrazines and pyridines, cyclohexanoic acids, substituted benzoic and phenylalkanoic acids.

Apart from these natural substances numerous xenobiotic substances were identified and attributed to the groups of pharmaceuticals, plasticizers, pesticides or chlorinated aromatics. Significant representatives of this substance group were bisphenol A, MCPP, DEET, NBBS and other sulfonamide plasticizers, phthalate plasticizers, 2,4-dichlorobenzoic acid, propyphenazone, ibuprofen and clofibric acid.

In comparision to the seepage water samples not all identified substances were recovered in the leakage water sample due to degradation processes or dilution to very low concentrations (see Tab. 3). Degradation processes were initiated by the change of environmental conditions from anaerobic conditions within the deposit to more aerobic conditions after release from the deposit. Quantitative data of selected contaminants (see Tab. 4) were used to discriminate substances affected by degradation processes and persistent compounds. The concentrations of limonene, 1,8-cineole, fenchone and borneol fell below the detection limit and bisphenol A as well as camphor were detected in the leakage water sample with significantly lower concentrations. On the contrary, several compounds appearing only in the leakage water indicated transformation processes occurring outside of the deposit. Examples include campherchinone, thiophenol, thiocresol and cycloalkanoic acids. Furthermore, the concentration of indole and skatole increased in the leakage water suggested an accelarated degradation of nitrogen-containing amino acids under the more aerobic conditions.

Several xenobiotics identified in both the seepage water and the leakage water on the same concentration level were not only persistent but also specific for the waste deposit-derived emission. Hence these stable source indicators (e.g. DEET, NBBS and propyphenazone) appear to be appropriate tracers for the subsequent route of the contamination derived from the waste deposit landfill.

### 2.1.2 Monitoring of groundwater contaminations caused by a leaking waste deposit landfill using organic tracer compounds‡

#### Introduction

For a long time landfilling has been the major disposal method for solid waste of municipial as well as of industrial origin. In case of careless installation practises (e.g. insufficient bottom sealing) or damages during the use seepage water derived from waste deposit landfills can enter the pedosphere. In particular the groundwater can be affected heavily by these emissions. In this case the impact on the groundwater quality depends on the chemical composition of the leachate, on their toxicological potential, as well as on the corresponding transport and degradation processes within the aquifer.

Generally, with respect to groundwater contamination the unambigious correlation of pollutants and responsible emission sources is an important aspect. Hence, detailed chemical information on emission sources and their contribution to the groundwater pollution in terms of spatial distribution and long term effects is an important tool for evaluating the environmental consequence of a contamination.

Seepage water of deposit landfills are characterized by very complex mixtures in particular of organic contaminants (see also Öman and Hynning 1993, Paxeus 1999, Schwarzbauer et al. 2002). Beside natural components anthropogenic substances partially with toxicological or ecotoxicological properties are common constituents often in high concentrations. Consequently, they also enter the aquatic environment in case of landfill leakages. However, specific anthropogenic compounds as well as characteristic patterns of technical mixtures are generally appropiate to attribute the environmental contamination to isolated sources of emission. With respect to landfill leachates numerous of the xenobiotical constituents seem to be suitable to act as tracer compounds which can be used to monitor leakage emissions.

However, environmental investigations on landfill derived groundwater contamination used frequently inorganic parameters for monitoring proposes (e.g. Murray et al. 1981; Looser et al. 1999; Vilomet et al., in press). These parameters include concentration of nonmetal anionic species (chloride, nitrate, sulfate, phosphate, borate etc.), heavy metals (Ni, Cr, Fe etc.) as well as alkali and alkaline earth metals (K, Na, Mg, Ca etc.).

‡ Mainly adapted/reprinted from Heim et al. 2004

On the contrary few studies focused on the distribution of organic compounds derived from landfill leachates (e.g. Reinhard et al. 1984, Albaiges et al. 1986, Rügge et al. 1995, Holm et al. 1996). The information provided by these studies are often limited by two different aspects: (i) The source specifity of many of the contaminants investigated are low (e.g. aromatic hydrocarbons, fatty acids) and, therefore, a superimposition of further emsisson sources cannot be excluded. (ii) The knowlegde on the degradation processes of the compounds studied within the individual compartments are often fairly scarce. Thus, microbial degradation or dilution/adsorption effects as major reasons for decreasing concentrations cannot be differentiated.

The following study was based on a formerly reported investigation which analysed the chemical composition of corresponding seepage and leakage water of a leaking waste deposit landfill and, further on, characterized appropriate source specific organic marker compounds (chapter 2.1.1 or Schwarzbauer et al. 2002). This continuative work applied these specific marker substances in order to trace the spatial distribution of the deposit landfill derived contamination of the underlying aquifer and, additionally, to monitor long time effects.

## Samples

The area of investigation comprised a groundwater system, which was presumably contaminated by emissions of a waste deposit landfill as the result of a leak discovered in the bottom sealing. Additionally, it was found that waste water was seeping down into the drainage water of an old lead/zinc mine situated below the deposit. Since 1999 technical measures were performed in order to diminish the resulting emissions.

For this investigation two different sets of samples were taken. In June, August and October of 2002 groundwater was sampled from five wells situated nearby the deposit landfill (W1 to W6). The sampling locations are illustrated in Fig. 1. Wells 1 to 5 were sampled at two different aquifer horizons named u(pper) and l(ower).

A second set of samples was taken from the central effluent of the mining drainage system. This drainage system dewaters not only the underlying mining system but also the complete groundwater system that was suspected to be affected by the landfill leachate. The sampling location is situated approx. 2 km apart from the landfill as indicated in Fig. 1. Surface water samples were taken monthly from June 2002 to May 2003. In addition an undergound water sample (RS) was taken within the mining system situated at an unaffected location upstream the leakage. This sample represents an uncontaminated reference material.

Details of quantitative analyses are given in Table 1 and Figure 1. The limit of quantification (LOQ) for underivatized compounds was in the range of 0.5 ng/L, but for methylated substances the LOQ was approx. 2 ng/L. No attempts were made to quantify components at concentrations of less than 1 ng/L or 5 ng/L, respectively.

**Tab. 1:** Compounds selected for quantification including the ion fragments used.

| Nr | Compounds | Ion fragments (m/z) |
|---|---|---|
| 1 | N,N-Diethyltoluamide, DEET | 91, 190 |
| 2 | N-Butylbenzenesulfonamide, NBBS | 141, 170 |
| 3 | Propyphenazone | 215, 230 |
| 4 | Nicotine | 84, 162 |
| 5 | Ibuprofen [m)] | 177,220 |
| 6 | Clofibric acid [m)] | 215,230 |
| 7 | Mecoprop [m)] | 169,228 |
| 8 | Bisphenol A | 213,228 |

[m)] =detected as methylesters

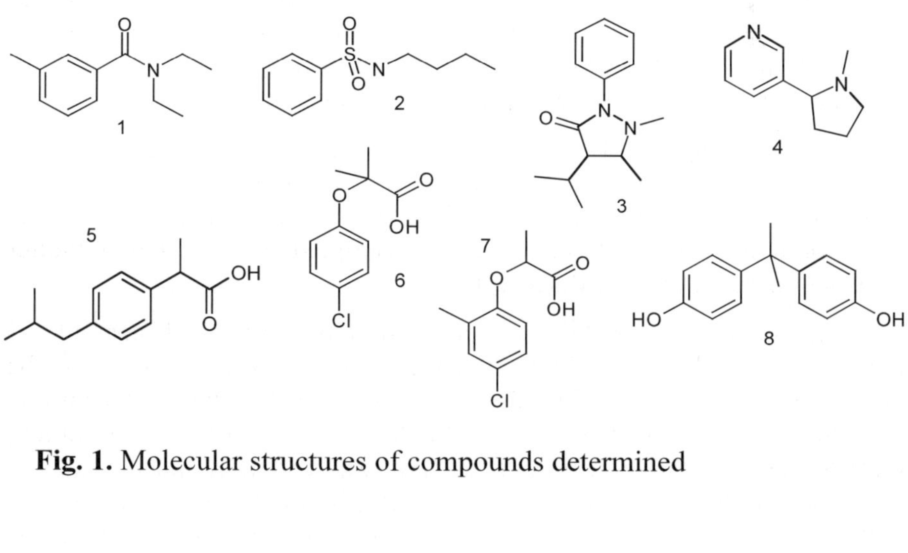


**Fig. 1.** Molecular structures of compounds determined

## Results and discussion

As formerly described (Schwarzbauer et al. 2002) a set of several organic substances seemed to be appropriate marker compounds to trace the emission derived from the deposit landfill leakage within the groundwater system in terms of spatial distribution and long time effects. In detail the pharmaceuticals propyphenazone and clofibric acid, the plasticizer N-butylbenzenesulfonamide (NBBS), the pesticides N,N-diethyltoluamide (DEET) and mecoprop were characterized as seepage water constituents with a high source specifity. They occurred in high concentration and without any compositional alteration not only in the seepage but also in leakage water. This leakage water was not directly collected but obtained after the leachate passed a short soil passage. Hence, an appropriate environmental stability has to be stated. In addition to these marker compounds three further source specific contaminants were included into the quantitative analyses. For nicotine, ibuprofene and bisphenol A a significant degradation with different transformation rates was observed after release of the deposit landfill (Schwarzbauer et al. 2002). Hence, these compounds were determined in order to contrast environmental stable and instable source indicators. Consequently, nicotine and ibuprofen were not detected in any sample investigated indicating a rapid degradation of these compounds in the surrounding aquifer. On the contrary bisphenol A was quantified in higher concentrations in the well samples suggesting a slower degradation rate. Its occurrence is discussed in detail in the following sections.

The origin of the marker compounds exclusively from the waste deposit landfill was confirmed by the analysis of the reference sample RS which revealed no contamination by the substances in question.

### *Spatial distribution*

For monitoring the spatial distribution of the landfill derived contamination the selected marker compounds were quantified in 5 groundwater wells at two different levels as indicated in Fig. 2. Selected quantitative data are also presented in Fig. 2. NBBS was excluded in this marker analyses due to an artificial contamination of the water samples during the pumping operation by plasticizers released by plastic fittings.

With respect to the lateral distribution all concentrations at sampling sites W1 and W2 were below the quantitation limit. The most abundant compound propyphenazone was detected in W3(l),W4 and W5 with increasing concentrations between <1 and 14000 ng/L towards the emission source. DEET, mecoprop and clofibric acid appeared with lower concentrations only in samples W4 and W5. However, their decreasing concentration profiles apart from the waste deposit also reflected a lateral expansion of approx. 200 to 500 m contrary to the groundwater flow direction as indicated in Fig.1.

With respect to the vertical distribution a comparison of the concentrations detected in the upper and lower aquifer layers indicated marginal differences. However, a slight accumulation of the contaminants in the upper layer can be stated.

The environmental stable marker compounds DEET and propyphenazone reached concentration levels in W5 which are diluted by the factor of approx. 100 to 1000 as compared to seepage and leakage water of the waste deposit (Schwarzbauer et al. 2002). However, various variations of the concentration data measured in june, august and octiober were observed. Data obtained from well W5 differed not significantly within this time period. On the contrary concentrations of all contaminants detected in well W3 and especially W4 in august were minimum values, whereas august and october data were clearly higher. Nevertheless the relative composition of the marker compounds did not vary confirming their applicability as source specific tracer substances.

*Long time effects*

In order to document long time effects of the waste deposit derived emissions a monthly sampling campaign was initiated. The unique hydrogeochmical conditions as specified in the experimental section allowed to monitor the total groundwater discharge by measuring only one sampling site at the gallery outlet of the mining drainage system. In addition the effluent run off is generally rather constant. Therefore, concentration data correspond very well to the load of the organic compounds.

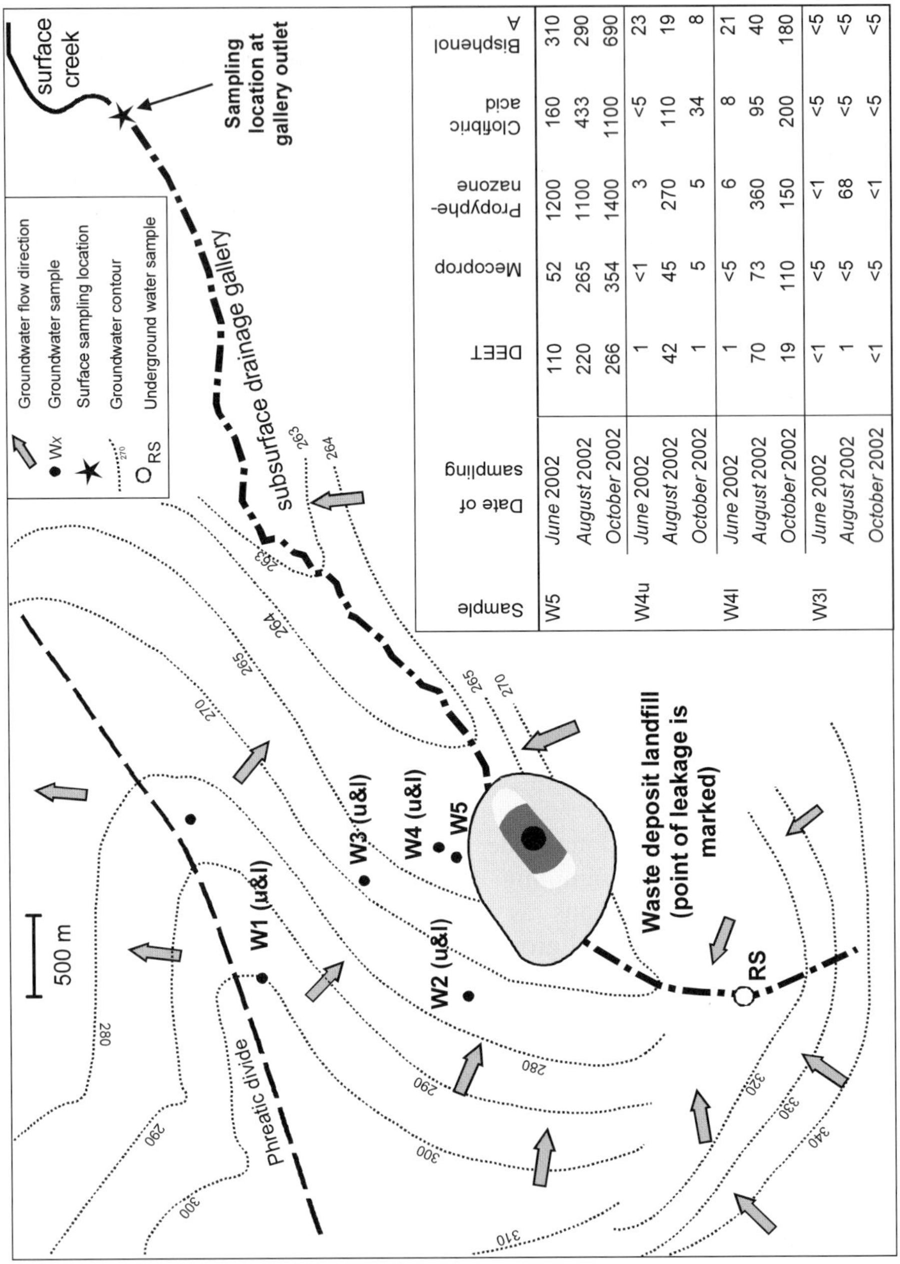


| Sample | Date of sampling | DEET | Mecoprop | Propyphenazone | Clofibric acid | Bisphenol A |
|---|---|---|---|---|---|---|
| W5 | *June 2002* | 110 | 52 | 1200 | 160 | 310 |
| | *August 2002* | 220 | 265 | 1100 | 433 | 290 |
| | *October 2002* | 266 | 354 | 1400 | 1100 | 690 |
| W4u | *June 2002* | 1 | <1 | 3 | <5 | 23 |
| | *August 2002* | 42 | 45 | 270 | 110 | 19 |
| | *October 2002* | 1 | 5 | 5 | 34 | 8 |
| W4l | *June 2002* | 1 | <5 | 6 | 8 | 21 |
| | *August 2002* | 70 | 73 | 360 | 95 | 40 |
| | *October 2002* | 19 | 110 | 150 | 200 | 180 |
| W3l | *June 2002* | <1 | <5 | <1 | <5 | <5 |
| | *August 2002* | 1 | <5 | 68 | <5 | <5 |
| | *October 2002* | <1 | <5 | <1 | <5 | <5 |

**Fig. 2.:** Sampling locations and quantitative data of selected marker compounds analysed in groundwater samples. All data are given in ng/L.

Regarding the time depending concentration profiles only a very slight decrease of the contamination was observed interrupted by individual maxima e.g. in march 2003. The tendency to a lower contamination might be attribute to the technical prevention measures performed since 1999.

Comparing the groundwater data of W5 with the concentration values in Tab. 2 dilution factors between 1 and 100 can be deduced. Noteworthy, the relative ratios of the environmental stable marker compounds still remained constant.

**Tab. 2:** Concentration (ng/L) of enviromental stable and instable marker compounds at the gallery outlet representing the long time effects of the discharge from a waste deposit landfill leakage.

| Date of sampling 2002<br>**Compound** | 04.06.02 | 01.07.02 | 05.08.02 | 11.09.02 | 01.10.02 | 14.11.02 | 10.12.02 |
|---|---|---|---|---|---|---|---|
| DEET | 10 | 5 | 5 | 1 | 2 | 3 | 3 |
| NBBS | 30 | 23 | 26 | <1 | 19 | 71 | 20 |
| Propyphenazon | 920 | 460 | 100 | 74 | 110 | 200 | 350 |
| Nicotin | <1 | <1 | <1 | <1 | <1 | <1 | <1 |
| Ibuprofene | <5 | <5 | <5 | <5 | <5 | <5 | <5 |
| Clofibric acid | <5 | <5 | 55 | <5 | 5 | <5 | <5 |
| Mecoprop | <5 | <5 | 17 | <5 | <5 | <5 | <5 |
| Bisphenol A | 340 | <5 | 123 | <5 | 70 | <5 | 320 |

| Date of sampling 2003<br>**Compound** | 13.01.03 | 19.02.03 | 10.03.03 | 06.05.03 | 26.05.20 |
|---|---|---|---|---|---|
| DEET | 3 | 3 | 18 | 1 | <1 |
| NBBS | 6 | 17 | 116 | 6 | <1 |
| Propyphenazon | 300 | 380 | 820 | 76 | 15 |
| Nicotin | <1 | <1 | <1 | <1 | <1 |
| Ibuprofene | <5 | <5 | <5 | <5 | <5 |
| Clofibric acid | <5 | 6 | <5 | <5 | <5 |
| Mecoprop | <5 | <5 | <5 | <5 | <5 |
| Bisphenol A | <5 | <5 | <5 | <5 | <5 |

With respect to bisphenol A an increased degradation has to be stated as revealed by the low concentrations between the quantitation limit and 34 ng/L. This might be caused by a longer duration of stay of the substances within the groundwater prior to the discharge via the gallery outlet. However, the bisphenol A analyses demonstrate the importance of enviromental stability as an essential property of marker compounds.

## Conclusions

Groundwater and drainage effluent samples were analysed by means of an organic marker compound approach in order to monitor a waste deposit landfill derived contamination. Considering the quantitative data presented the following two main results can be summarized:

- The applicability of the organic marker approach was confirmed by the continiously constant relative composition of the specific substances.
- Using the marker approach the spatial distribution, the dilution factors as well as long time emission behaviour were sufficiently determined and monitored.

Generally, source specific and environmental stable organic substances are powerful indicators not only for landfill leachates affecting groundwater areas. However, intense screening analyses combined with detailed information on the environmental behaviour are essential requirements for a successful isolation and application of appropriate marker compounds.

## 2.2. Chloronaphthalene related contamination of a groundwater system

### 2.2.1 Groundwater contamination by chlorinated naphthalenes and related substances caused by activities of a former military base‡

#### Introduction

Groundwater contaminations are generally caused either by non point sources (e.g. agricultural activities) or by point-source emissions for example of industrial sites, fuel depots and installations, dry cleaners or military bases (e.g. Albaiges et al., 1986; Guardiola et al., 1989; Cozzarelli et al., 1994; Furlong et al., 1997; Krueger et al. 1998; Suzuki et al., 1998). In contrast to diffuse emissions the contaminations resulting from point sources are usually characterised by high amounts of a limited number of substances.

Organic pollutants introduced to the underground can undergo various types of transport or modification/degradation processes. Depending on the environmental conditions as well as on the chemical properties of the contaminants either significant degradation, due to microbial activity, or stability over long periods of time (years, decades) may be observed.

Innovative remediation strategies, e.g. Natural Attenuation or Enhanced Natural Attenuation have been discussed intensively during the last years (e.g. Hart, 1996; Renner, 1998; Doll and Püttmann, 1999). With respect to these remediation activities information on occurrence and rate of microbial degradation is of fundamental importance (Sturchio et al., 1998). The groundwater-contaminated site presented in this study is currently monitored routinely for major known contaminants (chloronaphthalenes, PAH). Therefore the focus of our investigation was to identify possible degradation products of these contaminants and to quantify their abundance. Additionally the inorganic aquifer geochemistry was investigated for correlations between microbiological degradation and changes in concentrations of electron acceptors.

The investigated groundwater contamination was caused by long-term emissions from a former military base beginning in the late 1950s. The base was located in the Rhineland, Western Germany, and was run by the

‡ Mainly adapted/reprinted from Vinzelberg et al., 2005

British air force. For impregnation of woody material, coal tar based additives and technical formulations of mono- and dichlorinated naphthalenes were applied, causing continuous emission into the soil and the corresponding groundwater. The general setting is a quaternary unconfined sand-gravel aquifer of about 25 m thickness, followed by low permeable tertiary fine sands. The unsaturated zone is about 5 m thick. The hydraulic gradient is relatively low ($2 \cdot 10^{-4}$), so that groundwater flow velocity is about 30 m/a. Thus far, contaminants have spread in the groundwater over an area of approximately 400 x 200 m.

This study presents for the first time a comprehensive investigation of a high concentration groundwater contamination dominated by chloronaphthalenes; including contaminants, metabolites and isotope analysis. The GC/MS-analysis were performed in order to evaluate the state of pollution as well as to reveal molecular markers for a microbial activity with respect to the main organic pollutants.

## Samples

In 2002 twelve groundwater samples were taken from 6 different wells (see Fig. 1). The wells were chosen to represent different areas of the groundwater contamination: Sample A was taken in a region upstream of the potential emission source, sample B was located directly in the emission area, whereas samples C to F were situated downstream of the local contamination. With respect to changing chemical conditions within different vertical layers (e.g. varying oxidation-zones) half of the samples were taken from the upper (samples a-f) and the lower (samples A-F) part of the aquifer, respectively. The position of the investigated wells relative to the emission source as well as the depth of the sample points are shown in Tab. 2. Each well was pumped at all horizons simultaneously for 30 minutes at 2 $m^3/h$ to avoid vertical mixing.

Additionally, common inorganic parameters were determined and are presented also in Tab. 2.

The limit of quantification (LOQ) of the organic contaminants was in the range of approx. 0.5 ng/L. No attempts were made to quantify components at concentrations of less than 1 ng/L. Details of quantification and recovery rates are given in Tab. 1.

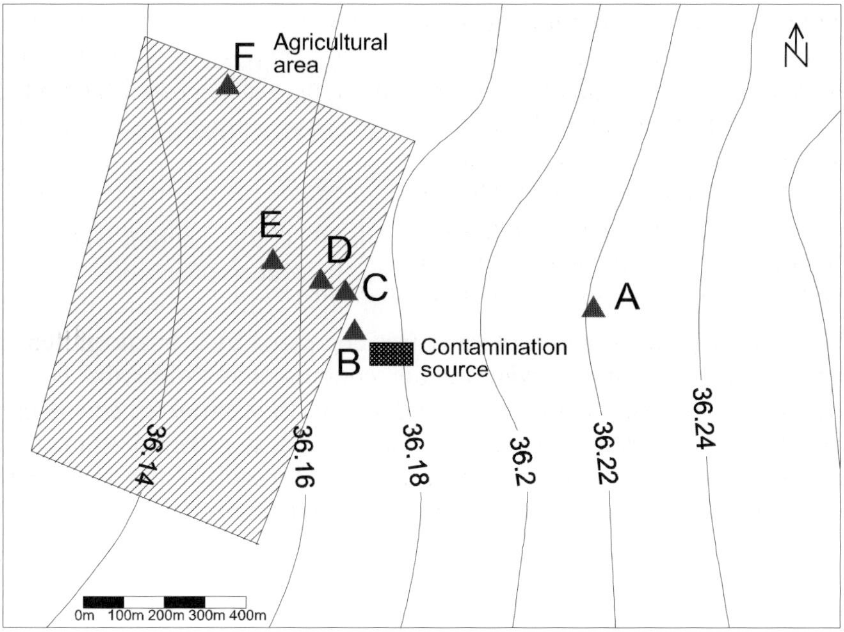


**Fig. 1:** Position of the sample wells and equipotential lines (height is given in meters above sealevel)

**Tab. 1:** Recovery rates and ion fragments used for quantitation of selected contaminants

| Compounds | Detected ion fragments [m/z] | Recovery rates[%] |
|---|---|---|
| *Chlorinated compounds* | | |
| 1-Chloronaphthalene | 162, 164 | 78 |
| Dichloronaphthalenes | 196, 198 | 80 |
| 1-Chloro-4-naphthol | 178, 180 | 65 |
| Chloronaphthoic acid | 220, 222 | 65* |
| Chlorobenzo(b)thiophene | 168, 170 | 78** |
| *PAC* | | |
| Acenaphtene | 76, 154 | 82 |
| Mehylnaphthalenes | 142, 115 | 88 |
| Dimethylnaphthalenes | 141, 156 | 80 |
| *Pesticides* | | |
| Atrazine | 200, 202 | 17 |
| Bromacine | 205, 207 | 76 |

* Recovery was taken from 4-chloronaphthol;

** Recovery and response factor were taken from 1-chloronaphthalene

## Synthesis of reference substances

1-Chloronaphthoic acid was synthesized using a direct acylation approach coupled with a haloform reaction. In detail, 1.7 g (1.05 mmol) of 1-chloronaphthalene was added to 1.4 g (10.5 mmol) of aluminium trichloride in 20 mL of 1,2-dichloroethane and cooled down to 4°C. Following, 0.8 g (1.02 mmol) of acetyl chloride was added slowly over the course of 2 hours. After 24 h of stirring at 50 °C the reaction was stopped by adding of 10 mL of water and the product (acetyl-1-chloronaphthene) was extracted three times with 10 mL of ether. Subsequently, the methyl group of the crude reaction product was separated by a haloform reaction using elemental bromine under alkaline conditions as followed: To 30 mL of a 16 % sodium hydroxide solution 1.25 mL (5.0 mmol) of bromine was added. The temperature did not exceed 10 °C during the addition. To the cold reaction solution (0 °C) approx. 2.0 g (1.0 mmol) of the crude acetyl-1-chloronaphthalene were added and the reaction mixture was stirred for 1 hour at room temperature. After extraction of the bromoform product with 20 mL of ether in triplicate, the organic layers were combined and 20 mL of a saturated sodium thiosulfate solution and 1 mL of concentrated sulfuric acid were added. The raw product was extracted with ether and purified by column chromatography on silica gel. Only one GC separable isomer was obtained. Following, the 1-chloronaphthoic acid was methylated to reveal the corresponding methyl ester. The methylation was carried out either with diazomethane in a microanalytical batch or in larger quantities with methanol under acidic conditions.

The approach used for the synthesis is outlined in Fig. 5. Overall yields were about 55%, and a purity of more than 95% was obtained. The product was characterized by mass spectrometry and infrared spectroscopy (see also Fig. 5). FTIR-spectra were acquired on a Nicolet 505 spectrometer with a resolution of 4 $cm^{-1}$. Details of GC-MS analysis are given above.

## Results and discussion

Initially, the groundwater system investigated was characterized by analyses of major cations and anions as given in Tab. 2. The upper aquifer layer represents an oxic environment with oxygen concentrations between 4.6 and 8.9 mg/L, whereas the lower layer shows concentrations between 0.4 and 1.7 mg/L.

**Tab. 2:** Samples and corresponding inorganic parameter provided by the engineering consultant GeoBit (Aachen).

| | | | **Sample** | **a** | **A** | **b** | **B** | **c** | **C** | **d** | **D** | **e** | **E** | **f** | **F** |
|---|---|---|---|---|---|---|---|---|---|---|---|---|---|---|---|
| | | **Detection limit** | **Sampling depth (m below surface)** | 8 | 28 | 8 | 23 | 7 | 26 | 7 | 26 | 7 | 26 | 8 | 30 |
| **Parameter** | **Unit** | | **Method** | | | | | | | | | | | | |
| Acid capacity | mg/L | 0.1 | DIN 38409 H7-1-2 | 3.9 | 6.3 | 7.9 | 7.0 | 6.3 | 6.5 | 1.5 | 6.6 | 1.9 | 6.7 | 1.7 | 5.8 |
| Ammonia-N | mg/L | 0.02 | DIN 38406 E5-1 | | | | 0.03 | 0.02 | 0.1 | | 0.07 | | 0.10 | | 0.27 |
| Nitrite-N | mg/L | 0.01 | DIN EN 26777 (D10) | | 0.02 | | | | | | | | | | |
| Nitrate-N | mg/L | 0.25 | DIN EN ISO 10304-1 (D19) | 5.3 | 0.6 | 11.9 | | 2.3 | | 39.2 | | 30.7 | | 5.3 | |
| Sulphate | mg/L | 1 | DIN EN ISO 10304-1 (D19) | 117 | 168 | 150 | 150 | 161 | 144 | 127 | 169 | 138 | 188 | 85.8 | 200 |
| Sulfid | mg/L | 0.02 | DIN EN ISO 10304-1 (D19) | | | | | | | | | | | | |
| Chloride | mg/L | 1 | DIN 38405 D27 | 34.9 | 58.7 | 5.1 | 36.9 | 20.9 | 37.3 | 64.6 | 37.1 | 78.0 | 46.6 | 16.1 | 40.4 |
| Sodium | mg/L | 0.05 | DIN EN ISO 10304-1 (D19) | 28.5 | 27.8 | 19.5 | 20.6 | 18.8 | 19.9 | 23.2 | 20.5 | 26.8 | 22.9 | 8.50 | 15.8 |
| Potassium | mg/L | 0.05 | DIN EN ISO 11885 (E22) | 6.21 | 4.90 | 6.89 | 6.57 | 7.90 | 5.08 | 1.13 | 5.70 | 4.06 | 5.21 | 6.01 | 2.99 |
| Calcium | mg/L | 0.02 | DIN EN ISO 11885 (E22) | 122 | 207 | 180 | 195 | 150 | 176 | 158 | 182 | 166 | 187 | 77.8 | 200 |
| Magnesium | mg/L | 0.02 | DIN EN ISO 11885 (E22) | 19.9 | 21.2 | 53.4 | 32.5 | 45.8 | 32.9 | 12.2 | 37.1 | 14.9 | 45.9 | 8.37 | 32.3 |
| Iron (II) | mg/L | 0.1 | Special method | | | | 2.7 | 2.4 | 7.4 | | 5.5 | | 5.3 | | 4.8 |
| Iron | mg/L | 0.01 | DIN EN ISO 11885 (E22) | | | 0.21 | 3.70 | 2.57 | 8.00 | 0.20 | 5.95 | 0.04 | 5.63 | 0.08 | 5.14 |
| Manganese | mg/L | 0.001 | DIN EN ISO 11885 (E22) | 0.29 | 1.92 | 0.01 | 1.91 | 2.98 | 1.55 | 0.04 | 2.18 | 0.06 | 1.67 | 0.02 | 1.24 |
| Phosphor (norm. $PO_4$) | mg/L | 0.025 | DIN EN ISO 11885 (E22) | 0.22 | 0.26 | 0.50 | 0.27 | 0.25 | 0.42 | 0.16 | 0.27 | 0.19 | 0.40 | 0.18 | 0.57 |
| Methane | µg/L | 1 | GC-FID | | | | | | | | 5 | | 6 | | 20 |

The elevated nitrate concentrations in samples d and e were attributed to agricultural input, in particular of fertilizers. Sulphide was absent in all samples. Methane concentrations reached a maximum of 20 μg/L in sample F. Therefore methane generation caused by microbiological activation could be excluded. Interestingly, no significant alteration of the quantitative distribution of anions was observed regarding the local emission situated near sample B. Similarly, with respect to cationic species no significant differences of concentrations were observed for either highly polluted or low contamination samples. Hence, evidence neither of aerobic nor anaerobic degradation processes within the plume near sampling site B was provided by these data.

Therefore, we focused our investigation on comprehensive organic-geochemical analyses of the contamination looking for contaminants as well as metabolites. Qualitative GC/MS analysis revealed a variety of organic compounds in the groundwater samples. The spectra of compounds detected in the contaminated samples are illustrated in Fig. 2 supplemented by Tab. 3. Based on the screening results abundant or structurally significant contaminants were quantified. Results are listed in Tab. 4.

In detail, several alkylated and nonalkylated polycyclic aromatic compounds (PAHs) were detected and dominated by acenaphthene with a maximum concentration of 24 μg/L in sample B. Total concentrations of methyl- and dimethylnaphthalenes reached 5.1 μg/L and 24 μg/L, respectively. Tricyclic aromatic compounds and oxygenated PAH species were detected comprising e.g. phenanthrene, 9H-fluorenone and acenaphthenone. Methylated and dimethylated benzo(b)thiophenes as well as dibenzothiophene were also present.

The most abundant contaminant was 1-chloronaphthalene with concentrations up to 750 μg/L followed by dichlorinated naphthalenes with concentrations up to 98 μg/L in total. The isomeric pattern of the dichloronaphthalenes as illustrated in Fig. 3 is very similar to technical mixtures like Halowax 1000, 1031 or Basileum SP-70. Chlorinated naphthalenes are known environmental contaminants. However, mainly the occurrence of the higher chlorinated congeners has been reported (Falandysz, 1998, and references cited therein). Only few investigations detected lower chlorinated compounds in the aquatic environment (Falandysz, 2003; Schwarzbauer et al., 2001).

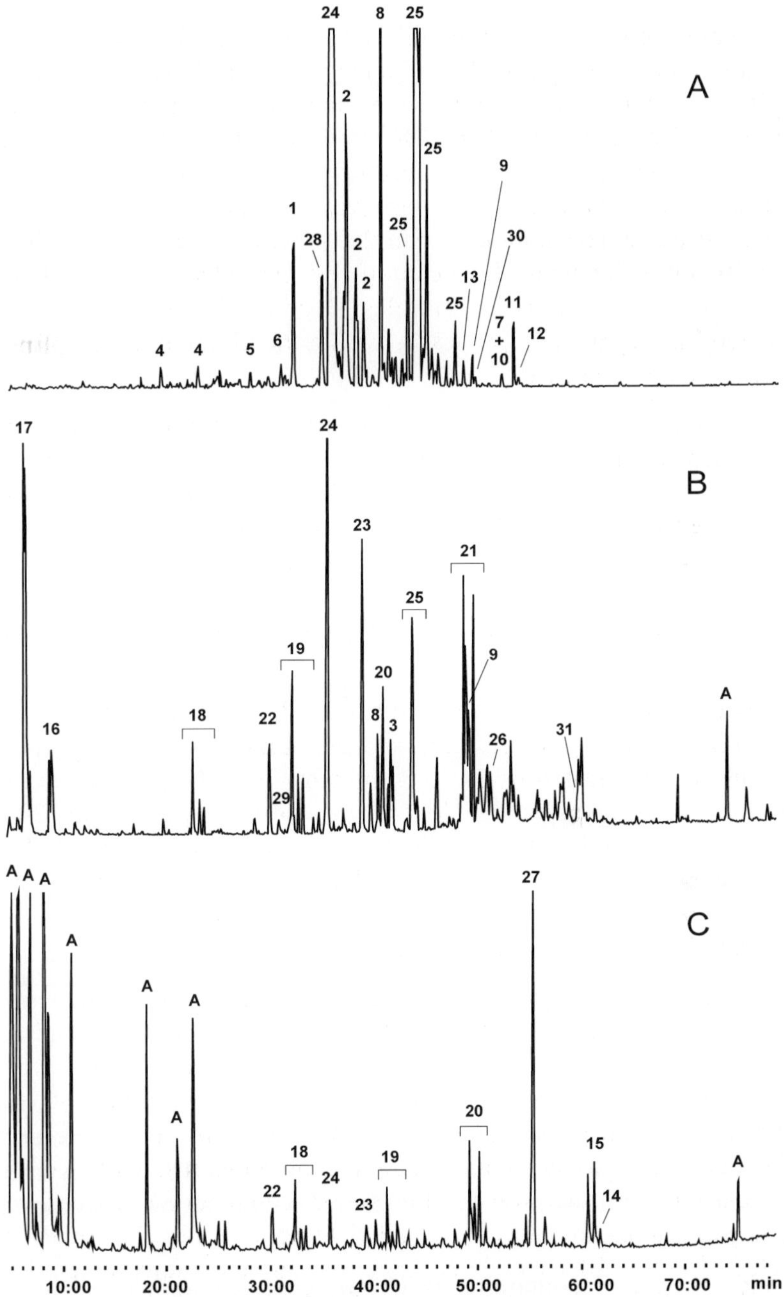


**Fig. 2:** GC/MS chromatograms of the pentane extract (A), dichloromethane extract (B) and dichloromethane extract after acidifaction to pH 2 (C) of sample B. The individual compounds identified are labelled according to Table 3.

**Tab. 3:** Organic contaminants identified in the groundwater samples. Numbers are used in Fig. 2.

| No. | Compound | No. | Compound |
|---|---|---|---|
| | *Polycyclic aromatic compounds* | | *Alcohols, phenols and ketones* |
| | Naphthalene | 16 | Hexanol (2 isomers) |
| 1 | Methylnaphtalene (2 isomers) | 17 | Hexanone (2 isomers) |
| 2 | $C_2$-Naphthalene (5 isomers) | | |
| 3 | Naphthol (2 isomers) | 18 | Dodecanol (3 isomers) |
| | | 19 | Tetradecanol (3 isomers) |
| 4 | Methylindane (2 isomers) | 20 | Hexadecanol (3 isomers) |
| 5 | $C_2$-Indane | 21 | Octadecanol (3 isomers) |
| 6 | Methylbenz(b)thiophene (2 isomers) | 22 | *tert*-Butylphenol |
| | $C_2$-Benz(b)thiophene (3 isomers) | 23 | Di-*tert*-butylphenol |
| 7 | Dibenzothiophene | | |
| | | | *Halogenated compounds* |
| 8 | Acenaphthene | | |
| 9 | Acenaphtenone [a)] | 24 | 1-Chloronaphthalene |
| | | 25 | Dichloronaphthalene (6 isomers) |
| 10 | 9H-Fluorenone | 26 | 1-Chloro-4-naphthol |
| | | 27 | Chloronaphthoic acid (2 isomers) [m)] |
| 11 | Phenanthrene | | |
| 12 | Anthracene | 28 | Chlorobenzo(b)thiophene [a)] |
| 13 | Dihydrophenanthrene | | |
| | | 29 | Chlorohydroxybenzaldehyde [a)] |
| | Methyldibenzofurane | | |
| | | | *Pesticides* |
| | *Carboxylic acid and esters* | | |
| | | 30 | γ-HCH, lindane |
| 14 | Octadecanoic acid [m)] | 31 | Bromacil |
| 15 | Triolein | | Atrazine |

a) = identified by comparison of the mass spectra with those of mass spectral data bases
m) = identified as methylester

Low chlorinated naphthalenes as well as selected PAHs are known as ingredients in coal tar based wood preservation agents (Pommer, 2001; Jacobsson and Asplund, 2000). Thus, we conclude that the occurrence of the described contaminants results from impregnation activities at the former military base.

Beside the main contaminants chlorobenzo(b)thiophene (one isomer) was also tentatively identified. Since reference material is commercial not available the identification based mainly on the gas chromatographic and mass spectral properties. First structural evidence for this compound was derived from the gas chromatographic analyses. The compound eluted a few seconds prior to 1-chloronaphthalene. This elution order is similar to the retention behaviour of the corresponding nonchlorinated substances, naphthalene and benzo(b)thiophene. The mass spectra as illustrated in Fig. 4 indicated the presence not only of one chlorine atom (deducible from the isotopic pattern of the molecular ion as well as from the loss of 35 m/z) but also of an aromatic moiety with a mass of 134 Dalton (indicated by the fragment 133). Thus, the analytical data strongly support the proposed molecular structure of chlorobenzo(b)thiophene. Interestingly, this compound is not only structurally related to chloronaphthalene but also appeared in the groundwater samples closely associated with the chlorinated naphthalenes. Although the concentrations are much lower as compared to chloronaphthalene (up to 5 μg/L), the distribution suggests this substance to be part of the formulation used for impregnation. Nevertheless, it has not been described as an ingredient of technical chloronaphthalene compositions so far.

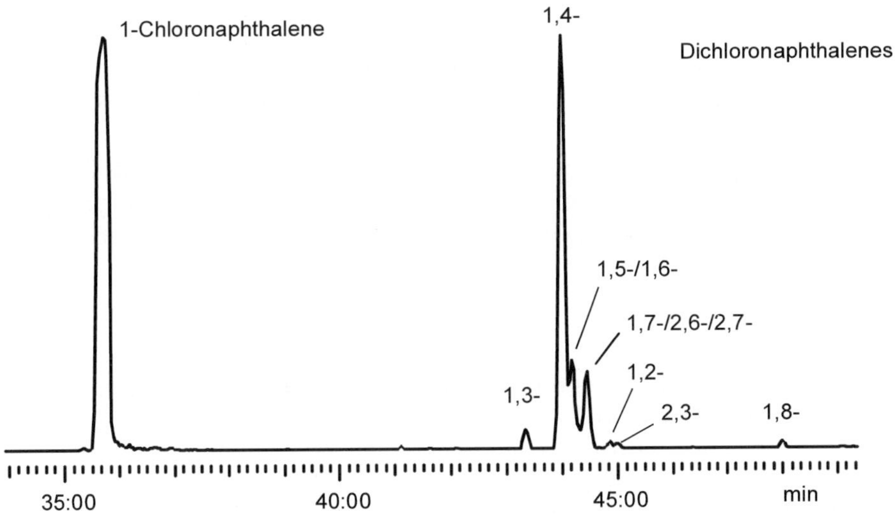


**Fig. 3:** Congener pattern of low chlorinated naphthalenes as identified in the groundwater samples.

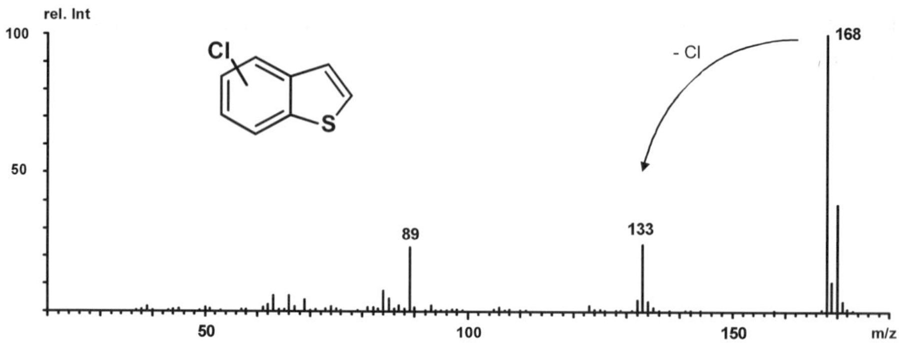


**Fig. 4:** Mass spectra of the tentatively identified chlorobenzo(b)thiophen.

The horizontal distribution of the contaminants is consistent with a point source emission localised near sampling site B. The vertical distribution characterized by elevated concentrations in the lower aquifer layer indicates an accumulation at the bottom of the groundwater system. Keeping in mind that the input of the contaminants took place in the unsaturated zone at the surface, the question arises why the upper part of the aquifer is only slight contaminated as compared to the lower part. Since no significant signals of intense degradation processes could be detected from the inorganic parameters, natural attenuation in form of "microbiological cleanup" of the upper part of the aquifer seems unlikely. Hence, it has to be assumed that the original contaminants reached the lower part of the aquifer by means of gravitational separation of the heavier plume (compare densities of 1-chloronaphthalene: 1.19 g/cm).

Following the groundwater flow direction from sampling site C to F, rapidly decreasing concentrations were observed in the lower layer indicating a distinct spatial distribution of the plume. At sampling site A, which is located upstream of the point source, only a negligible contamination was observed. The decreasing concentration profile and the distinct distribution of the contamination can be the result either of immobilisation caused by retention processes (e.g. sorption) or of efficient microbial degradation within the lower part of the aquifer.

The strong decrease of concentrations from sample B to C might suggest strong influence of sorption to organic matter in particular, if the relative high $K_{OC}$-values of mono- and dichlorinated naphthalenes (between 3.9 and 4.8, Jakobsson and Asplund, 2000) and the PAHs identified are considered (approx. 3.5 to 4.5, Schwarzenbach et al., 1993).

**Tab. 4:** Quantitative results of selected organic contaminants. All data are given in ng/L.

| **Upper layer** / **Compounds** | **a** | **b** | **c** | **d** | **e** | **f** |
|---|---|---|---|---|---|---|
| 1-Chloronaphthalene | 60 | 40 | 100 | | | |
| Dichloronaphthalenes | 30 | 20 | 10 | | | |
| 1-Chloro-4-naphthol | | | | | | |
| Chloronaphthoic acid | 910 | | | | | |
| Chlorobenzo(b)thiophene | | | | | | |
| Acenaphtene | | | | | | |
| Methylnaphthalene | | | | | | |
| Dimethylnaphthalene | | | | | | |
| Atrazine | 3400 | 3000 | 1700 | 1500 | 300 | |
| Bromacil | 3300 | 670 | 130 | | | |

| **Lower Layer** / **Compounds** | **A** | **B** | **C** | **D** | **E** | **F** |
|---|---|---|---|---|---|---|
| 1-Chloronaphthalene | 580 | 750000 | 75000 | 18000 | 1700 | 90 |
| Dichloronaphthalenes | 310 | 98000 | 9800 | 340 | 120 | |
| 1-Chloro-4-naphthol | | 29000 | 7000 | | | |
| Chloronaphthoic acid | | 81000 | 3900 | | 2500 | |
| Chlorobenzo(b)thiophene | | 7900 | 540 | 70 | | |
| Acenaphtene | | 24000 | 4300 | 1600 | | |
| Methylnaphthalene | | 5100 | 1600 | 730 | | |
| Dimethylnaphthalene | | 24000 | 2400 | 210 | | |
| Atrazine | 1300 | | 260 | 140 | | |
| Bromacil | 5000 | 2900 | | | | |

Furthermore, the qualitative screening analyses revealed specific contaminants that were structurally related to chloronaphthalene. These compounds, 1-Chloro-4-naphthol and one isomer of 1-chloronaphthoic acids, have not been described as constituents of technical mixtures of chlorinated naphthalenes. Nevertheless, they show similar distributions as compared to the concentration profile of 1-chloronaphthalene. With respect to their molecular structures they can probably be considered as molecular markers for an anaerobic transformation of chloronaphthalene. 1-Chloro-4-naphthol and 1-chloronaphthoic acid (identified as methylester after derivatisation by co-injection of authentic reference material) were detected with concentrations up to 29 µg/L and 103 µg/L respectively. To our knowledge, monochlorinated naphthoic acids have formerly not been described as environmental contaminants. The identification of both substances was based on the comparison of gas chromatographic and mass spectral properties with those of authentic reference material. Since 1-chloronaphthoic acids and their corresponding methyl esters are commercially not available a reference substance was synthesized according to the synthesis pathway depicted in Fig. 5. Interestingly, not only the mass spectral but also the gaschromatographic properties of the isomer obtained by synthesis matches perfectly the properties of the natural occurring substance.

The mass spectrum of the methylated compound is presented in Fig. 5 indicating the loss of a methoxy group ($M^+$ – m/z 31) and a methyl acetate moiety ($M^+$ - m/z 59) followed by the loss of a chlorine atom (forming m/z 126) as the major mass spectral fragmentation. Since mass spectra do not allow determination of the substitution pattern on aromatic compounds, we used infrared data to obtain more information on the position of the carboxylic group. Beside characteristic bands representing aromatic and aliphatic hydrogen atoms, the carboxylic group as well as aromatic carbon atoms the IR spectra exhibits two bands in the diagnostic area between 700 and 900 $cm^{-1}$ (see Fig. 5). The absorption at 766 $cm^{-1}$ represents four neighbored hydrogen atoms attached to an aromatic ring system. An identical absorption pattern was observed in the IR spectra of the educt, 1-chloronaphthalene, measured under the same conditions (data not shown). Therefore, this band clearly reflects the unsubstituted ring of naphthalene, and, consequently, the position of the carboxylic group has to be attributed to the chlorosubstituted ring system. The distinct substitution position within this ring (ortho, meta or para position next to the chloro atom at C1) cannot be clarified by interpretation of the IR data. The characteristic bands for an isolated hydrogen substituent representing a meta substitution are normally of minor diagnostic value. The absorption band characterizing ortho or para substitution (two neighbored hydrogen substituents) appears normally in a region between 800 $cm^{-1}$ and 840 $cm^{-1}$. Since the second absorption in the diagnostic region at 787 $cm^{-1}$ does not fit, we were not able to reveal the position of the carboxylic group.

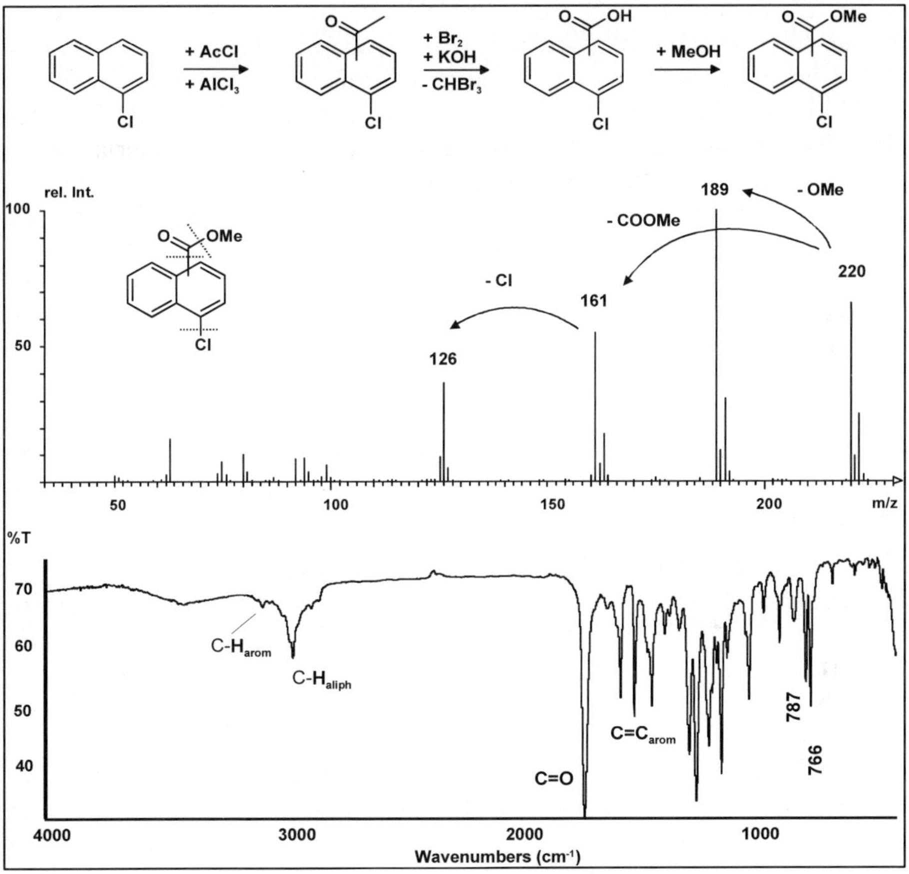


**Fig. 5:** Synthetic pathway, mass spectra and fragmentation behaviour of chloronaphthoic acid. The substitution position of the carboylic group cannot be determined solely based on mass spectral data.

The available knowledge about transformation processes of chlorinated naphthalenes is very limited. Only few investigations on the aerobic metabolism have been conducted and have revealed hydroxylation to be the most important initial transformation reaction (Jacobson and Asplund; 2000, and references cited therein). Information on anaerobic degradation of chlorinated naphthalenes is absent. However, the degradation processes of chemical very similar compounds are better understood. For example, the transformation of non-chlorinated naphthalene to naphthoic acid as well as phenanthrene to penthanthrene carboxylic acid under sulfate-reducing conditions has been described (Zhang and Young, 1997; Meckenstock et al., 2000; Griebler et al., 2004). Furthermore, anaerobic carboxylation of benzo(b)thiophene, benzene and phenol has been reported (e.g. Annweiler et al., 2001; Coates et al., 2002; Schink et al., 2000). Additionally, the hydroxylation of chlorinated and non-chlorinated aromatics by methanotrophic microbes forming phenolic compounds has been reported (Higgins et al., 1980). These studies covered e.g. naphthalene, benzene, chlorinated benzenes and biphenyls (e.g. Colby et al., 1977; Green and Dalton, 1989; Brusseau et al., 1990; Adriaens and Grbic-Galic, 1994; Coates et al., 2002).

Considering the microbial and environmental conditions described for the anaerobic degradation of aromatic substances, as well as the hydroxylated and carboxylated transformation products a similar metabolism of chloronaphthalene by anaerobic bacteria forming 1-chloro-4-naphthol and 1-chloronaphthoic acid can be assumed. Therefore, the occurrence of these proposed transformation products with elevated concentrations in the lower aquifer layers might indicate a microbial degradation of 1-chloronaphthalene in analogy to naphthalene. Interestingly, the occurrence of 1-chloro-4-naphthol suggested a different metabolic pathway as compared to degradation processes leading to the carboxylated species. This indicates that the microbial degradation of chloronaphthalenes probably proceeded through multiple pathways in this particular aquifer. However, it has to be noted that hydroxylated and carboxylated aromatic compounds, as the result of similar transformation processes of other main pollutants (acenaphthene, alkylated naphthalenes etc.), have not been observed. Furthermore, following the concentration profiles from B to F, no significant accumulation of the transformation products as the result of ongoing microbial activities has been detected in parallel to decreasing 1-chloronaphthalene concentrations (see Fig. 6). These observations dis-agree with the presence of extensive degradation processes with respect to 1-chloronaphthalene.

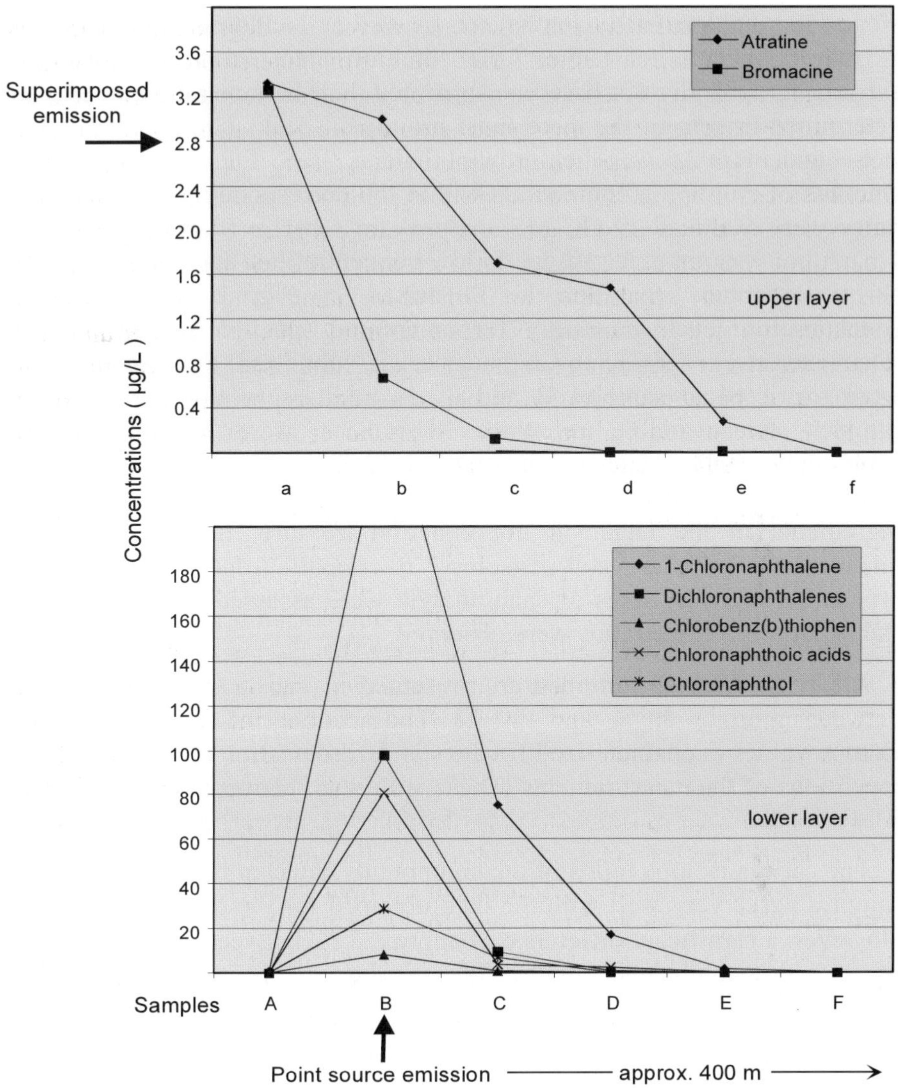


**Fig. 6:** Concentration profiles of the pesticides atrazine and bromacine (upper graphic) as well as 1-chloronaphthalene and related compounds (lower graphic). Both profiles represent contaminations of different emission sources.

In order to obtain more information on the proposed metabolism of 1-chloronaphthalene, compound specific stable carbon isotope analyses were performed. These analyses included the isotopic characterization of mono- and dichlorinated naphthalenes as well as 1-chloronaphthoic acid as a major metabolite. Since the dichlorinated isomers were gas chromatographically not base line separated their isotope ratio values were determined in sum of the three most prominent peaks (isomers: 1,4-, 1,5-/1,6- and 1,7-/2,6-/2,7-dichloronaphthalene, see Fig. 3). Significant amounts of co-eluting compounds within the corresponding retention time range were excluded by GC-MS analyses for samples B and C. 1-Chloro-4-naphthol was not detected due to low concentrations and insufficient gas chromatographic separation as illustrated in Fig. 2. Regarding the concentration levels necessary for compound specific isotope analyses (Dempster et al., 1997; Schwarzbauer et al., submitted) the measurements were restricted to samples B and C. In addition to the environmental samples the available reference substances were analysed. These compounds represent technical synthesized material and might give an idea on the original isotope ratios of the chloronaphthalene related ingredients in the technical impregnation mixture. In order to avoid misinterpretation as a result of isotopic fractionation due to the analytical procedure, recovery rates of each analyte were assessed. No shifts of the stable carbon isotope ratios were observed.

All $\delta\ ^{13}C$-values determined are presented in Tab. 5 / Fig. 7 and range between approx. -20 ‰ and -30 ‰. The accuracy of the stable carbon isotope values is characterized by the standard deviation obtained by three repetitions of the measurements. These values ($\delta 1$) covered the range from 0.1 to 0.4 ‰.

The carbon isotope ratios of the compounds detected in the groundwater samples agree very well with the data obtained for the reference material. However, a significant difference of approx. 3 ‰ between the $\delta\ ^{13}C$-values of 1-chloronaphthalene and 1-chloronaphtoic acid was observed in samples B and C. This difference might be interpreted as an indicator of microbial transformation, typically, with a $^{13}C$-depleted product (1-chloronaphthoic acid) and an isotopically heavier educt (1-chloronaphthalene). Considering the isotopic values of educt and product an enrichment factor $\varepsilon$ (also called isotope separation factor) between -2.0 ‰ and -3.0 ‰ can be estimated based on the differences of the $\delta^{13}C$-values observed (according to Whiticar, 1999). These data correspond well with values published for aromatic hydrocarbons in comparable groundwater systems ($\varepsilon$ = -1.1 ‰ to –3.6 ‰, sulfate reducing conditions; Schmidt et al., 2004).

**Tab. 5**: Stable carbon isotope ratios of 1-chloronaphthalene, dichloronaphthalenes and 1-chloronaphthoic acid as detected in the groundwater samples B and C. In addition the $\delta^{13}C$-values of corresponding reference compounds determined prior and after a recovery experiment are given.

[a] Data on chloronaphthoic acid are measured for the corresponding methyl esters and are corrected for the effect of the derivatisation

| Samples (n = number of repetítions) | | 1-Chloronaphthaline | Dichloronaphthalines | Chloronaphthoic acid methylester [a] |
|---|---|---|---|---|
| Sample B (n=3) | $\delta^{13}C$-values (‰ PDB) | **-23.1** | **-23.9** | **-25.8** |
| | stand. dev. (‰ PDB) | 0.2 | 0.2 | 0.2 |
| Sample C (n=3) | $\delta^{13}C$-values (‰ PDB) | **-23.6** | **-25.11** | **-26.7** |
| | stand. dev. (‰ PDB) | 0.2 | 0.4 | 0.1 |
| Reference substance (n=3) | $\delta^{13}C$-values (‰ PDB) | **-24.8** | **-24.6** | **- 25.4** |
| | stand. dev. (‰ PDB) | 0.2 | 0.3 | 0.2 |
| Recoveredsubstance (n=3) | $\delta^{13}C$-values (‰ PDB) | **-24.8** | **-24.6** | **- 25.8** |
| | stand. dev. (‰ PDB) | 0.4 | 0.3 | 0.2 |

However, comparing the $\delta^{13}C$-values of 1-chloronaphthalene in samples B and C only a slight shift of approx. 0.5 ‰ was determined. Comparing this data shift with published data of analogous compounds a low degree of degradation of 1-chloronaphthalene in the aquifer investigated is evident. E.g. naphthalene degradation via carboxylation reactions in a natural groundwater system was reported to lead to isotopic shifts of up to 3 ‰ corresponding to a degradation rate of approx. 90 % (Griebler et al., 2004).

Regarding the very slow groundwater flow of approx. 30 m/a the distance between sampling site B and C of approx. 60 m has to be related to a time period of at least 2 years. In case of an unretarded horizontal flow the corresponding isotopic shift between sampling site B and C has to be attributed also to degradation processes within a minimum of 2 years. Thus, only a very slow metabolism can be deduced from the isotopic data. This assumption implies that microbial degradation processes leading to the carboxylated metabolites has been initiated already at sampling point B and continued during the transport to sampling point C. Considering the very similar pattern of the relevant inorganic electron acceptors (see

Tab. 1), similar redox conditions for sulfate reducing bacteria at both sampling points can be deduced and, consequently, a similar microbial activity can be assumed for the whole section between the two sampling points.

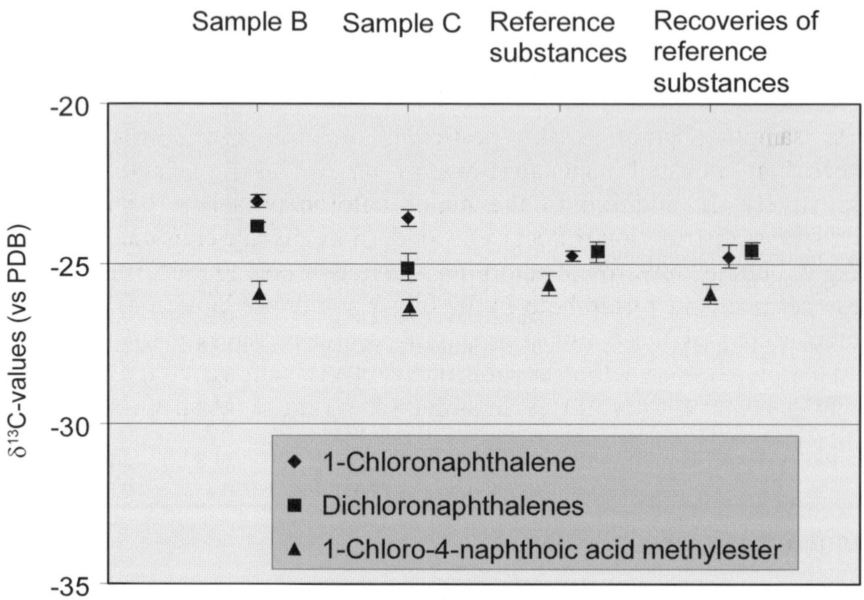


**Fig. 7:** Stable carbon isotope ratios of 1-chloronaphthalene, dichloronaphtalenes and 1-chloronaphthoic acid methylester as detected in the groundwater samples B and C. In addition the $\delta^{13}$C-values of corresponding reference compounds determined prior and after a recovery experiment are presented.

The low degree of degradation as well as the slow metabolism process as indicated by the isotopic values contrast with the rapid decrease of the 1-chloronaphthalene concentrations. Therefore, microbial degradation contributed only to a minor extent to the concentration profile observed in the aquifer layers. However, low transformation rates do not agree with the high concentration of the metabolite chloronaphthoic acid as compared to the amounts of 1-chloronaphthalene. In order to explain this apparent contradiction the limited water solubility of 1-chloronaphtahlene (in the range of 2 mg/L) has to be considered. The concentrations measured in sample B were in this range, corresponding to a very concentrated solution. Therefore, the availability of 1-chloronaphthalene was apparently limited by desorption from soil matter and it has to be assumed that the main portion of the contamination is fixed on the immobile particulate matter. On the contrary, the transformation product 1-chloronaphthoic acid is characterized by significantly higher water solubility as the result of its carboxylic group. Accordingly, it accumulated dominantly in the water phase. In conclusion, the similar concentration levels of 1-chloronaphthalene and 1-chloronaphthoic acid in samples B,C and D are the result of the accumulation of the transformation product due to a slow metabolism over a long period of time accompanied by an ongoing but slow dissolution of the particle associated transformation educt.

At sampling point B the pesticides atrazine and bromacine were detected at elevated concentrations of up to 3.4 µg/L and 5.0 µg/L, respectively. In addition to the major chloronaphthalene contamination their spatial distribution reflects a second, more diffuse emission source for these pollutants upwards of sampling location A (see Fig. 6). Interestingly, these pollutants occurred both in the lower and upper layers of the aquifer. The emission of these substances was obviously not related to the long term impregnation activities but likely to an intensive application of pesticides in the early 90'ies in order to minimise weed growth on the military base.

## Summary and conclusions

Qualitative and quantitative as well as stable carbon isotope analyses were performed and discussed in terms of environmental stability of the groundwater residues described. Anaerobic microbial degradation of the main pollutant 1-chloronaphthalene was evident from identification of 1-chloro-4-naphthol and 1-chloronaphthoic acid as biotic transformation products. While this observation indicates microbial involvement in degradation of chloronaphthalenes, several aspects of our analysis are inconsistent with an extensive degree of microbial degradation:

- Only a slight isotopic shift accompanied by rapidly decreasing concentrations was determined for 1-chloronaphthalene and its degradation products between sampling locations B and C.
- No accumulation of the biotic transformation products was detected in parallel to decreasing concentrations of the educt as the result of ongoing microbial activities.
- No corresponding metabolites derived from other, chemically related main pollutants (e.g. acenaphthene, methylnaphthalenes) were identified as the result of comparable transformation processes.

- Electron acceptors showed no change along the flow path.

Therefore, microbial degradation contributed only to a minor extent to the concentration profile of 1-chloronaphthalene observed in the aquifer. Consequently, the significant attenuation of 1-chloronaphthalene levels observed within a few meters from the point source has to be attributed mainly to sorption processes. For the polycyclic aromatic contaminants a similar environmental behaviour can be assumed (Stupp and Püttmann, 2001).

Finally, to the best of our knowledge, the detection of chlorobenzo(b)thiophene and chloronaphthoic acid reported here is the first observation of these compounds at a contaminated site.

In addition to the major contamination a second, more diffuse emission source for the pesticides bromacine and atrazine pollutants upwards of sampling location A was pointed out.

# 3 Occurrence of low molecular weight anthropogenic compounds in riverine water and corresponding sediments

## 3.1 Lippe river

### 3.1.1 The anthropogenic contribution to the organic load of the Lippe river (Germany) - Qualitative characterisation of low-molecular weight organic compounds‡

**Introduction**

Rivers are sensitive and complex ecosystems hosting a diverse spectrum of organisms. Human activities have resulted in increasing river pollution due to input of various contaminants into the river systems. A huge proportion of this contamination is caused by low-molecular weight organic compounds, like pesticides, plasticizers, ingredients of personal care products etc. Some of them, like PCB, PAH, DDT have already been studied in detail and their behaviour in aqueous systems is well known. During the last decade, screening analyses revealed furthermore an enormous diversity of detectable low-molecular weight organic contaminants in wastewater effluents and river water (e.g. Franke et al., 1995; Paxéus, 1996; Espadaler et al., 1997; Castillo and Barceló, 1999). Many of them had rarely been noticed so far. It will be an important task for the future to study their occurrence and fate in natural environments.

The present study is focussed on screening analyses of water samples which were taken from the longitudinal section of a river. A selection of the identified compounds is presented here as well as available information about their natural occurrence and/or industrial application. Some contaminants of special interest are discussed in detail. Besides, in order to reconstruct input pathways of frequently detected compounds

‡ Mainly adapted/reprinted from Dsikowitzky et al., 2004a

along the river course, various input sources of organic matter along the river were sampled. In part II of our study, the distribution of specific contaminants within the river course will be studied on the basis of quantitative data and a calculation of the organic loads.

The Lippe river (North Rhine-Westphalia, Germany) was chosen for this approach because it is located in a densely populated area with many diverse anthropogenic input sources, e.g. chemical industry, mining, agriculture and sewage. The vincinity of the less densely populated upper reaches of the Lippe river is a rural region which is mainly used for agricultural purposes. For part I and part II of our study, altogether four sample sets of river water were taken semi-annually at 19 sampling sites on a longitudinal section of the Lippe river.

## Samples

Samples of river water were taken on August, 16-17, 1999 at 19 sampling sites on a longitudinal section of the Lippe river, North-Rhine Westphalia, Germany (see Fig. 1). They were scooped up from below the water surface at midstream and bottled in pre-cleaned glass flasks. Filled sample flasks were sealed free of air bubbles with glass stoppers and stored in the darkness at a temperature of approximately 4 °C.

On August, 22, 1999 the effluent of a sewage treatment plant (STP) in Hamm and the effluent of a pharmaceutical plant in Bergkamen were sampled. Tributaries like the Alme river (see Figure 1) and the Quabbe Brook were sampled on August, 29, 2000. Water samples from the Seseke river (see Figure 1) were taken on March, 8, 2001.

## Results and Discussion

### *Identified organic compounds in Lippe river water samples*

The identified low-molecular weight compounds ranged from nonpolar constituents like aliphatic hydrocarbons to polar constituents like *n*-carboxylic (fatty) acids and phenols. Table 1 shows a selection of the detected compounds which are subdivided and arranged either by their structural properties or their applications. Compounds which represent a typical laboratory contamination (phthalates) as well as structural uninteresting compounds with only sporadic occurrence in the river (e.g. xylenes and aliphatic aldehydes) were not included.

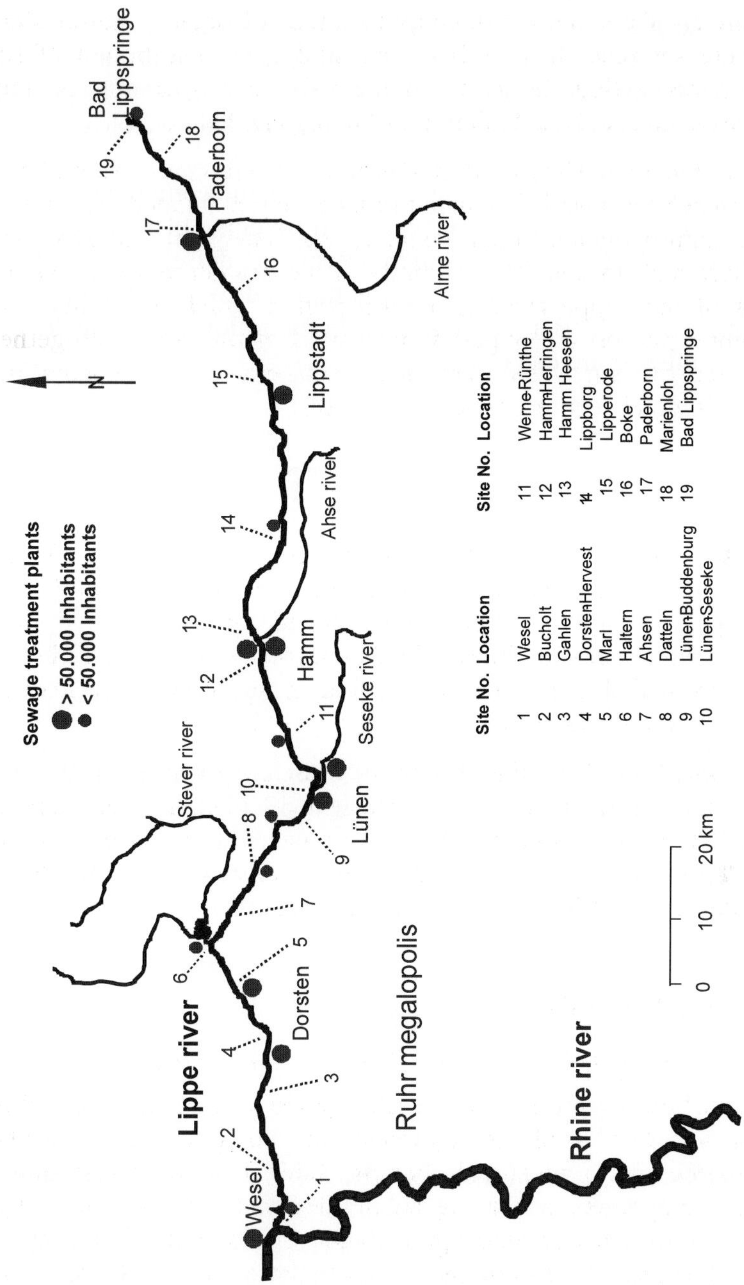


**Fig. 1:** Map showing the Lippe river system (North-Rhine Westphalia, Germany) with sampling locations (sites 1 to 19) and sewage treatment plants.

Available informations about the natural occurrence and/or industrial application of the selected compounds are summarised in Table 2. Most of them are of anthropogenic origin (marked with 'A') or have a known industrial application (marked with 'I').Compounds which we related to an anthropogenic origin have structural features which appear rarely in nature (e.g.ethylhexyl-groups, chlorinated organic compounds) or they do not naturally occur in the investigated region (like caffeine). Their occurrence in the aquatic environment is definetely attributed to anthropogenic activities. Only a few compounds are related to a biogenic input (marked with 'B') (see Table 2). In the following, some compounds of special interest are discussed.

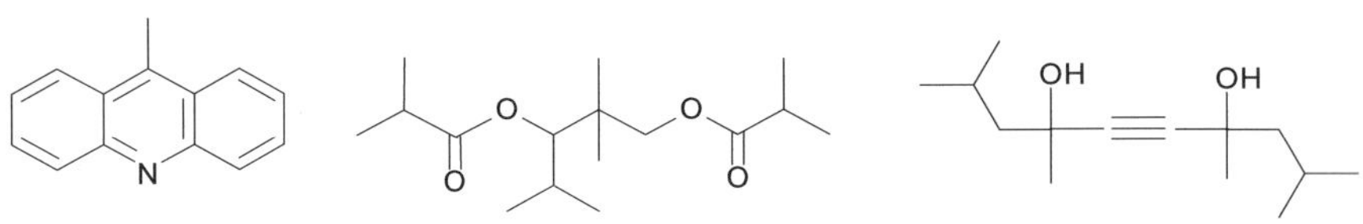

9-Methylacridine (No. 8) TXIB (No. 21) TMDD (No. 26)

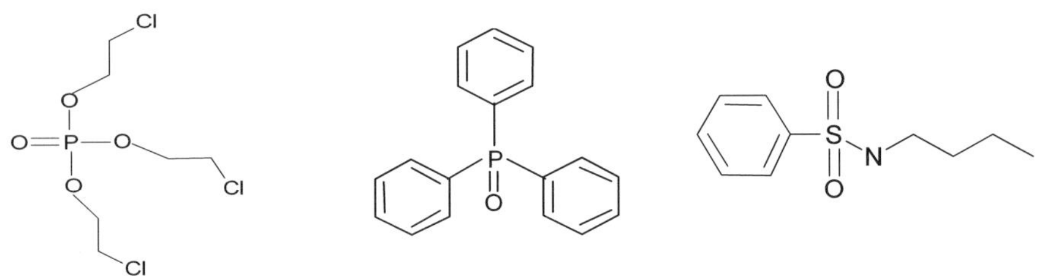

TCEP (No. 34) Triphenylphosphinoxide (No. 37) NBBS (No. 47)

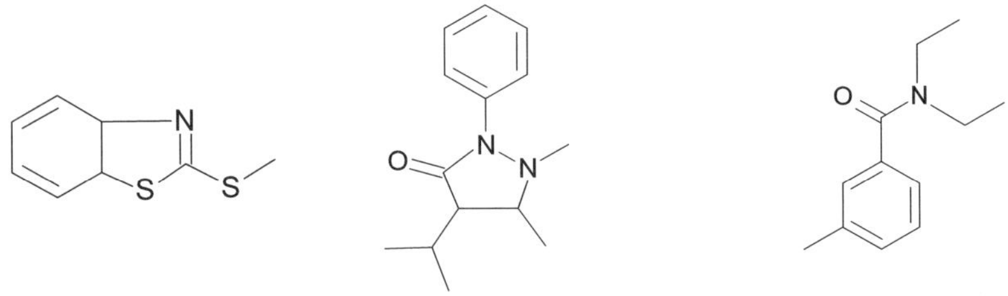

MTB (No. 56) Propyphenazone (No. 64) DEET (No. 68)

**Fig. 2:** Chemical structures of some identified compounds of special interest (full chemical names are listed in Table 1).

**Tab. 1:** Organic compounds in Lippe river water samples, sampling in August 1999 (site numbers see Figure 1)

| No | Sampling locations / Compounds | 1 | 2 | 3 | 4 | 5 | 6 | 7 | 8 | 9 | 10 | 11 | 12 | 13 | 14 | 15 | 16 | 17 | 18 | 19 |
|---|---|---|---|---|---|---|---|---|---|---|---|---|---|---|---|---|---|---|---|---|
| | *Aliphatics* | | | | | | | | | | | | | | | | | | | |
| 1 | Homologues series of *n*-alkanes ($C_{11}$-$C_{18}$, $C_{24}$, $C_{25}$) | + | + | + | + | + | + | + | + | + | + | + | + | + | + | + | + | + | + | + |
| 2 | Squalene | + | + | + | + | + | + | + | + | + | + | + | + | | + | | + | + | + | + |
| | *Polycyclic aromatic compounds, PACs* | | | | | | | | | | | | | | | | | | | |
| 3 | Naphthalene | + | + | + | + | | + | | + | + | + | + | + | | | + | | + | + | + |
| 4 | Phenanthrene | + | | + | | + | + | | + | | + | | + | + | + | + | + | + | + | + |
| 5 | Fluorene | | | | | | | + | | | | | | | | | | | | |
| 6 | 9-Fluorenone | + | | + | | + | + | | + | | | + | + | | + | + | | | + | + |
| 7 | Acridine | | | | | | + | | | | | | | | | | | | | |
| 8 | 9-Methylacridine | + | + | + | + | + | + | + | + | + | + | + | + | + | + | + | + | + | + | |
| | *Aliphatic carboxylic acids* | | | | | | | | | | | | | | | | | | | |
| 9 | Homologues series of *n*-carboxylic acids ($C_8$-$C_{18}$) | + | + | + | + | + | + | + | + | + | + | + | + | + | + | + | + | + | + | + |
| | *Aromatic carboxylic acids* | | | | | | | | | | | | | | | | | | | |
| 10 | Benzoic acid [(m)] | + | + | + | + | | + | + | + | + | + | + | + | + | + | + | + | + | | + |
| 11 | 4-*tert*-Butylbenzoic acid [(m)] | + | + | + | | + | + | + | + | + | + | + | + | + | + | + | + | + | + | + |
| 12 | 3,5-Di-*tert*-butyl-4-hydroxybenzoic acid [(m)] | | | | + | + | + | + | + | + | + | | | + | + | + | + | + | + | |
| 13 | Phenylacetic acid [(m)] | | + | + | | | | | | | + | | | | | | + | + | | |
| 14 | 3-Phenylpropionic acid [(m)] | + | + | + | | | + | | + | | + | | + | | | | | | | |
| 15 | Phthalic acid* | + | + | | | + | | | | | | + | | | | | | | | |
| | *Esters* | | | | | | | | | | | | | | | | | | | |
| 16 | 2-Butenedioic acid dibutylester | + | | + | + | + | + | | + | | + | + | + | + | + | + | + | + | + | + |

| No | Sampling locations / Compounds | 1 | 2 | 3 | 4 | 5 | 6 | 7 | 8 | 9 | 10 | 11 | 12 | 13 | 14 | 15 | 16 | 17 | 18 | 19 |
|---|---|---|---|---|---|---|---|---|---|---|---|---|---|---|---|---|---|---|---|---|
| 17 | Di-*iso*-propyl-adipate* | + | | | | | | | + | | | | + | | | | | | | |
| 18 | Di-*n*-butyladipate | + | | | + | | | | | | | + | + | | | | | | + | |
| 19 | Bis(2-ethylhexyl)-adipate | + | | + | + | + | + | + | + | | + | + | + | + | + | + | | | | + |
| 20 | Isopropyl myristate | | | | | | | | | | | | | | | | | | | + |
| 21 | TXIB, 2,2,4-Trimethyl-1,3-pentanedioldi-*iso*-butyrate | + | + | | + | + | | | + | | + | + | + | + | + | + | + | + | + | + |
| 22 | Triethyl citrate | | + | + | + | + | + | + | + | + | + | + | | | | | + | | + | |
| 23 | 2-(2-Butoxyethoxy)-ethylacetate | + | | | | | | + | | + | + | + | + | | | + | + | + | + | + |
| | *Aldehydes and ketones* | | | | | | | | | | | | | | | | | | | |
| 24 | Vanillin | + | + | + | + | + | + | + | + | + | + | + | + | + | + | + | + | + | + | + |
| 25 | 4-Oxoisophorone | | | | | | + | + | | + | | + | | | | | + | | | |
| | *Alcohols and ethers* | | | | | | | | | | | | | | | | | | | |
| 26 | TMDD, 2,4,7,9-Tetramethyl-5-decyne-4,7-diol | + | + | + | + | + | + | + | + | + | + | + | + | + | + | + | + | + | + | + |
| 27 | Propylenoxide dimer* (2 isomers) | + | + | + | + | + | + | + | | + | + | + | + | + | + | + | + | | | |
| 28 | Propylenoxide trimer* (3 isomers) | (+) | (+) | (+) | (+) | (+) | (+) | (+) | (+) | (+) | (+) | + | (+) | (+) | (+) | (+) | + | (+) | + | (+) |
| | *Phenols* | | | | | | | | | | | | | | | | | | | |
| 29 | Ionol, BHT, 2,6-Di-*tert*-butyl-4-methyl-phenol | | + | + | + | | + | | | | + | + | | | | | + | | + | + |
| 30 | BHA, *tert*-Butyl-hydroxyanisole* | | + | + | + | + | + | + | + | | + | | + | | | + | + | + | + | + |
| | *P-containing organic compounds* | | | | | | | | | | | | | | | | | | | |
| 31 | Tri-*iso*-butyl phosphate | + | + | + | + | + | + | + | + | | + | + | + | + | | | + | + | + | |
| 32 | Tri-*n*-butyl phosphate | + | + | + | + | + | + | + | + | | + | + | | + | | | + | | + | |
| 33 | Triethyl phosphate | | | | | | + | | | | | + | | | | | | | | |

| No | Sampling locations / Compounds | 1 | 2 | 3 | 4 | 5 | 6 | 7 | 8 | 9 | 10 | 11 | 12 | 13 | 14 | 15 | 16 | 17 | 18 | 19 |
|---|---|---|---|---|---|---|---|---|---|---|---|---|---|---|---|---|---|---|---|---|
| 34 | TCEP, Tris-2-chloroethyl phosphate | + | + | + | + | + | + | + | + | + | + | + | + | + | + | + | + | + | + | |
| 35 | Triphenyl phosphate | + | | | | + | + | + | + | | | | | + | | + | | | | |
| 36 | TCPP, Tris(chloropropyl) phosphate (2 isomers) | (+) | (+) | (+) | | + | (+) | + | (+) | | (+) | | | | | | (+) | | | |
| 37 | Triphenylphosphinoxide | + | + | + | + | | | | | + | + | | | | | | | | | |
| | *Chlorinated organic compounds* | | | | | | | | | | | | | | | | | | | |
| 38 | Hexachlorobutadiene | + | + | + | + | | | | | | | | | | | | | | | |
| 39 | Chlorobutanol | | | | | | | | | | | | | | | | | | + | |
| 40 | 1,2,3-Trichloropropene* | | | | | | | | | | | | | | | + | | | | |
| 41 | Dichlorobenzene | | | | | | + | | | | | + | | | | | | | | |
| 42 | Dichloroaniline | | | | + | | | | | | | | | | | | | | | |
| 43 | Dichlorobenzophenone | | | | + | | | | | | | | | | | | | | | |
| 44 | Bis(chloropropyl)-ethers (2 isomers) | + | + | + | + | | | | | | | | | | | | | | | |
| | *S-containing organic compounds* | | | | | | | | | | | | | | | | | | | |
| 45 | Di-*iso*-propyldisulfide | | + | | + | | + | + | | + | + | + | | | + | + | | + | + | + |
| 46 | Dipropyltrisulfide* | | + | | + | | | | | + | + | + | | | | | + | + | | + |
| 47 | NBBS, N-Butylbenzenesulfonamide | + | + | + | + | + | + | + | + | + | + | + | + | + | + | + | + | + | + | |
| | *N-containing organic compounds* | | | | | | | | | | | | | | | | | | | |
| 48 | Toluenesulfonic acid N-ethylamide | + | + | + | | + | + | + | + | | + | + | + | | | | + | | + | |
| 49 | TAED, N,N,N`,N`-Tetraacetylethylene diamine* | | | | | | | | | | | + | | | | | | | | |
| 50 | Methylbenzonitrile* | | | | | | | | | | | | | | | | | + | | |
| 51 | Phenylisocyanate | | | | | | + | | | | | + | | | | | | | | |
| 52 | Methylphenylisocyanate* | | | | | | | | | | | + | | | | | | | | |
| 53 | Cyclohexylisocyanate* | | | | | | + | | | | | + | | | | | | | | |

| No | Sampling locations / Compounds | 1 | 2 | 3 | 4 | 5 | 6 | 7 | 8 | 9 | 10 | 11 | 12 | 13 | 14 | 15 | 16 | 17 | 18 | 19 |
|---|---|---|---|---|---|---|---|---|---|---|---|---|---|---|---|---|---|---|---|---|
| | *Heterocyclic organic compounds* | | | | | | | | | | | | | | | | | | | |
| 54 | 2,5-Dihydro-2,5-dimethylfurane* | | + | | | + | + | + | | | + | + | | + | | | + | | | |
| 55 | Benzothiazole | | + | | | | | | + | | | | + | | | | | | | |
| 56 | MTB, 2-Methylthiobenzothiazole | | + | + | + | + | + | + | + | + | + | + | + | + | + | + | + | | + | |
| | *Terpenoids* | | | | | | | | | | | | | | | | | | | |
| 57 | Acetylcedrene* | | | | | | | | | | + | | | | | | | | | |
| 58 | Camphor | | + | | | | | | | | | | | | | | | | | |
| 59 | Epicamphor* | | + | | | | | | | | | | | | | | | | | |
| 60 | Menthylacetate* | | | | | | + | | | | | | | | | | | | | |
| 61 | Dehydroabietic acid (m) | | + | + | + | + | + | + | + | | + | | | | + | + | + | + | + | + |
| | *Physiological effective compounds* | | | | | | | | | | | | | | | | | | | |
| 62 | Ibuprofen (m) | | + | | | | | | + | + | | | + | | | | | | + | |
| 63 | Carbamazepine | + | + | + | + | + | + | + | + | + | + | + | + | + | | + | + | | + | + |
| 64 | Propyphenazone | + | + | + | + | + | + | + | + | + | | + | + | + | | + | | | | |
| 65 | Clofibric acid (m) | + | + | + | + | + | + | + | + | + | + | + | + | + | + | + | + | + | + | |
| 66 | Diclofenac | | | | | | | + | + | | + | | + | + | + | | | | + | |
| 67 | Caffeine | | + | + | | | | + | + | + | | | | + | | | | + | + | |
| 68 | DEET, N,N-Diethyl-*m*-toluamide | + | + | + | | | + | + | + | + | + | + | + | + | + | + | + | + | + | |
| | *Polycyclic musks* | | | | | | | | | | | | | | | | | | | |
| 69 | HHCB, Galaxolide | + | + | + | + | + | + | + | + | + | + | + | + | + | + | + | + | + | + | + |
| 70 | AHTN, Tonalide | + | + | + | + | + | + | + | + | + | + | + | + | + | + | + | + | + | + | + |
| 71 | ADBI, Celestolide | | | | + | | + | | | | + | | | | | | + | | + | |
| 72 | AHMI, Pantolide | | | | + | | + | | | | + | | | | | | + | | + | |

* Identified by comparison of mass spectral data with those of mass spectral data bases
(m) detected as methyl ester
(+) not all isomers could be detected

**Tab. 2:** Occurrence and application of the identified organic compounds

| No. | Compounds | | Occurrence and application |
|---|---|---|---|
| | *Aliphatics* | | |
| 1 | Homologues series of *n*-alkanes ($C_{11}$-$C_{18}$, $C_{24}$, $C_{25}$) | BI | occurrence in every natural material, technical application,e.g. for fat synthesis, short-chain *n*-alkanes are used as solvents and fuels |
| 2 | Squalene | BI | natural occurrence e.g. in shark liver, mainly used for synthesis of squalane |
| | *Polycyclic aromatic compounds* | | |
| 3 | Naphthalene | BI | in exhaust fumes, used for the synthesis of dyes and pesticides |
| 4 | Phenanthrene | | |
| 5 | Fluorene | | |
| 6 | 9-Fluorenone | BI | dehydrogenating agent for the Oppenauer oxidation |
| 7 | Acridine | BI | used for the synthesis of dyes and pharmaceutics |
| 8 | 9-Methylacridine | A | industrial application unknown |
| | *Aliphatic acids* | | |
| 9 | Homologues series of *n*-carboxylic acids ($C_8$-$C_{18}$) | BI | occurrence in all organisms, basis product for the synthesis of soaps, detergents, plasticizers and other industrial products |
| | *Aromatic carboxylic acids* | | |
| 10 | Benzoic acid (m) | BI | natural occurrence in resins and berrys, used as preservating agent for food |
| 11 | 4-*tert*-Butylbenzoic acid (m) | A | anticorrosive agent for bore and grinding oils |
| 12 | 3,5-Di-*tert*-butyl-4-hydroxybenzoic acid (m) | A | |
| 13 | Phenylacetic acid (m) | BI | in ethereal oils, used for synthesis of pharmaceutics |
| 14 | 3-Phenylpropionic acid (m) | BI | used as fixating agent in perfumes |
| 15 | Phthalic acid* or Phthalic anhydride* | BI | used for the synthesis of dyes and phthalates |

| No. | Compounds | | Occurrence and application |
|---|---|---|---|
| | *Esters* | | |
| 16 | 2-Butenedioic acid dibutylester | | |
| 17 | Di-*iso*-propyladipate* | | |
| 18 | Di-*n*-butyladipate | | |
| 19 | Bis(2-ethylhexyl)-adipate | BI | plasticizer (Katase & Kim, 1999) |
| 20 | Isopropyl myristate | BI | additive for cosmetic products (Ash & Ash, 1991) |
| 21 | TXIB, 2,2,4-Trimethyl-1,3-pentanedioldi-*iso*-butyrate | A | plasticizer (Ash & Ash, 1991) |
| 22 | Triethyl citrate | I | natural occurrence in wine, used as plasticizer, additive for cosmetic products and foods |
| 23 | 2-(2-Butoxyethoxy)-ethylacetate | A | |
| | *Aldehydes and ketones* | | |
| 24 | Vanilline | BI | in vanilla pods, wood and some foods, used as additive for cosmetic products |
| 25 | 4-Oxoisophorone | BI | used by the perfume industry (Papa & Sherman, 1981) |
| | *Alcohols and ethers* | | |
| 26 | TMDD, 2,4,7,9-Tetra-methyl-5-decyne-4,7-diol | A | dispersing agent in dyes and cleansing products (Ash & Ash, 1991) |
| 27 | Propylenoxide dimer* (2 isomers) | A | |
| 28 | Propylenoxide trimer* (3 isomers) | A | |
| | *Phenols* | | |
| 29 | Ionol, BHT, 2,6-Di-*tert*-butyl-4-methyl-phenol | A | technical antioxidant |
| 30 | BHA, *tert*-Butyl-hydroxyanisole* | A | antioxidant for foods |

| No. | Compounds | | Occurrence and application |
|---|---|---|---|
| | *P-containing compounds* | | |
| 31 | Tri-*iso*-butyl phosphate | A | plasticizer (Koch, 1995) |
| 32 | Tri-*n*-butyl phosphate | A | plasticizer, flame retardant |
| 33 | Triethyl phosphate | A | flame retardant (Ash & Ash, 1991) |
| 34 | TCEP, Tris-2-chloro-ethyl phosphate | A | plasticizer, flame retardant |
| 35 | Triphenyl phosphate | A | flame retardant |
| 36 | TCPP; Tris(chloro-propyl) phosphate (2 isomers) | A | flame retardant |
| 37 | Triphenylphosphinoxide | | product of the Wittig reaction |
| | *Chlorinated compounds* | | |
| 38 | Hexachlorobutadiene | A | solvent, liquid in hydraulic systems (Koch, 1995) |
| 39 | Chlorobutanol | A | plasticizer and preservating agent |
| 40 | 1,2,3-Trichloropropene* | A | - |
| 41 | Dichlorobenzene | A | solvent (Koch, 1995) |
| 42 | Dichloroaniline | A | precursor of some dyes, pesticides and pharmaceutics |
| 43 | Dichlorobenzophenone | A | degradation product of DDT |
| 44 | Bis(chloropropyl)ethers (2 isomers) | A | |
| | *S-containing compounds* | | |
| 45 | Di-*iso*-propyldisulfide | B | formation by blue-green algae (Microcystis flos-aquae) in fresh waters (Hofbauer & Jüttner, 1988), natural occurrence of Di-iso-propyldisulfide in onions (Boelens et al., 1971) |
| 46 | Dipropyltrisulfide* | B | |
| 47 | NBBS, N-Butylben-zenesulfonamide | A | plasticizer (Ash & Ash, 1991) |
| | *N-containing compounds* | | |
| 48 | Toluenesulfonic acid N-ethylamide | A | |

| No. | Compounds | | Occurrence and application |
|---|---|---|---|
| 49 | TAED, N,N,N`,N`-Tetraacetylethylene-diamine* | A | activator of bleaching agents in washing powders |
| 50 | Methylbenzonitrile* | I | used for organic synthesis |
| 51 | Phenylisocyanate | A | used for the synthesis of pharmaceutics and plasticizers |
| 52 | Methylphenylisocyanate | A | |
| 53 | Cyclohexylisocyanate* | A | |
| | *Heterocyclic compounds* | | |
| 54 | 2,5-Dihydro-2,5-dimethylfurane* | | |
| 55 | Benzothiazole | I | solvent for synthesis at high temperatures |
| 56 | MTB, 2-Methylthio-benzothiazole | A | originates during the production of rubber (Takada and Eganhouse 1998) |
| | *Terpenoids* | | |
| 57 | Acetylcedrene* | I | natural occurrence in ethereal oils, used by the perfume industry |
| 58 | Camphor | I | natural occurrence in ethereal oils, used as plasticizer |
| 59 | Epicamphor* | I | - |
| 60 | Menthylacetate* | BI | natural occurrence in ethereal oils, used by the perfume industry |
| 61 | Dehydroabietic acid (m) | BI | natural occurrence in resins and barks (e.g. Corin et al., 2000) |
| | *Physiological effective compounds* | | |
| 62 | Ibuprofen (m) | A | analgesic |
| 63 | Carbamazepine | A | antiepileptic drug |
| 64 | Propyphenazone | A | analgesic |
| 65 | Clofibric acid (m) | A | active metabolite of clofibrate, a blood lipid regulator |
| 66 | Diclofenac | | analgesic |
| 67 | Caffeine | A | in coffee and tea, used as a pharmaceutical drug |
| 68 | DEET, N,N-Diethyl-*m*-toluamide | A | insect repellent |

| No. | Compounds | | Occurrence and application |
|---|---|---|---|
| | *Polycyclic musks* | | |
| 69 | HHCB, Galaxolide | A | used as fragrances in cosmetics and detergents |
| 70 | AHTN, Tonalide | A | used as fragrances in cosmetics and detergents |
| 71 | ADBI, Celestolide | A | used as fragrances in cosmetics and detergents |
| 72 | AHMI, Pantolide | A | used as fragrances in cosmetics and detergents |

A anthropogenic marker
B known biogenic or geogenic source
I industrial application or industrial by-product
BI industrial application or industrial by-product, biogenic origin possible
* Identified by comparison of mass spectral data with those of mass spectral data bases
$^{(m)}$ detected as methyl ester

Plasticizers and flame retardants

TXIB (2,2,4-trimethyl-1,3-pentanedioldi-*iso*-butyrate, No. 21 according to Table 1, chemical structure see Figure 2) appeared in most of the investigated Lippe river water samples (see Table 1). Its occurrence as contaminant in Elbe river water and sediments of the German Bight has been reported by Franke et al. (1995) and Schwarzbauer et al. (2000), respectively. According to a study of the Danish EPA (2001), TXIB has a relatively low water solubility (1-2 mg/L, estimated $K_{OW}$ 4.1) and a low toxicity.

Various alkyl phosphates (tri-*iso*-butyl phosphate, tri-*n*-butyl phosphate, triethyl phosphate and triphenyl phosphate, Nos. 31,32,33 and 35) and TCEP (tris-2-chloroethyl phosphate, No. 34, chemical structure see Figure 2) were detected along the Lippe river (see Table 1). These compounds are flame retardants and weak plasticizers which are frequently used as plastic additives as well as flame inhibitors in pieces of upholstered furniture, carpets, casings of electric devices etc. They are well known contaminants in river water, sewage effluents: and river sediments (e.g. Ishikawa et al., 1985; Paxeus, 1996; Metzger and Möhle, 2001). Because of its persistence, the industrial usage of TCEP was stopped some years ago. Nevertheless, it is still found in aquatic systems (Metzger and Möhle, 2001).

Two isomers of tris(chloropropyl) phosphate could be identified in the Lippe river water, to be precise mainly in samples from the lower reaches (see Table 1). Tris(1-chloro-2-propyl)phosphate (TCPP) was discovered by other authors in water, sediments and air samples (Kawagoshi et al., 1999; Fries and Püttmann, 2001; Ingerowski et al., 2001; Schwarzbauer et al., 2002). It is predominately used as a flame retardant in rigid polyurethane foam. Due to information of a manufacturer, almost 23.000 tons of TCPP were sold in Western Europe in 1998. Details about pathways of TCPP into aquatic environments have not been clarified so far.

NBBS (N-butylbenzenesulfonamide, No. 47, chemical structure see Figure 2) was identified in all water samples, except at the river source (see Table 1). This is surprising, because the usage in Germany was stopped when neurotoxic effects of this plasticizer became known (Huppert and Würtele, 1998; Nerurkar et al., 1993). The occurrence of NBBS in German sewage effluents and Elbe river water has been described by Huppert and Würtele (1998) and Franke et al. (1995), respectively.

Ingredients of personal care products

Vanillin (No. 24) was detected abundantly in Lippe river samples, whereas acetylcedrene and menthylacetate (Nos. 57 and 60) were identified in one sample only. The three compounds are present in nature but are as well synthezised industrially as cosmetic and perfume additives (see Table 2). Probably, the ubiquitous occurrence of vanillin results from the biodegradation of wood in the river. Contrary, the occurrence of acetylcedrene and menthylacetate can be attributed to anthropogenic input because they are components of cedar and peppermint oil, respectively. Both plants are not frequent at the vicinity of the river. Most of the investigated Lippe river water samples contained triethyl citrate (No. 22). This chemical is found in low concentrations in wine and is used as plasticizer as well as food and cosmetic additive (see Table 2). 4-Oxoisophorone (No. 25) was detected in five of nineteen water samples and is used as additive in perfumes (Papa and Sherman, 1981). This fragrance was prevalent in Lippe river sediments (Kronimus et al., 2004) and was also identified in sediments of the Havel and Spree rivers (Ricking et al., 2003). Four polycyclic musks (Nos. 69 to 72) which are well known contaminants in the aquatic environment (e.g. Rimkus, 1999) and are used as fragrances in washing powders and cosmetics were identified (see Table 1). Their occurrence and distribution in Lippe river water and sediments was reported in Dsikowitzky et al. (2002).

Physicological effective compounds

The occurrence of pharmaceutical drugs in the aquatic environment has recently been the subject of several studies (e.g. Ternes, 1998; Halling-Sørensen et al., 1998; Kolpin et al., 2002). In Lippe river water, five pharmaceutics were identified. Ibuprofen (No. 62) and diclofenac (No. 66), which are analgesics, were only present in a few samples (see Table 1). Both compounds have already been detected in fresh waters and sewage effluents (e.g. Ternes, 1998; Buser et al., 1998a; Buser et al.,1999) and are easily degradable (Buser et al., 1999; Buser et al., 1998b). Most samples from the course of the Lippe river contained the pharmaceutics carbamazepine (No. 63), propyphenazone (No. 64, chemical structure see Figure 2) and clofibric acid (No. 65) (see Table 1). Carbamazepine is a frequently used antiepileptic drug and is a prevalant contaminant in surface waters and sewage effluents (e.g.Ternes,1998). Propyphenazone, an analgesic and structural relative of phenazone, has rarely been noticed so far but is prevalent in Berlin surface waters and groundwater (Heberer et al., 1998). Clofibric acid is the active metabolite of the blood lipid regulators clofibrate, etofyllinclofribate and etofibrate. It is formed via hydrolysis very soon after ingestion and is excreted primarily as glucuronide of the acidic metabolites. The presence of clofibric acid in sewage effluents indicates the hydrolysis of the conjugate during wastewater treatment (Ternes, 1998).

Caffeine (No. 67) is a component of several pharmaceutics as well as beverages and was detected in some Lippe river water samples (see Table 1). It was ubiquitous in river water samples from nineteen different German rivers, in concentrations up to 1 µg/L (Prösch and Puchert, 1998). Because of its frequent occurrence in sewage effluents, it has the potential to be used as anthropogenic marker compound (Standley et al., 2000). The insect repellent DEET (N,N-diethyl-*m*-toluamide, No. 68, chemical structure see Figure 2) was present in most samples, even at the source region. This compound was already detected in numerous German and American rivers (Franke et al., 1995; Kolpin et al., 2002). The input of DEET into the aquatic environment mainly takes place via sewage effluents and shows a strong seasonality with the highest concentrations during the summer (Knepper et al., 1996). With respect to toxicological effects, does the uptake of a single dosis DEET via the skin (400 mg/kg) lead to increased 8-hydroxy-2'-deoxyguanosine (a biomarker of oxidative DNA damage ) in rat urine (Abu-Quare and Abou-Donia, 2000).

Solvents and products of industrial synthesis

Triphenylphosphinoxide (No. 37, chemical structure see Figure 2) was identified in six samples from the lower reaches of the Lippe river (see Table 1). It is a product of the Wittig-synthesis, which is carried out for the industrial production of olefinic compounds like carotinoids, vitamin D, squalene etc. Besides it originates from the desoxygenation of peroxides and other oxides with triphenylphosphine. To our knowledge, triphenylphosphinoxide has not been described as environmental contaminant so far. Hexachlorobutadiene (No. 38) was identified in four water samples from the area of the river mouth (see Table 1). Its occurrence in Lippe river sediments was reported by Kronimus et al. (2004). The main input pathways of hexachlorobutadiene into the aquatic environment are effluents of the organic chemical industry (Verschueren, 1996; Lee and Fang, 1997). Hexachlorobutadiene is very volatile and can exist vaporised in the atmosphere or adsorbed to airborne particulate matter (Choudhary, 1995). In the aquatic environment, it is preferentially adsorbed to particulate matter (log $K_{OW}$ 4.78; Verschueren, 1996). In sediments of the Great Lakes, U.S., a heavily indiustrialised area, concentrations of up to 120 μg/kg were determined (Choudhary, 1995). Data about the occurrence, biological effects and properties of hexachlorobutadiene are summarised in Verschueren (1996). The chemical was included into the hazardous substance priority list of the European Union (Decision No. 2455/2001/EG). Bis(chloropropyl)ethers (No. 44) were also exclusively identified in water samples from the lower reaches of the Lippe river (see Table 1). They are known as by-products of various technical synthesis (e.g. the aqueous chlorination of propene). Two isomers could also be identified in sediments from the Lippe river mouth (Kronimus et al., 2004). 1-Chloro-2-propyl-2-chloro-1-propylether was listed as a priority pollutant by the American Environmental Protection Agency. As early as 1988 the mutagenic and cancerogenic potential of this isomer became known (McGregor et al., 1988).

MTB (2-methylthiobenzothiazole, No. 56, chemical structure see Figure 2) was found in most water samples (see Table 1). It does not have an industrial application but originates as a by-product during the synthesis of rubber. MTB is a component of tyes and consequently, high concentrations of this compound were measured in the vincinity of highways (ARGE Elbe, 2000). Besides that, it is found as a degradation product of the pesticide 2-thiocyanomethylthiobenzothiazole in tannery wastewaters (Reemtsma et al., 1995). MTB was detected in the Elbe river system (Germany) and in estuarine and marine waters of the North Sea (Bester et al., 1997). The partition coefficient (log $K_{OW}$ 3.1) of MTB suggests a high tendency for bioaccumulation and adsorption to particulate matter (Schmegel, 1995). The compound is not completely removable during biological wastewater treatment and is of concern because of its limited biodegradability and potential toxicity (Reemtsma et al., 1995).

Compounds with natural occurrence

*N*-alkanes (No. 1) and *n*-carboxylic acids (No. 9) occur in all natural materials and are as well used for numerous industrial syntheses (see Table 2). Thus, their appearance in Lippe river water can be attributed to various sources. This is also the case for vanillin (see section 'Perfumes, odors and additives for cosmetics' ). In contrast, di-*iso*-propyldisulfide (No. 45) and dipropyltrisulfide (No. 46) which were detected in several water samples (see Table 1) are clearly related to natural sources. It is known that they are formed by blue-green algae (*Microcystis flos-aquae*) in fresh waters (Hofbauer and Jüttner, 1988) and the industrial application is unknown.

Miscellaneous

9-Methylacridine (No. 8, chemical structure see Figure 2) was detected in every Lippe river water sample, except at the river source (see Table 1). There are no information available about an industrial application of this chemical. 9-Methylacridine is a common contaminant in effluents of sewage treatment plants but is not present in the corresponding influents. Therefore, Grohmann et al. (1998) suggest that it is formed during wastewater treatment. If further research reveals the ubiquitous occurrence of 9-methylacridine in sewage effluents, it is probably a useful anthropogenic marker.

TMDD (2,4,7,9-tetramethyl-5-decyne-4,7-diol, No. 26, chemical structure see Figure 2) was identified in all Lippe river water samples. It is used as dipersing agent for dyes and cleansing products (Ash and Ash, 1991) and was so far not known as contaminant in aqueous systems. However, TMDD-concentrations of up to 69 μg/L were detected in mineral waters in beverage carton packages. TMDD is a surfactant in printing inks and migrates during the production process into the inner polyethylene film and subsequently into the product (Kleinschnitz and Schreier, 1998).

*Potential sources of the identified organic contaminants*

In order to get an idea about the input pathways of the detected compounds into the Lippe river, some tributaries and potential sources of organic contaminants were sampled. Random samples were taken from the Alme river and the Quabbe Brook which are located at the less densely populated upper reaches of the Lippe river. Additionally, the Seseke river was investigated, a dirty water course which is heavily polluted with sewage effluents. Analyses of effluents from the municipial sewage treatment plant (STP) in the city of Hamm and a pharmaceutical plant were also carried out. The compound spectra which was identified in the Lippe river (see Table 1) was used as a basis for investigating the source samples. The results are summarised in Table 3.

Many of the organic contaminants which were found in Lippe river water were also present in the source samples (see Table 3). The sewage effluent sample and the Seseke river showed the best accordance with the compound spectrum of the Lippe river. However, also in the two tributaries from the rural upper reaches of the river, numerous specific contaminants like 9-methylacridine (No. 8), alkyl phosphates (Nos. 31, 32) and chlorinated alkyl phosphates (Nos. 34, 36) appeared. In the effluent of a pharmaceutical plant, only a few Lippe river contaminants like *n*-alkanes (No. 1), naphthalene (No. 3), TXIB (No. 21) and caffeine (No. 67) were detected (see Table 3). Therein, mainly structural relatives of androstanone like 3β-hydroxy-5β-androstan-17-one, 3α-hydroxy-5β-androstan-17-one and androstan-5β-3,17-dione were present. These compounds are probably by-products of the synthesis of hormone preparations. Some polycyclic aromatic compounds, halogenated compounds and terpenoids were not detected in the source samples (see the underlined compounds in Table 3) and probably have another origin. Representative sampling of various input sources have to be carried out to prove the origin of these compounds. Hexachlorobutadiene (No. 38) and bis(chloropropyl)ethers (No. 44) appear exclusively at the lower reaches of the Lippe river (see Table 1), downstream the chemical plants in Marl. They are attributed to inputs of the chlorochemical industry (see section 3.1). Hence, this suggests their input by an industrial point source.

**Tab. 3:** Organic compounds in tributaries and wastewater effluents at the Lippe river

| No | Compound | | STP Hamm [a] | Pharmaceutical plant [b] | Seseke river [c] | Alme river [d] | Quabbe Brook [e] |
|---|---|---|---|---|---|---|---|
| | *Aliphatics* | | | | | | |
| 1 | Homologues series of *n*-alkanes ($C_{11}$-$C_{18}$, $C_{24}$, $C_{25}$) | BI | + | + | + | + | + |
| 2 | Squalene | BI | + | | + | + | + |
| | *Polycyclic aromatic compounds* | | | | | | |
| 3 | Naphthalene | BI | | + | + | + | + |
| 4 | Phenanthrene | | | | | + | |
| 5 | Fluorene | | | | | | |
| 6 | 9-Fluorenone | BI | | | | | |
| 7 | Acridine | BI | | | | | |
| 8 | 9-Methylacridine | A | + | | + | + | + |
| | *Aliphatic acids* | | | | | | |
| 9 | Homologues series of *n*-carboxylic acids ($C_8$-$C_{18}$) | BI | + | | | + | + |
| | *Aromatic carboxylic acids* | | | | | | |
| 10 | Benzoic acid (m) | BI | + | | | | + |
| 11 | 4-*tert*-Butylbenzoic acid (m) | A | + | | | | + |
| 12 | 3,5-Di-*tert*-butyl-4-hydroxybenzoic acid (m) | A | + | | | | |
| 13 | Phenylacetic acid (m) | BI | + | | | | |
| 14 | 3-Phenylpropionic acid (m) | BI | | | | | |
| 15 | Phthalic acid* or Phthalic anhydride* | BI | + | | + | | + |

| No | Compound | | STP Hamm [a)] | Pharmaceutical plant [b)] | Seseke river [c)] | Alme river [d)] | Quabbe Brook [e)] |
|---|---|---|---|---|---|---|---|
| | *Esters* | | | | | | |
| 16 | 2-Butenedioic acid dibutylester | | | | | | |
| 17 | Di-*iso*-propyladipate* | | | | + | + | |
| 18 | Di-*n*-butyladipate | | + | | + | + | |
| 19 | Bis(2-ethylhexyl)adipate | BI | | | | + | |
| 20 | Isopropyl myristate | BI | | | | | |
| 21 | TXIB, 2,2,4-Trimethyl-1,3-pentanedioldi-*iso*-butyrate | A | + | + | | + | + |
| 22 | Triethyl citrate | I | + | | + | | + |
| 23 | 2-(2-Butoxyethoxy)ethylacetate | A | | | | | |
| | *Aldehydes and ketones* | | | | | | |
| 24 | Vanilline | BI | | | | | + |
| 25 | 4-Oxoisophorone | BI | + | | + | + | + |
| | *Alcohols and ethers* | | | | | | |
| 26 | TMDD, 2,4,7,9-Tetramethyl-5-decyne-4,7-diol | A | + | | + | + | + |
| 27 | Propylenoxide dimer* (2 isomers) | A | | | + | | |
| 28 | Propylenoxide trimer* (3 isomers) | A | | | + | + | |
| | *Phenols* | | | | | | |
| 29 | Ionol, BHT, 2,6-Di-*tert*-butyl-4-methylphenol | A | + | | + | | + |
| 30 | BHA, *tert*-Butylhydroxyanisole* | A | + | | + | + | + |

| No | Compound | | Sampling Location: STP Hamm [a)] | Pharmaceutical plant [b)] | Seseke river [c)] | Alme river [d)] | Quabbe Brook [e)] |
|---|---|---|---|---|---|---|---|
| | *P-containing compounds* | | | | | | |
| 31 | Tri-*iso*-butyl phosphate | A | + | | + | + | + |
| 32 | Tri-*n*-butyl phosphate | A | | | + | + | + |
| 33 | Triethyl phosphate | A | + | | | | |
| 34 | TCEP, Tris-2-chloroethyl phosphate | A | + | | + | + | + |
| 35 | TCPP, Triphenyl phosphate | A | | | | | |
| 36 | Tris(chloropropyl) phosphate (2 isomers) | A | + | | + | + | + |
| 37 | Triphenylphosphinoxide | | | | + | | |
| | *Chlorinated compounds* | | | | | | |
| 38 | Hexachlorobutadiene | A | | | | | |
| 39 | Chlorobutanol | A | | | | | |
| 40 | 1,2,3-Trichloropropene* | A | | | | | |
| 41 | Dichlorobenzenes | A | | | + | | |
| 42 | Dichloroaniline | A | | | | | |
| 43 | Dichlorobenzophenone | A | | | | | |
| 44 | Bis(chloropropyl)ethers (3 isomers) | A | | | | | |
| | *S-containing compounds* | | | | | | |
| 45 | Di-*iso*-propyldisulfide | B | | | | + | + |
| 46 | Dipropyltrisulfide* | B | | + | | | |
| 47 | NBBS, N-Butylbenzene-sulfonamide | A | + | | + | | |

| No | Compound | | STP Hamm [a] | Pharmaceutical plant [b] | Seseke river [c] | Alme river [d] | Quabbe Brook [e] |
|---|---|---|---|---|---|---|---|
| | *N-containing compounds* | | | | | | |
| 48 | Toluenesulfonic acid N-ethylamide | A | + | | + | | |
| 49 | TAED, N,N,N`,N`-Tetraacetyl ethylenediamine* | A | + | | + | | + |
| 50 | Methylbenzonitrile* | I | | | | | |
| 51 | Phenylisocyanate | A | | | + | | + |
| 52 | Methylphenylisocyanate* | A | | | | | |
| 53 | Cyclohexylisocyanate* | A | | | | | |
| | *Heterocyclic compounds* | | | | | | |
| 54 | 2,5-Dihydro-2,5-dimethylfuran* | | + | | | | |
| 55 | Benzothiazole | I | | | + | | |
| 56 | MTB, 2-Methylthiobenzothiazole | A | + | | + | | + |
| | *Terpenoids* | | | | | | |
| 57 | Acetylcedrene* | I | + | | | | |
| 58 | Camphor | I | | | + | | + |
| 59 | Epicamphor* | I | | | + | | |
| 60 | Menthylacetate* | BI | | | | | |
| 61 | Dehydroabietic acid [m] | BI | | | | | |
| | *Physiological effective compounds* | | | | | | |
| 62 | Ibuprofen [m] | A | + | | | | + |
| 63 | Carbamazepine | A | + | | + | + | + |
| 64 | Propyphenazone | A | + | | + | | |
| 65 | Clofibric acid [m] | A | + | | | | + |
| 66 | Diclofenac | | + | | + | | + |

| No | Compound | | STP Hamm [a)] | Pharmaceutical plant [b)] | Seseke river [c)] | Alme river [d)] | Quabbe Brook [e)] |
|---|---|---|---|---|---|---|---|
| 67 | Caffeine | A | | + | + | | + |
| 68 | DEET, N,N-Diethyl-*m*-toluamide | A | + | | + | | + |
| | *Polycyclic musks* | | | | | | |
| 69 | HHCB, Galaxolide | A | + | | + | + | + |
| 70 | AHTN, Tonalide | A | + | | + | + | + |
| 71 | ADBI, Celestolide | A | + | | + | | |
| 72 | AHMI, Pantolide | A | + | | + | | |

a) Effluent of the sewage treatment plant in Hamm, sampling on 22/8/2000
b) Effluent of a pharmaceutical plant, sampling on 22/8/2000, without 3rd fraction
c) Water sample from the Seseke river, a tributary of the Lippe river, after wastewater treatment plant at the river mouth, sampling on 8/3/2001
d) Water sample from the Alme river, a tributary of the Lippe river, sampling on 29/8/2000
e) Water sample of the Quabbe Brook, a tributary of the Lippe river, sampling on 29/8/2000
underlined: compound was only detected in Lippe river water samples
* Identified by comparison of mass spectral data with those of mass spectral data bases
(+) not all isomers could be detected
(m) detected as methyl ester
A anthropogenic marker
B known biogenic or geogenic source
I industrial application or industrial by-product

*Anthropogenic marker compounds in the Lippe river*

Several compounds of anthropogenic origin like 9-methylacridine (No. 8), TXIB (No. 21), triethyl citrate (No. 22), TMDD (No. 26) , tri-*iso*-butyl phosphate (No. 31), TCEP (No. 34), NBBS (No. 47), MTB (No. 56), some pharmaceutics (Nos. 63 – 65), DEET (No. 68) and two polycyclic musks (Nos. 69 and 70) were ubiquitous in Lippe river water (see Table 1), in sewage effluent and also in samples from the tributaries (see Table 3). Interestingly, most of these compounds were even present in the small Quabbe Brook which is located in a rural area. Even here an input of sewage effluents from small villages takes place. The listed compounds are obviously typical sewage derived contaminants in the area of the Lippe river. They have the potential to be generally used as anthropogenic markers if the following conditions according to Takada and Eganhouse (1998) are fulfilled: i) source specificity, ii) widespread and massive usage and iii) environmental persistence.

9-Methylacridine (No. 8), tri-*iso*-butyl phosphate (No. 31), TCEP (No. 34), NBBS (No. 47), MTB (No. 56), carbamazepine (No. 63) and clofibric acid (No. 65), DEET (No. 68) and the polycyclic musks were identified by other authors in sewage effluents (see section 3.1). This speaks in favour of the source specificity and the widespread massive usage of these compounds at least in Germany. To our knowledge, the plasticizer TXIB (No. 21), triethyl citrate (No. 22), the dispersing agent TMDD (No. 26) and the pharmaceutical drug propyphenazone (No. 64) have not been identified in sewage effluents so far. Because of their application, their input via sewage effluents into the environment and their widespread usage are likely. The third condition, the environmental persistence of the marker candidates has to be proved by degradation experiments and/or quantitative examinations. The latter will be discussed in part II of our study (see chapter 3.1.2).

## Summary and conclusions

Screening analyses of water samples from a longitudinal section of the Lippe river, Germany, revealed the presence of a wide spectrum of low molecular weight organic compounds ranging from nonpolar constituents like aliphatic hydrocarbons to polar constituents like *n*-carboxylic acids and phenols. Most of the identified compounds could be attributed to anthropogenic input. Amongst them are components of household products (e.g. plasticizers, flame retardants, synthetic fragrances), by-products of industrial synthesis and pharmaceutical drugs.

Some of the organic compounds which were frequently detected in Lippe river water were also present in the effluent of a municipial sewage treatment plant and in Lippe river tributaries. These compounds are obviously typical sewage derived contaminants in the area of the Lippe river. Their ubiquitous occurrence in German rivers and sewage effluents, which was confirmed by other authors, suggests their use as markers. Further quantitative analyses are underway to examine the persistence of these compounds, which is the condition for their general use as anthropogenic markers for sewage effluents.

# 3.1.2 The anthropogenic contribution to the organic load of the Lippe river (Germany) - Part II: Quantification of specific organic contaminants‡

## Introduction

Part I of our study (Dsikowitzky et al. 2004a; chapter 3.1.1) revealed the presence of a wide spectrum of low-molecular weight organic compounds which were detected in Lippe river water samples. For a more detailed study, fourteen contaminants of special interest were quantified in four water sample sets and are presented here. These were compounds which have rarely been noticed as aquatic contaminants so far (e.g. TXIB (2,2,4-trimethyl-1,3-pentanedioldi-*iso*-butyrate), TMDD (2,4,7,9-tetra-methyl-5-decyne-4,7-diol), triphenylphosphinoxide) or which were interesting because of their application (e.g. pharmaceuticals).

The water course of the Lippe river can be divided into the more rural upper reaches (sites 19 to 14) and the densely populated and highly industrialised lower reaches (sites 13 to 1). According to the population density, there is an increasing frequency of sewage treatment plants below sampling site 13 (see Fig. 1). Besides there are coal mines as well as numerous industrial plants located in the area of the lower reaches.

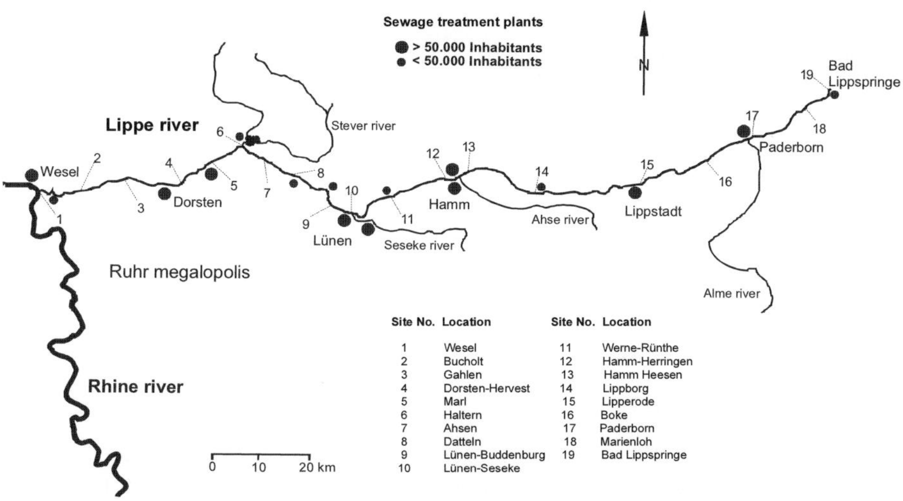


**Fig. 1:** Map showing the Lippe river system (North-Rhine Westphalia, Germany) with sampling locations (sites 1 to 19) and sewage treatment plants.

‡ Mainly adapted/reprinted from Dsikowitzky et al., 2004b

## Samples

See Chapter 3.1.1

Quantitative data were obtained by integration of selected ion chromatograms extracted from the total ion current. The ions used for quantification are presented in Table 1. For GC/MS-analyses of reference compounds the limit of quantitation (LOD) was 0.5 ng (signal to noise ratio 10:1). In Lippe river water samples concentrations of 10 ng/L resulted in similar signal to noise ratios due to matrix effects. Therefore these values were defined as limits of quantitation in the present study.

**Tab. 1:** Ions used for quantitation and recoveries of selected compounds from high-purity water (n = 4; Triphenylphosphinoxide n =2; Clofibric acid n = 3). The data represent average recovery rates and relative standard deviations. The chemical names of the pharmaceutical drugs and HHCB can be found in chapter 3.1.2 and Dsikowitzky (2002).

| Compound | Quantitation Ions [m/z] | Recovery rate [%] |
|---|---|---|
| TXIB, 2,2,4-Trimethyl-1,3-pentanedioldi-*iso*-butyrate | 71 | 92 ± 30 |
| Triethyl citrate | 157 | 96 ± 3 |
| TMDD, 2,4,7,9-Tetramethyl-5-decyne-4,7-diol | 109 | 36 ± 13 |
| Hexachlorobutadiene | 225;227 | 53 ± 17 |
| TCEP, Tris-2-chloroethyl phosphate | 249;251 | 85 ± 34 |
| Triphenylphosphinoxide | 199;277 | 85 ± 22 |
| NBBS, N-Butylbenzenesulfonamide | 77;170 | 102 ± 15 |
| MTB, 2-Methylthiobenzothiazole | 148;181 | 84 ± 10 |
| Carbamazepine | 193 | 65 ± 23 |
| Propyphenazone | 215;230 | 89 ± 34 |
| Clofibric acid | 128;130 | 97 ± 23 |
| Diclofenac | 214;242 | 90 ± 36 |
| Caffeine | 109;194 | 58 ± 16 |
| HHCB, Galaxolide | 243;258 | 94 ± 7 |

## Results and Discussion

*Compound concentrations in water*

Sustainable information (e.g. industrial application, occurrence in the environment) about the contaminants which were chosen for quantification are summarised in part I of our study. The quantitative data of the first sample set (taken in August 1999) are shown in Table 2. In order to register temporal changes and/or general distribution pattern, the selected compounds were also quantified in the other water samples sets (Tables 3 to 5).

The bulk of the quantitative data show that concentrations of the compounds differed on a large scale. Carbamazepine and diclofenac, two pharmaceutical drugs, had the highest concentrations (up to 2000 ng/L and up to 1800 ng/L, respectively). The highest concentration of the dispersing agent TMDD was 1500 ng/L. In contrast, concentrations of hexachlorobutadiene and MTB were often below the limit of quantitation (< 10 ng/L). Maximum values were 80 ng/L and 60 ng/L, respectively. Concentrations of the plasticizers TXIB, triethyl citrate, TCEP and NBBS and of the pharmaceutical drugs propyphenazone and clofibric acid as well as of triphenylphosphinoxide, caffeine and HHCB (a polycyclic musk) ranged between these two extrema.

*Calculation of organic loads*

It has to be taken into account that changing river runoff influences compound concentrations in water. To this end, the organic loads of each compound were balanced on the basis of river runoff on the day of sampling. Runoff data for the sampling sites are shown in Table 6. The organic load $L_{org}$ [g/d] was calculated with the compound concentration in water C [ng/L] and the median river runoff MQ [$m^3/s$]:

$$L_{org} = C \times MQ$$

$L_{org}$ describes the amount of the quantified compound which was transported within the aqueous phase on the day of sampling. Changing river runoff does not affect $L_{org}$ so that general spatial distributions of the investigated compounds along the river water course can be recognised.

For the present study, random water samples (see Section 2.1) from a longitudinal section of the Lippe river were taken. The interpretation of the determined pattern only makes sense if compound input does not vary within a short time scale. TXIB, TMDD, TCEP, NBBS, HHCB and the pharmaceutical drugs are contaminants which are derived from effluents of municipal sewage treatment plants (see Part I of our study and references therein). The application of triethyl citrate as plasticizer and additive in foods and personal care products suggests the same source. Caffeine has already been used as marker for municipal sewage effluents (Standley et al., 2000). Compound concentrations in sewage influents often fluctuate according to the human day and night rhythm (e.g. synthetic fragrances, Simonich et al., 2002). During wastewater treatment the influent is collected in big basins and mixed, a continuous process which takes several hours up to more than a day. Therefore, short variations of influent

**Tab. 2:** Concentrations of selected compounds [ng/L] in water samples from the Lippe river (site numbers see Fig. 1), sampling in August 1999.

| | **Sampling location** | **11** | **12** | **13** | **14** | **15** | **16** | **17** | **18** | **19** |
|---|---|---|---|---|---|---|---|---|---|---|
| **No.** | **Compounds** | | | | | | | | | *river source* |
| **1** | **TXIB** | 30 | 60 | <10 | 30 | 10 | 40 | 50 | 70 | 40 |
| **2** | **Triethyl citrate** | 20 | <10 | <10 | <10 | <10 | <10 | <10 | <10 | <10 |
| **3** | **TMDD** | 750 | 310 | 360 | 320 | 250 | 1500 | 20 | 120 | 10 |
| **4** | **Hexachlorobutadiene** | <10 | <10 | <10 | <10 | <10 | <10 | <10 | <10 | <10 |
| **5** | **TCEP** | 220 | 120 | 80 | 140 | 20 | 40 | 10 | 20 | <10 |
| **6** | **Triphenylphosphinoxide** | <10 | <10 | <10 | <10 | <10 | <10 | <10 | <10 | <10 |
| **7** | **NBBS** | 40 | 10 | 60 | 60 | 10 | 40 | 10 | 40 | <10 |
| **8** | **MTB** | 10 | 30 | <10 | 10 | 10 | 30 | <10 | 10 | <10 |
| **9** | **Carbamazepine** | 790 | 730 | 650 | <10 | 330 | 540 | <10 | 290 | 60 |
| **10** | **Propyphenazone** | 40 | 50 | 10 | <10 | <10 | <10 | <10 | 10 | <10 |
| **11** | **Clofibric acid** | 30 | 60 | 30 | 10 | 20 | 30 | <10 | 20 | <10 |
| **12** | **Diclofenac** | <10 | 160 | 70 | 30 | <10 | <10 | <10 | 50 | <10 |
| **13** | **Caffeine** | <10 | <10 | 30 | <10 | <10 | <10 | 10 | 20 | <10 |
| **14** | **Galaxolide** | 120 | 90 | 100 | 50 | 50 | 90 | 50 | 140 | 10 |

| | Sampling location | 1 | 2 | 3 | 4 | 5 | 6 | 7 | 8 | 9 | 10 |
|---|---|---|---|---|---|---|---|---|---|---|---|
| **No.** | **Compounds** | *river mouth* | | | | | | | | | |
| **1** | **TXIB** | 100 | <10 | <10 | 30 | 30 | 10 | 20 | 180 | <10 | 20 |
| **2** | **Triethyl citrate** | <10 | 10 | 20 | 10 | 10 | <10 | 10 | 10 | <10 | 20 |
| **3** | **TMDD** | 310 | 580 | 620 | 100 | 660 | 390 | 510 | 520 | 300 | 490 |
| **4** | **Hexachlorobutadiene** | 10 | 30 | 40 | 10 | <10 | <10 | <10 | <10 | <10 | <10 |
| **5** | **TCEP** | 200 | 180 | 170 | 190 | 240 | 150 | 200 | 200 | 210 | 210 |
| **6** | **Triphenylphosphinoxide** | 90 | 110 | 100 | 60 | <10 | <10 | <10 | <10 | 30 | 40 |
| **7** | **NBBS** | 70 | 120 | 130 | 30 | 60 | 20 | 90 | 140 | 60 | 100 |
| **8** | **MTB** | <10 | 30 | 50 | <10 | 10 | <10 | 60 | 40 | 10 | 20 |
| **9** | **Carbamazepine** | 950 | 1100 | 950 | 1300 | 1200 | 1300 | 760 | 860 | 570 | 1100 |
| **10** | **Propyphenazone** | 40 | 60 | 70 | 60 | 60 | 40 | 60 | 50 | 70 | 70 |
| **11** | **Clofibric acid** | 50 | 70 | 60 | 50 | 50 | 40 | 50 | 60 | 40 | 70 |
| **12** | **Diclofenac** | <10 | <10 | <10 | <10 | <10 | <10 | 230 | 120 | <10 | 130 |
| **13** | **Caffeine** | <10 | 130 | 100 | <10 | <10 | <10 | 60 | 60 | 20 | <10 |
| **14** | **Galaxolide** | 80 | 70 | 120 | 60 | 160 | 170 | 170 | 180 | 100 | 110 |

**Tab. 3:** Concentrations of selected compounds [ng/L] in water samples from the Lippe river (site numbers see Fig. 1), sampling in February 2000.

| | **Sampling location** | **11** | **12** | **13** | **14** | **15** | **16** | **17** | **18** | **19** |
|---|---|---|---|---|---|---|---|---|---|---|
| **No.** | **Compounds** | | | | | | | | | *river source* |
| **1** | **TXIB** | 40 | 10 | 10 | 20 | 10 | 10 | 20 | 10 | 10 |
| **2** | **Triethyl citrate** | 20 | 40 | 10 | 10 | 10 | <10 | 10 | 10 | <10 |
| **3** | **TMDD** | 30 | 100 | 20 | 20 | 30 | 70 | <10 | 30 | 30 |
| **4** | **Hexachlorobutadiene** | <10 | <10 | <10 | <10 | <10 | <10 | <10 | <10 | <10 |
| **5** | **TCEP** | 10 | 20 | 10 | <10 | <10 | 10 | <10 | <10 | <10 |
| **6** | **Triphenylphosphinoxide** | <10 | <10 | <10 | <10 | <10 | <10 | <10 | <10 | <10 |
| **7** | **NBBS** | 10 | 30 | 10 | 10 | 10 | <10 | 10 | 10 | <10 |
| **8** | **MTB** | <10 | <10 | <10 | <10 | <10 | <10 | <10 | <10 | <10 |
| **9** | **Carbamazepine** | 180 | 410 | 180 | 50 | 110 | 180 | 60 | 70 | <10 |
| **10** | **Propyphenazone** | 10 | 20 | <10 | <10 | <10 | <10 | <10 | <10 | <10 |
| **11** | **Clofibric acid** | 30 | 30 | <10 | <10 | <10 | 30 | <10 | <10 | <10 |
| **12** | **Diclofenac** | 70 | 150 | <10 | <10 | <10 | 80 | <10 | <10 | <10 |
| **13** | **Caffeine** | 120 | 170 | 40 | <10 | 30 | 30 | 30 | 40 | <10 |
| **14** | **Galaxolide** | 50 | 50 | 10 | 20 | 30 | 20 | 20 | 20 | <10 |

| | Sampling location | 1 | 2 | 3 | 4 | 5 | 6 | 7 | 8 | 9 | 10 |
|---|---|---|---|---|---|---|---|---|---|---|---|
| No. | Compounds | *river mouth* | | | | | | | | | |
| 1 | **TXIB** | 20 | 20 | 10 | 20 | 20 | 20 | 20 | 20 | 10 | 30 |
| 2 | **Triethyl citrate** | 40 | 40 | 60 | 50 | 50 | 40 | 50 | 50 | 40 | 80 |
| 3 | **TMDD** | 90 | 80 | 40 | 50 | 60 | 50 | 80 | 80 | 50 | 110 |
| 4 | **Hexachlorobutadiene** | 10 | 10 | 20 | 20 | <10 | <10 | <10 | <10 | <10 | <10 |
| 5 | **TCEP** | 20 | 20 | 30 | 30 | 20 | 30 | 20 | 10 | 30 | 30 |
| 6 | **Triphenylphosphinoxide** | 70 | 60 | 130 | 100 | 20 | 20 | 10 | 10 | 10 | 30 |
| 7 | **NBBS** | 20 | 40 | 60 | 50 | 40 | 40 | 30 | 20 | 20 | 60 |
| 8 | **MTB** | 10 | <10 | <10 | 10 | 10 | <10 | <10 | 10 | <10 | 10 |
| 9 | **Carbamazepine** | 260 | 380 | 450 | 590 | 410 | 430 | 470 | 270 | 350 | 520 |
| 10 | **Propyphenazone** | 10 | 20 | 20 | 30 | 30 | 20 | 20 | 10 | 20 | 30 |
| 11 | **Clofibric acid** | 60 | 140 | 30 | 80 | 130 | 170 | <10 | 90 | 90 | 70 |
| 12 | **Diclofenac** | 90 | 110 | 120 | 150 | <10 | 200 | <10 | 160 | <10 | 120 |
| 13 | **Caffeine** | 100 | 90 | 240 | 190 | 160 | 170 | 130 | 120 | 90 | 320 |
| 14 | **Galaxolide** | 70 | 80 | 70 | 80 | 50 | 60 | 110 | 90 | 90 | 100 |

**Tab. 4:** Concentrations of selected compounds [ng/L] in water samples from the Lippe river (site numbers see Fig. 1), sampling in August 2000.

| | **Sampling location** | **11** | **12** | **13** | **14** | **15** | **16** | **17** | **18** | **19** |
|---|---|---|---|---|---|---|---|---|---|---|
| **No.** | **Compounds** | | | | | | | | | *river source* |
| **1** | **TXIB** | 20 | 10 | 10 | 10 | <10 | 10 | 10 | 10 | 20 |
| **2** | **Triethyl citrate** | 10 | 20 | 10 | 20 | 10 | 10 | <10 | 10 | <10 |
| **3** | **TMDD** | 70 | 120 | 110 | 120 | 60 | 110 | 20 | 30 | 40 |
| **4** | **Hexachlorobutadiene** | <10 | <10 | <10 | <10 | <10 | <10 | <10 | <10 | <10 |
| **5** | **TCEP** | 30 | 60 | 40 | 40 | 30 | 60 | 10 | 20 | <10 |
| **6** | **Triphenylphosphinoxide** | 10 | 10 | 20 | 10 | <10 | 10 | <10 | <10 | <10 |
| **7** | **NBBS** | 20 | 50 | 30 | 50 | 20 | 60 | 20 | 40 | <10 |
| **8** | **MTB** | 10 | 10 | <10 | 10 | <10 | 20 | <10 | 10 | <10 |
| **9** | **Carbamazepine** | 530 | 1300 | 1000 | 990 | 530 | 1100 | 380 | 630 | <10 |
| **10** | **Propyphenazone** | 10 | 30 | 10 | 20 | 10 | 10 | <10 | 10 | <10 |
| **11** | **Clofibric acid** | 50 | 180 | 10 | 150 | 20 | 130 | <10 | 30 | <10 |
| **12** | **Diclofenac** | <10 | 400 | 200 | 570 | 220 | 290 | 110 | 650 | <10 |
| **13** | **Caffeine** | <10 | 40 | 40 | 50 | <10 | <10 | 30 | 40 | <10 |
| **14** | **Galaxolide** | 30 | 30 | 90 | 50 | 10 | 30 | 20 | 20 | <10 |

| No. | Sampling location / Compounds | 1 *river mouth* | 2 | 3 | 4 | 5 | 6 | 7 | 8 | 9 | 10 |
|---|---|---|---|---|---|---|---|---|---|---|---|
| 1 | **TXIB** | 30 | 90 | 50 | 50 | 50 | <10 | 20 | 20 | 10 | 10 |
| 2 | **Triethyl citrate** | 60 | 70 | 50 | 40 | 30 | 50 | 40 | 30 | 40 | 100 |
| 3 | **TMDD** | 290 | 220 | 210 | 110 | 60 | 120 | 120 | 110 | 130 | 170 |
| 4 | **Hexachlorobutadiene** | 60 | 50 | 80 | 70 | <10 | <10 | <10 | <10 | <10 | <10 |
| 5 | **TCEP** | 120 | 170 | 200 | 150 | 120 | 140 | 90 | 60 | 110 | 140 |
| 6 | **Triphenylphosphinoxide** | 170 | 130 | 190 | 110 | 20 | 20 | 30 | 10 | 20 | 30 |
| 7 | **NBBS** | 100 | 110 | 100 | 80 | 70 | 90 | 80 | 50 | 70 | 120 |
| 8 | **MTB** | 20 | 10 | 20 | 10 | <10 | 10 | 10 | 10 | 20 | 20 |
| 9 | **Carbamazepine** | 1300 | 1100 | 1700 | 1400 | 840 | 1200 | 1000 | 760 | 1100 | 1700 |
| 10 | **Propyphenazone** | 40 | 60 | 70 | 50 | 60 | 40 | 40 | 20 | 30 | 50 |
| 11 | **Clofibric acid** | 170 | 150 | 290 | 20 | 110 | 160 | 110 | 70 | 80 | 110 |
| 12 | **Diclofenac** | 360 | 440 | 800 | 270 | 220 | 130 | <10 | 120 | 590 | 1000 |
| 13 | **Caffeine** | 210 | 360 | 80 | 70 | 50 | <10 | 30 | 20 | <10 | 30 |
| 14 | **Galaxolide** | 60 | 90 | 120 | 90 | 60 | 80 | 40 | 40 | 90 | 160 |

**Tab. 5:** Concentrations of selected compounds [ng/L] in water samples from the Lippe river (site numbers see Fig. 1), sampling in March 2001.

| | Sampling location | 11 | 12 | 13 | 14 | 15 | 16 | 17 | 18 | 19 |
|---|---|---|---|---|---|---|---|---|---|---|
| No. | Compounds | | | | | | | | | *river source* |
| 1 | TXIB | 20 | 20 | 20 | 10 | 10 | 10 | 10 | 10 | <10 |
| 2 | Triethyl citrate | 50 | 40 | 10 | 10 | 10 | 10 | <10 | 10 | <10 |
| 3 | TMDD | 230 | 260 | 170 | 70 | 130 | 140 | 30 | 20 | <10 |
| 4 | Hexachlorobutadiene | <10 | <10 | <10 | <10 | <10 | <10 | <10 | <10 | <10 |
| 5 | TCEP | 20 | 20 | 10 | <10 | 20 | 10 | <10 | <10 | <10 |
| 6 | Triphenylphosphinoxide | <10 | <10 | <10 | <10 | <10 | <10 | 220 | <10 | <10 |
| 7 | NBBS | 60 | 50 | 20 | 20 | 40 | 40 | 20 | 10 | <10 |
| 8 | MTB | 10 | 20 | <10 | <10 | <10 | <10 | <10 | <10 | <10 |
| 9 | Carbamazepine | 750 | 720 | 440 | 610 | 680 | 830 | 350 | 420 | <10 |
| 10 | Propyphenazone | 20 | <10 | <10 | <10 | <10 | <10 | <10 | <10 | <10 |
| 11 | Clofibric acid | 70 | 80 | 20 | 20 | 40 | 90 | 30 | 20 | <10 |
| 12 | Diclofenac | 170 | <10 | <10 | <10 | <10 | 460 | 390 | 400 | <10 |
| 13 | Caffeine | 120 | 190 | 40 | 70 | 90 | 80 | 50 | 10 | <10 |
| 14 | Galaxolide | 80 | 90 | 90 | 60 | 20 | 30 | 10 | 20 | <10 |

| | Sampling location | 1 | 2 | 3 | 4 | 5 | 6 | 7 | 8 | 9 | 10 |
|---|---|---|---|---|---|---|---|---|---|---|---|
| No. | Compounds | *river mouth* | | | | | | | | | |
| 1 | **TXIB** | 20 | 20 | 30 | 40 | 40 | 60 | 40 | 20 | 20 | 20 |
| 2 | **Triethyl citrate** | 80 | 80 | 70 | 70 | 110 | 80 | 110 | 150 | 90 | 130 |
| 3 | **TMDD** | 190 | 200 | 130 | 120 | 420 | 380 | 230 | 670 | 270 | 410 |
| 4 | **Hexachlorobutadiene** | 30 | 30 | 30 | 40 | <10 | <10 | <10 | <10 | <10 | <10 |
| 5 | **TCEP** | 30 | 40 | 20 | 30 | 40 | 30 | 40 | 50 | 40 | 40 |
| 6 | **Triphenylphosphinoxide** | 910 | 560 | 680 | 750 | 40 | <10 | 80 | 90 | 50 | 80 |
| 7 | **NBBS** | 80 | 70 | 60 | 60 | 60 | 60 | 90 | 90 | 80 | 90 |
| 8 | **MTB** | <10 | <10 | <10 | <10 | <10 | <10 | 20 | 30 | <10 | <10 |
| 9 | **Carbamazepine** | 1700 | 1400 | 1000 | 1500 | 1100 | 1400 | 2000 | 1700 | 1300 | 1700 |
| 10 | **Propyphenazone** | 40 | 40 | 20 | 40 | 40 | 30 | 40 | 90 | 20 | 30 |
| 11 | **Clofibric acid** | 100 | <10 | 80 | 200 | 150 | 160 | 90 | 80 | 140 | 110 |
| 12 | **Diclofenac** | 380 | 130 | 950 | 1200 | 1700 | 1800 | 60 | 470 | 1500 | 550 |
| 13 | **Caffeine** | 90 | 110 | 200 | 200 | 80 | <10 | 420 | 270 | 60 | 80 |
| 14 | **Galaxolide** | 130 | 120 | 140 | 120 | 170 | 150 | 230 | 250 | 200 | 240 |

**Tab. 6:** river runoff data from several gauges along the Lippe river.

| Site No. | Gauge | river runoff [$m^3/s$] 16/17 Aug 1999 | 23/24 Feb 2000 | 29/30 Aug 2000 | 7/8 Mar 2001 |
|---|---|---|---|---|---|
| 1 | Fusternberg [a)] | 23.0 | 69.7 | 20.0 | 29.3 |
| 2 | Fusternberg [a)] | 23.0 | 69.7 | 20.0 | 29.3 |
| 3 | Schermbeck [b)] | 24.8 | 69.7 | 20.0 | 29.3 |
| 4 | Schermbeck [b)] | 24.8 | 69.7 | 20.0 | 29.3 |
| 5 | Hüls [a)] | 19.0 | 59.1 | 18.0 | 21.6 |
| 6 | Haltern [b)] | 19.9 | 59.1 | 18.0 | 21.6 |
| 7 | Leven [b)] | 16.1 | 48.7 | 14.1 | 19.1 |
| 8 | Vinnum [a)] | 14.0 | 50.0 | 10.3 | 14.2 |
| 9 | Lünen [a)] | 13.0 | 41.0 | 11.5 | 16.0 |
| 10 | Lünen [a)] | 13.0 | 41.0 | 11.5 | 16.0 |
| 11 | Rünthe [a)] | 9.0 | 39.6 | 9.7 | 9.6 |
| 12 | Hamm downstream weir [c)] | 10.0 | 36.7 | 10.0 | 10.0 |
| 13 | Hamm upstream weir [c)] | 12.0 | 57.1 | 14.2 | 24.4 |
| 14 | Kesseler [b)] | 10.9 | 48.3 | 12.3 | 19.9 |
| 15 | Lippstadt Pegel 1 Nordumflut [b)] | 7.9 | 33.9 | 8.9 | 12.5 |
| 16 | Bentfeld [b)] | 5.2 | 24.4 | 5.3 | 9.1 |
| 17 | Neuhaus [b)] | 0.8 | 4.3 | 0.8 | 1.1 |
| 18 | Lippspringe [b)] | 0.4 | 1.3 | 0.5 | 0.7 |
| 19 | river source | - | - | - | - |

a) Preliminary runoff data; Emschergenossenschaft Lippeverband. Germany
b) Environmental Protection Agency of North-Rhine Westphalia. Germany (LUA NRW)
c) Wasserschiffahrtsamt Duisburg

compositions are equalised and random sampling is an adequate method to investigate the input of sewage derived contaminants. Hexachlorobutadiene and triphenylphosphinoxide, products of industrial synthesis, are likely derived from industrial effluents. In this case, the applied method is only sufficient if their emission takes place as a steady input.

*Spatial distribution of the calculated loads along the river*

Increasing loads of a specific compound along the river document the input of this compound e.g. by effluents of sewage treatment plants. Decreasing loads in contrast indicate a removal from the aqueous phase caused by degradation of the compound or its partitioning into other phases. Degradation of compounds either occurs on the biological or physico-chemical pathway. Partitioning into other phases can take place as volatilization or adsorption to particulate matter. Because input and removal of compounds are concurrent opposing processes, increasing loads along the river can be interpreted as an over-compensation of removal processes by a higher input. The nature of the relevant removal processes depends a lot on the physico-chemical properties of the compounds (an example is discussed in Dsikowitzky et al., 2002). Distribution of organic loads along the river showed distinctive patterns, depending upon the input situation and extent of removal which take place. Figures 2 to 15 depict the calculated organic loads of the compounds which were plotted for all sample sets from the river source to the mouth.

It is obvious that the spatial distribution of hexachlorobutadiene (Fig. 14) and triphenylphosphinoxide (Fig. 15) differ fundamentally from that of the others. Both appear exclusively at the lower reaches of the river. In contrast, the other compounds were present in most samples from along the water course. They are all characterised by low loads at the upper reaches (sites 19 to 14) and higher loads at the lower reaches (sites 13 to 1). Some of them show this trend in a very clear way with steadily increasing loads along the river. These are triethyl citrate (Fig. 2), TCEP (Fig. 3), NBBS (Fig. 4), carbamazepine (Fig. 5), propyphenazone (Fig. 6) and HHCB (Fig. 7). They were classified as Type 1 compounds. The others follow the same trend but their loads tend to to vary strongly and show a more erratic distribution. These are TXIB (Fig. 8), TMDD (Fig. 9), MTB (Fig. 10), clofibric acid (Fig. 11), diclofenac (Fig. 12) and caffeine (Fig. 13) and are named as Type 2 compounds. Hexachlorobutadiene and triphenylphosphinoxide were classified as Type 3 compounds. In the following, spatial distributions of the calculated loads will be discussed according to these three categories.

*Type 1 compounds*

The pharmaceutical drug carbamazepine (Fig. 5) is a good example for Type 1 compounds. In the area of the river source (sites 19 to 17), low loads (< 40 g/d) were determined. Downstream they increased (up to 655 g/d) and stayed roughly at the same level until site 11. At site 10, a sudden increase is observed. Downstream this location, the loads decreased but then showed an upward trend towards site 4 (up to 3795 g/d). In the area of the river mouth (sites 3 to 1), in most cases lower loads were determined.

The steady input of carbamazepine via effluents of sewage treatment plants (see Dsikowitzky et al., 2004a) causes its prevalent occurrence in the Lippe river. Increasing loads indicate that the input rate is generally higher than removal rate. This accumulation in the aqueous phase results from the low tendency of the compounds to partition into other phases and/or its relative persistence against degradation. Therefore, its distribution witnesses the input of sewage effluents into the Lippe river. In Figure 1 it can be seen that areas with many sewage treatment plants correspond with high loads of carbamazepine (Fig. 5). Carbamazepine loads do not increase steadily along the water course but show variations (which are higher than the relative standard deviation, Table 1). This illustrates the dynamic character of the occurring processes.

There is not much known about the physico-chemical properties of carbamazepine. In a quantitative study of influents and effluents of a German municipal sewage treatment plant it was determined that only 7 % of the drug was degraded during wastewater treatment (Ternes, 1998), suggesting its persistence in the aqueous phase. Volatilization, adsorption or degradation – whatever contributes to the removal of carbamazepine is compensated by a much larger input via sewage effluents.

The pharmaceutical drug propyphenazone (Fig. 6) had lower loads than carbamazepine (up to 181 g/d) but showed a similar distribution. Propyphenazone has so far rarely been detected in aquatic environments. Its application as pharmaceutical drug indicates the input via sewage effluents into the Lippe river. The accumulation in the aqueous phase implies a low tendency to partition into other phases and its persistence. The same can be deduced for the plasticizers TCEP (highest load 408 g/d at site 4, see Fig. 3) and NBBS (highest load 362 g/d at site 3, see Fig. 4) which behaved similarly. Both have been detected in sewage effluents so far (Metzger and Möhle, 2001; Huppert et al., 1998). The plasticizer and food additive triethyl citrate and the synthetic fragrance HHCB follow the same trend, with highest loads of 362 g/d at site 3 and 482 g/d at site 4, respectively. Interestingly, their loads were lower in the summer than in February/March, probably as a result of seasonally changing input or degradation.

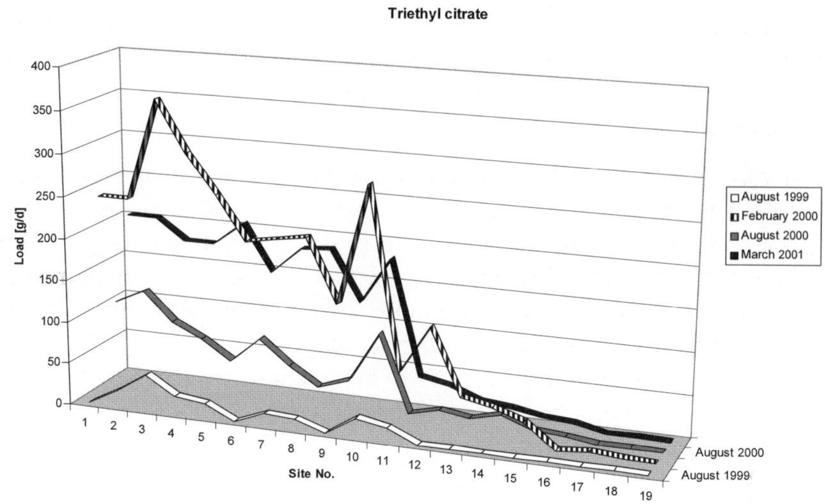


**Fig. 2:** Temporal and spatial distribution of the plasticizer triethyl citrate (loads in g/d) along a longitudinal section of the Lippe river.

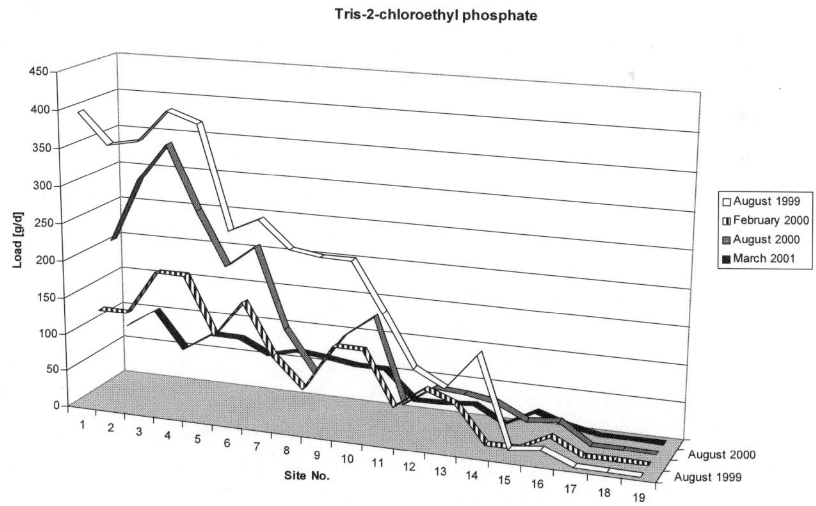


**Fig. 3:** Temporal and spatial distribution of the plasticizer TCEP ( tris(2-chloroethyl)phosphate, loads in g/d) along a longitudinal section of the Lippe river.

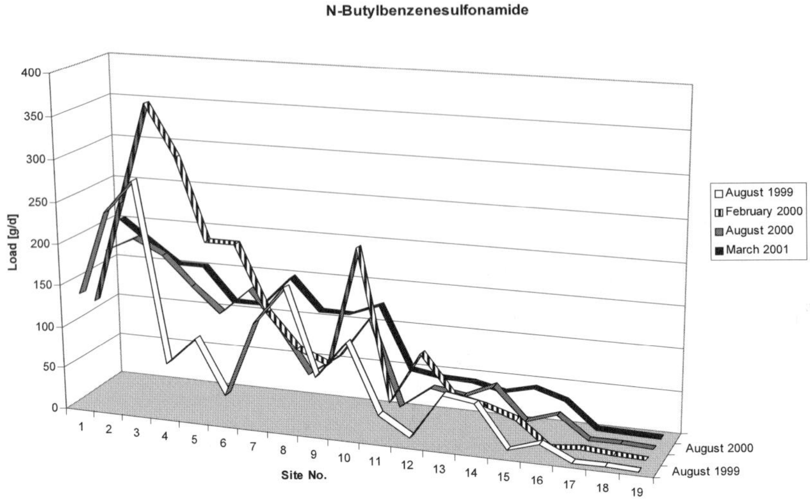


**Fig. 4:** Temporal and spatial distribution of the plasticizer NBBS (N-butylbenzenesulfonamide, loads in g/d) along a longitudinal section of the Lippe river.

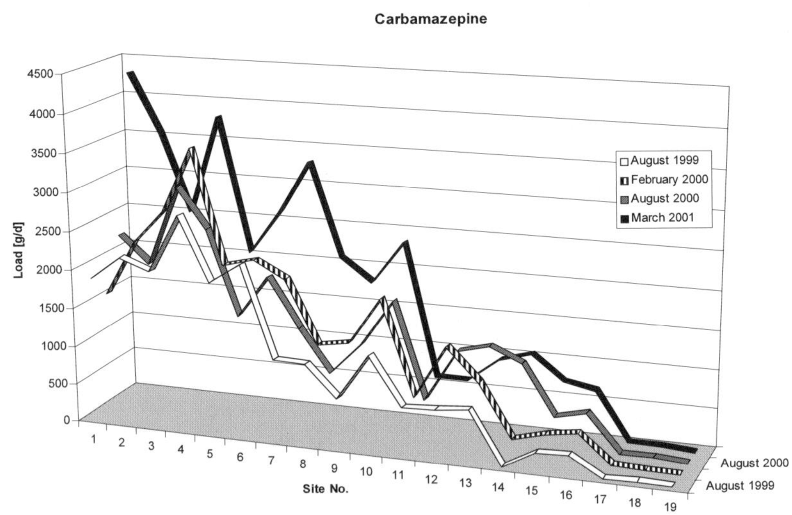


**Fig. 5:** Temporal and spatial distribution of the pharmaceutical drug carbamazepine (loads in g/d) along a longitudinal section of the Lippe river.

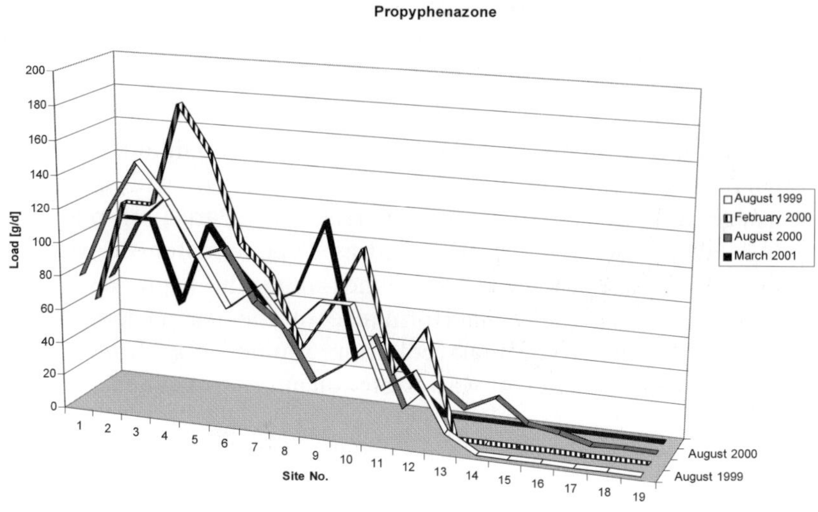


**Fig. 6:** Temporal and spatial distribution of the pharmaceutical drug propyphenazone (loads in g/d) along a longitudinal section of the Lippe river.

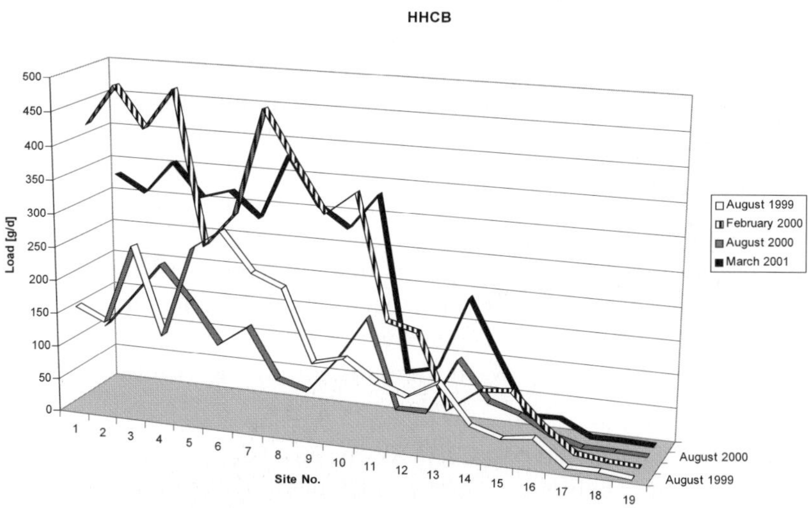


**Fig. 7:** Temporal and spatial distribution of the polycyclic musk HHCB (loads in g/d) along a longitudinal section of the Lippe river.

The occurrence and distribution of the polycyclic musks HHCB and AHTN in Lippe river water and sediments has been discussed in detail in Dsikowitzky et al. (2002).

Type 1 compounds have the potential to be used as anthropogenic markers for sewage effluents according to Takada and Eganhouse (1998) if the following criteria are fulfilled: i) source specifity, ii) widespread and massive usage and iii) environmental persistence. Part I of our study revealed that all compounds we classified as Type 1 match the first two criteria. Increasing loads of these compounds along the Lippe river water course speak in favour for the performance of the last criterion. Such compounds can be used to track pollution plumes e.g. from sewage treatment plants with insufficient wastewater cleaning capacities.

*Type 2 compounds*

Similar to Type 1 the loads of Type 2 compounds were low at the upper reaches of the Lippe river and higher at the lower reaches. Calculated loads of the plasticizer TXIB (Fig. 8) were 83 g/d or lower from the source to site 14. Downstream site 14, loads of up to 218 g/d were determined. In contrast to Type 1, TXIB loads were characterised by strong variations along the water course. Often high loads at one site were followed by low loads at the next site. The dispersing agent TMDD (Fig. 9), the pharmaceutical drugs clofibric acid (Fig. 11) and diclofenac (Fig. 12) as well as caffeine (Fig. 13) showed a similar behaviour. Interestingly, the loads of clofibric acid and diclofenac were generally lower in August 1999 and caffeine loads were lower in the summer than in the other sample sets. Although MTB (Fig. 10) roughly had a similar spatial distribution, it is a special case. Concentrations of MTB in water were low (in most cases < 20 ng/L) and often below the limit of quantitation (< 10 ng/L). It is a component of tyres and could therefore occur ubiquitous in surface waters. However, its distribution in the Lippe river can not be interpreted because its concentrations were to low to be satisfyingly determined with the applied method.

TMDD and TXIB have rarely been noticed as environmental contaminants so far. Their application and frequent detection in Lippe river water samples suggests their input via sewage effluents. Clofibric acid, diclofenac and caffeine were already found in several European rivers and in sewage effluents (e.g. Ternes, 1998; Buser et al., 1998a; Prösch and Puchert, 1998; Standley et al., 2000). They are polar compounds with a good water solubility and diclofenac is easily degradable by photolysis (Buser et al., 1998b).

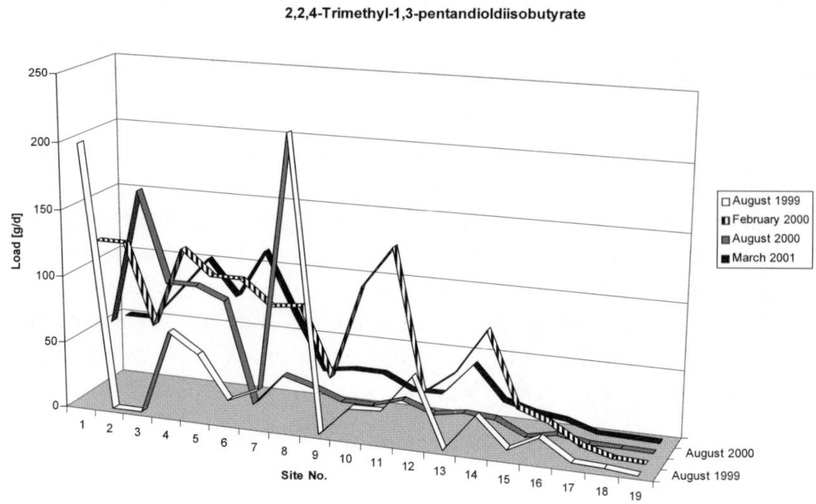


**Fig. 8:** Temporal and spatial distribution of the plasticizer TXIB (2,2,4-trimethyl-1,3-pentanedioldi-*iso*-butyrate, loads in g/d) along a longitudinal section of the Lippe river.

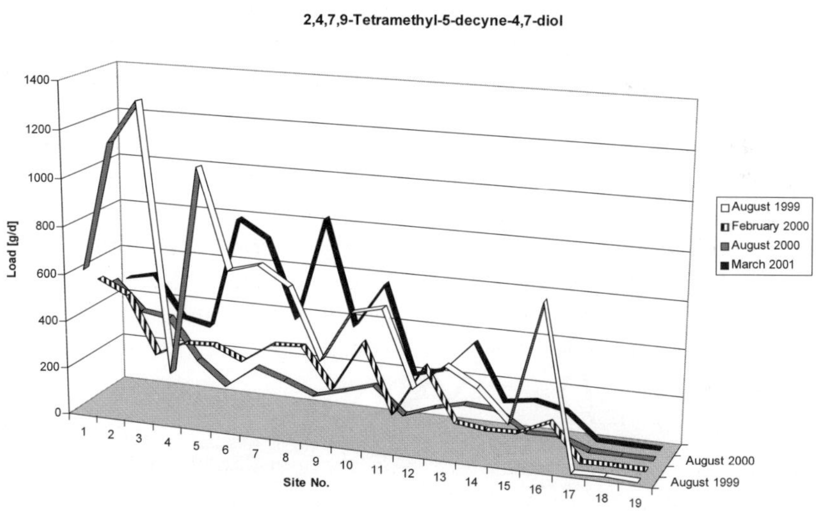


**Fig. 9:** Temporal and spatial distribution of the surfactant TMDD (2,4,7,9-tetramethyl-5-decyne-4,7-diol, loads in g/d) along a longitudinal section of the Lippe river.

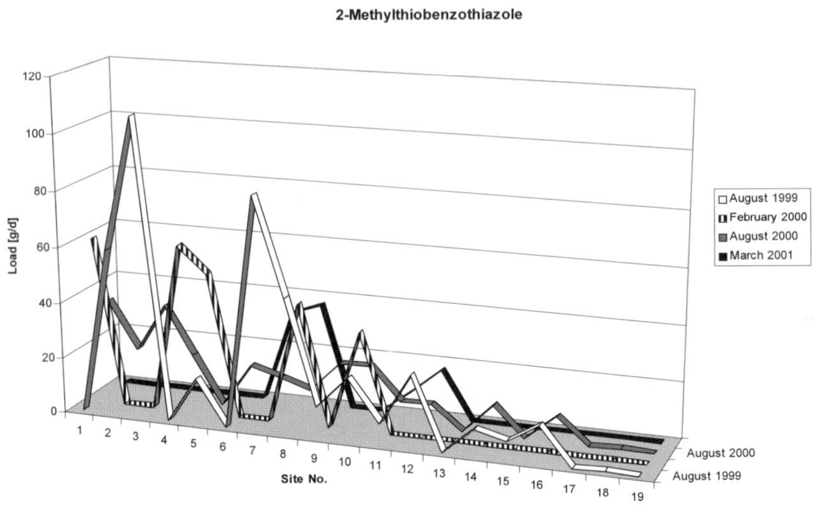


**Fig. 10:** Temporal and spatial distribution of the MTB (2-methylthiobenzothiazole, loads in g/d) along a longitudinal section of the Lippe river.

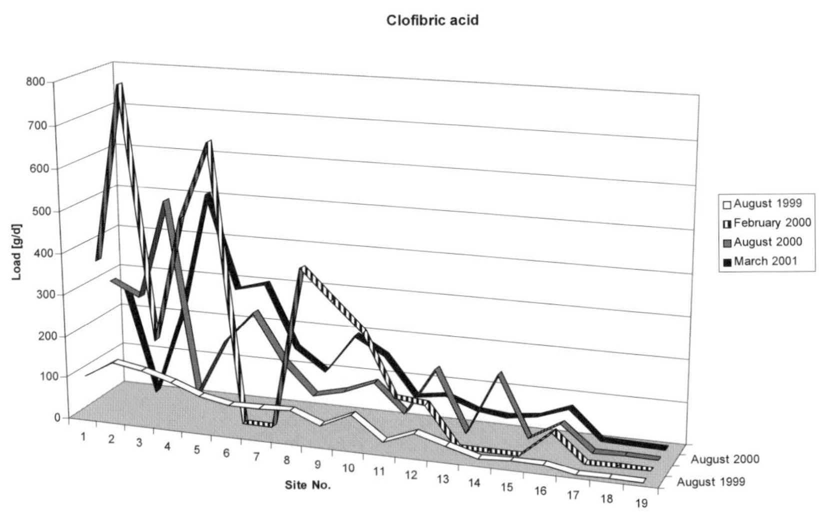


**Fig. 11:** Temporal and spatial distribution of the pharmaceutical drug clofibric acid (loads in g/d) along a longitudinal section of the Lippe river.

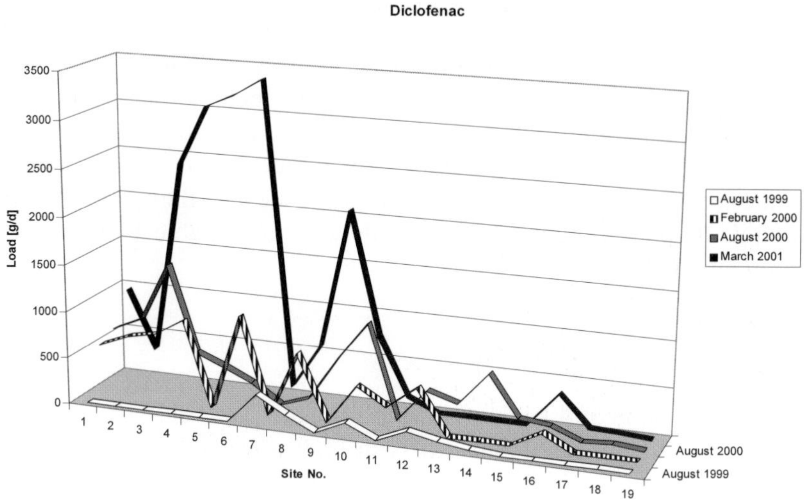


**Fig. 12:** Temporal and spatial distribution of the pharmaceutical drug diclofenac (loads in g/d) along a longitudinal section of the Lippe river.

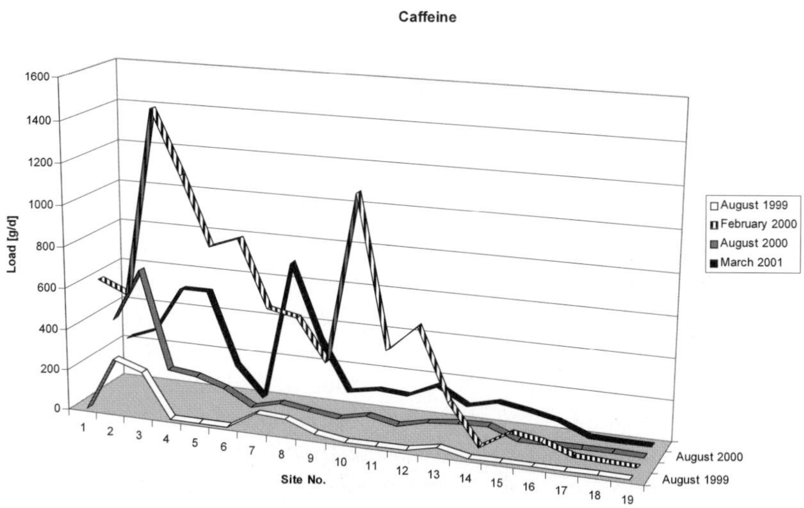


**Fig. 13:** Temporal and spatial distribution of caffeine (loads in g/d) along a longitudinal section of the Lippe river.

The presence of the investigated Type 2 compounds in the Lippe river is due to their steady input via sewage effluents. Varying loads, especially the sudden decrease of loads within a short flow distance implies their rapid removal from the aqueous phase. This happens either by partition processes or degradation. Especially the good water solubility of caffeine and the pharmaceutics and thus their low tendency to be adsorbed to particulate matter suggests their degradability. Because they are not stable in the aqueous phase, Type 2 compounds are not suitable to be used as anthropogenic markers.

*Type 3 compounds*

Hexachlorobutadiene and triphenylphosphinoxide were classified as Type 3 compounds and appeared only at restricted areas of the Lippe river. Hexachlorobutadiene, a product of industrial synthesis, was exclusively detected downstream site 5 (loads of up to 138 g/d, Fig. 14). Hence, its presence can be traced back to an industrial point source between sites 5 and 4. Indeed, on of the biggest chemsites of Europe is located in this area which comprises e.g. petrochemical plants and an own sewage treatment plant. Hexachlorobutadiene has a low water solubility (Verschueren, 1996). It is very volatile (Choudhary, 1995; Verschueren, 1996) and has a high tendency to be adsorbed to particulate matter (log $K_{OW}$ 4.8; Verschueren, 1996). Volatilization and adsorption to particulate matter can therefore cause its removal from the aqueous phase. Inspite of these possible removal pathways it remained relatively stable in the aqueous phase and was detected in the whole area of the Lippe river mouth.

Triphenylphosphinoxide occurred sporadically at the upper reaches (loads < 20 g/d, Fig. 15). Downstream the Seseke river mouth, loads of up to 132 g/d were determined whereas downstream site 5 the highest loads of up to 2302 g/d were observed. Similar to hexachlorobutadiene, a sudden increase of loads below sampling site 5 suggests its input from a point source. Triphenylphosphinoxide is a product of industrial synthesis and likely originates from the same industrial source as hexachlorobutadiene. It was not detected in effluent of the sewage treatment plant in Hamm (see Part I), hence its presence at the upper reaches of the river is probably due to additional industrial point sources. Triphenylphosphinoxide has not been noticed as a contaminant so far and its behaviour in the environment is unknown.

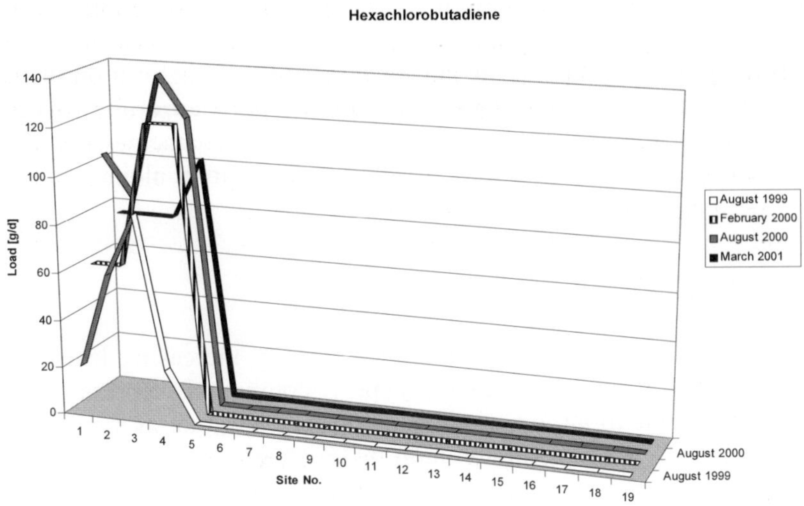


**Fig. 14:** Temporal and spatial distribution of hexachlorobutadiene (loads in g/d) along a longitudinal section of the Lippe river.

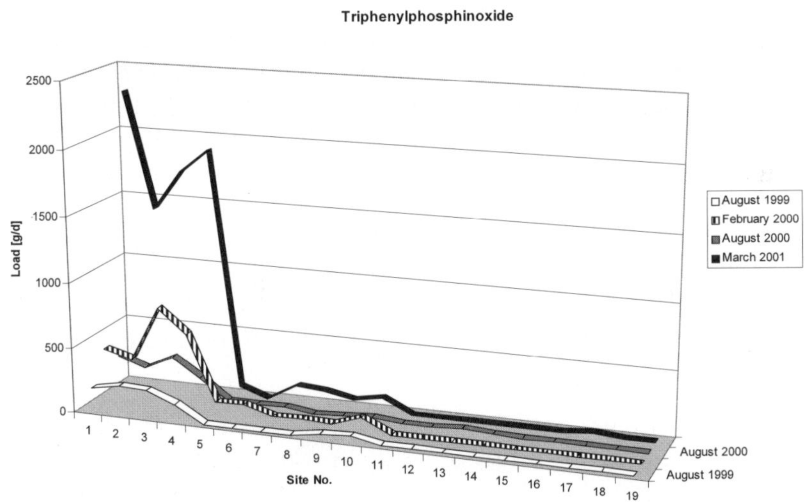


**Fig. 15:** Temporal and spatial distribution of triphenylphosphinoxide (loads in g/d) along a longitudinal section of the Lippe river.

Both Type 3 compounds were detected in all four sample sets suggesting their permanent input into the river. Their stability in the aqueous phase seems to be sufficient for tracing back the location of their input. Thus, they are suitable markers for identifying their origin. Because the composition of industrial wastewaters differs fundamentally the two compounds are only useful to track specific point sources along the Lippe river.

**Summary and Conclusions**

Type 1 compounds were characterised by steadily increasing loads from the river source to the river mouth. Type 2 compounds represented the same trend but, due to large variations along the river, not as significantly as Type 1. Both types have in common that a steady input along the river by effluents of sewage treatment plants is taking place. Increasing loads of Type 1 compounds along the river indicate that the input rate is always higher than removal rate. This accumulation in the aqueous phase indicates the persistence of Type 1 compounds and their low tendency to partition into other phases. Their distribution along the river witnesses therefore the input situation of pollutants into the Lippe river in a very concrete fashion. The erratic distribution pattern of Type 2 compounds results from partial compensation of input by effective removal processes. The fact that loads are decreasing within a short flow distance implies that these processes take place within a short period of time. Consequently, it can be deduced that Type 2 compounds are rapidly degraded in the aqueous phase and/or tend to partition into other phases (e.g. by volatilization or adsorption to particulate matter). The third type of compounds occurred only in restricted sections of the river implying a punctual input pathway (point source). Type 1 compounds which remain relatively stable in the aqueous phase are useful molecular markers for anthropogenic input of organic pollutants. They can be applied to track plumes from discrete pollution sources (e.g. sewage effluents).

## 3.1.3 Anthropogenic organic contaminants in sediments of the Lippe river, Germany ‡

### Introduction

The Lippe river, a tributary to the Rhine river, is situated in North Rhine Westfalia, Germany (Fig. 1). Because of several municipal communities and industrial settlings along the river, the Lippe river is characterized by many qualitatively different anthropogenic effluents within a short river length. Therefore this riverine system is very suitable to target numerous anthropogenic contaminants within a restricted area of investigation. Generally, the population density increases from the upper reaches of the Lippe river system towards the downstream areas. Additionally, the main impact of industrial emissions is also located in the area between the city of Hamm and the confluence with the Rhine river.

Only a few publications have considered the anthropogenic organic contamination of the Lippe river. Friege et al. (1991) investigated the herbicide pollution of the Lippe water and Schöberl and Spilker (1996) determined linear alkylbenzenesulfonates (LAS) in a dated sediment core from the Lippe river. The historical pollution with polycyclic aromatic hydrocarbons and several chlorinated compounds were also investigated on sediment cores derived from a corresponding riparian wetland (Klös and Schoch, 1993). Polychlorinated biphenyls (PCB) and tetrachlorobenzyltoluenes (TCBT) were investigated in Lippe river sediments by Poppe et al. (1991). Dsikowitzky et al. (2002) investigated the occurrence and distribution of polycyclic musks in water and sediments of the Lippe river.

All investigations described focused mainly on a preselected set of contaminants, especially priority pollutants. Our effort was to perform a more comprehensive analysis of the anthropogenic organic contamination of Lippe river sediments. Consequently, a non-target-screening approach was applied to sediment samples derived from nine locations along the longitudinal profile of the river. In order to gather time-dependent influences the sampling campaign was performed four times between August 1999 and March 2001 about every six months. Focal points of the present investigation were to reveal a variety of specific organic compounds in order to target sources and patways with assistance of the concept of anthropogenic molecular markers. Besides, unknown and rarely detected contaminants were also considered.

---

‡ Mainly adapted/reprinted from Kronimus et al., 2004

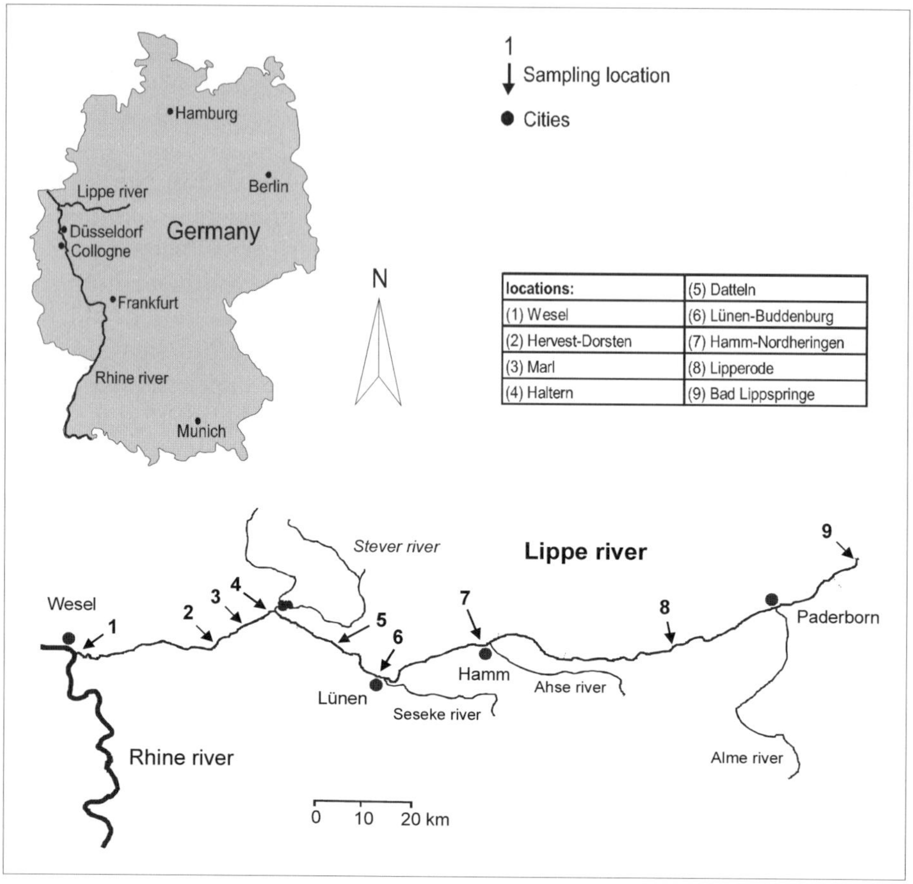


**Fig. 1:** Geographical situation and sampling locations (arrows)

Anthropogenic molecular markers are either xenobiotics or natural compounds which are discharged into environmental compartments by human activities. Additionally they have to comply with three conditions: 1) source-specifity 2) massive and widespread use 3) environmental persistence (Takada and Eganhouse, 1998). Such compounds can be used to trace pathways of effluents from specific sources as well as to determine and to distinguish between several anthropogenic sources (e.g. domestic wastes, industrial effluents) Examples for municipal anthropogenic markers are synthetic fragrances, tenside related compounds like LAB (linear alkylbenzenes) and anthropopgenic fecal steroids (cholesterol and coprostanol). These and further examples are described in more detail elsewhere (Takada and Eganhouse, 1998, Ricking et al., 2003).

## Samples

All sampling locations are presented in Fig. 1. The sampling campaigns were performed in August 1999, February 2000, August 2000 and March 2001 including nine sediment samples, respectively. Only two sampling locations (No. 8 and 9) were selected within the upper reaches of the river considering the lower concentrations of anthropogenic contaminants expected as the result of the emission situation described above.

**Tab. 1:** Recoveries and associated standard derivations of the quantified compounds

| | Ions (m/z) | Recovery rates (%) ±Std. dev. (%) |
|---|---|---|
| *Chlorinated compounds* | | |
| Tetrachlorobenzyltoluenes | 283.0, 285.0 | 21 ± 1 |
| Dichlorobenzenes | 145.9, 147.9 | 19 ± 3 |
| Trichlorobenzenes | 179.9, 181.9 | 32 ± 5 |
| Tetrachorobenzenes | 213.9, 215.9 | 42 ± 12 |
| Pentachlorobenzene | 250.0, 252.0 | 59 ± 14 |
| Hexachlorobenzene | 283.9, 285.9 | 81 ± 10 |
| Methyltriclosan | 302.1, 304.1 | 35 ± 7 |
| Bis(1-chloro-2-propyl)ether | 45.1, 121.1 | 43 ± 7 |
| (1-Chloro-2-propyl-2-chloro-1-propyl)ether | 45.1, 121.1 | 43 ± 7 |
| Pentachlorobutadiene | 225.9, 227.9 | 32 ± 8 |
| Hexachloro-1,3-butadiene | 259.9, 261.9 | 32 ± 8 |
| Octachlorostyrene | 307.9, 309.9 | 88 ± 14 |
| *Nitrogen containing compounds* | | |
| Bis(4-octylphenyl)amine | 322.4 | 24 ± 3 |
| N- Formylpiperidine | 112.2, 113.2 | 25 ± 4 |
| 3,6-Dichlorocarbazole | 235.0, 237.0 | 33 ± 7 |
| *Fragrances* | | |
| 4-Oxoisophorone | 68.2, 152.2 | 52 ± 5 |
| HHCB | 243.3, 258.3 | 43 ± 16 |
| AHTN | 243.3, 258.3 | 46 ± 17 |
| *Metalorganic compounds* | | |
| Tetrabutyl tin | 289.2 291.2 | 39 ± 9 |

All samples were taken from stagnant water zones near the riverside at a maximum depth of 5 cm with high-grade steel vessels and filled in glass vessels with teflon seals. The samples were stored at +4°C until extraction. Further details of the applied sampling techniques have been described previously (Schwarzbauer et al., 2001).

Quantification limits (LOQ) were calculated at 0.5 ng/g dry matter (signal to noise ratio of approx. 10:1 in real samples), whereas detection limits were specified at 0.1 ng/g dry matter. All concentrations are recovery corrected and normalized on dry weight bases.

## Results and discussion

The GC/MS non-target screening analyses revealed numerous compounds of which selected ones are subsequently presented and discussed in the following mainly with respect to their source specifity and their spatial distribution within the riverine system. Concentrations of anthropogenic marker compounds as well as of still unnoticed contaminants will effectively characterize anthropogenic emissions and provide a more detailed view on the pollution of the Lippe river. All quantitative data are shown in Tab. 2 a-d.

### *Municipal markers*

Municipal anthropogenic markers are characterized by an affinity to municipal waste discharged into natural systems by aquatic transport. Effluents from municipal waste water treatment plants are the ordinary pathway of dissolved municipal waste compounds into riverine systems. Useful anthropogenic municipal markers are environmentally stable ingredients in detergents and body care products because of their significant content in municipal waste water. Some of the most common anthropogenic markers (e.g. linear alkylbenzenes, nonylphenols) belong to this group of contaminants.

Also synthetic fragrances are appropriate municipal marker compounds, especially the polycyclic musks, which are used in detergents, body lotions, shampoos, perfumes and after shaves (Eschke et al., 1994; Geyer et al., 2000). So far, the occurrence of seven different polycyclic musk species have been reported (Rimkus, 1999). These fragrances were multiple detected in aquatic systems as well as in organisms (Bester et al., 1998a; Dsikowitzky et al., 2002; Eschke et al. 1994, 1995a/b; Franke et al., 1999; Heberer et al., 1998; Schwarzbauer et al., 2000). Because of their ubiquity in aqueous environments and their bioaccumulation tendency there is concern about the toxicity of polycyclic musks (Kevekordes et al., 1997; Balk and Ford, 1999a/b; Kallenborn et al., 1999).

**Tab. 2a:** Quantitative analytical data of the campaign of August 1999. All data in ng/g
n.d. (not detected) means the compound was below detection limit (<0.1 ng/g)
< 0.5 means the compound was below limit of quantification and above detection limit (0.1 to 0.5 ng/g)

| **Sampling location** | **1** | **2** | **3** | **4** | **5** | **6** | **7** | **8** | **9** |
|---|---|---|---|---|---|---|---|---|---|
| *Chlorinated compounds* | | | | | | | | | |
| Tetrachlorobenzyltoluenes | 260 | 40 | 40 | 40 | 100 | 60 | 20 | n.d | n.d |
| Dichlorobenzenes (3 isomers) | 27 | 39 | 8 | n.d. | n.d. | 58 | 5 | n.d. | n.d. |
| Trichlorobenzenes (3 isomers) | 11 | 10 | 0.5 | n.d. | n.d. | <0.5 | 2 | n.d. | n.d. |
| Tetrachorobenzenes (3 isomers) | 11 | 6 | n.d. | n.d. | n.d. | n.d. | 3 | n.d. | n.d. |
| Pentachlorobenzene | 6 | 2 | <0.5 | n.d. | n.d. | n.d. | 1 | n.d. | n.d. |
| Hexachlorobenzene | 27 | 17 | 1 | <0.5 | 1 | <0.5 | 2 | n.d. | 1 |
| Methyltriclosan | 220 | 49 | 32 | 34 | n.d. | 28 | 18 | 29 | <0.5 |
| Bis(1-chloro-2-propyl)ether (1-Chloro-2-propyl-2-chloro-1-propyl)ether | 18 | 31 | n.d. | n.d. | n.d. | n.d. | n.d. | n.d. | n.d. |
| Hexachloro-1,3-butadiene | 18 | 58 | <0.5 | n.d. | <0.5 | n.d. | n.d. | n.d. | n.d. |
| Petachlorobutadiene (1 Isomer) | 13 | 8 | n.d. | n.d. | n.d. | n.d. | n.d. | n.d. | n.d. |
| Octachlorostyrene | 4 | n.d. | n.d. | n.d. | n.d. | n.d. | n.d. | n.d. | n.d. |
| *Nitrogen containing compounds* | | | | | | | | | |
| Bis(4-octylphenyl)amine | 31 | 8 | 5 | 1 | <0.5 | 18 | <0.5 | <0.5 | <0.5 |
| N- Formylpiperidine | 250 | 12 | 40 | 445 | n.d. | n.d. | n.d. | n.d. | n.d. |
| 3,6-Dichlorocarbazole | n.d. | n.d. | n.d. | n.d. | n.d. | n.d. | n.d. | 50 | n.d. |
| *Fragrances* | | | | | | | | | |
| 4-Oxoisophorone | 18 | 35 | 32 | 39 | 57 | 12 | 1 | 28 | 17 |
| HHCB | | | | | | | | | |
| AHTN | | | | | | | | | |
| *Metalorganic compounds* | | | | | | | | | |
| Tetrabutyl tin | 15 | 50 | 10 | 12 | 37 | 26 | n.d. | n.d. | n.d. |

**Tab. 2b:** Quantitative analytical data of the campaign of February 2000. All data in ng/g
n.d. (not detected) means the compound was below detection limit (<0.1 ng/g)
< 0.5 means the compound was below limit of quantification and above detection limit (0.1 to 0.5 ng/g

| **Sampling location** | **1** | **2** | **3** | **4** | **5** | **6** | **7** | **8** | **9** |
|---|---|---|---|---|---|---|---|---|---|
| *Chlorinated compounds* | | | | | | | | | |
| Tetrachlorobenzyltoluenes | 210 | n.d. | 100 | 70 | 570 | 2400 | 210 | n.d. | n.d. |
| Dichlorobenzenes (3 isomers) | 4 | <0.5 | 7 | 33 | 120 | 17 | 3 | n.d. | n.d. |
| Trichlorobenzenes (3 isomers) | < 1 | n.d. | <0.5 | 1 | 9 | n.d. | n.d. | <0.5 | <0.5 |
| Tetrachorobenzenes (3 isomers) | n.d. | n.d. | n.d. | n.d. | 1 | n.d. | n.d. | n.d. | n.d. |
| Pentachlorobenzene | 3 | n.d. | 1 | <0.5 | 4 | 4 | <0.5 | n.d. | <0.5 |
| Hexachlorobenzene | 25 | 2 | 3 | 1 | 6 | 3 | 1 | <0.5 | 1 |
| Methyltriclosan | 250 | n.d. | 70 | 30 | 450 | 280 | 16 | 50 | n.d. |
| Bis(1-chloro-2-propyl)ether (1-Chloro-2-propyl-2-chloro-1-propyl)ether | 26 | 7 | n.d. | n.d. | n.d. | n.d. | n.d. | n.d. | n.d. |
| Hexachloro-1,3-butadiene | 13 | 10 | n.d. | n.d. | n.d. | n.d. | n.d. | n.d. | n.d. |
| Petachlorobutadiene (1 Isomer) | 10 | <0.5 | n.d. | n.d. | n.d. | n.d. | n.d. | n.d. | n.d. |
| Octachlorostyrene | 6 | n.d. | n.d. | n.d. | n.d. | n.d. | n.d. | n.d. | n.d. |
| *Nitrogen containing compounds* | | | | | | | | | |
| Bis(4-octylphenyl)amine | 11 | n.d. | 8 | 6 | <0.5 | <0.5 | n.d. | n.d. | n.d. |
| N- Formylpiperidine | n.d. | n.d. | n.d. | 29 | 9 | n.d. | n.d. | n.d. | n.d. |
| 3,6-Dichlorocarbazole | n.d. | n.d. | n.d. | n.d. | n.d. | n.d. | n.d. | n.d. | n.d. |
| *Fragrances* | | | | | | | | | |
| 4-Oxoisophorone | 7 | n.d. | <0.5 | n.d. | n.d. | n.d. | <0.5 | 5 | n.d. |
| HHCB | <0.5 | <0.5 | <0.5 | <0.5 | <0.5 | n.d. | n.d. | <0.5 | <0.5 |
| AHTN | 23 | 1 | <0.5 | <0.5 | 2 | 1 | n.d. | n.d. | <0.5 |
| *Metalorganic compounds* | | | | | | | | | |
| Tetrabutyltin | 25 | <0.5 | 11 | 7 | 49 | <0.5 | n.d. | n.d. | n.d. |

In this study the quantitatively most important polycyclic musk fragrances , 7-acetyl-1,1,3,4,4,6-hexamethyl-1,2,3,4-tetrahydronaphthalene (AHTN, Tonalide®) [Structure #1; Appendix] and 1,2,3,4,6,7,8-hexahydro-4,6,6,7,8,8-hexacyclopenta[g]-2-benzopyrane (HHCB, Galaxolide®) [Structure 2; Appendix] were considered. Generally low concentrations between the limit of detection (LOD) and approx. 90 ng/g were detetcted without a distinct trend along the longitudinal profile of the river. These results are in accordance with the observations by Dsikowitzky et al. (2002).

A further synthetic fragrance was detected in the sediment samples investigated. 4-Oxoisophorone [Structure #3; Appendix] is an ingredient mainly in perfumes (Papa and Sherman, 1981) and was formerly reported as a contaminant in sediments of the Havel and Spree rivers (Ricking et al., 2003) as well as in water samples of the Lippe river (Dsikowitzky, 2002). Noteworthy, to our knowledge there is no information available concerning fate and stability of 4-oxoisophorone in the environment so far. The distribution patterns obtained characterized this fragrance as an almost ubiquitous contaminant released from diffusive sources with maximum concentrations of about 90 ng/g. However, no significant accumulation was observed.

Antiseptics may also serve as municipal molecular markers. Triclosan [Structure #4; Appendix] is a common antiseptic used in toothpastes, detergents, socks, underwear, plastic products, shampoos, deodorants and soaps (Adolfsson-Erici et al., 2002; Lindström et al., 2002). With respect to these specific applications and to its environmental stability triclosan can be regarded as a municipal molecular marker. In the Lipper river sediments investigated methyltriclosan [Structure #5; Appendix], a methylated transformation product of the antiseptic, was identified exclusively. This methylated metabolite was formerly identified in lakes and rivers in Switzerland together with its primary substrate triclosan (Lindström et al., 2002).

However, as an unambiguous transformation product methyltriclosan can also act succesfully as an anthropogenic marker of municipal effluents. In sediments of the Lippe river methyltriclosan was detected with concentrations of up to 450 ng/g.

In order to visualize the contribution of municipial sewage effluents as revealed by the marker compounds introduced Figure 2 illustrates the spatial and temporal distribution of 4-oxoisophorone and methyltriclosan for all four sampling campaigns. Emissions along the longitudinal profile are characterized by a high variation and sligthly higher concentrations downstream of Hamm. With respect to the seasonal variations no preferential trends were observed.

**Tab. 2c:** Quantitative analytical data of the campaign of August 2000. All data in ng/g
n.d. (not detected) means the compound was below detection limit (<0.1 ng/g)
< 0.5 means the compound was below limit of quantification and above detection limit (0.1 to 0.5 ng/g

| **Sampling location** | **1** | **2** | **3** | **4** | **5** | **6** | **7** | **8** | **9** |
|---|---|---|---|---|---|---|---|---|---|
| Chlorinated compounds | | | | | | | | | |
| Tetrachlorobenzyltoluenes | <0.5 | 1 | <0.5 | n.d. | 5 | <0.5 | <0.5 | n.d. | n.d. |
| Dichlorobenzenes (3 isomers) | 7 | 66 | 8 | 4 | 32 | <0.5 | n.d. | n.d. | n.d. |
| Trichlorobenzenes (3 isomers) | <0.5 | 4 | <0.5 | <0.5 | <0.5 | n.d. | 1 | n.d. | n.d. |
| Tetrachorobenzenes (3 isomers) | n.d. | n.d. | n.d. | n.d. | n.d. | n.d. | n.d. | n.d. | n.d. |
| Pentachlorobenzene | 1 | 1 | 1 | n.d. | <0.5 | <0.5 | <0.5 | n.d. | n.d. |
| Hexachlorobenzene | 2 | 4 | <0.5 | n.d. | 1 | <0.5 | <0.5 | n.d. | <0.5 |
| Methyltriclosan | 1 | 6 | <0.5 | n.d. | 34 | 20 | 4 | n.d. | n.d. |
| Bis(1-chloro-2-propyl)ether (1-Chloro-2-propyl-2-chloro-1-propyl)ether | 3 | 49 | n.d | n.d | n.d | n.d | n.d | n.d | n.d |
| Hexachloro-1,3-butadiene | 3 | 26 | n.d. | n.d. | n.d. | n.d. | n.d. | n.d. | n.d. |
| Petachlorobutadiene (1 Isomer) | <0.5 | n.d. | n.d. | n.d. | n.d. | n.d. | n.d. | n.d. | n.d. |
| Octachlorostyrene | <0.5 | n.d. | n.d. | n.d. | n.d. | n.d. | n.d. | n.d. | n.d. |
| *Nitrogen containing compounds* | | | | | | | | | |
| Bis(4-octylphenyl)amine | 3 | <0.5 | n.d. | <0.5 | 5 | 6 | <0.5 | n.d. | n.d. |
| N- Formylpiperidine | n.d. | n.d. | n.d. | n.d. | n.d. | n.d. | n.d. | n.d. | n.d. |
| 3,6-Dichlorocarbazole | n.d. | n.d. | n.d. | n.d. | n.d. | n.d. | n.d. | n.d. | n.d. |
| *Fragrances* | | | | | | | | | |
| 4-Oxoisophorone | 6 | n.d. | 4 | 8 | 10 | <0.5 | <0.5 | <0.5 | <0.5 |
| HHCB | 5 | <0.5 | 7 | 3 | <0.5 | <0.5 | 10 | <0.5 | 0.5 |
| AHTN | 29 | 1 | 13 | 14 | 1 | 4 | 4 | n.d. | <0.5 |
| *Metalorganic compounds* | | | | | | | | | |
| Tetrabutyltin | 2 | 1 | 1 | 2 | 15 | 12 | n.d. | n.d. | n.d. |

*Industrial and agricultural markers*

In contrast to the municipal marker compounds industrial ones are often discharged from point sources and are spatially restricted. In the present study various anthropogenic markers were identified representing different industrial emission sources.

**Tab. 2d:** Quantitative analytical data of the campaign of March 2001. All data in ng/g
n.d. (not detected) means the compound was below detection limit (<0.1 ng/g)
< 0.5 means the compound was below limit of quantification and above detection limit (0.1 to 0.5 ng/g)

| Sampling location | 1 | 2 | 3 | 4 | 5 | 6 | 7 | 8 | 9 |
|---|---|---|---|---|---|---|---|---|---|
| *Chlorinated compounds* | | | | | | | | | |
| Tetrachlorobenzyltoluenes | 4 | <0.5 | <0.5 | <0.5 | n.d. | <0.5 | n.d. | n.d. | n.d. |
| Dichlorobenzenes (3 isomers) | 22 | 50 | 30 | 53 | 8 | 7 | 10 | 1 | <0.5 |
| Trichlorobenzenes (3 isomers) | 6 | 13 | <0.5 | 1 | n.d. | <0.5 | <1.5 | <0.5 | n.d. |
| Tetrachorobenzenes (3 isomers) | 3 | 7 | n.d. | n.d. | n.d. | n.d. | <0.5 | n.d. | <0.5 |
| Pentachlorobenzene | 4 | 3 | <0.5 | <0.5 | n.d. | <0.5 | 1 | <0.5 | <0.5 |
| Hexachlorobenzene | 18 | 6 | 2 | <0.5 | n.d. | <0.5 | 1 | 1 | 1 |
| Methyltriclosan | 32 | 2 | 20 | 4 | <0.5 | 2 | <0.5 | <0.5 | n.d. |
| Bis(1-chloro-2-propyl)ether (1-Chloro-2-propyl-2-chloro-1-propyl)ether | 11 | 34 | n.d. | n.d. | n.d. | n.d. | n.d. | n.d. | n.d. |
| Hexachloro-1,3-butadiene | 17 | 34 | n.d. | n.d. | n.d. | n.d. | n.d. | n.d. | n.d. |
| Petachlorobutadiene (1 Isomer) | 9 | 7 | n.d. | n.d. | n.d. | n.d. | n.d. | n.d. | n.d. |
| Octachlorostyrene | 1 | <0.5 | n.d. | n.d. | n.d. | n.d. | n.d. | n.d. | n.d. |
| *Nitrogen containing compounds* | | | | | | | | | |
| Bis(4-octylphenyl)amine | n.d. | n.d. | n.d. | n.d. | n.d. | n.d. | n.d. | n.d. | n.d. |
| N- Formylpiperidine | n.d. | n.d. | n.d. | n.d. | n.d. | n.d. | n.d. | n.d. | n.d. |
| 3,6-Dichlorocarbazole | n.d. | n.d. | n.d. | n.d. | n.d. | n.d. | n.d. | n.d. | n.d. |
| *Fragrances* | | | | | | | | | |
| 4-Oxoisophorone | 8 | 1 | 24 | 4 | 2 | 2 | 20 | 4 | n.d. |
| HHCB | 20 | 1 | 56 | 1 | 3 | 4 | 41 | n.d. | n.d. |
| AHTN | 90 | 2 | 45 | 4 | 2 | 2 | 15 | n.d. | n.d. |
| *Metalorganic compounds* | | | | | | | | | |
| Tetrabutyltin | 11 | 4 | 34 | 35 | 4 | 4 | n.d. | n.d. | n.d. |

Polychlorinated biphenyls (PCBs) were detected as ubiquitous contaminations. Since PCBs are used in a wide field of applications they are not suitable as specific anthropogenic markers. In contrast, the PCB substituents tetrachlorobenzyltoluenes (TCBT) [Structure #6; Appendix] can serve as marker compounds of industrial emissions because of their higher specifity due to their limited technical applications. In Europe

TCBTs are commercially available mainly as a technical isomeric mixture named ‚Ugilec 141'. The application of Ugilec 141 in Germany is restricted to the usage as a hydraulic fluid in mining industry. An isomer-specific determination of more than 70 isomers of Ugilec 141 was performed by Ehmann and Ballschmitter (1989). The biodegradability of TCBTs by comparison with PCBs is discussed elsewhere (Haelst et al., 1995). A second technical formulation named ‚Ugilec T' is a mixture of 40% trichlorobenzenes and 60% Ugilec 141. In Germany, this formulation is exclusively used as a dielectric fluid in transformers. Thus, the appearence of TCBTs indicates contaminations either from mining activities or derived from transformers. These different emission sources can be differentiated by analysing accompanied trichlorbenzenes.

The contamination of sediments from the Lippe river with Ugilec 141 is evident by comparing the elution patterns of TCBTs in the isomeric mixture with those in the sediments (Figure 3). Since the distribution patterns of trichlorobenzenes do not correlate with those of Ugilec 141, the TCBT contaminations can be clearly declared as effluents from mining industry (see Table 2). In all sampling campaigns TCBTs are absent at the upper reaches. However, Ugilec 141 was detected downstream from sampling location 7 (Hamm) with concentrations of up to 2400 ng/g. Hence, the occurrence of TCBT downstream of Hamm clearly reflects the contamination of the lower part of the river by effluents from mining activities (see Figure 4). This observation is in very good accordance with coal mining activities in the Ruhr Basin, which is partially drained by the Lippe river.

A second marker compound is tetrabutyl tin reflecting another industrial emission. Organotins are well known contaminants, especially of the aquatic environment (e.g. Chau and Wong, 1991; Hattori et al., 1984; Maguire, 1987). An important class of organotins are butylized tins. Mono-, di- and tributyl tins used as hydrophobation additives, plastic stabilizers and biozides are the most important class of organotins, whereas tetrabutyl tin [Structure #7; Appendix] is the synthetic precurser. This compound was detected downstream of sampling location 6, which is situated close to the Seseke river (a tributary to the Lippe river) discharging into the Lippe river (see Figure 4). Considering a large organotin producing plant situated nearby a small tributary of the Seseke river, contaminations from effluents of this plant and a subsequent transport through the Seseke river into the Lippe river is evident from the analytical data.

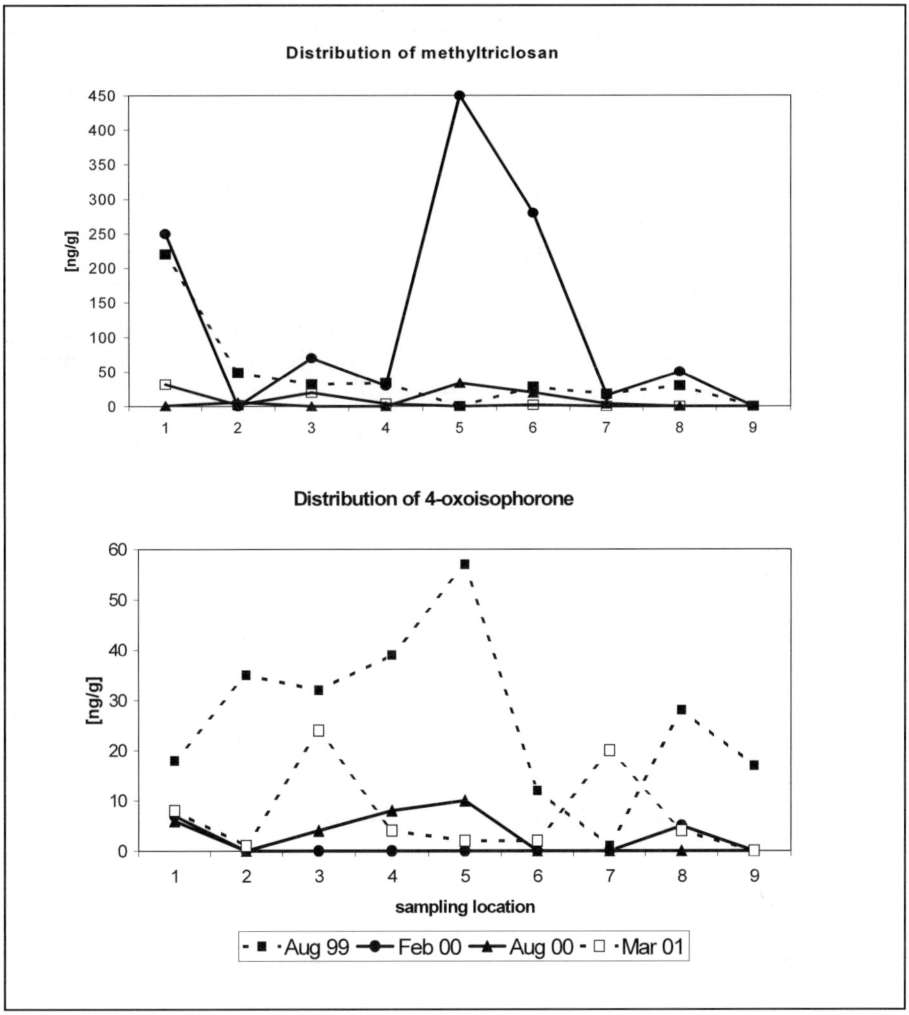


**Fig. 2:** Distribution patterns of methyltriclosan and 4-oxoisophorone as selected municipal molecular markers

As a third set of industrial marker compounds several chlorinated compounds were detected some of which are characterized by high source specifities. Dichlorinated dipropylethers detected in the sediment samples investigated represent a first group of halogenated marker compounds. Since bischloropropylethers are known by-products from various technical syntheses (e.g. the aquous chlorination of propene) the potential sources of

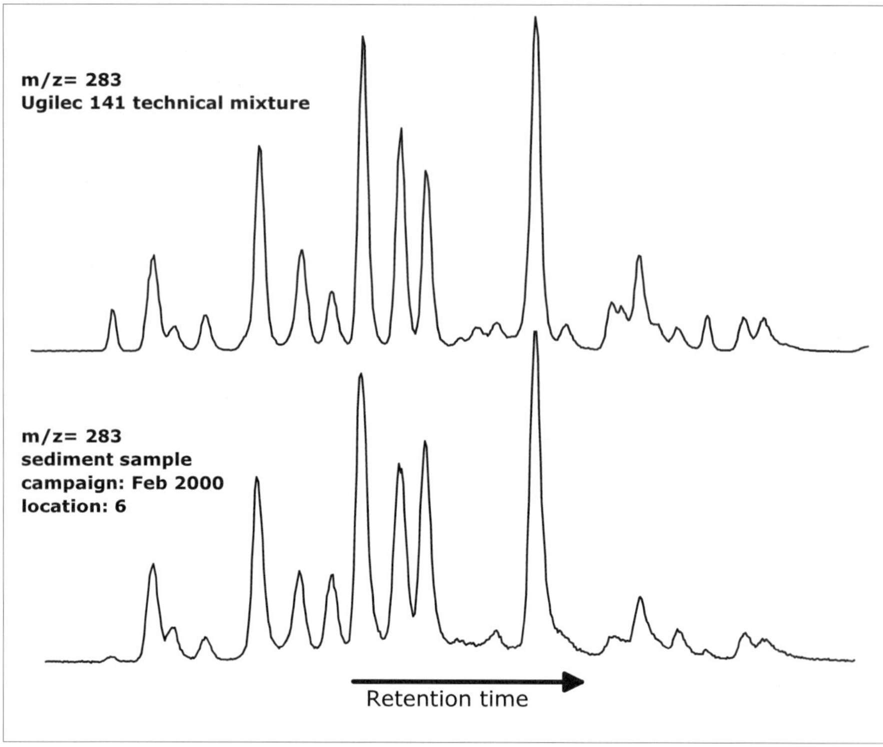


**Fig. 3:** Ion chromatograms of tetrachlorobenzyltoluenes (TCBT) in the technical mixture ‚Ugilec 141‘ and in a selected sample.

these contaminants are chemical plants which perform technical organochlorine syntheses.In this study bis-(1-chloro-2-propyl)ether [Structure #8; Appendix] and (1-chloro-2-propyl-2-chloro-1-propyl)ether [Structure #9; Appendix] were quantified cummulatively as they were chromtographically not fully separable. The dichlorinated haloethers were detected at sampling locations 1 and 2 exclusively with concentrations ranging from 7 to 34 ng/g. It has to be noted that (1-chloro-2-propyl-2-chloro-1-propyl)ether listed as priority pollutant by the EPA is toxic for aquatic species and seems to have mutagenic and cancerogenic properties (McGregor et al., 1988).

Additonally, hexachloro-1,3-butadiene (HCBD) [Structure #10; Appendix] was also identified in selected sediment samples of the Lippe river. This substance is used as a heat transfer agent, solvent and biozide, but the compound is also environmentally relevant as a by-product from the technical synthesis of volatile chlorinated organics like tri- and tetrachloroethene (Booker and Pavlostathis, 2000). Anerobic transformation experiments revealed 15 metabolites of HCBD including one pentachlorobutadiene isomer (Booker and Pavlostathis, 2000). HCBD and the pentachlorobutadiene isomer were detected almost exclusively in samples from locations 1 and 2 with concentrations up to 34 ng/g. In two cases HCBD was detected upstream of location 2, but with concentrations below the limit of quantification.

Also octachlorostyrene [Structure #11; Appendix] was detected exclusively in samples from locations 1 and 2 with concentrations up to 6 ng/g. The occurrence of octachlorostyrene (OCS) in lacustrine sediments was correlated with electrolytical chlorine production performed with graphite anodes (Kaminsky and Hites, 1984). The formation of OCS in electrolytical processes other than chlorine gas production are also conceivable (e.g. manganese production). Therefore OCS can be used as a marker compound to prove effluents from chlorochemical industry or at least from chemical industry at all.

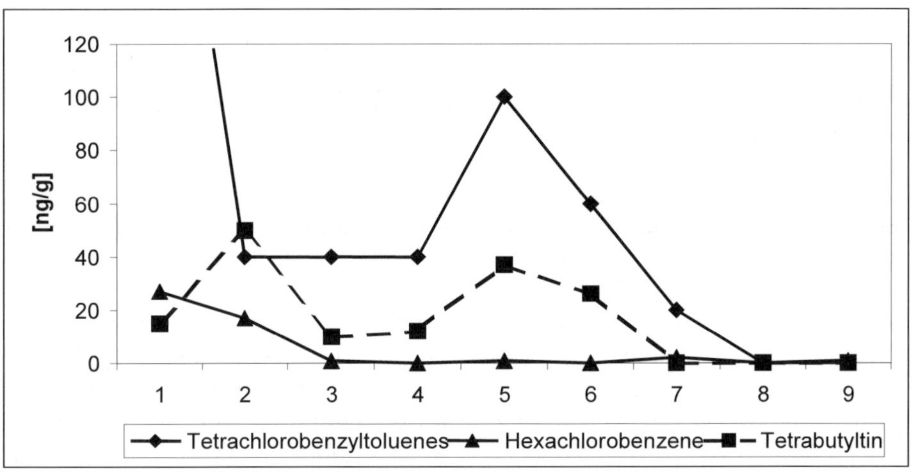


**Fig. 4:** Distribution patterns of the three gathered industrial marker compound in samples from the campaign of August 1999

Considering the spatial distribution of the chlorinated marker compounds (bischloropropylethers, chlorinated butadienes and OCS) a significant contribution of effluents from chlorochemical industry has to be stated for samples from locations 1 and 2. In fact there is a settlement of several chemical plants nearby sampling location 2. Hence, it is obvious that from this industrial area effluents with production residues are discharging into the Lippe river. On the contrary a significant temporal variation was not observed.

Further chlorinated compounds detected in Lippe river sediments could not be attributed clearly to distinct emission sources. Within the group of chlorinated benzenes, tri-, tetra- and pentachlorinated isomers did not show siginicant distribution patterns. In contrast, samples from the upper areas (sampling locations 8 and 9) were significantly less contaminated by dichlorobenzenes. Among other applications dichlorobenzenes are ingredients of toilet cleaners. Therefore, their distribution pattern in Lippe river sediments may reflect partially the pollution from municipal effluents.

Hexachlorobenzene (HCB) [Structure #12, Appendix] occurred ubiquitous in the sediment samples analysed. This compound was used as a pesticide in the past. However, this application is prohibited since several years in Germany. A striking fact is that in samples from locations 1 and 2 hexachlorobenzene reached concentrations up to 16 times higher than at local concentration maximums in upstream samples (see Figure 4). These increasing concentrations correlate with the appearance of OCS, HCB and Bischloropropylethers. Hence a significant proportion of the HCB contamination in the samples 1 and 2 can be attributed to chemical industrial effluents. Due to its environmental persistence the occurrence of HCB at lower concentrations can be interpreted as residues of former application of HCB as herbicide (see Figure 4). Therefore this contamination reflects mainly agricultural activities.

*Nitrogen containing compounds*

In addition to the marker compounds described above three nitrogen containing compounds were identified and quantified, which are still unnoticed riverine contaminants.

N-Formylpiperidine (NFP) [Structure #13; Appendix] was detected in several sediment samples taken during the first two sampling campaigns only. It is used in multiple ways e.g. as solvent, as complexing agent with lewis acids, as formylation agent of organometallic compounds, as stabilizer for polyester manufacturing, as anticorrosive agent for steel against $SO_2$, as development accelerator for chromophototermography and as a phase transfer agent. Therefore, the occurrence of N-formylpiperidine in Lippe river sediments seems to be associated with effluents from chemical or steel industry. However, Paxéus and Schröder (1996) detected

N-formylpiperidine in municipal wastewater influents to wastewater treatment plants whereas in the effluents the compound was absent. Hence, the origin of NFP is not clarified so far. This compound was also detected in sediments from the Havel and Spree rivers, Germany (Ricking et al., 2003).

Two nitrogen organic compounds were identified, which were to our knowledge formerly not described as riverine contaminants, namely 3,6-dichlorocarbazole [structure #14; appendix] and bis(4-octylphenyl)amine [structure #15; appendix]. The mass spectra of these substances are presented in Figure 5.

Bis(4-octylphenyl)amine was detected in samples take during the first three sampling campaigns in Lippe river sediments. However within the fourth campaign the compound was absent in all sediments. The concentrations were up to 30 ng/g. Interestingly there are lower concentration levels within the upper reaches. As illustrated in Figure 5 the mass spectral properties of bis(4-octylphenyl)amine are restricted mainly to the loss of pentyl moieties. This fragmentation reflects the structure of the branched side chains as the result of the synthesis pathway. The technical synthesis is performed by reaction of di-*iso*-butylene with diphenylamin forming a 4,4'-alkylsubstituted diphenylamin with a methyl substitution at the 3-position of the side chain. Bis(4-octylphenyl)amine is used as a high-temperature antioxidant additive in lubricants with concentration between 0.5% and 2%. There are no information about the pathways into aquatic systems yet.

3,6-Dichlorocarbazole was detected in only one sample from the first campaign. The origin and environmental pathway of this chlorinated carbazole are completely unknown so far. Its identification is based on the comparison of mass spectral and gas chromatographic properties with those of reference material. The synthesis performed according to Moskalev et al. (1985) as well as the mass spectra of 3,6-dichlorocarbazole are presented in Figure 5. In recent investigations of dated sediment cores taken from floodplains of the Lippe river both compounds described were also detected (Heim et al., 2004).

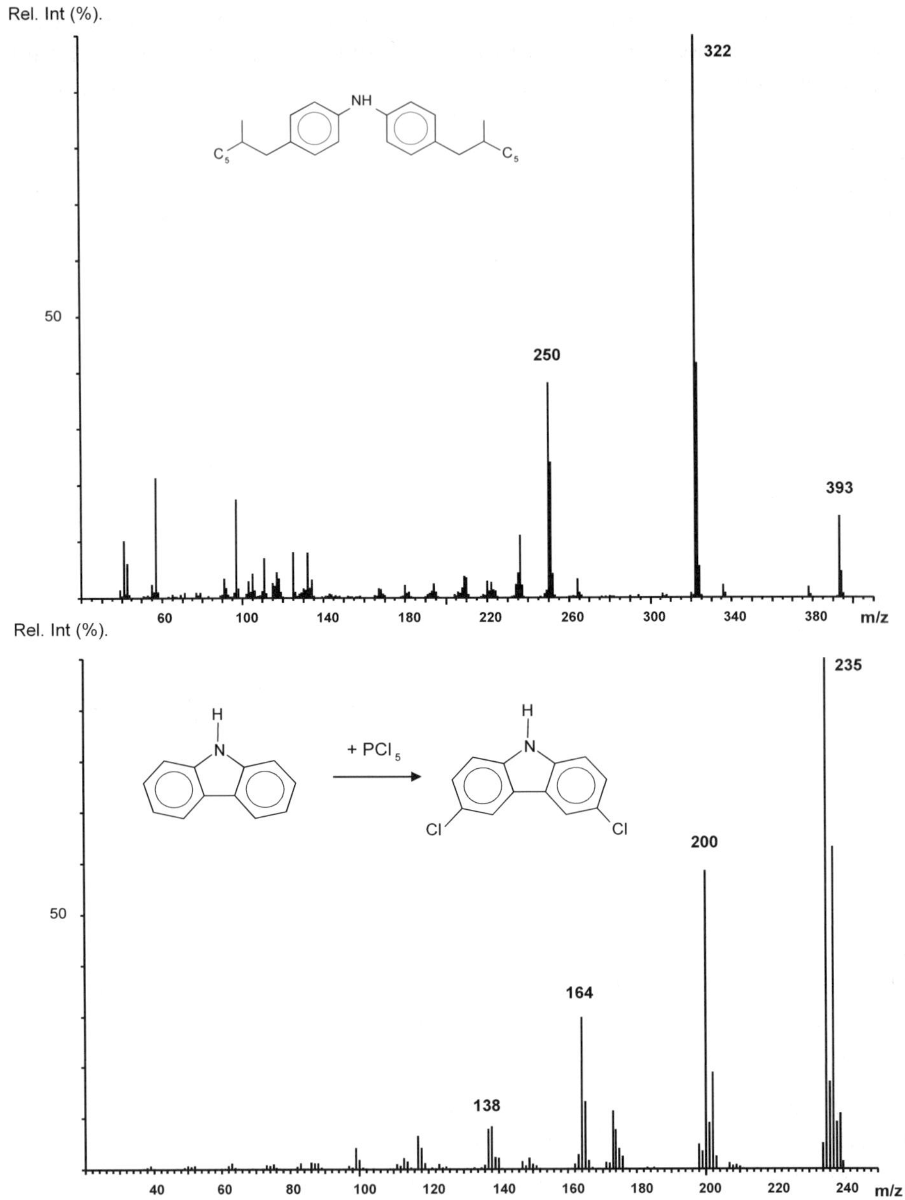


**Fig. 5:** Mass spectra of bis(4-octylphenyl)amin (A) and 3,6-dichlorocarbazol (B). In the insets the molecular structures as well as the synthesis of 3,6-dichlorocarbazol are presented.

The occurence of all three nitrogen containing compounds discussed above were not reproducible through all sampling campaigns. Although there were only surface sediments sampled, each sample did not contain particulate matter deposited during one season only. Generally low sedimentation rates in riverine systems imply thin seasonal sediment layers which prevent sampling of sediments deposited during one season exclusively. According to that the unreproducibility of the above mentioned time dependent occurrences are either caused by degradation or remobilization processes.

## Conclusions

GC-MS analyses have been applied to sediment samples of the Lippe river, a tributary to the Rhine river, in order to gather the pollution situation considering molecular source parameters as well as seasonal and spatial distribution patterns. Detailed screening analyses revealed various lipohilic organic compounds discharged from 1) municipal sources characterized by diffusive effluents, 2) point sources of three different industrial branches characterized by spatially limited appearance, 3) diffusive agricultural sources. The anthropogenic markers of industrial origin were assigned to mining activities, chlorochemical as well as organotin industry. Specific sources for selected nitrogen containing compounds comprising N-formylpiperidin, 3,6-dichlorocarbazole and bis(4-octylphenyl)amine, which are obviously of anthropogenic origin, could not be clarified. To our knowledge, the latter two were detected for the first time in environmental samples.

The spatial distribution of the anthropogenic markers derived from each industrial branch was reproducible through all sampling campaigns. However, no preferred seasonal trends according to the distribution patterns of each compound were observed. The pollution of the upper reaches was on average much smaller than in the downstream area.

The resulting emission situation of the Lippe river is visualized in Figure 6 as revealed by the above presented analyses of marker compounds including their linkage to individual emission sources.

The spatial distribution of the anthropogenic markers derived from each industrial branch was reproducible through all sampling campaigns. However, no preferred seasonal trends according to the distribution patterns of each compound were observed. The pollution of the upper reaches was on average much smaller than in the downstream area.

The resulting emission situation of the Lippe river is visualized in Figure 6 as revealed by the above presented analyses of marker compounds including their linkage to individual emission sources.

**Fig. 6:** Map of the Lippe river with the emission situation

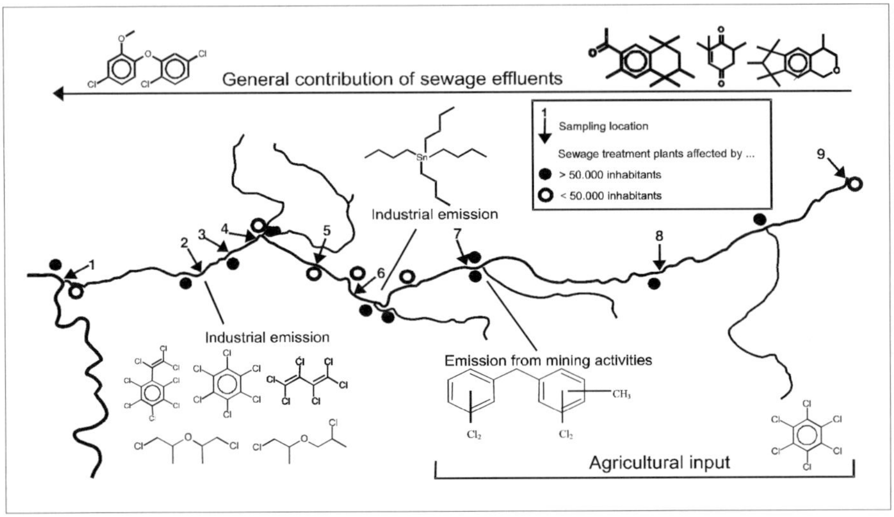

## Appendix

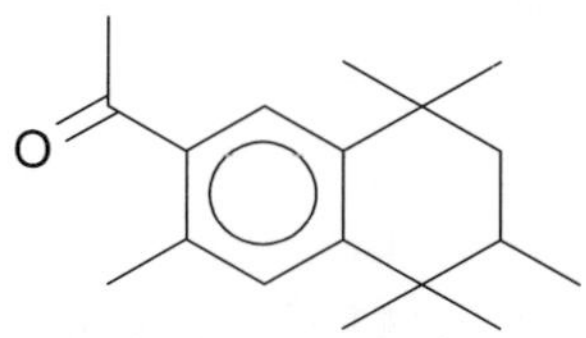


**1**

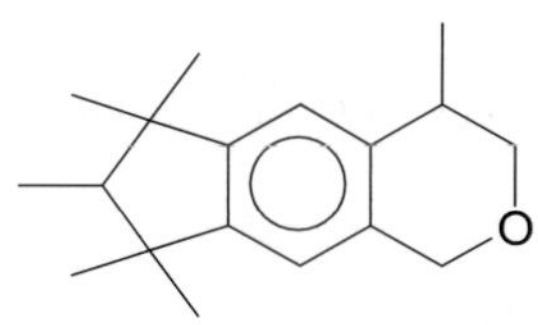


**2**

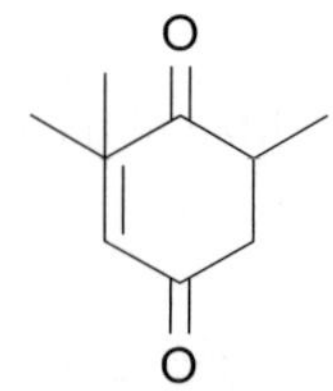


**3**

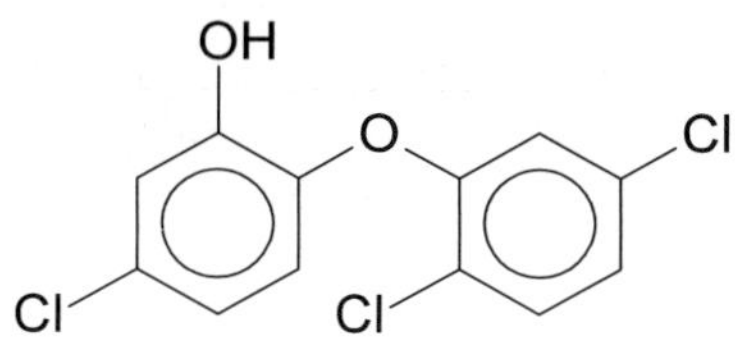


**4**

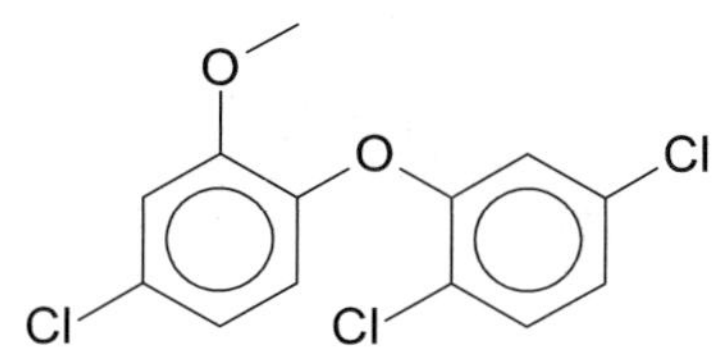


**5**

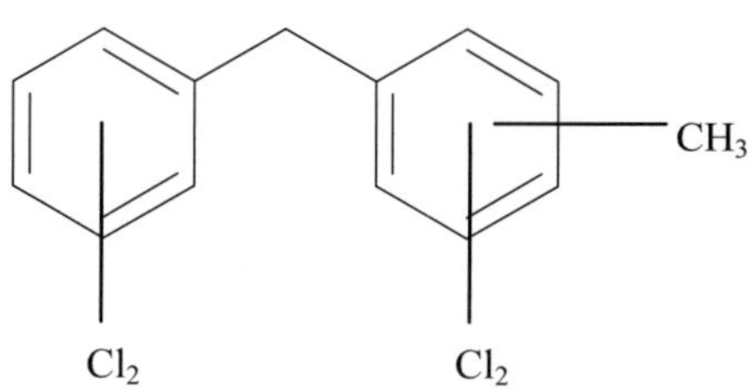


**5**

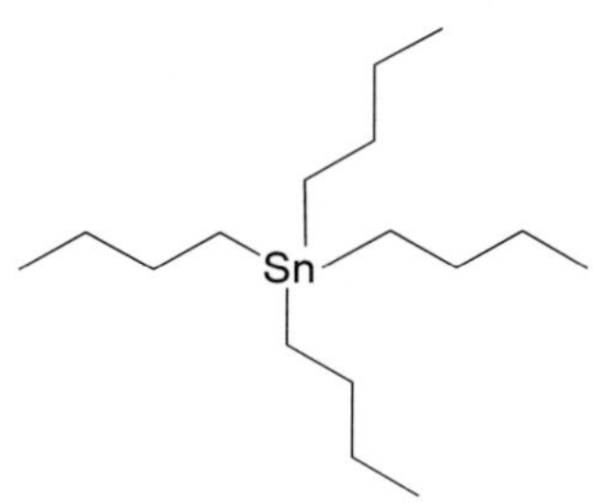


**7**

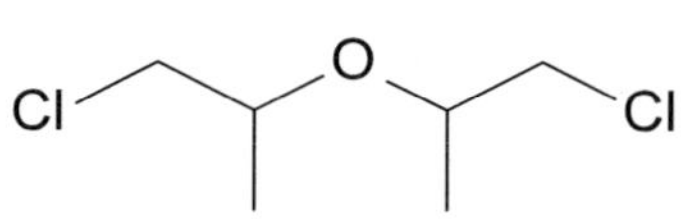


**8**

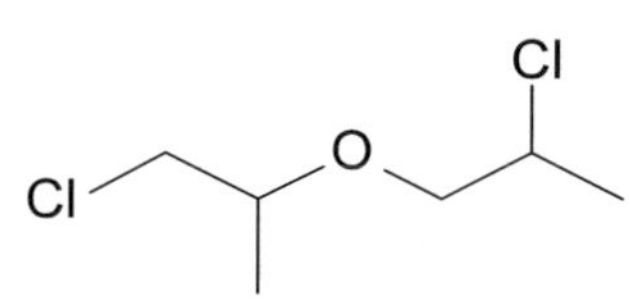


**9**

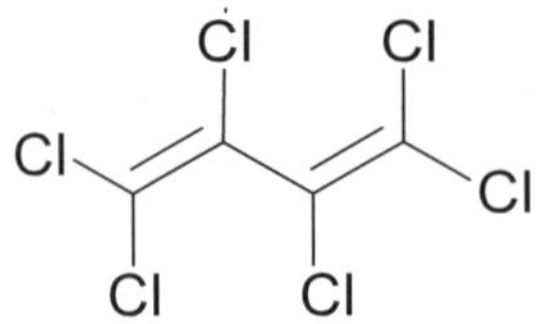

**10**

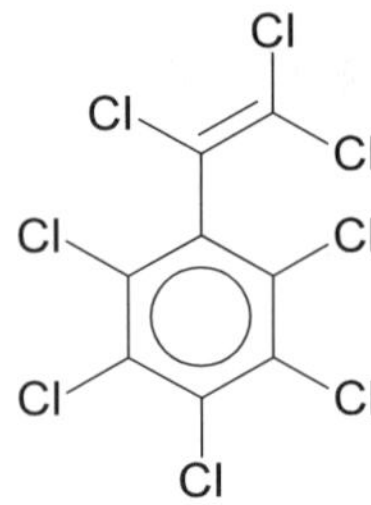

**11**

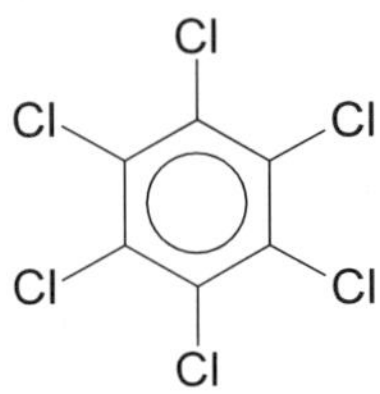

**12**

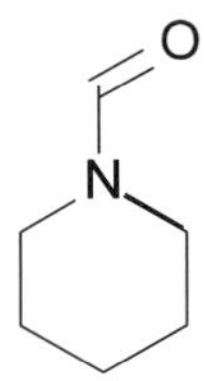

**13**

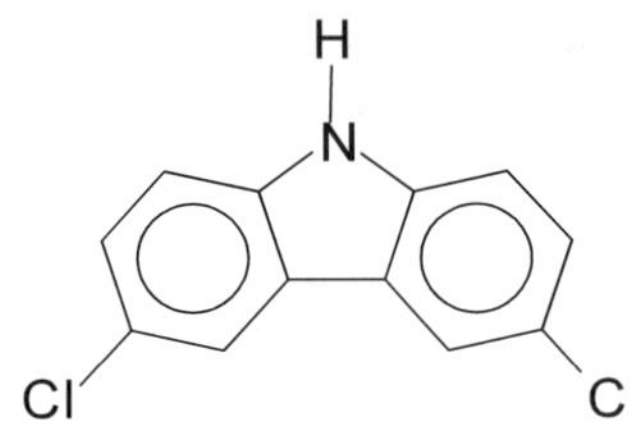

**14**

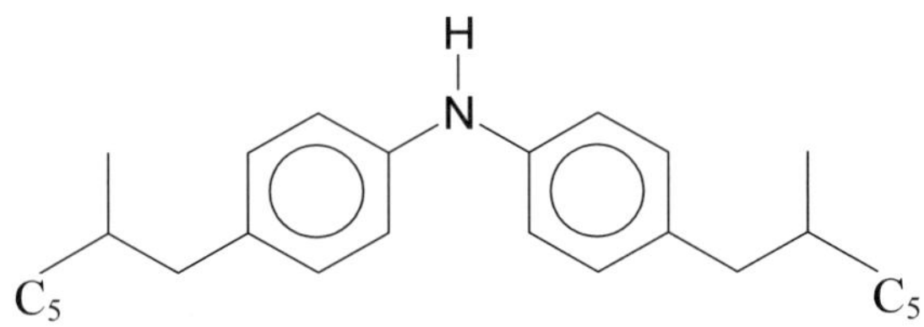

**15**

## 3.2. Rhine river

### 3.2.1 Lipophilic organic contaminants in the Rhine river, Germany ‡

#### Introduction

The Rhine river system, a major riverine system in Europe with a drainage area of approx. 224500 $km^2$ and and annual discharge of more than 70 $km^3$ (Flintrop et al. 1996), flows through numerous highly populated and industrialized areas in Switzerland, Germany and the Netherlands which produce a huge pollutant load. Since the Rhine river is also an important reservoir in terms of drinking water abstraction and other kinds of water consumption, many scientific investigations focussed on its contamination level. With respect to the organic contamination e.g. pesticides, petrogenic pollutants, phosphorous and sulfur containing contaminants as well as volatile organic compounds were considered (e.g. Meijers and van der Leer 1976; Frimmel et al. 1989; Reupert and Plöger 1989; Brauch 1993; van Urk et al. 1993; Lenhart and Lemm 1993; Schullerer and Frimmel 1993, Miermans et al. 1999).

However, analyses on organic substances in river water traditionally focused on selected pollutants or groups of contaminants, in particular on common priority pollutants which are monitored routinely. The environmental occurrence of further contaminants, e.g. pharmaceuticals, personal care products or chelating agents, which were unnoticed for a long time, has received increasing attention within the last decade. Consequently, also the knowledge about the environmental behaviour of a rising number of riverine compounds has been extended continously. Of major interest are ecotoxicological and toxicological effects caused by the individual compounds. However, investigations regarding the Rhine water contamination in the Netherlands, at the downstream end of the river basin, revealed that many organic substances responsible for toxic effects remain unidentified (Hendriks et al. 1994; Lahr et al. 2003).

Therefore, the main issue of the presented study is to provide a comprehensive view on the state of pollution of Rhine river water with respect to lipophilic low molecular weight organic contaminants. The investigation is spatially restricted to the Rhine river section of North Rhine – Westfalia (Germany) which is affected dominantly by emissions

---

‡ Mainly adapted/reprinted from Schwarzbauer and Heim, 2005c

from the Lower Rhine – Ruhr megalopolis, one of the most populated and industialized areas in Europe. Within the river section investigated several polluted tributaries are discharging into the Rhine river including in particular the Emscher, Lippe, Sieg, Wupper, Erft and Ruhr rivers (see Fig. 1).

## Samples

River water samples were taken on March, 9, 2001 at 8 sampling locations along the Rhine river (Germany) between Koblenz and Venlo (see Figure 1). 2 L - water samples were scooped up from below the water surface at a distance of approx. 2m from the river side and bottled in pre-cleaned glass flasks.

## Results and discussion

### *Screening analyses*

All contaminants identified in Rhine water extracts by GC/MS screening analyses are summarized in Tab. 2. They are arranged either by structural properties or by technical/commercial application.

A huge proportion of the contaminants were of anthropogenic origin. Several of these compounds are known as riverine pollutants and are consequently reported in various environmental investigations. However, numerous compounds identified are obviously man made chemicals but not or rarely reported as riverine contaminants. Thus, the knowledge about their usage, emission pathway and environmental behaviour is limited.

Domestic sewage emission are indicated by several specific groups of compounds including e.g. fragrances, pharmaceuticals, polymer additives like plasticizers, ingredients of personal care products or washing agents. In the Rhine water samples we identified in detail tonalide, galaxolide and 4-oxoisophorone (fragrances), carbamazepine, clofibric acid and mephobarbital (pharmaceuticals), DEET (repellent), Mecoprop and isoproturon (pesticide), TAED (bleach activator in washing agents), 4-methoxycinnamic acid 2-ethylhexyl ester (UV-protector) as well as caffeine (food constituent). With one exception the occurrence of these compounds in riverine systems is reported sporadically (Franke et al. 1995, Heberer et al. 1999b, Öllers et al. 2001, Dsikowitzky et al. 2002, Ricking et al. 2003, Buerge et al. 2003, Dsikowitzky et al. 2004a/b), but the sedative mephobarbital is still unnoticed as river water contaminant.

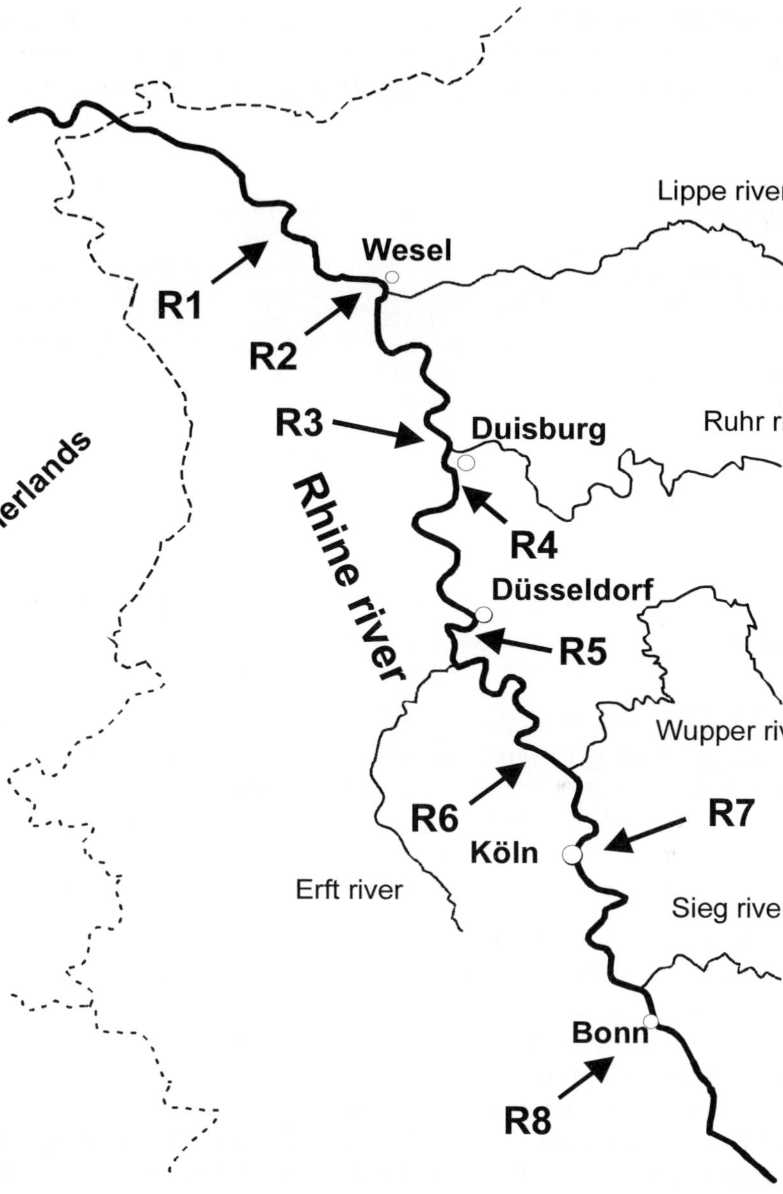


**Fig. 1:** Sampling locations along the Rhine river.

**Tab. 1:** Selected contaminants determined in Rhine water extracts and ion fragments used for quantification

| Compounds | Ion fragments | Compounds | Ion fragments |
|---|---|---|---|
| *Nitroaromatics* | | *Halogenated aromatics* | |
| Nitrobenzene | 77, 123 | 2-/4-Bromophenol | 172, 174 |
| | | 2,4-Dibromophenol | 250, 252 |
| *Technical additives, plasticizers* | | 2,4,6-Tribromophenol | 330, 332 |
| Di-iso-propylnaphtalenes, DIPNs | 197, 212 | 4-Bromo-2-chlorophenol | 206, 208 |
| NBBS | 141, 170 | 3-(Bromo-4-methoxyphenyl)-propionic acid methyl ester | 272, 274 |
| Triethylcitrate | 157, 203 | | |
| TPDB | 71, 111 | | |
| TMDD | 109, 151 | *Pharmaceuticals, pesticides and alkaloids* | |
| Triethylphosphate | 99, 155 | Carbamazepine | 193, 236 |
| Tri-n-butylphosphate | 99, 155 | Coffeine | 109, 170 |
| Tri-iso-butylphosphate | 99, 155 | DEET | 119, 190 |
| Tris(2-chloroethyl)phosphate | 249, 251 | Isoproturon | 146, 161 |
| Tris(chloropropyl)phosphates | 99, 125 | | |
| Tris(butoxyethoxy)phosphate | 199, 299 | *Sulfones* | |
| | | Methylphenylsulfone | 141, 156 |
| Dimethylphthalate | 163, 194 | Chlormethylphenylsulfone | 175, 190 |
| Diethylphthalate | 149, 222 | | |
| Di-iso-butylphthalate | 149, 223 | *Personal care products and detergents related compounds* | |
| Di-n-butylphthalate | 149, 223 | | |
| Bis(2-ethylhexyl)phthalate, DEHP | 149, 279 | | |
| | | 4-Oxoisophorone | 96, 152 |
| *Miscellaneous* | | TAED | 72, 127 |
| 4-iso-Propylcyclohexanone | 98, 154 | | |

The huge group of technical additives comprises on the one hand widely used polymer plasticizers and lubricants such as phosphates and phthalates, NBBS, TPDB and triethylcitrate. Although phosphates and phthalates are well known technical constituents released in major amounts into the aquatic environment (e.g. Huhn 1978, Bohlen et al. 1990), the diversity of individual compounds as described in this study has been reported rarely. On the other hand further more specific technical additives were detected including TMDD used as dispersing additives in pigment paints, the organic solvent NMP, triacetine used as plasticizer with fungistatic activities, and an isomeric mixture of di-*iso*-propylnaphthalenes (DIPNs) which are utilized as constituents of coating agents, adhesives, PCB substitutes, as well as heat transfer and insulating oils. It has to be noted that the occurrence of DIPNs, triacetine, TMDD and TPDB in the aquatic environment is rarely reported (Petermann and Delfino, 1990, Franke et al. 1995, Dsikowitzky et al. 2004a).

Further on, a wide variety of esters were detected containing specific molecular moieties characterizing man made chemicals. E.g. the 2-ethylhexyl group represents a molecular substructure frequently used in the technosphere dominantly as the corresponding alcohol or acid, but the natural occurrence of this molecular moiety is very scarce. Thus, the detected 2-hydroxypropylester of 2-ethylhexanoic acid represents a specific anthropogenic contaminant. The *iso*-propyl and butyl esters identified also reflect mainly the emission of technical contaminants chiefly derived from migration processes of polymer additives.

Halogenated compounds were detected in higher concentrations in all Rhine water samples investigated. Surprisingly, the halogenated aromatics were highly dominated by brominated compounds comprising brominated phenols, (methoxyphenyl)propionic acids and hydroxymethylacetophenones. Information on the latter two substance groups are presented in detail in the following chapters.

Noteworthy, a technical application and a subsequent emission of mono- to tribrominated phenols to the aquatic environment has not been reported so far. In contrast, brominated phenols and anisols are well-known organohalogens derived from biogenic formation, but exclusively detected in the marine environment (Ballschmiter 2003). Thus, the origin of brominated phenols in the Rhine water samples is still ambiguous. In any case, the prevalence of brominated substances as compared to chlorinated contaminants in the riverine environment is unusual.

Within the group of sulfur containing compounds the sulfones are remarkable. Both methylphenylsulfone and its chlorinated analogue as well as dimethylsulfone are, to our knowledge, formerly not reported as riverine pollutants. Both aromatic sulfones are described to be natural degradation products of pesticides applied to agricultural soils. Chlorophenylmethylsulfone is the hydrolysis product of the insectide fonofos (O-ethyl-S-phenylethyldithiophosphonate) (Somasundaram et al. 1990), whereas phenylmethylsulfone represents a metabolite of the herbicide planavin (N,N-dipropyl-2,6-dinitro-4-methylsulfonaniline) (Tomkins et al. 1998). The emission pathway of these compounds entering the aquatic environment has to be attributed to the water exchange with river associated polluted groundwater or to soil errosion and subsequent deposition on the water surface. Also an industrial emission by waste discharge of synthesis residues has to be considered (Take et al., 1993).

**Tab. 2:** Organic contaminants identified in Rhine water samples

*Isoprenoides*

Limonene [b)]
Menthol [b)]
Squalene [b)]
Dihydroactinidiolide [a)]

*Polycyclic aromatic compounds*

Dibenzofurane [b)]
Dibenzophenone [b)]
Fluorenone [b)]
Fluoranthene [b)]

*Alcohols and ethers*

3-Phenoxypropan-1-ol [a)]
1-Tetradecanol [b)]
1-Hexadecanol [b)]
1-Octadecanol [b)]

*Aldehydes and ketones*

Benzaldehyde [b)]
4-iso-Propylcyclohexanone [b)]

*Carboxylic acids*

Benzoic acid [m)]
$C_7$ to $C_{18}$ carboxylic acids [m)]
Linoleic acid [b)]
Palmitoleinic acid [b)]
Oleic acid [b)]
Acetic acid [m)]
Methoxypropionic acid [m)]

*Esters and amides*

Isopropylmyristate [b)]
Adipinic acid dibutylester [a)]
Isopropylpalmitate [b)]
Butanedioic acid diethylester [a)]
2-Ethylhexanoic acid 2-hydroxy-propyl-ester [a,m)]
Butandioic acid di-*iso*-butylester [a)]
Acetamide [b)]

*Technical additives, plasticizers and solvents*

Di-*iso*-propylnaphthalenes (5 isomers), DIPNs [b)]
N-Butylbenzenesulfonamide, NBBS [b)]
Triethylcitrate [b)]
Triacetine [b)]
2,4,4-Trimethylpentane-1,3-dioldi-*iso*-butyrate, TPDB [b)]
2,4,7,9-Tetramethyl-5-decyne-4,7-diol, TMDD [b)]
N-Methylpyrrolidone, NMP [b)]
Triethylphosphate [b)]
Tri-*n*-butylphosphate, TBP [b)]
Tri-*iso*-butylphosphate [b)]
Tris(2-chloroethyl)phosphate, TCEP [b)]
Tris(chloropropyl)phosphates (2 isomers), TCPP [b)]
Tris(butoxyethoxy)phosphates [b)]
Dimethylphthalate [b)]
Diethylphthalate [b)]
Di-*n*-butylphthalate, DnBP [b)]
Di-*iso*-butylphthalate, DiBP [b)]
Bis(2-ethylhexyl)phthalate, DEHP [b)]

*Halogenated aromatics*

2-Bromophenol [b)]
4-Bromophenol [b)]
2,4-Dibromophenol [b)]
2,4,6-Tribromophenol [b)]
4-Bromo-2-chlorophenol [b)]
3-(Bromo-4-methoxyphenyl)propionic acid [m,b)]
3-(Bromo-3-methoxyphenyl)propionic acid [m,b)]
3-(Bromo-2-methoxyphenyl)propionic acid (2 isomers) [m,b)]
3-(Dibromo-4-methoxyphenyl) propionic acid [m,b)]
3-(Dibromo-3-methoxyphenyl) propionic acid [m,b)]

**Table 2 continued**

*Nitrogen containing compounds*

3,3-Trimethyl-2-oxoindol [a)]
N-Acetylmorpholine [a)]
N-Ethylaniline [b)]
Nitrobenzene [b)]

*Sulfones and benzothiazoles*

2-Methylthioacetic acid [a)]
3-Methylthiopropionic acid [a)]
2-Methylthioacetic acid ethylester [a)]
Dimethylsulfone [a)]
Methylphenylsulfone [a)]
Chloromethylphenylsulfone [a)]
Benzthiazol [b)]
2-Thiomethylbenzothiazol [b)]

3-(Dibromo-2-methoxyphenyl) propionic acid (2 isomers) [m,b)]
Bromohydroxymethylacetophenone[m,b)]
Dibromohydroxymethylacetophenone[m,b)]

*Personal care products, washing agents and related compounds*

4-Oxoisophorone [b)]
Galaxolide [b)]
Tonalide [b)]
N,N,N',N'-Tetraacetylethylene diamine, TAED [b)]
4-Methoxycinnamic acid 2-ethylhexylester [b)]

*Pharmaceuticals, pesticides and alkaloids*

Carbamazepin [b)]
Mephobarbital [a)]
Clofibric acid [b,m)]
Coffeine [b)]
N,N-Diethyltoluamide, DEET [b)]
Mecoprop [b)]
Isoproturon

[a)] = identified by comparison of mass spectral data with those of mass spectral data bases
[b)] = identified by comparision of gas chromatographic and mass spectral data with those of reference compounds
[m)] = detected as methylester

The occurrence of dimethylsulfone in Rhine water samples can be attributed either to industrial sources or to municipial effluents due to its application as technical additive (plasticizer, dispersion agent) and solvent.

In addition methylthio substituted carboxylic acids are still unnoticed riverine contaminants. Methylthioacetic acid ethylester is a known volatile constituent of melones, passion fruits, further tropical fruits as well as wine (e.g. Wong and Ong, 1993, Beaulieu and Grimm 2001, Jordan et al. 2002, Moreira et al. 2002). Therefore its application as flavour constituent in fragrances seemed to be possible. Further technical usage also of methylthio acetic and propionic acids are unknown. On the contrary methylthiobenzothiazol and its metabolite benzothiazole are well known riverine contaminants as the result of their usage as biocorrosion inhibitors, coating agents of metallic surfaces and dominantly as vulcanisation accelerator in the rubber industry (Rennie 1988; Fiehn et al. 1998).

Further compounds detected in the water samples are nonspecific in terms of source elucidation. These compounds might derive either from natural or anthropgenic orgins. Examples include the isoprenoides, PACs, as well as the listed alcohols, ketones, aldehydes and carboxylic acids.

*Identification of brominated contaminants*

During the GC/MS screening analyses several still unknown contaminants were detected. Consequently, structure elucidating analyses were performed on two different groups of compounds both appearing in the third extracts. Thus, polar and acidic properties have to be stated for these substances. The identification of the individual compounds was based mainly on the interpretation of their mass spectral properties and subsequent verification of the proposed molecular structures by synthesized reference material.

The first group of contaminants comprises several isomers of mono- and dibrominated (methoxyphenyl)propionic acids. Their occurrence and isomeric distribution is illustrated in Fig. 2. All position isomers with respect to the methoxy group were present. However, for each methoxy substitution (2-, 3- and 4-substitution) only one isomer considering the bromination position was observed in most cases. Just one further monobrominated isomer was detected in the water samples which was not attributed to synthesized reference material. With respect to its mass spectral properties (as discussed above) the occurrence of a further 3-(bromo-2-methoxyphenyl)propionic acid methylester is likely. Interestingly, a very similar isomer pattern of reference substances was obtained by nonselective bromination of the corresponding nonhalogenated compounds (see also Fig. 5). Therefore, a technical preparation of these contaminants can be suggested.

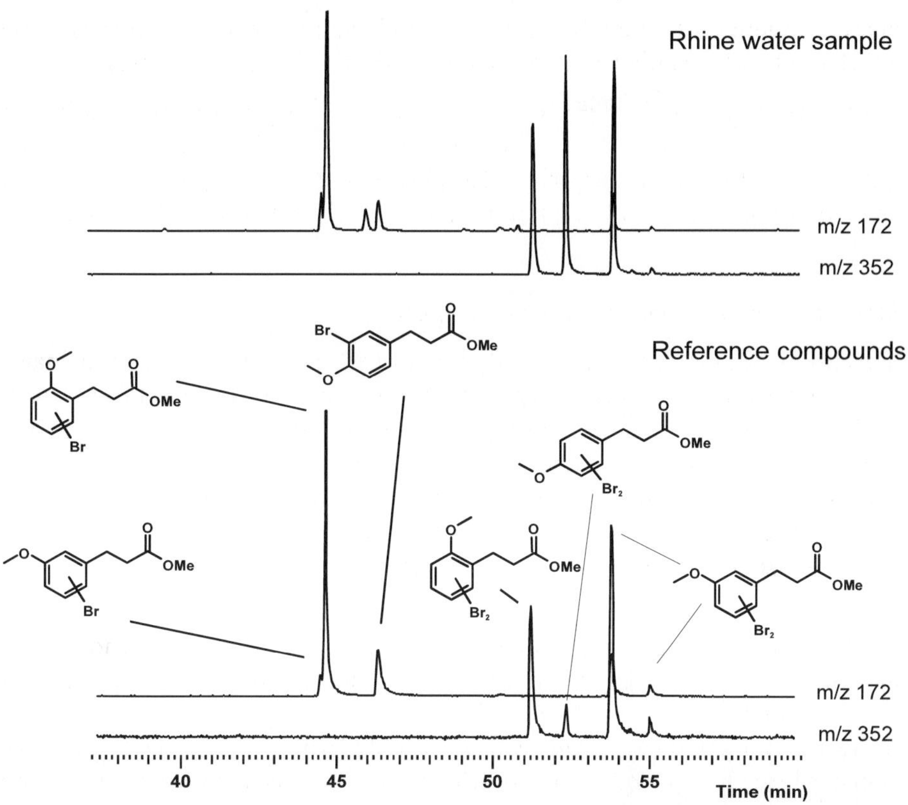


**Fig. 2:** Ion chromatograms of a third extract derived from a Rhine water sample. Mono- and dibrominated (methxyphenyl)propionic acids appeared as methylated derivatives. The elution pattern is compared with those of authentical reference compounds derived from chemical syntheses. It has to be noted, that the relative concentrations are not reflected by the peak areas or peak heights due to different relative proportions of the ion fragments 172 and 352 m/z within the individual mass spectra (see Fig. 3 and 4).

The mass spectra of these halogenated acids, detected and identified as methylesters, are presented in Fig. 3 and 4. Mono- or dibromo substitution are reflected by the isotopic pattern of the molecular ions. Interestingly, the mass spectra differ significantly depending on the substitution position of the methoxy group. 2- and 4-substituted compounds are characterized dominantly by the loss of formic acid methylester and $CH_2COOCH_3$-moieties forming the ion fragments 212 and 199 m/z or 277 and 290 m/z, respectively.

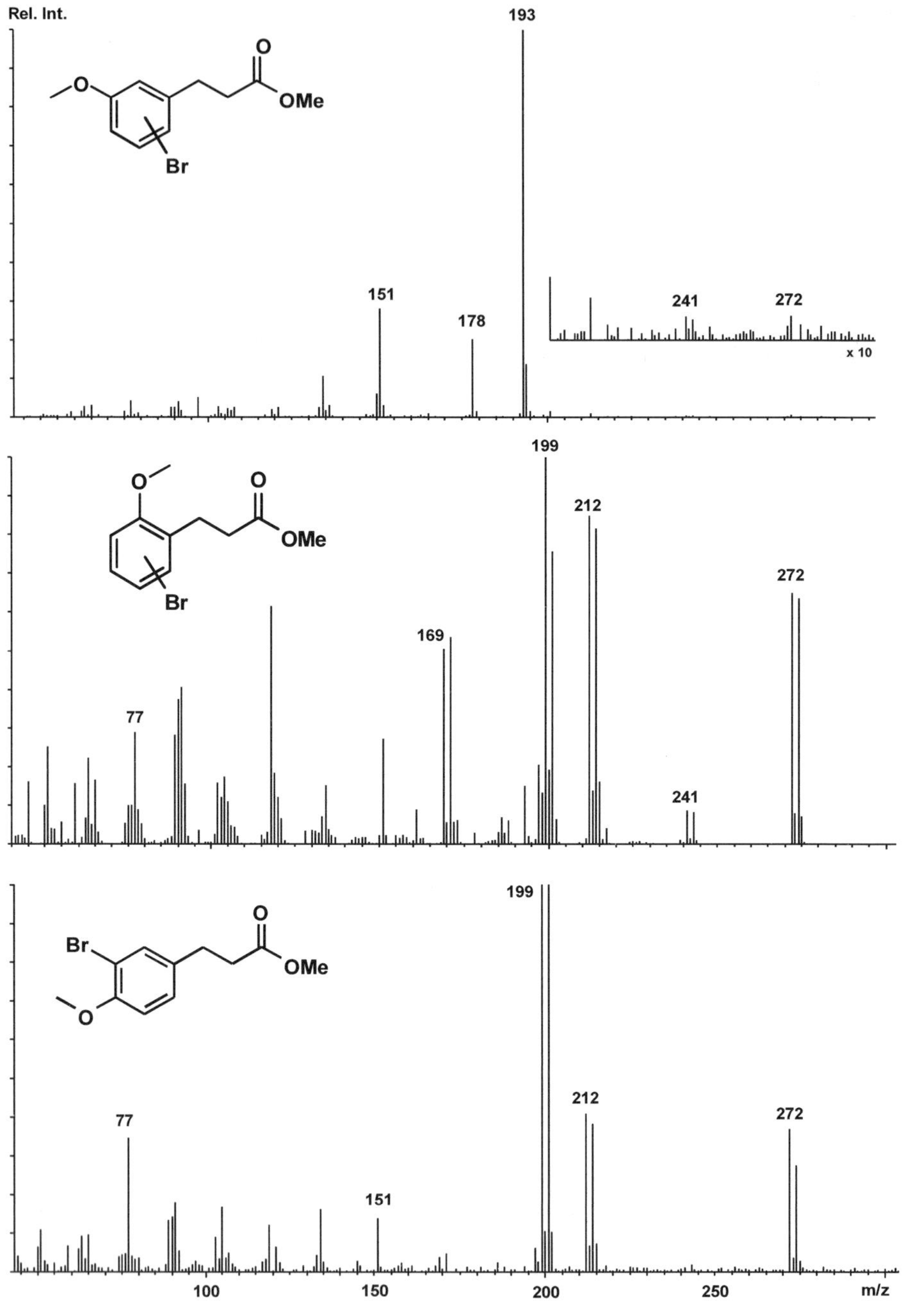


**Fig. 3:** Mass spectra of brominated 3-(methoxyphenyl)propionic acid methylesters identified in Rhine water samples.

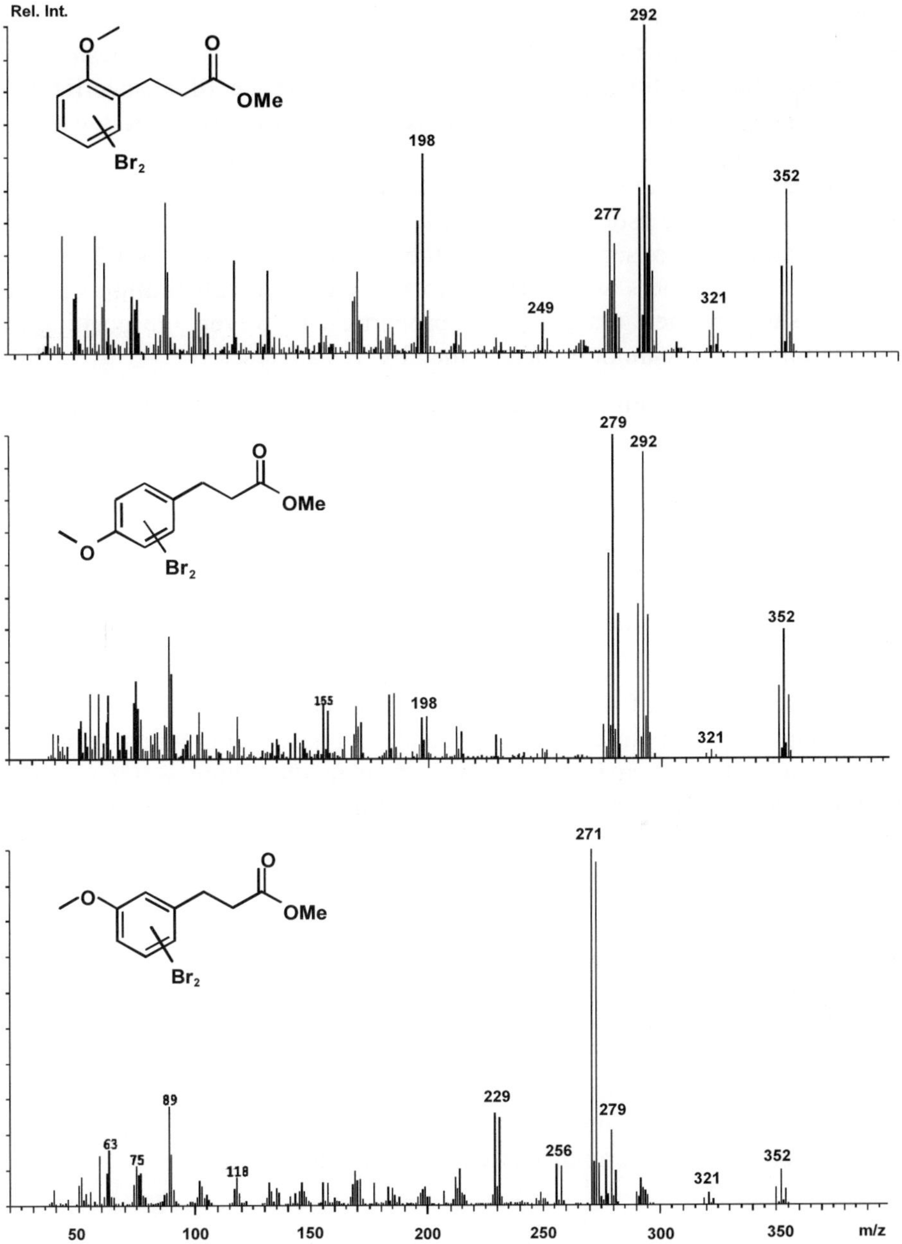


**Fig. 4:** Mass spectra of dibrominated 3-(methoxyphenyl)propionic acid methylesters identified in Rhine water samples.

Also the loss of methyl groups was observed ($M^+$ - 15 m/z) to a minor extend. 2- and 4-methoxy substituted isomers can be distingueshed by slight variation of the intensities and the occurrence of a specific ion fragment (229 and 169 m/z, respectively) appearing only in the mass spectra of halogenated (2-methoxyphenyl)propionic acid methylesters. The mass spectra of 3-methoxy substituted contaminants differed significantly as compared to the isomers discussed so far. Dominantly the loss of a bromine atom was observed forming the ion fragments 193 and 271 m/z. Therefore, all isomers with respect to the methoxy substitution can be distinguished by their mass sepctral properties. However, the position of the bromine atoms cannot be determined by MS analyses.

In Fig. 5 the synthetic pathway as well as the IR spectroscopic properties of one reference substance, 3-(bromo-4-methoxyphenyl)propionic acid methylester, are presented. However, it has to be noted that also IR spectroscopic analyses are not suitable for the determination of the bromine substitution position.

Information on the concentrations of a representative isomer detected along the riverine section investigated is presented and discussed in the following subchapter.

A second group of still unknown contaminants included mono- and dibrominated hydroxymethylacetophenones. These compounds appeared only unsteadyly and in minor concentration in the Rhine water samples investigated. Hence, they were not considered in quantitative analyses.

However, they represent a group of still unnoticed riverine contaminants, interestingly, substituted with bromine atoms. Thus, also structure elucidating analyses were performed on this substances. These compounds were identified by comparison of their analytical data with those of reference material in case of the mono-substituted contaminants. The identification of the dibrominated compounds is based on their mass spectral properties as well as on their gas chromatographic elution in comparison with the monobrominated compounds. In the following the mass spectral, gas chromatographic and IR spectroscopical properties are illustrated (Fig. 6 to 9) and discussed shortly.

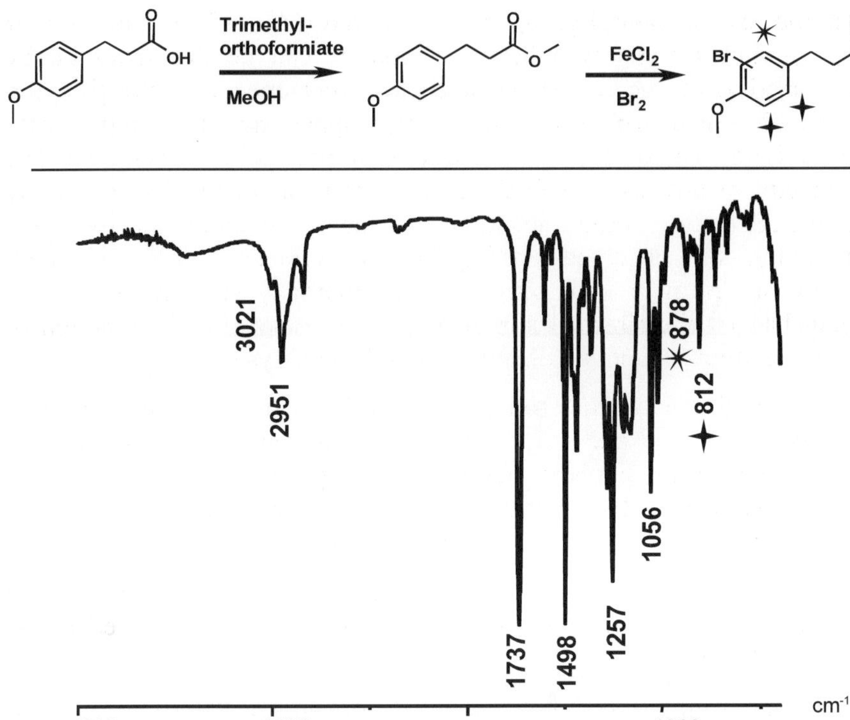


**Fig. 5:** Synthetic pathway and FTIR spectra of 3-(bromo-4-methoxyphenyl)propionic acid methylester

The mass spectra of the mono- and dibrominated hydroxymethylacetophenones are very similar and dominantly characterized by the loss of a methylgroup forming the base peaks $M^+$ - 15 m/z (213 m/z and 301 m/z , respectively). The molecular ion is always present with relative intensities between 20 % and 40 %. At the most abundant fragments the isotopic pattern of bromine substitutions can be observed. Minor fragmentations form the ions m/z 185 and 263, respectively, representing the loss of the actyl moieties.

Interestingly, the dibrominated compound could be methylated by derivatisation with trimethylsulphonium hydroxid (TMSH) forming the corresponding methylether. The mass spectra are presented in Fig. 7. In contrast, the monobrominated phenol remained unaffected as the result of less acidity due to the lower degree of bromine substitution.

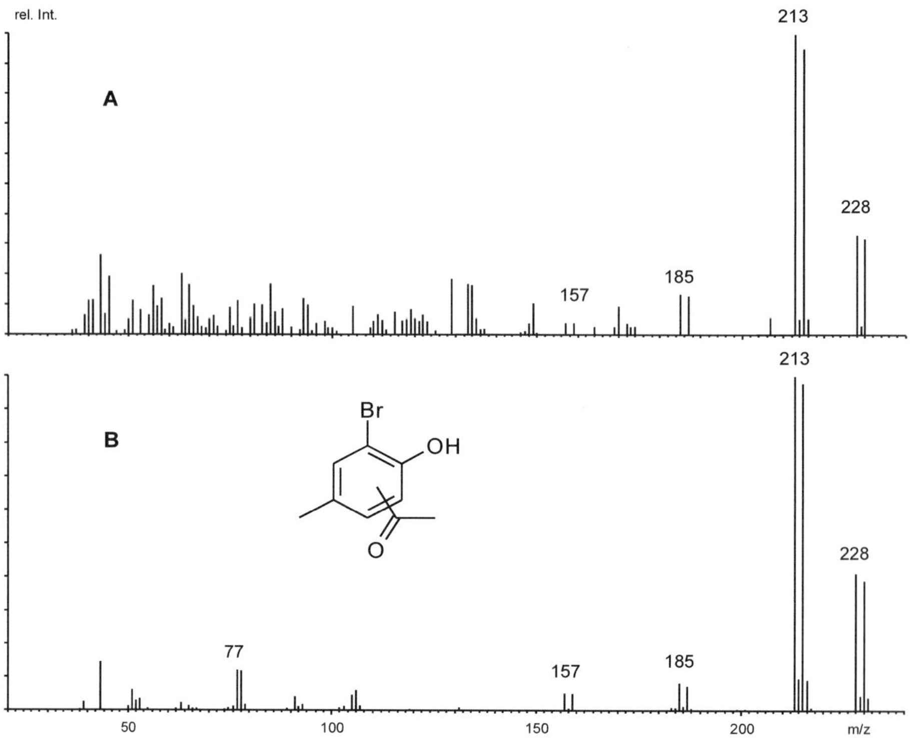


**Fig. 6:** Mass spectra of a brominated contaminant identified in Rhine water samples (A) and of the reference substance obtained by acylation of 2-bromo-4-methylphenol.

With respect to the gas chromatographic behaviour it has to be noted that only a single peak was detected for both mono- and dibrominated compounds, respectively. However, the identification of the substances cannot be reduced to the occurrence of one isolated isomer, because the substitutional isomers do not differ significantly with respect to their gas chromatographic retention time. This was observed by gas chromatographic comparison of two isomers synthesized (see Fig. 8). The very similar retention times of both isomers (acylated 2-bromo- and 3-bromomethylphenol) have to be attributed to only minor differences in the resulting dipole moment. Consequently, also for the naturally occurring isomers the gas chromatographic separation has to be assumed to be insufficient. Therefore, an isomer specific detection cannot be achieved under the chromatographic conditions used.

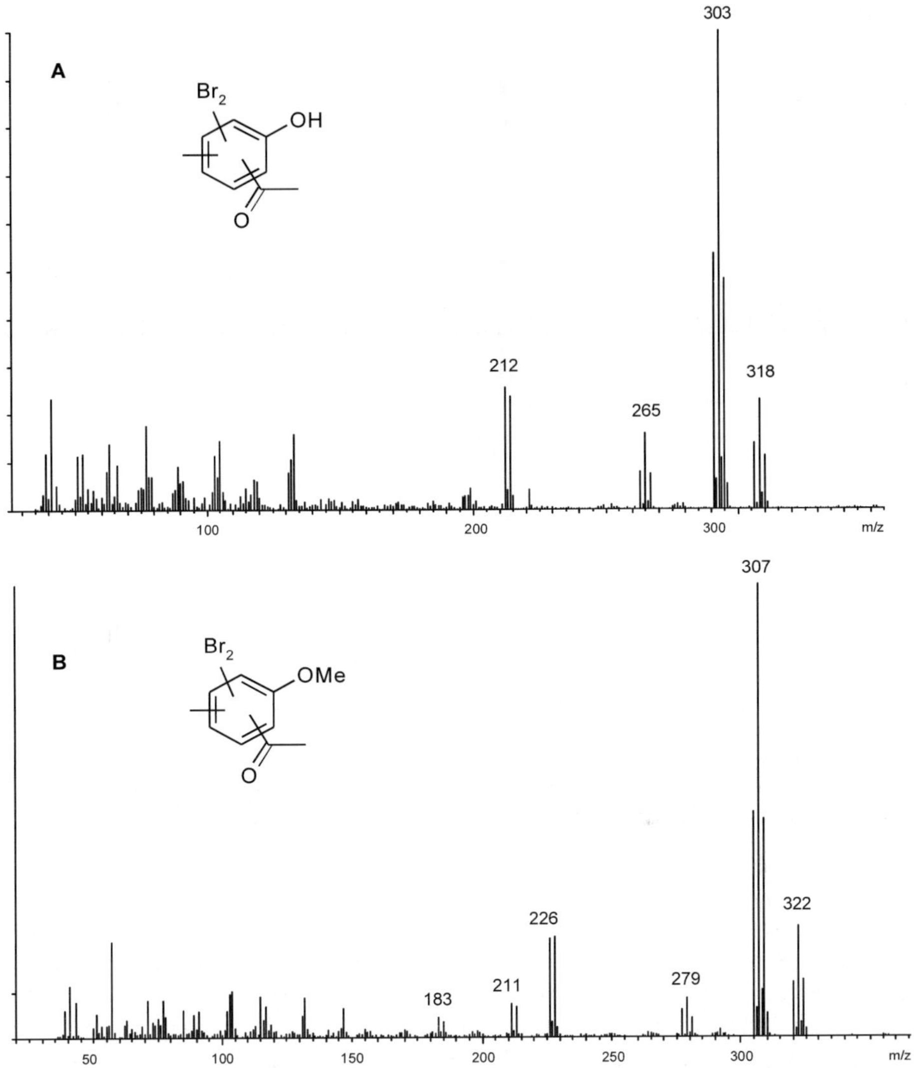


**Fig. 7:** Mass spectra of an unknown isomer of dibromohydroxymethylacetophenone detected in Rhine water samples prior (A) and after (B) derivatisation with TMSH.

FTIR spectroscopic analyses of the reference substances verified the molecular structures as illustrated in Fig. 9. Additionally, the synthetic pathway is depicted.

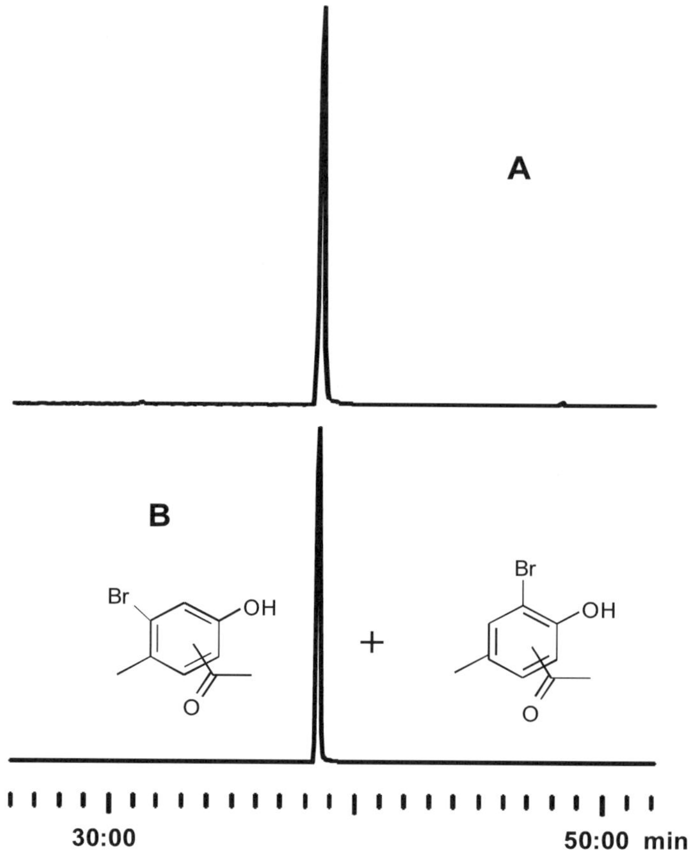


**Fig. 8:** Ion chromatogramms (m/z 228) of the third extract of a Rhine water sample (A) and of reference substances synthesized (B).

Several characteristic functional groups or specific molecular moieties are characterized by the IR-spectra including the carbonyl group (1638 cm-$^{1}$), the OH-absorptions (around 3500 cm$^{-1}$), the methyl as well as the aromatic C-H bonds (2963 cm$^{-1}$ and 3073 cm$^{-1}$, respectively). Absorptions linked to the bromine-carbon bond (detectable at 1075 – 1030 cm$^{-1}$ and 680 – 515 cm$^{-1}$ ) as well as bands reflecting the substitution pattern at the phenyl ring (between 700 cm$^{-1}$ and 900 cm$^{-1}$) cannot be attributed definitely to individual signals due the high complexicity of the spectra in the range between 1500 cm$^{-1}$ and 500 cm$^{-1}$.

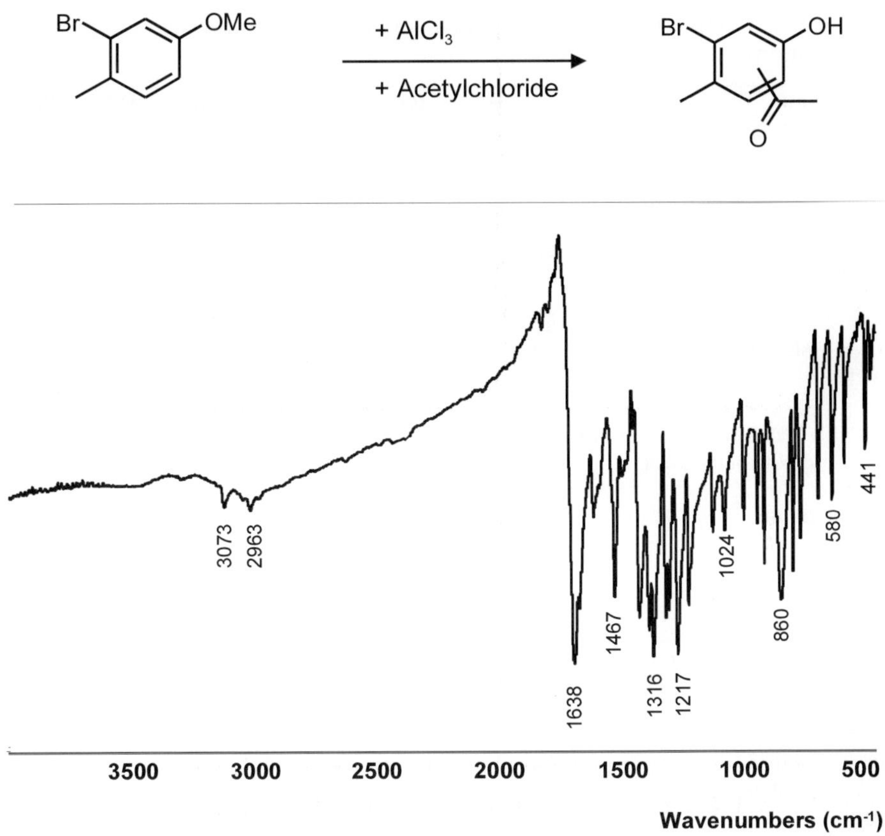


**Fig. 9:** Synthetic pathway and FTIR spectra of acetylated 3-bromo-4-methylphenol

*Quantitation*

Numerous contaminants were quantified selected either by their specific chemical structure, their ecotoxicological properties or their technical application. All concentration data are presented in Tab. 3. Additionally the confluences of major tributaries are marked.

The concentration levels detected ranged between the quantitation limit and 10000 ng/L. Major contaminants were bromophenols as well as the phthalates DEHP, DiBP and DnBP.

**Tab. 3:** Quantitative data of selected contaminants in Rhine water samples. All concentrations are given in ng/L.

| **Compounds** **Sample** | **R1** | **R2** | **R3** | **R4** | **R5** | **R6** | **R7** | **R8** |
|---|---|---|---|---|---|---|---|---|
| *Nitroaromatics* | | | | | | | | |
| Nitrobenzene | <5 | <5 | <5 | <5 | <5 | <5 | 120 | 220 |
| *Sulfones* | | | | | | | | |
| Methylphenylsulfon | 65 | 41 | 27 | 39 | 26 | 26 | 14 | 17 |
| Chlormethylphenylsulfon | 69 | 37 | 29 | 30 | 29 | 25 | <5 | <5 |
| *Personal care products, detergent related compounds* | | | | | | | | |
| 4-Oxoisophorone | 19 | 29 | 23 | 16 | 13 | 30 | 17 | 22 |
| TAED | 640 | 320 | 460 | 490 | 370 | 610 | 470 | 570 |
| *Technical additives plasticizers* | | | | | | | | |
| Di-*iso*-propylnaphthalenes | 20 | 21 | 13 | 12 | 11 | 30 | 75 | 120 |
| NBBS | 190 | 110 | 120 | 130 | 110 | 190 | 92 | 110 |
| Triethylcitrate | 200 | 100 | 120 | 140 | 96 | 150 | 89 | 100 |
| TPDB | 50 | 25 | 7 | 28 | 53 | 71 | 72 | 49 |
| TMDD | 990 | 1000 | 720 | 770 | 2300 | 1600 | 1000 | 840 |
| Triethylphosphate | <5 | <5 | <5 | 17 | 44 | 79 | 18 | 12 |
| TBP | 190 | 680 | 190 | 180 | 1100 | 1300 | 1500 | 340 |
| Tri-*iso*-butylphosphate | 320 | 62 | 13 | 20 | 220 | 290 | 370 | 49 |
| TCEP | 210 | 130 | 100 | 120 | 120 | 220 | 150 | <5 |
| TCPP | 360 | 240 | 200 | 260 | 550 | 1000 | 450 | 180 |
| Tris(butoxyethoxy)phosphate | <5 | <5 | <5 | <5 | 2200 | 5800 | 1900 | 440 |
| Dimethylphthalate | 240 | 190 | 29 | 41 | 28 | 75 | 68 | 52 |
| Diethylphthalate | 400 | 580 | 150 | 170 | 210 | 450 | 310 | 340 |
| DiBP | 630 | 2400 | 410 | 570 | 650 | 840 | 920 | 1200 |
| DnBP | 3200 | 6300 | 190 | 280 | 400 | 1300 | 1700 | 1100 |
| DEHP | 6000 | 9800 | 1200 | 1800 | 2300 | 3600 | 3100 | 3200 |
| *Pharmaceuticals, pesticides, alkaloids* | | | | | | | | |
| Carbamazepine | 650 | 410 | 600 | 540 | 410 | 560 | 310 | 290 |
| DEET | 27 | 15 | 15 | 19 | 19 | 30 | 22 | 14 |
| Coffeine | 730 | 370 | 630 | 560 | 450 | 670 | 450 | 620 |
| Isoproturon | 110 | 45 | 80 | 110 | 83 | 120 | 120 | 160 |
| *Halogenated aromatics* | | | | | | | | |
| 2-/4-Bromophenol | 4000 | 5800 | 1000 | 1600 | 550 | 1600 | 450 | 900 |
| 2,4-Dibromophenol | 180 | 110 | <5 | <5 | <5 | 55 | <5 | <5 |
| 2,4,6-Tribromophenol | 45 | 56 | <5 | 13 | <5 | <5 | <5 | <5 |
| 4-Bromo-2-chlorophenol | 230 | 140 | <5 | <5 | <5 | <5 | <5 | <5 |
| 3-(Bromo-4-methoxyphenyl)-propionic acid | 390 | 380 | 780 | 950 | 420 | 1100 | <5 | 300 |
| *Miscellaneous* | | | | | | | | |
| 4-iso-Propylcyclohexanone | 22 | 26 | 34 | 23 | 25 | 49 | 39 | 28 |
| *Tributaries* | | Sieg river ↑ | Wupper river ↑ | Erft river ↑ | | Ruhr river ↑ | Lippe river ↑ | |

Higher concentrations were also observed for diethylphthalate, selected phosphates (tris(butoxyethoxy)phosphate, TBP, TCPP), 3-(bromo-4-methoxyphenyl)propionic acid, coffeine, carbamazepine, TMDD and TAED with concentrations up to 1000 ng/L.

With respect to the longitudinal profile investigated different concentration trends were observed. A first group of contaminants were characterized by a nearly constant concentration profile between Koblenz (R8) and the border to the Netherlands (R1). This group includes NBBS, 4-oxoisophorone, TAED, triethylcitrate, TCEP, DEET, caffeine isoproturon, and 4-*iso*-propylcyclohexanone. Nearly all these compounds are typical constitutents of sewage effluents and represent domestic waste water contributions (Ricking et al. 2003, Takada and Eganhouse 1998). Therefore an input of these compounds by sewage treatment plants situated along the river system has to be expected. Also, the contributions of the tributaries should increase the concentrations of these compounds due to a higher sewage waste water contamination of most of these smaller riverine systems dewatering densely populated regions. Therefore, the constant concentration profiles reflect a steady state of accumulating input and several output processes like adsorption to the particulate matter or degradation. Further substances derived from municipial effluents like TMDD or TPDB and, additionally, the phthalates are characterized by highly variable concentration profiles without a clear trend of increasing or decreasing values. Accordingly, these more dynamic profiles are the result of emissions from multiple sources superimposed by intensive output processes as described above. However, a close correlation with the confluence of tributaries cannot be observed.

As a third type of concentration profiles the quantitation of 3-(bromo-4-methoxyphenyl)propionic acid and all phosphates exluding TCEP revealed a broad maximum with elevated concentrations between Cologne (sampling site R7) and Duisburg (sampling site R3). These contaminants are emitted to the Rhine river as the result of huge industrial and municipal effluents dominantly derived from the Lower Rhine - Ruhr District. Although significant degradation or adsorption processes seemed to be likely, the contribution derived from this highly industrialized and populated region exceed the output processes.

In contrast an emission of nitrobenzene derived from sources upstream of Koblenz was deduced from the concentration profile of this compound. At sampling sites R7 and R8 nitrobenzene was determined with decreasing concentrations and was not detected further downstream. A similar concentration profile was analysed for the di-*iso*-propylnaphthalenes.

Elevated concentrations at sampling sites R7 and R8 were followed by a significantly lower concentration level downstream of Cologne. A major emission source for these technical additives upstream of Koblenz has to be assumed.

A fifth group of compounds reflecting steadily increasing concentration towards the border to the Netherlands comprised the brominated phenols, carbamazepine and aromatic sulfones. Whereas the pharmaceutical and the sulfones were characterized by constantly rising values likely as the result of constant diffuse emissions wihtout remarkable output from the water body, the concentrations of brominated phenols rapidly increased at sampling site R2. Hence, a superposition of a slowly increasing contamination by discharges derived from a point source has to be stated for the brominated aromatics probably situated close to the Lippe river inflow. This observation contradicts a natural source of mono- and dibromophenols which has been discussed above. An anthropogenic origin has to be suggested for these compounds.

## Summary and conclusions

The qualitative GC/MS analyses revealed a wide variety of organic contaminants in the Rhine water samples. Only a minor proportion was attributed to biogenic sources. Within the group of predominant anthropogenic contaminants, only a few compounds were characterized as frequently detected or priority pollutants (phthalates, caffeine). Numerous compounds exhibiting physiological or ecotoxicological properties were identified which are only rarely reported or still unnoticed riverine contaminants (e.g. DEET, mephobarbital). Furthermore, a large number of rarely noticed riverine components were attributed to the group of technical additives (TPDB, TCEP, DIPN). Information on environmental behaviour or ecotoxicological effects are still limited for most of these substances.

Additionally, several brominated compounds were identified for the first time as environmental contaminants. Their analytical properties (mass spectral, gas chromatographic and IR spectroscopical data) revealed the occurrence of mono- and dibrominated (methoxyphenyl)propionic acids and hydroxymethylacetophenones. Interestingly, the Rhine water samples investigated were dominated by brominated compounds as compared to chlorinated substances. This is an unusual state of riverine pollution with respect to halogenated compounds.

Quantitative analyses characterized up to five groups of pollutants with respect to their concentration profiles as the result of input and output processes. In particular the spatial distribution and the intensity of emission sources on the one hand and the environmental stability as well as the tendency to adsorb on the particulate matter on the other hand determined the quantitative occurrence of individual compounds. Nevertheless, accumulation tendencies or elevated concentrations were observed for several contaminants with potential harmful effects (e.g. carbamazepine, brominated phenols, aromatic sulfones and phthalates).

In conclusion the qualitative and quantitative GC/MS analyses constituted a comprehensive characterization of the low molecular weight organic contamination of Rhine river water. In contrast monitoring analyses on preselected pollutants fail to identify and quantify a wide spectrum of riverine compounds which remain unnoticed, although ecotxicological or harmful effects cannot be excluded. Consequently, non-target screening analyses as well as structure elucidating analyses represent an important skill within environmental investigations on riverine systems.

## 3.3 Havel and Spree rivers

### 3.3.1 Molecular markers of anthropogenic activity in sediments of the Havel and Spree rivers (Germany)‡

**Introduction**

Anthropogenic molecular markers are divided into two principal classes: 1) natural products derived from anthropogenic activities (e.g., coprostanol), and 2) industrially synthesized products. Due to special conditions in the aquatic environment of the Havel and Spree rivers (Pachur et al., 1993, Schwarzbauer et al., 2001), which are characterized by high inputs of anthropogenic contaminants into an eutrophic to hypertrophic riverine system with very slow flowing conditions, this region is to be considered as ideal for searching for anthropogenic markers. The sampling locations in Berlin are characterized by an anoxic sediment-water interface, short residence times in the water phase due to rapid sorption to suspended matter, and sedimentation rates of up to 3 cm/year (Pachur et al., 1993).

In order to identify a wide variety of organic contaminants and to differentiate between point and non-point sources, 28 recent surface sediment samples from the Havel and Spree river, representing the sedimentary record from 1979/80 up to 1995, were analysed by means of a non-target screening approach with subsequent quantification of selected groups of compounds.

The main focus of this study was to identify organic substances in Havel and Spree river sediments that could trace anthropogenic emissions of specific sources within the system. Therefore, selected contaminants with respect to their source specificity and to their environmental stability are discussed in order to characterize new potential anthropogenic marker compounds.

‡ Mainly adapted/reprinted from Ricking et al. 2003

## Samples

All sampling locations are given in Table 1 and are illustrated in Figure 1. The sediment samples were taken in 1993/4. Surface sediment samples (S1-S11, S13, S14, H1-H4, H6, H9, T1, T4-T6) were collected with an Ekman-Birge-grab sampler yielding up to 20 cm of the top layer.

Layers from sediment cores of up to 2 m length were obtained from selected sampling locations in the urban area of Berlin and in Brandenburg at Lake Quenz (S12, H5, H7, H8, T2, T3) using a deep-freezing method (Schwarzbauer et al., 2001). The sample from station H8 represents a sediment layer of 1984/85 and the sediment sample from station T2b is dated back to 1979/80, as determined by gamma spectrometric measurements of Pb-210 and Cs-137 of sediment core slices.

**Tab. 1:** Samples of Havel and Spree river sediments. Sampling devices: EB = Eckman-Birge-grab sampler; NF = liquid nitrogen deep freeze method.

| River | Sample | Sampling site | Date of sampling | Sampling device | TOC (%) | Water depth (m) |
|---|---|---|---|---|---|---|
| **Spree** | | | | | | |
| | S1 | Spree, km 145, inflow of industrial area | 07-94 | EB | 55.6 | 0.50 |
| | S2 | Spree, km 145 | 07-94 | EB | 8.2 | 0.70 |
| | S3 | Spree km 133, upriver Spremberg | 07-94 | EB | 7.9 | 2.50 |
| | S4 | Spree km 165, upriver Cottbus | 07-94 | EB | 8.2 | 0.80 |
| | S5 | Spreewald, tributary to Spree | 07-94 | EB | 12.2 | 1.50 |
| | S6 | Spreewald, tributary to Spree | 07-94 | EB | 13.1 | 1.40 |
| | S7 | Spree, km 238, outflow Spreewald | 07-94 | EB | 0.9 | 1.50 |
| | S8 | Lake Zeuthen | 07-94 | EB | 14.7 | 2.80 |
| | S9 | Dahme | 03-93 | EB | 13.8 | 3.10 |
| | S10 | Lake Dämeritz | 03-93 | EB | 14.7 | 2.80 |
| | S11 | Lake Müggel | 07-94 | EB | 13.0 | 4.80 |
| | S12 | Lake Rummelsburg | 08-94 | NF | 13.6 | 3.80 |
| | S13 | Spandau Shipping Canal, km 0.6 | 07-94 | EB | 9.2 | 2.30 |

| River | Sample | Sampling site | Date of sampling | Sampling device | TOC (%) | Water depth (m) |
|---|---|---|---|---|---|---|
| | S14 | Spandau Shipping Canal, km 2.1 | 07-94 | EB | 9.3 | 2.20 |
| **Havel** | | | | | | |
| | H1 | Great Lake Labus | 07-94 | EB | 9.1 | 3.90 |
| | H2 | Havel, km 160, upriver Berlin | 08-94 | EB | 4.4 | 0.60 |
| | H3 | Lake Niederneuendorf | 03-93 | EB | 10.3 | 3.50 |
| | H4 | Lake Tegel | 03-93 | EB | 9.9 | 2.30 |
| | H5 | Spandau North harbour | 08-94 | NF | 6.2 | 2.70 |
| | H6 | Pichelsdorf Gmünd | 03-93 | EB | 15.1 | 3.40 |
| | H7 | Lower Havel river, km 9.5 | 08-94 | NF | 12.0 | 4.10 |
| | H8 | Lake Quenz; core 0.35-0.37 m depth | 08-94 | NF | 10.4 | 3.00 |
| | H9 | Greater Lake Wend | 08-94 | EB | 10.3 | 3.50 |
| **Teltow Canal** | | | | | | |
| | T1 | Teltow Canal, km 3, Berlin-Grünau;Factory outlet | 07-94 | EB | 7.1 | 0.20 |
| | T2a | Teltow Canal, km 2.5; core 0.00-0.05 m depth | 07-94 | NF | 12.0 | 3.80 |
| | T2b | Teltow Canal, km 2.5; core 0.35-0.40 m depth | 07-94 | NF | 10.7 | 3.80 |
| | T3 | Teltow Canal, km 2, core 0.00-0.20 m depth | 06-95 | NF | 11.7 | 3.60 |
| | T4 | Teltow Canal, harbour Lichterfelde | 07-94 | EB | 11.9 | 2.70 |
| | T5 | Lake Machnow | 03-93 | EB | 4.3 | 2.80 |
| | T6 | Lake Griebnitz | 03-93 | EB | 7.5 | 2.60 |

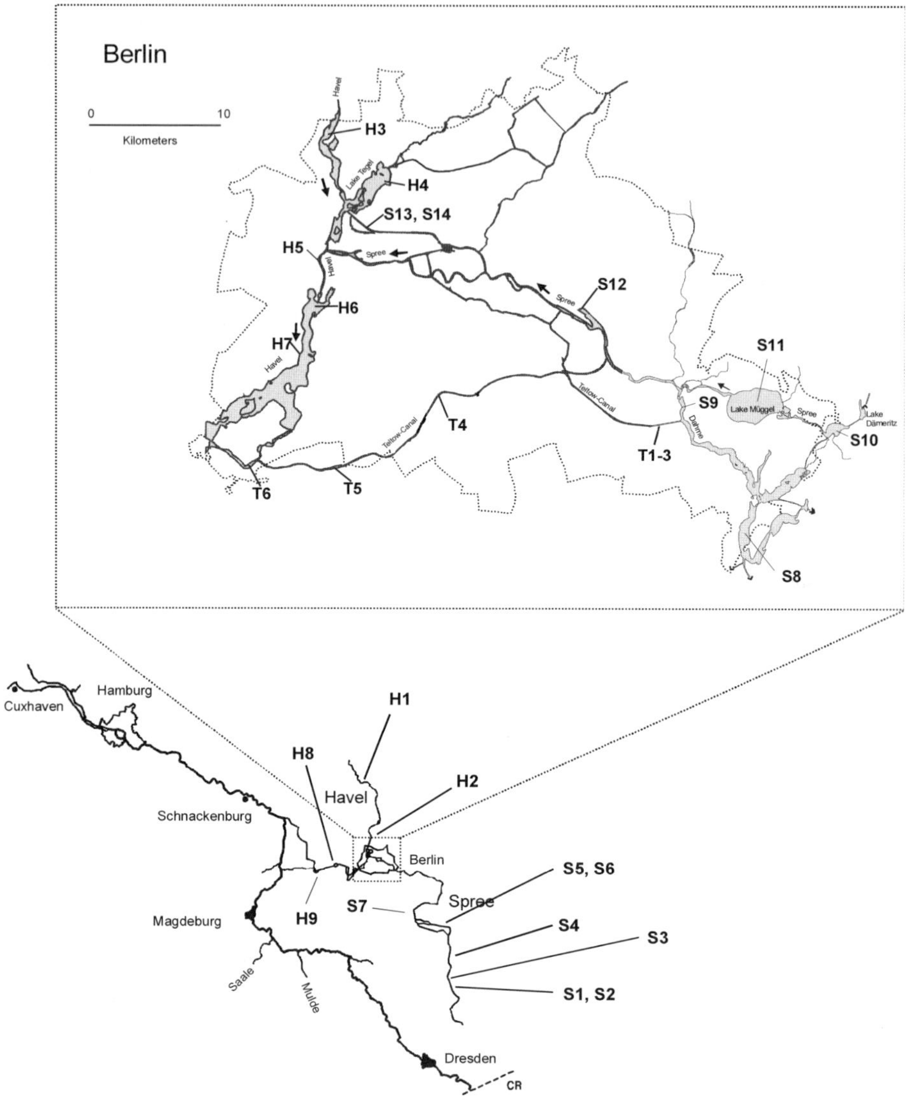


**Fig. 1** Sampling locations in the Havel and Spree river area. Further informations are provided in Table 1.

In contrast to the qualitative non-target screening analyses, which were applied to the complete sample set, the quantitative analyses were restricted to selected sediment samples including H3, H6, S2, S9, S10, S13, S14,T1, T2, T5 and T6. These samples were selected with respect to their spatial distribution representing well the contamination of the Havel and Spree river sediments in the urban area of Berlin. The quantitative data were not normalized to TOC content as no significant correlation with the data has been found.

The limit of determination was calculated at 0.1 ng/g dry matter (signal to noise ratio 5:1 in real samples); no attempt was made to quantify components with concentrations less than 0.5 ng/g dry matter. Concentrations are recovery corrected (see Table 2) and normalized on a dry weight basis.

**Tab. 2:** Selected contaminants and ion fragments used for quantification as well as determined recovery rates

| **Compounds** | **Detected ions** (m/z) | **Recovery rate** (%) |
|---|---|---|
| Tributyl phosphate | 99, 155 | 78 |
| Tritolyl phosphate (4 isomers) | 368 | 69 |
| Alkylsulfonic acid phenyl esters | 94 | 77-82 |
| Alkylsulfonic acid cresyl esters | 108 | 77-82 |
| Dibenzyl ether | 92, 107 | 84 |
| Benzophenone | 105, 182 | 90 |
| Benzyl benzoate | 105, 212 | 80 |
| 4-Methoxycinnamic acid 2-ethylhexyl ester | 161, 178 | 75 |
| Formyl piperidine | 84,113 | 38 |
| Acetyl piperidine | 84,127 | 47 |
| N-Ethyl-N-phenyl formamide | 106, 149 | 57 |
| N-Ethyl-N-phenyl acetamide | 106, 163 | 57 |
| Nitrobenzene | 77,123 | 41 |
| 4-Ethylnitrobenzene | 105,151 | 56 |
| Butyltriethyl tin | 177,179 | 63 |
| Dibutyltriethyl tin | 177,179 | 75 |
| Tributylethyl tin | 261,263 | 78 |
| Tetrabutyl tin | 289, 291 | 84 |

In the case of N-ethylformamide the response factor and the recovery rate of the related compound, N-ethylacetamide, was used.

## Results and Discussion

GC/MS non-target screening analysis revealed a wide variety of compounds. The chemical structures of selected compounds are given in the appendix and the numbers are given in brackets behind the compounds name.

The compounds identified included terpenoids, hopanoids, steranes, alkylbenzenes, PAHs, aromatic ethers, aliphatic aldehydes and ketones, phenylaldehydes and -ketones, amides, phenols, ethers and vitaminoids. In addition, several groups of well-known anthropogenic markers, including fecal steroids (coprostanol, epicoprostanol, coprostanone), detergent related products ($C_{10}$-$C_{13}$ LABs, nonylphenols), PAHs and oxygenated compounds (e.g., benzanthrone, fluorenone) and halogenated aromatics were present in nearly all samples investigated (see Table 3). Superimposed on this diffuse contamination, a strong local emission of an industrial point source was reflected by specific markers including brominated, chlorinated and mixed halogenated aromatics, as well as several chlorinated and brominated pesticides (Schwarzbauer et al., 2001).

In addition to this published data we identified several non-halogenated contaminants, which are also specific for this local emission due to their obviously non-biogenic structures and the point source related spatial distribution. In detail, 4-nitrobenzoic acid [# 1] was only detected at sampling locations T1 and T2 situated nearby the industrial point source (see Table 4). Also in samples T1 and T2 the concentrations of 4-ethylnitrobenzene [# 2], formylpiperidine [# 3] and acetylpiperidine [# 4] increased significantly up to 61, 39 and 28 ng/g, respectively (see Table 3), compared to stations T5 and T6, which are assumed to be contaminated only by diffuse sources, representing therefore background samples with respect to the industrial point source.

In contrast, the concentrations of structurally related compounds, nitrobenzene, N-ethyl-N-phenylformamide [# 5] and N-ethyl-N-phenylacetamide [# 6] did not peak at T1 and T2. Therefore, the distribution of these structurally related compounds indicated no source specificity.

Beside industrial emissions sewage input mainly characterizes anthropogenic contamination of riverine systems, attributed to different groups of fragrances, personal care products, plasticizers or technical additives (see Table 4).

**Tab. 3:** Commonly observed lipophilic organic contaminants in sediments of the Havel and Spree rivers.

| **Alkanes** | **Steranes and steroids** |
|---|---|
| Homol. series of *n*-alkanes ($C_{11}$-$C_{34}$) | Cholestane (3 isomers) |
| 2,2,4,4,6,8,8-Heptamethylnonane | $C_{27}$-Diasterane (2 isomers) |
| | Cholestene |
| *Terpenoides and degradation products* | Cholestadiene (2 isomers) |
| Limonene | Cholestatriene (3 isomers) |
| *p*-Cymol | Ergostane (2 isomers) |
| Camphene | $C_{28}$-Diasterane (2 isomers) |
| Camphor | Ergostene |
| α-Ionone | Ergostadiene (2 isomers) |
| β-Ionone | Ergostatriene (2 isomers) |
| Dihydroactinidiolide | Stigmastane(3 isomers) |
| Loliolide | $C_{29}$-Diasterane (2 isomers) |
| Cadalene | Stigmastene (2 isomers) |
| Dehydroabietic acid methyl ester | Stigmastadiene |
| Tetrahydroabietic acid methyl ester | Stigmastatriene (2 isomers) |
| Dehydroabietic acid $^{(m)}$ | Cholestanone |
| Abietic acid $^{(m)}$ | Coprostanone |
| Dihydroabietic acid $^{(m)}$ | Cholestanol |
| Tetrahydroabietic acid $^{(m)}$ | Epicholestanol |
| Retene | Coprostanol |
| Tetrahydroretene | Epicoprostanol |
| Dehydroabietin | Cholesterol |
| Dehydroabietane | 3β,5β-Ergostanone |
| Pristane | 3β,5α-Ergostanone |
| Phytane | 3β,5β-Ergostanol |
| Phytene | 3β,5α-Ergostanol |
| Phytadiene | 3β,5α-Ergostenol |
| 6,10,14-Trimethylpentadecan-2-one | 3β,5β-Stigmastanone |
| Phytol | 3β,5α-Stigmastanone |
| Dihydrophytol | 3β,5β-Stigmastanol |
| Phytolic acid $^{(m)}$ | 3β,5α-Stigmasterol |
| Squalene | |

**Hopanoids**

α-Trisnorhopane
α, β-Norhopane
β, α-Norhopane
β, β-Norhopane
α, β-Hopane
α, β-Homohopane
β, α-Homohopane
β, β-Homohopane
α, β-Bishomohopane (2 isomers)
α, β-Trishomohopane (2 isomers)
α, β-Tetrakishomohopane (2 isomers)
α, β-Pentakishomohopane (2 isomers)

**Alkylbenzenes**

Toluene
*o-/p-/m*-xylene
$C_3$-benzenes (7 isomers)
$C_4$-benzenes (15 isomers)
Indan
Methylindan (2 isomers)

**Sulphur containing PAH**

Benz(b)thiophene
Methylbenz(b)thiophene (4 isomers)
Dibenzothiophene
Methyldibenzothiophene (3 isomers)
$C_2$-Dibenzothiophene (6 isomers)
Benz(b)naphtho(2,1-d)thiophene
Benz(b)naphtho(1,2-d)thiophene
Benz(b)naphtho(2,3-d)thiophene
Phenanthro(9,10-b)thiophene
Methylbenzonaphthothiophene (4 isomers)

**PAH**

Naphthalene
1-/2-Methylnaphthalenes
1-Ethylnaphthalene
Dimethylnaphthalene (6 isomers)
$C_3$-naphthalene (15 isomers)
Biphenyl
3-/4-Methylbiphenyls
$C_2$-Biphenyl (3 isomers)
$C_3$-Biphenyl (3 isomers)
Diphenylmethane
Acenaphthylene
Acenaphthene
Methylacenaphthene (4 isomers)
Fluorene
9-/2-/1-Methylfluorenes
$C_2$-Fluorene (5 isomers)
1-/2-Phenylnaphthalenes
Phenanthrene
Anthracene
1-/2-/3-/4-/9-Methylphenanthrene
2-Methylanthracene
1,2-/1,3-/1,4-/1,7-/1,8-/1,9-/2,7-/2,9-/3,5-/3,10-/4,9-Dimethylphenanthrene
Ethylphenanthrene (2 isomers)
$C_3$-Phenanthrene (5 isomers)
4H-Cyclopenta(def)phenanthrene
Triphenylmethane
Fluoranthene
Acephenanthrylene
Pyrene
Benz(a)fluorene
Methylfluoranthene/-pyrene (6 isomers)
Dimethylfluoranthene/-pyrene (9 isomers)
Ethylfluoranthene/-pyrene
*o-/m-/p*-Terphenyl
Benz(ghi)fluoranthene

| **PAH (cont.)** | **Phenols** |
|---|---|
| Benz(c)phenanthrene | Phenol |
| Cyclopenta(cd)pyrene | 4-Cresol |
| Benz(a)anthracene | 2-/3-Cresol |
| Chrysene/Triphenylene | |
| Methyl-228 (4 isomers)* | **Ethers** |
| Dimethyl-228 (8 isomers and ethyls)* | Diphenyl ether |
| 1,2´-/2,2´-Binaphthyls | Methyl diphenyl ether (2 isomers) |
| Phenylphenanthrene/ | |
| -anthracene (3 isomers) | **Aliphatic aldehydes and Ketones** |
| Benz(x)fluoranthene, (x=j,k,b) | Homologous series of *n*-aldehydes ($C_8$-$C_{32}$) |
| Benz(e)pyrene | Homologues series of 2-alkanones ($C_8$-$C_{17}$) |
| Benz(a)pyrene | |
| Perylene | **Phenones and phenylaldehydes** |
| Methyl-252 (5 isomers)* | Benzaldehyde |
| Indeno(1,2,3-cd)pyrene | Methylbenzaldehyde (2 isomers) |
| Benz(ghi)perylene | Dimethylbenzaldehyde |
| Dibenzanthracene | Acetophenone |
| Benz(b)chrysene | |
| Picene | **Esters** |
| Dibenzo(def,mno)chrysene | Complex mixture of wax esters ($C_{26}$- $C_{32}$) |
| Coronene | |
| | Methyl myristate |
| **Oxygen containing PAH** | Isopropyl myristate |
| Dibenzofurane | Butyl myristate |
| Methyldibenzofurane (3 isomers) | Methyl pentadecanoate |
| $C_2$-Dibenzofurane (5 isomers) | Methyl palmitate |
| Benz(b)naphtho(2,1-d)furane | Ethyl palmitate |
| Benz(b)naphtho(1,2-d)furane | Isopropyl palmitate |
| Benz(b)naphtho(2,3-d)furane | Methyl heptadecanoate |
| Methylbenzonaphthofurane (4 isomers) | Methyl stearate |
| $C_2$-Benzonaphthofurane (5 isomers) | Ethyl stearate |
| | |
| **Hydrogenated PAH** | **Aromatic alkanoic acids** |
| Tetraline | Benzoic acid [(m)] |
| 1,1,6-Trimethyltetraline | Phenylacetic acid [(m)] |
| 3,3,7-Trimethyl-1,2,3,4-tetrahydrochrysene | |

| Aliphatic alkanoic acids | Vitaminoids and metabolites |
|---|---|
| Homologous series of *n*-carboxylic acids ($n$-$C_{12}$ – $n$-$C_{28}$) (m) | 2-Methyl-1,4-naphthochinone |
| | Phthiokol |
| Homologous series of *iso*-carboxylic acids ($iso$-$C_{15}$ – $iso$-$C_{26}$) (m) | α-Tocopherol |
| | 4,8,12,16-Tetramethylheptadecan-4-olide |
| Homologous series of *anteiso*-carboxylic acids ($anteiso$-$C_{15}$ –$anteiso$-$C_{26}$) (m) | 3,4-Dimethyl-2,5-furandione |
| | |
| Palmitoleic acid (m) | **Amides** |
| Linoleneic acid (m) | *n*-Pentadecanamide |
| Linoleic acid (m) | *n*-Hexadecanamide |
| Oleic acid (m) | *n*-Heptadecanamide |
| Arachidonic acid (m) | *n*-Octadecanamide |

(m) = detected as their methylated derivatives.

Two groups of plasticizers, alkylsulfonic acid esters (ASPE) [# 7], as well as tributyl (TBP) [# 8] and tritolyl phosphates (TTP) [# 9], although known as environmental contaminants, have not previously been used for tracing sewage-derived contamination. Alkylsulfonic acid phenyl and cresyl esters are used as technical additives and plasticizers (Mesamoll™) in PVC and related polymers, nitrocellulose lacquer, and other applications (Franke et al., 1998a). They have been in use for more than 50 years (Haslam et al., 1951). Alkylsulfonic acid aryl esters were further detected in sediment samples of the Elbe river system with elevated concentrations near industrial point sources (Franke et al., 1998a). The published data reflect the ubiquitous distribution of these plasticizers in the Elbe river system and the adjacent North Sea. Therefore, ASPE seem to be suitable as anthropogenic markers for the Elbe river (Schwarzbauer et al., 2000). Taking their widespread use in PVC production into account, a world-wide contamination and subsequent application as anthropogenic markers can be suggested.

In Havel and Spree river sediment samples a background concentration of ASPE of 1500 - 2500 ng/g was peaked by concentrations up to 8860 ng/g, indicating a distinct emission source at station H6.

Current investigations on dated sediment cores sampled at stations T2 and H8 reveal the occurrence of ASPE with isomeric patterns according to the surface sediment samples in sediment layers of 1979/80 and 1984/85, respectively.

**Tab. 4:** Potential anthropogenic marker compounds identified in sediment samples of the Havel / Spree rivers. Quantitative data are given in ng/g dry matter.

| Compounds / Stations | H3 | H6 | S2 | S9 | S10 | S13 | S14 |
|---|---|---|---|---|---|---|---|
| *Nitro compounds* | | | | | | | |
| Nitrobenzene | | 4 | | 2 | | | 10 |
| 4-Ethylnitrobenzene | | | | 4 | | 4 | 6 |
| 4-Nitrobenzoic acid (m) | | | | | | | |
| *N-substituted amides* | | | | | | | |
| Formyl piperidine | | 8 | | 15 | 1 | 8 | 4 |
| Acetyl piperidine | | 5 | | 5 | | 4 | 2 |
| N-Ethyl-N-phenylformamide | | 13 | | | | | |
| N-Ethyl-N-phenylacetamide | | 18 | | 0.5 | | | 0.5 |
| *Detergents related substances* | | | | | | | |
| 4-iso-Nonylphenol (10 isomers) | + | | + | (+) | | + | + |
| Phenyldecanes to –tridecanes (LABs) | + | + | + | + | | (+) | + |
| *Plasticizers* | | | | | | | |
| Tributyl phosphate | 41 | 46 | 82 | 44 | 15 | 31 | 28 |
| Tritolyl phosphate (4 isomers) | 17 | 20 | 94 | 100 | 12 | 100 | 140 |
| Alkylsulfonic acid phenyl esters | 1800 | 6600 | 1500 | 1200 | 1600 | 2000 | n.q. |
| Alkylsulfonic acid cresyl esters | 660 | 2290 | 340 | 320 | 380 | 240 | n.q. |
| *Fragrances and UV-protectors* | | | | | | | |
| 4-Methoxycinnamic acid 2-ethylhexyl ester | 3 | 4 | | | | | 1 |
| Galaxolide | + | + | + | + | + | + | + |
| Tonalide | + | + | + | + | + | + | + |
| 4-Oxoisophorone | + | + | + | + | + | + | + |
| *Technical additives, solvents* | | | | | | | |
| Dibenzyl ether | 1 | 3 | | 3 | 2 | 4 | 2 |
| Benzophenone | 3 | 1 | | 2 | 2 | 2 | 1 |
| Benzyl benzoate | | 0.5 | | 0.5 | 0.5 | | |
| Thymol | + | + | + | + | | + | + |
| *Organotin compounds* | | | | | | | |
| MBT (as butyl triethyl tin) | 410 | 710 | n.a. | 550 | 380 | n.a. | n.a. |
| DBT (as dibutyl diethyl tin) | 1400 | 4900 | n.a. | 2500 | 1100 | n.a. | n.a. |
| TBT (as tributyl ethyl tin) | 860 | 2700 | n.a. | 760 | 1000 | n.a. | n.a. |
| Tetrabutyl tin | 9 | | n.a. | | 15 | n.a. | n.a. |

| *Tab 4. cont.* **Stations**<br>**Compounds** | **T1** | **T2** | **T5** | **T6** |
|---|---|---|---|---|
| *Nitro compounds* | | | | |
| Nitrobenzene | | 2 | | |
| 4-Ethylnitrobenzene | 34 | 61 | 2 | 1 |
| 4-Nitrobenzoic acid [(m)] | + | + | | |
| *N-substituted amides* | | | | |
| Formyl piperidine | 39 | 23 | | |
| Acetyl piperidine | 28 | 15 | 1 | |
| N-Ethyl-N-phenylformamide | | | 2 | |
| N-Ethyl-N-phenylacetamide | | | 0.5 | |
| *Detergents related substances* | | | | |
| 4-iso-Nonylphenol (10 isomers) | + | + | (+) | + |
| Phenyldecanes to –tridecanes (LABs) | (+) | (+) | + | (+) |
| *Plasticizers* | | | | |
| Tributyl phosphate | 170 | 140 | 54 | 46 |
| Tritolyl phosphate (4 isomers) | 390 | 320 | 36 | 36 |
| Alkylsulfonic acid phenyl esters | 2990 | + | 1310 | 1460 |
| Alkylsulfonic acid cresyl esters | 480 | + | 220 | 460 |
| *Fragrances and UV-protectors* | | | | |
| 4-Methoxycinnamic acid 2-ethylhexyl ester | | | 1 | |
| Galaxolide | + | + | + | + |
| Tonalide | + | + | + | + |
| 4-Oxoisophorone | + | + | + | + |
| *Technical additives, solvents* | | | | |
| Dibenzyl ether | 8 | 21 | 0.5 | 1 |
| Benzophenone | 0.5 | 4 | 2 | 3 |
| Benzyl benzoate | 0.5 | 0.5 | 0.5 | |
| Thymol | + | + | + | + |
| *Organotin compounds* | | | | |
| MBT (as butyl triethyl tin) | n.a. | n.a. | 110 | 230 |
| DBT (as dibutyl diethyl tin) | n.a. | n.a. | 350 | 450 |
| TBT (as tributyl ethyl tin) | n.a. | n.a. | 150 | 570 |
| Tetrabutyl tin | n.a. | n.a. | | 2 |

[(m)] = detected as methylated compounds after derivatisation
\+ = detected, but not quantified; n.a. = not analysed

The observed pattern was very similar as compared to the isomeric composition of LABs, attributed to a preferential microbial degradation of The isomeric distribution of ASPE in sediments from different sampling locations and in selected sediment layers was virtually identical (see Figure 2), suggesting their persistence in anaerobic environments. external relative to internal substituted isomers (Takada and Eganhouse, 1998). The observed pattern of ASPE indicated a very similar environmental behaviour as compared to LABs, that are well established molecular marker. In addition, information on the aerobic catabolism of ASPE by soil bacteria presented also the observed pattern as demonstrated for various soil samples (Schmidt et al., 2000).

Tributyl (TBP) and tritolyl phosphates (TTP), two persistent and acute harmful compounds to aquatic biota (Stuer-Lauridsen et al., 2001), are also used as PVC plasticizers and as additives in nitro-based paint, hydraulic fluids, gasoline, and as flame retardants (Saeger *et al.*, 1979). Background concentrations in the sediment samples investigated were in the range of 15 - 80 ng/g TBP and 10 - 100 ng/g TTP. Near the industrial point source at location T1 the concentrations peaked at 170 and 390 ng/g, respectively (see Table 4). Although the potential biodegradation in water and activated sludge slurries has been demonstrated (e.g., Saeger *et al.*, 1979), during the current investigation TBP and TTP have been detected at elevated concentrations at station T2a in a sediment layer dated back to 1979/80, illustrating their environmental stability. In addition to our results for the Havel and Spree river phosphate plasticizers were also detected in comparable environmental systems, e.g. the Mulde and Elbe river as well as the Hamburg Harbour (Schwarzbauer, 1997). Fries and Püttmann (2001) published data for the Rhine, Elbe, Main and Odra river in the concentration range of 17 – 1510 ng/l (tributyl phosphate, tris(2-chloroethyl) phosphate and tris(2-butoxyethyl) phosphate) with the highest levels detected in waste water samples.

HHCB (Galaxolide®) [# 10] and AHTN (Tonalide®) [# 11] are persistent polycyclic synthetic musk substitutes (PMF), with widespread application as fragrances in soaps, perfumes, air fresheners, detergents, fabric softeners and other household cleaning products (Rimkus, 1999). They are characterized by low water solubilities (Balk and Ford, 1999a/b; 1.25–1.75 mg/l), and log $K_{ow}$ values in the range of 5.7-5.9 (Rimkus, 1999). We detected HHCB and AHTN accompanied by 4-oxoisophorone [# 12], a constituent of perfumes (Ullmann, 2001), in all samples within our research area, reflecting a widespread application and subsequent emission (see Table 4).

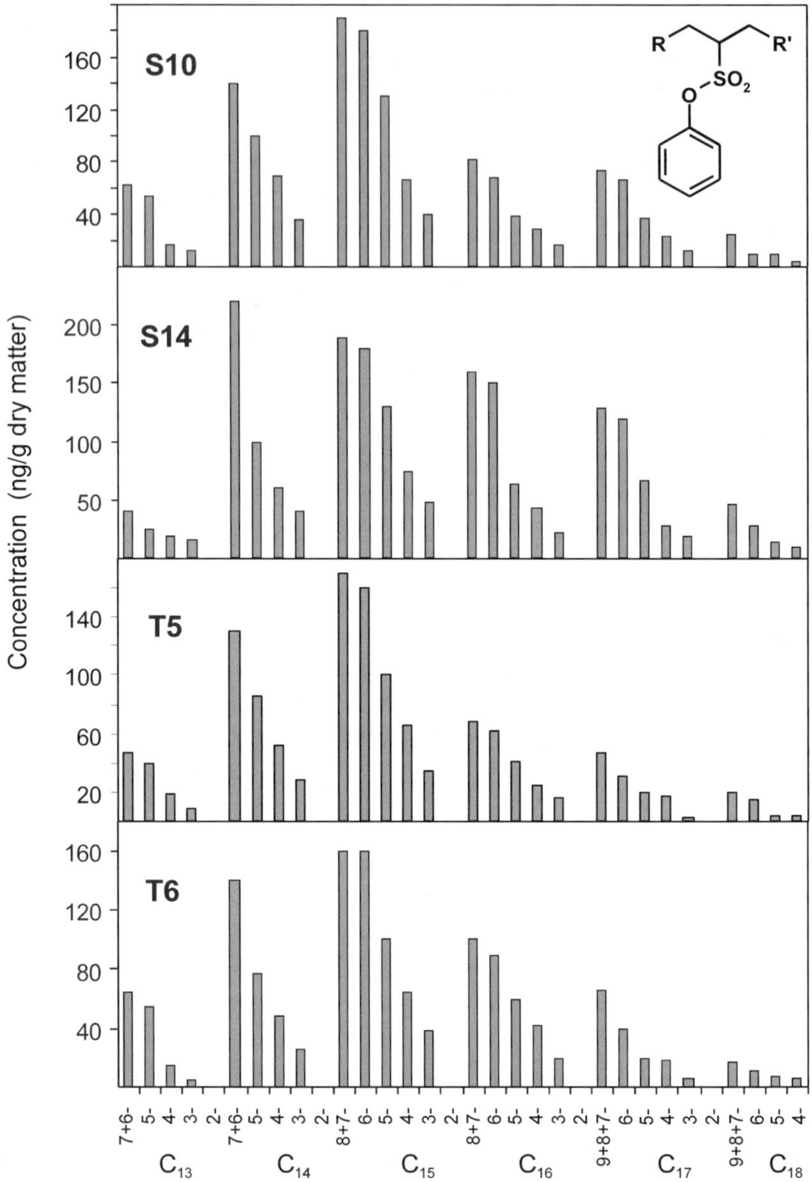


**Fig. 2:** Concentrations of alkylsulfonic acid phenylesters (molecular structure is given in the upper box) in sediment samples of the Spree river (S10 and S14) and the Teltow Canal (T5 and T6). On the x-axis the homologues series with aliphatic side chain length ranging from $C_{13}$ to $C_{18}$ and the individual positional isomers of each homologues groups (7-/6-, 5-, 4- 3- ....= position of substitution at the side chain) are noted according to Franke et al., 1998a. All concentrations are given in ng/g dry matter.

Their occurrence in riverine sediments characterized by significant anthropogenic contaminations was also observed in Mulde, Elbe and Lippe river sediments (Schwarzbauer, 1997; Dsikowitzky et al., 2002) and in marine sediments (Schwarzbauer et al., 2000). Investigations that have examined the environmental stability of musk fragrances indicate decomposition during atmospheric transport (Aschmann et al., 2001), significant degradation rates during wastewater treatment (Simonich et al., 2000) and substantial metabolism in aquatic biota (Balk and Ford, 1999). However, information on the lifetime of HHCB and AHTN in aquatic sediments are to our knowledge not available. In addition, recent analyses of a sediment core obtained near sampling location T3 indicate a high accumulation potential for a time period of approximately 20 years with concentrations of more than 200 ng/g of HHCB and AHTN. High concentrations in the non-extractable fraction illustrate their environmental stability (Schwarzbauer et al., 2002). Characterizing the source specificity of musk fragrances in the Havel and Spree river system, Heberer et al. (1999b) and Fromme et al. (1999) revealed municipal treatment plant effluents as the main source of contamination for the Berlin area containing up to 12500 μg/l HHCB. The emission was monitored by analysing water samples and eel samples, respectively. Very recent results indicate HHCB and AHTN as suitable tracers for the US and Canada, too (Osemwengie and Steinberg, 2001; Ricking et al., 2002b).

Information on 4-oxoisophorone with respect to its environmental fate and stability are to our knowledge not available.

Additional compounds associated with fragrance and cosmetic products could serve as potential tracers of sewage inputs due to their source specificity, although data on their persistence are to be elucidated. All the following compounds presented were widespread detected and mainly quantified.

The UV-protector 4-methoxycinnamic acid 2-ethylhexyl ester [# 13] , is used as an UV-B-absorber and sun screening agent (Bernerd et al., 2000). In Havel and Spree river sediments it has been detected at concentrations near the detection limit and up to 4 ng/g (see Table 4). Considering the occurrence of 4-methoxycinnamic acid 2-ethylhexyl ester also in sediment samples from the German Bight (Schwarzbauer et al., 2000), a long life time of this compound in the aquatic environment can be assumed, with respect of the prolonged residence time during the transport from the sewage treatment effluent into the estuarine sediments.

As a second marker candidate the compound benzyl benzoate [# 14], used in perfume manufacture to stabilize volatile odorants and in pharmaceuticals, was detected in six samples (H6, S9, S10, T1, T2, T5) with concentrations close to the detection limit (see Table 4). Although the source specifity is verified, the low concentrations limited the application of benzyl benzoate as an anthropogenic marker.

Also benzophenone [# 15] might act as an anthropogenic marker. It is mainly used along with substituted diphenyl ketones as UV-absorber in creams, as a fixative in perfumes and soap as well as an intermediate in pharmaceuticals and agricultural chemicals production. As well, only low concentrations were found in the Havel and Spree river sediments in the range of 0.5 to 3 ng/g, recommending an optimized analytical procedure.

Furthermore, dibenzyl ether [# 16] and thymol [# 17] are used in large quantities in soap perfumes, fragrances and as intermediate in menthol production. Both compounds were detected frequently in the sediment samples investigated, only thymol was not quantified (see Table 4).

To our knowledge no further information about the environmental persistence and the environmental fate for 4-methoxycinnamic acid 2-ethylhexyl ester, benzyl benzoate, benzophenone, dibenzyl ether and thymol are available. Nevertheless, due to their source specificity, these compounds are appropriate candidates for the application as anthropogenic markers in the case of sufficient persistence.

In selected samples of the Havel and Spree river sediments (H3, H6, S9, S10, T2 and T5) we detected high amounts of monobutyl tin (MBT) [# 18;] and dibutyl tin (DBT) [# 19;] in the range of 110-710 ng/g and 350-4900 ng/g dry matter, respectively (see Table 4). These compounds are used as stabilizers in plastics and PVC (Fent, 1996), and, thus, they are common contaminants of sewage inputs.

DBT and MBT are also primary degradation products of tributyl tin (TBT) [# 20;] which has been widely applied as antifouling agent. Considering the the degradation half-life time of TBT in water of approx.1-3 weeks (Fent, 1996) the superimposition of TBT emissions caused by shipping activities has to be excluded using DBT and MBT as sewage treatment plant marker. In addition the applicability of DBT and MBT as anthropogenic tracers in aquatic systems is restricted to the particulate matter due to the low environmental persistence in the water phase.

## Conclusions

Superimposed on diffuse anthropogenic contamination, a local emission from an industrial point source recently characterized by halogenated aromatics and pesticides has been reflected also by specific nitrogen containing compounds including 4-ethylnitrobenzene, formyl piperidine and acteyl piperidine.

In addition numerous compounds were detected which may serve as potential anthropogenic markers with respect to their source specifity and environmental persistence. Plasticizers (alkylsulfonic acid aryl esters, tributyl and tricresyl phosphates), synthetic fragrances (galaxolide, tonalide, 4-oxoisophorone) and additives of personal care products (4-methoxycinnamic acid 2-ethylhexyl ester, benzyl benzoate, dibenzyl ether, benzophenone) occurred due to sewage treatment plant effluents and reflect therefore an anthropogenic contamination of the particulate riverine matter, even in marine systems.

## Appendix

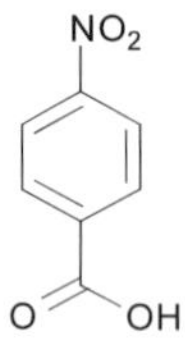
1

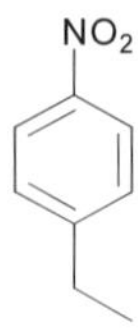
2

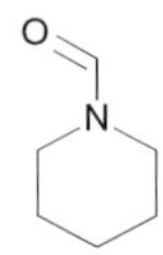
3

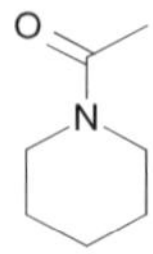
4

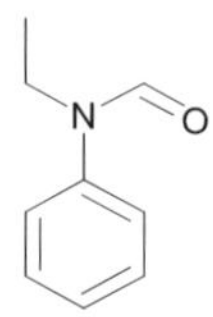
5

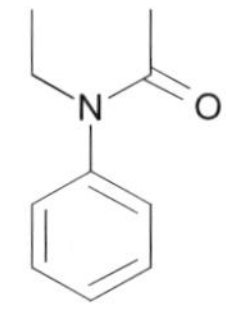
6

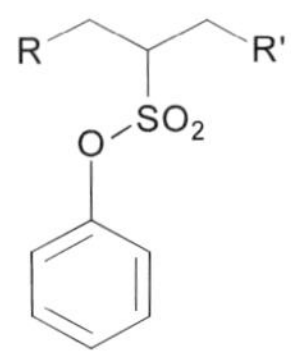
7

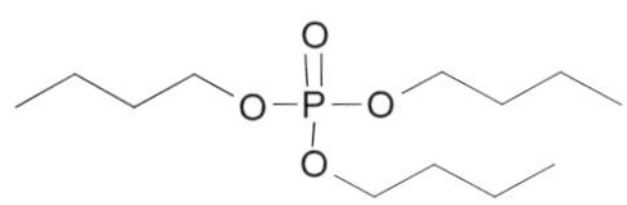
8

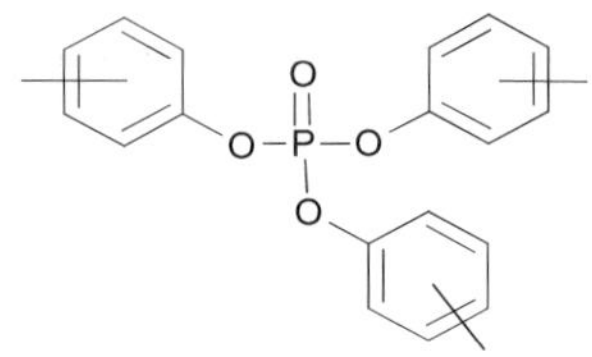
9

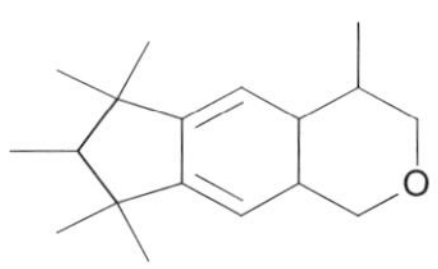
10

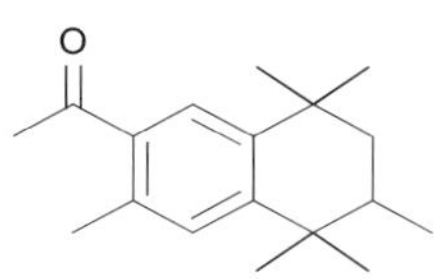
11

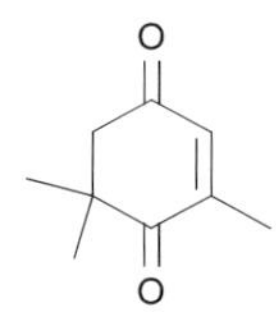
12

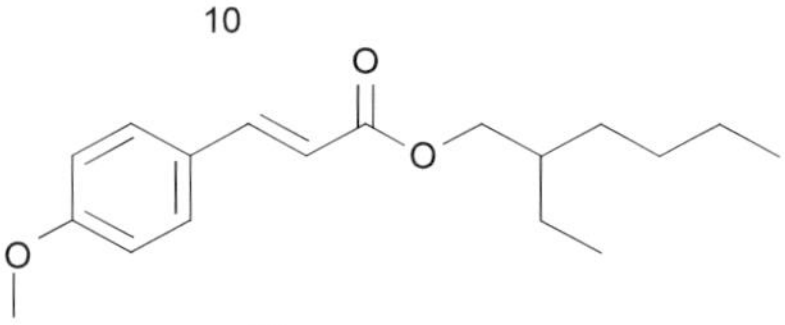
13

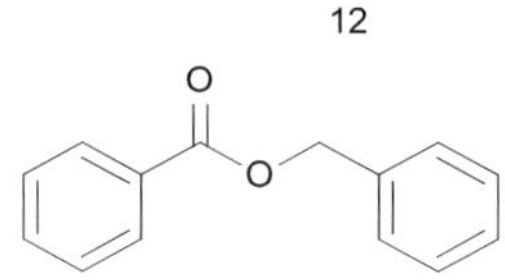
14

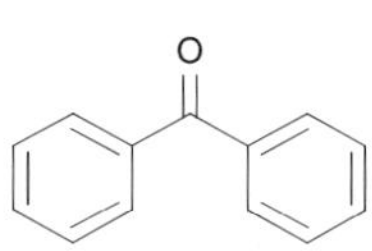
15

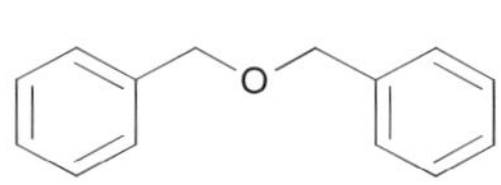
16

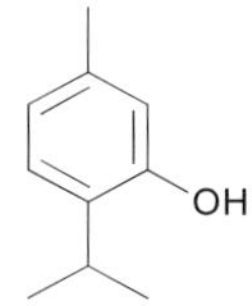
17


18

19

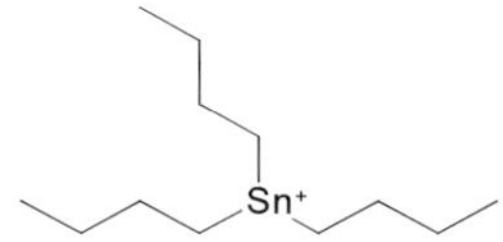
20

Molecular structures of the Havel/Spree contaminants discussed. These molecular marker compounds are numbered according to the numbers given in the text (#-numbers).

# 4 Fate and distribution of organic contaminants in riverine systems

## 4.1 Emission and subsequent distribution of musk fragrances in a riverine system

### 4.1.1 Polycylic aromatic musk compounds in sewage treatment plant effluents of Canada and Sweden ‡

**Introduction**

Polycyclic musk fragrances (PMF) are the second of two major classes of synthetic musks, after nitro musks. They are widely used in an array of fragrances for cosmetics and other personal and household care products and were commercially introduced in the 1950s (Daughton and Ternes, 1999). Worldwide, more than 5000 tons per year of synthetic musks are produced recently (Gebauer and Bouter, 1997). In Europe, the usage rate of the two most abundant compounds HHCB and AHTN exceeded 2000 tons $yr^{-1}$ (Balk & Ford, 1999a).

With respect to their environmental behaviour PMF are characterized as semi-volatile compounds, with a log $K_{ow}$ of 5.4 – 6.3 (Osemwengie and Steinberg, 2001). The structural feature of all polycyclic musks is an indane or tetraline skeleton, which is highly substituted mainly by methyl groups (Table 1). Polycyclic musks and nitrocyclic musks are classified as not readily biodegradable (RIZA 2001, cited in Simonich et al., 2002) and toxic (Swedish Society for Nature Conservation, 2000). After application, most of these materials are released as household effluents in sewage, and hence these compounds are detected in influent waters of sewage treatment plants (STP) or effluents of unprocessed sewage (Eschke et al., 1994; Simonich et al., 2000) and sewage sludge (Herren and Berset, 2000). They are not completely removed during wastewater treatment (e.g. Eschke et

‡ Mainly adapted/reprinted from Ricking et al., 2002b

al., 1994). Simonich et al. (2000) investigated concentrations of fragrance materials in sewage treatment plant influents and effluents in the U.S. and reported a mean HHCB and AHTN removal of 87.4 % and 86.2 %, respectively, during wastewater treatment. Very recently Simonich et al. (2002) published wastewater effluent concentrations in the µg/L range for the U.S and Europe, whereas it would appear that industrial sewage treatment plants are not emitting PMF in significant amounts (Verbruggen et al., 2000). Since residues of polycyclic musks are discharged via municipal treatment plant effluents into the receiving waters these compounds have been detected frequently in the aquatic environment (e.g., Franke et al., 1995; Müller et al., 1996; Heberer et al., 1999; Dsikowitzky et al., 2002).

Balk and Ford (1999a) observed that HHCB and AHTN were metabolised and excreted with a turnover rate of one third per day by bluegill sunfish (Lepomis macrochirus). The calculated lifetimes of HHCB and other fragrance materials in the atmosphere are a few hours or less due to daytime reaction with the OH radical (for a OH radical concentration of $2.0 \times 10^6$ molecule $cm^{-3}$) (Aschmann et al., 2001). These results indicate that HHCB and AHTN are altered by chemical and biological degradation processes after their release into the environment.

However, despite their moderate environmental stability Standley et al. (2000) reported their use as molecular tracers to track varying inputs of sewage treatments in river water and Ricking et al. (2003) in sediments. In surface water samples of urbanised areas they are found at concentrations up to the µg/L-level (Rimkus, 1999). Due to their high log $K_{ow}$ values, they are detected in sewage sludges, aquatic sediments and so far in organisms at the top of the aquatic food chain where they are found at concentrations of up to more than 100 mg/kg lipid in fish and mussels (Rimkus, 1999)

The ecotoxicological and toxicological risks of polycyclic musks are discussed continually. Considering the high volume emissions, the intensive usage, the bioaccumulation tendency and the widespread distribution, an ecological risk could be associated with these xenobiotics (Kevekordes et al., 1997; Balk and Ford, 1999b, Kallenborn et al., 1999)

Results of toxicity tests have revealed that direct exposure to HHCB and AHTN in fragrances or indirect exposure in the environment poses no significant risk to human health (Ford, 1998). It is nevertheless alarming that HHCB and AHTN in alcoholic solutions enter the human body via diffusion through the skin (Ford et al., 1999). Because of their lipophilicity, polycyclic musks are stored in the body and have been detected in human fat tissues and in human milk samples (Eschke et al.,

1995b). Possible pheromone-like and hormone-like interactions are expected by Kallenborn et al. (1999) and should not be neglected while estimating the ecotoxicological risk of the PMF.

The present chapter provides quantitative data on PMF in different wastewater treatment effluents in Canada and Europe. The intention of the study was to provide preliminary information on the variations within the emission pattern of polycyclic musks with respect to the discharge location. In that way water samples were analysed for an array of PMF (HHCB-Galaxolide® AHTN-Tonalide®, ATTI-Traseolide®, -AHMI-Phantolide®, -ADBI-Celestolide®, and DPMI-Cashmerane®) and the most prominent nitro musks (Musk Xylene-MX, and Musk Ketone–MK).

## Samples

*Canadian samples*

The Canadian samples were obtained in the provinces of Nova Scotia (Mill Cove Plant) and New Brunswick (Lancaster and Strawberry Marsh). All STPs receive mixed household/industry influents.

The Mill Cove plant is a conventional secondary sewage treatment plant with preliminary treatment, primary settling, activated sludge treatment, secondary settling and finally UV. Biosolids are treated by anaerobic digestion and transported to a lagoon. It has been in operation since 1969 and expanded in 1981, to treat 22.5 million litres per day. It is now capable of treating more than 31.5 million litres per day.

The Lancaster sample is from the Lancaster secondary sewage treatment plant of the city of Saint John, New Brunswick, but the plant treats only a fraction of the generated sewage. The Strawberry Marsh sample is from a similar plant of the city of the Miramichi in northern New Brunswick (previously called Newcastle).

*Swedish samples:*

The Swedish samples were taken from the provinces Uppsala (Enköping), Västra Götaland (Gässlösa, Nolhaga, Skene) and Gävleborg (Ljusne). All STPs receive mixed household/industry influents.

The Enköping sample was collected from the effluent of a STP with activated sludge treatment combined with $Fe_{III}$ pre-precipitation. The Ljusne sample was obtained from the effluents of a STP with avtivated sludge treatment/biosorption combined with Al post-precipitation. The Gässlösa, the Nolhaga and the Skene samples were obtained from effluents of STPs with trickling filtration/activated sludge and nitrogen removal combined with Al pre- and post-precipitation treatment, respectively.

The water samples were obtained according to the conventional sampling procedures. In Canada the water samples were taken in January 2002 directly out of the STP effluent water at a water depth of 1 m below water surface, stored at 4°C in the dark and shipped immediately to the laboratory in Aachen. The samples were composite samples where an aliquot was taken for analysis. The Swedish samples were collected during a survey for estrogenic compounds in June-December 1999 by means of a SPE extraction and stored frozen at –18°C until shipment to Berlin.

Limits of quantitative analyses were in the range of 0.5 ng/L (signal to noise ratio of 5:1 in real samples) in case of full scan analyses and for single-ion mode measurements. With respect to varying matrix influences no attempt was made to quantify components with concentrations less than 1 ng/L. The data on concentrations are recovery corrected according to spiking experiments with concentrations of 1 µg/L to 1.25 µg/L of the respective reference compounds (Dsikowitzky et al. 2002).

**Tab. 1:** Structure and physico-chemical parameters of musk fragrances. (Source: Simonich et al. 2000)

| Compounds structure | Trade name and CAS name including CAS number | $pK_{ow}$ | Water solubility (mg $L^{-1}$) | Ions for Quantification |
|---|---|---|---|---|
| 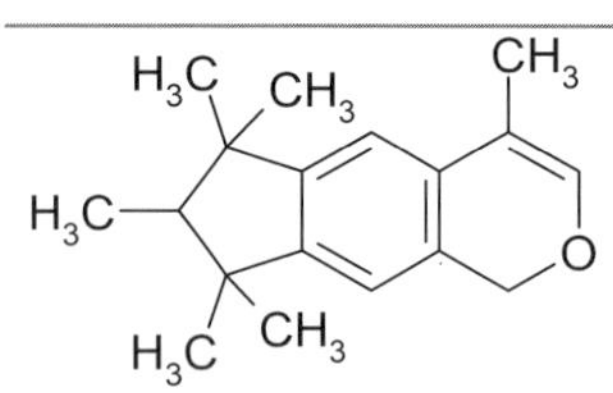 | Galaxolide<br>1,3,4,6,7,8-hexahydro-4,6,6,7,8,8-hexamethyl-cyclopenta-[g-2-benzopyran]<br>HHCB<br>1222-05-5 | 5.9 | 1.75 | 243, 258 |
| 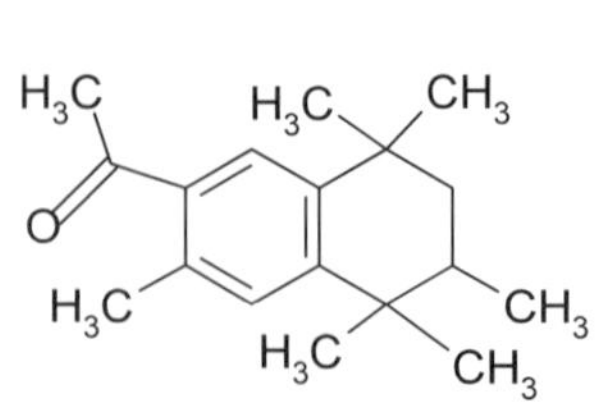 | Tonalide<br>1-(5,6,7,8-tetrahydro-3,5,5,6,8,8-hexamethyl-2-naphthenyl-ethanone<br>AHTN<br>1506-02-1 | 5.7 | 1.25 | 243<br>258, |

| Compounds structure | Trade name and CAS name including CAS number | $pK_{ow}$ | Water solubility ($mg\ L^{-1}$) | Ions for Quantification |
|---|---|---|---|---|
| 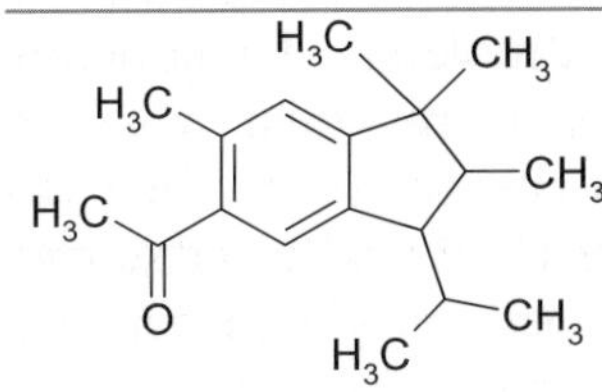 | Traseolide<br>1-[2,3-dihydro-1,1,2,6-tetramethyl-3-(1-methyl-ethyl)-ethanone<br>ATII<br>68140-48-7 | 6.3 | n.a. | 215, 216 |
| 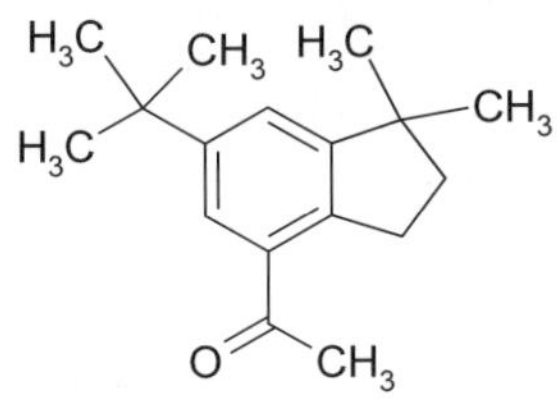 | Celestolide<br>1-[6-(1,1-dimethylethyl)-2,3-dihydro-1,1-methyl-1H-inden-4-yl]-ethanone<br>ADBI<br>13171-00-1 | 5.4 | n.a. | 229, 244 |
| 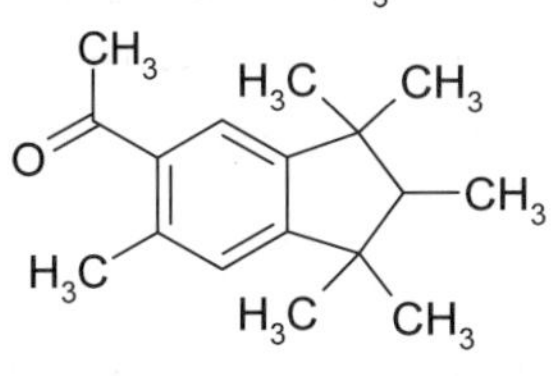 | Phantolide<br>1-(2,3-dihydro-1,1,2,3,3,6-hexamethyl-1H-inden-5-yl) ethanone<br>AHMI<br>15323-35-0 | 5.9 | n.a. | 229, 244 |
| 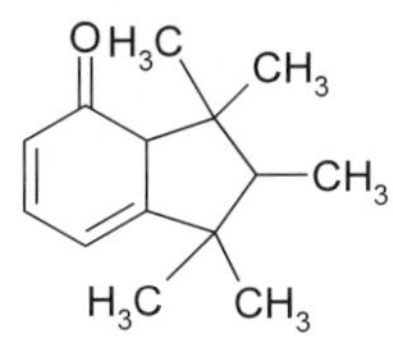 | Cashmeran<br>1,2,3,5,6,7-hexahydro-1,1,2,3,3-pentamethyl-4H-inden-4-one<br>DPMI<br>33704-61-9 | 5.9 | n.a. | 191, 192 |
| 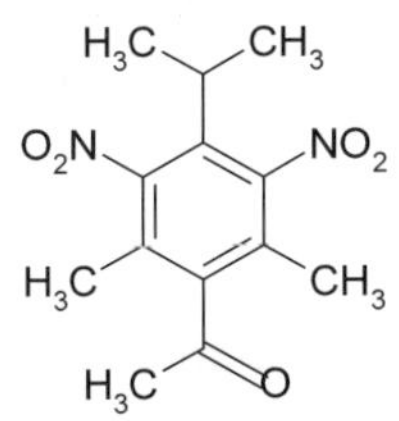 | Musk ketone<br>1-tert.-Butyl-3,5-dimethyl-2,6-dinitro-4-acetylbenzene<br>MK<br>81-14-1 | 4.3 | 1.9 | 279, 294 |
| 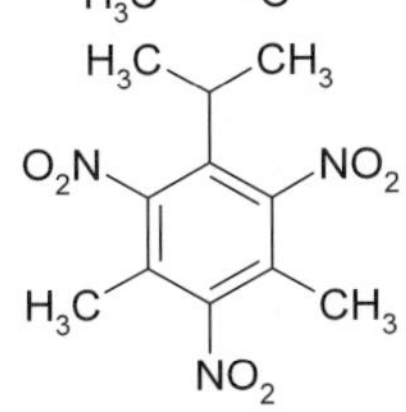 | Musk xylene<br>1-tert.-Butyl-3,5-dimethyl-2,4,6-trinitrobenzene<br>MX<br>81-15-2 | 4.8 | 0.49 | 282, 297 |

## Results and discussion

In all samples HHCB and AHTN (Table 1) were clearly detected, while ADBI, AHMI were also present, but close to the detection limit. The other PMF DPMI, ATTI,, and more surprisingly musk ketone and musk xylene were not detected in any sample analysed. These two compounds display the lowest log $K_{ow}$ of the target analytes and are expected to partition more into the water than sediments or particulates, in comparison to the other targets that would prefer sediments more. Therefore, their absence indicates their lower concentration and environmental use prior to discharge, or their higher degradation relative to HHCB and AHTN. The concentrations are given in Table 2 along with some published data for comparison. As a continuous decline of these substances can be observed in the aquatic environment for several years (Rimkus 1999), only the biotransformation products, especially for musk ketone (Kallenborn et al. 1999) are relevant and should be addressed in future works.

In the Canadian samples the concentrations ranged between HHCB 205 -1300 n/L, AHTN 110-520 ng/L, ADBI 4–19 ng/L, and AHMI 2–6 ng/L in the effluent water samples. The highest concentrations for all compounds were measured in the effluents from the Mill Cove Plant in Bedford, Nova Scotia. The HHCB/AHTN ratios were close in the range of 1.86 – 2.50.

In the Swedish samples the concentrations ranged between HHCB 157-423 ng/L, AHTN 42-104 ng/L, ADBI 2–8 ng(L, and AHMI 2–5 ng/L in the effluent water samples. The highest concentrations for HHCB and AHTN were measured in the effluents from the Gässlösa STP. The concentrations of ADBI and AHMI were close to the detection limit. The HHCB/AHTN ratios were in the range of 3.7 – 5.3.

In comparison the samples from Canada were contaminated at a higher level than the Swedish samples, by a maximum factor of ~ 10 for HHCB and AHTN. There was no correlation trend with the population and applied STP technology observed. Noteworthy in contrast to Paxéus 1996, no MK was detected in any of the samples.

Compared with published data the concentrations are in the moderate to high level relative to the Berlin area (Fromme et al. 2001); far below the maximum concentrations measured by Heberer et al. (1999); higher than the concentrations reported from the Lippe river and the first results presented by Osemwengie and Steinberg (2001). Furthermore the HHCB concentrations in this study were a factor of 4-10 lower compared to data reported by Paxeus (1996). These differences may be due to the semi-quantitative calculation by Paxéus.

With respect to the nitro musks we detected no compounds neither in Swedish nor in Canadian samples. This is slightly in agreement with Simonich et al. (2000, 2002) who reported concentrations of HHCB and AHTN dominating by far those of MK and MX. In contrast to published data for biota samples from the same Canadian province (Gatermann et al. 1999) which revealed musk ketone concentrations several-fold orders of magnitude higher than the concentrations for the predominant polycyclic compound (HHCB), we detected no nitro musks in the Canadian samples. Our water was collected from the effluents of a secondary treatment plant, while the mussels were collected close to numerous raw sewage effluents of the same harbour, but more than 10 km away from the STP. These preliminary results tend to demonstrate the advantage of the secondary STP, and need pursuing.

Ratios of dissolved HHCB/AHTN are included in Table 2. The ratios for the Canadian samples (1.9 to 2.5) matched closely those reported for the North Sea, and other remote sampling locations and the US (Bester et al. 1998a; Heberer et al. 1999; Osemwengie and Steinberg 2001), while those for the Swedish samples (3.7 to 5.3) were closer to highly contaminated samples from Berlin (Heberer et al. 1999; Fromme et al. 2001) and other STP-effluents (Dsikowitzky et al. 2002).

The median HHCB/AHTN ratio in Lippe river water was 2.9 (Dsikowitzky et al., 2002), close to the HHCB/AHTN ratio of 2.5, which was deduced from the 1995 European use volumes (Balk and Ford, 1999a). However, in water samples from the Elbe river, the Saale river and the Mulde river, Germany, lower median HHCB/AHTN ratios were determined, 1.4 to 1.8 (data compiled by Rimkus, 1999). Apparently, HHCB/AHTN ratios are subject to fluctuations, since Simonich et al. (2000) have shown that the percent removal of both compounds during wastewater treatment in the STPs analysed was very similar.

**Tab. 2:** Results of the analysis of PMF in samples from Canada and Sweden and data for comparison (data are given in ng/L).

| | Cashmeran (DPMI) | Celestolide (ADBI) | Phantolide (AHMI) | Galaxolide (HHCB) | Traseolide (ATTI) | Tonalide (AHTN) | Musk Xylene | Musk Ketone | Ratio HHCB/AHTN | References |
|---|---|---|---|---|---|---|---|---|---|---|
| Enköping (Sweden) (population 21000) | <1 | 7 | 4 | 336 | <1 | 90 | <1 | <1 | 3.7 | 1 |
| Skene (Sweden) (population 17280) | <1 | 3 | 2 | 218 | <1 | 42 | <1 | <1 | 5.1 | 1 |
| Gässlösa (Sweden) (population 79000) | <1 | 6 | 5 | 423 | <1 | 104 | <1 | <1 | 4.1 | 1 |
| Nolhaga (Sweden) (population equiv. 39500) | <1 | 2 | 2 | 157 | <1 | 42 | <1 | <1 | 3.8 | 1 |
| Ljusne (Sweden) (population 2596) | <1 | 8 | 3 | 407 | <1 | 77 | <1 | <1 | 5.3 | 1 |
| Strawberry Marsh ATP (Canada) (population 20000) | <1 | 7 | 2 | 480 | <1 | 220 | <1 | <1 | 2.2 | 1 |
| Lancaster (Canada) (population 90000) | <1 | 4 | 2 | 205 | <1 | 110 | <1 | <1 | 1.9 | 1 |
| Mill Cove Plant Bedford (Canada) (population 350000) | <1 | 19 | 6 | 1300 | <1 | 520 | <1 | <1 | 2.5 | 1 |
| Stockholm, Göteborg, Malmö (Sweden) | n.m. | n.m. | n.m. | 1000 - 6000 | n.m. | n.m. | n.m. | 1000-5000 | - | 2 |
| Lippe (Germany) | n.m. | <10-20 | <10-60 | 50 - 1400 | n.m. | 10 - 360 | n.m. | n.m. | 2.0-5.0 | 3 |
| Havel river - Berlin mean moderate to high level (Germany) | n.m. | <20-20 | <10-70 | 230 - 1590 | <10-70 | 70 - 530 | n.m. | n.m. | 3.0-3.3 | 4 |
| Elbe Estuary and German Bight (North Sea) (Germany) | n.m. | n.m. | n.m. | 0.09 - 95 | n.m. | 0.08 -67 | n.m. | n.m. | 1.1-1.9 | 5 |
| Berlin area (Germany) | n.m. | 20-410 | n.m. | 30 - 12500 | n.m. | 40 - 6800 | <10 | <5-390 | 0.8-4.6 | 6 |
| Hamm STP influent (Germany) | n.m. | 20 | 20 | 970 | n.m. | 320 | n.m. | n.m. | 3.0 | 3 |
| Hamm STP effluent (Germany) | n.m. | <10 | 60 | 1400 | n.m. | 360 | n.m. | n.m. | 3.9 | 3 |
| Sewer Schönerlinde (Germany) | n.m. | 410 | n.m. | 10800 | n.m. | 5800 | <10 | 320 | 1.9 | 6 |
| Sewage plant effluent (Germany) sample 06/24/97 | n.m. | n.m. | n.m. | 4200 | n.m. | 1900 | n.m. | n.m. | 2.2 | 7 |
| Sewage plant effluent (Germany) sample 07/29/97 | n.m. | n.m. | n.m. | 3700 | n.m. | 1700 | n.m. | n.m. | 2.2 | 7 |

| | Cashmeran (DPMI) | Celestolide (ADBI) | Phantolide (AHMI) | Galaxolide (HHCB) | Traseolide (ATTI) | Tonalide (AHTN) | Musk Xylene | Musk Ketone | Ratio HHCB/AHTN | References |
|---|---|---|---|---|---|---|---|---|---|---|
| Sewage pond influent (Germany) sample 06/24/97 | n.m. | n.m. | n.m. | 3600 | n.m. | 1500 | n.m. | n.m. | 2.4 | 7 |
| Sewage pond effluent (Germany) sample 07/29/97 | n.m. | n.m. | n.m. | 1700 | n.m. | 640 | n.m. | n.m. | 2.7 | 7 |
| STP effluent (USA) | <DL | 0.3-2.1 | 2.4-5 | 40.8-152 | <DL -126 | 36.8 - 92.2 | <DL -1.3 | <DL-27.5 | 1.1-2.1 | 8 |
| STP influent (USA) (n=12) | n.m. | n.m. | n.m. | 16600 ± 10400 | n.m. | 12500 ± 7350 | n.m. | n.m. | n.c. | 9 |
| STP effluent (primary gravitational settling and activated sludge (USA) (n=4) | n.m. | n.m. | n.m. | 2053 ± 1314 | n.m. | 1326 ± 270 | n.m. | n.m. | n.c. | 9 |
| STP effluent (primary gravitational settling and activated sludge (EU) (n=1) | n.m. | n.m. | n.m. | 4620 | n.m. | 1440 | n.m. | n.m. | 3.2 | 9 |
| STP effluent (primary gravitational settling and carousel) (EU) (n=2) | n.m. | n.m. | n.m. | 1065 | n.m. | 1235 | n.m. | n.m. | 0.9 | 9 |
| STP effluent (primary gravitational settling and oxidation ditch) (USA) (n=2) | n.m. | n.m. | n.m. | 1495 | n.m. | 1010 | n.m. | n.m. | 1.5 | 9 |
| STP effluent (primary gravitational settling and ticking filtering) (USA) (n=3) | n.m. | n.m. | n.m. | 2056 ± 655 | n.m. | 1555 ± 522 | n.m. | n.m. | n.c. | 9 |
| STP effluent (primary gravitational settling and ticking filtering) (EU (n=2) | n.m. | n.m. | n.m. | 2400 | n.m. | 1645 | n.m. | n.m. | 1.5 | 9 |
| STP effluent (primary gravitational settling and rotating biological contractor (USA) (n=1) | n.m. | n.m. | n.m. | 2210 | n.m. | 1710 | n.m. | n.m. | 1.3 | 9 |

References: 1 - this study; 2 - Paxéus 1996; 3 - Dsikowitzky et al. 2002; 4 - Fromme et al. 2001 ; 5 - Bester et al. 1998a; 6 - Heberer et al. 1999; 7 - Gattermnann et al. 2002 ; 8 - Osemwengie & Steinberg 2001; 9 - Simonich et al. 2002

(DL means Methods detection limit in Osemwengie & Steinberg 2001; The number means the average and the ± means the SD in Simonich et al. 2002; nm = not measured; nc = not calculated)

However, the type of treatment, may it be biological, chemical, or a combination could have a different outcome. Variations of HHCB/AHTN ratios were possibly a result of ongoing supply of HHCB and AHTN with variable ratios or a higher biodegradability of HHCB (Postma et al., 1996). Based on different influent concentrations in the US and Europe (see Table 2; Simonich et al., 2002) we speculate on processes during the transport within the aquatic system, as the ratios in effluent samples are different from those reported for aquatic systems.

Parallels of the Swedish samples were also tested for estrogenicity and also exposed fish were checked for hormonal effects, as musk substances could perhaps interfere with the receptor tests that were applied. There was no significant positive correlation ($r = -0.40$) between the sum of the musks and estrogenicity in the same five investigated Swedish municipal effluents, reported earlier (Svenson et al., 2002). The synthetic contraceptive ethinylestradiol and the natural estrogenic steroids, estradiol and estrone, have so far been found to dominate the municipal effluent estrogenicity (Desbrow et al. 1998, Belfroid et al., 1999, Baronti et al., 2000, A. Svenson unpublished data). As first published data on the estrogenicity of HHCB and AHTN indicated a weak estrogenic activity (Seinen et al., 1999), it is not likely that these substances contribute substantially to the estrogenicity in municipal effluents.

These musk substances had no mutagenicity in *Salmonella typhimurium* tests (Mersch-Sundermann et al., 1998), but caused an acute hepatic damage in rats (Steinberg et al., 1999). The very recently published calculated PEC (predicted environmental concentration) values, based on a conservative approach, are in the range of 3.0 µg/L for AHTN and 7.2 µg/L for HHCB (Salvito et al., 2002). These concentrations are less than a factor of 10 higher than the concentrations measured in our samples. The conservative approach disregards additive and synergistic effects due to other compounds emitted via STPs. It is therefore essential to monitor these substances and to study the environmental fate, especially in sediments, and potential toxic effects of single or mixed musk compounds on aquatic ecosystems.

## 4.1.2 Distribution of polycyclic musks in water and particulate matter of the Lippe river (Germany)‡

### Introduction

The distribution of organic compounds in aquatic environments is complex and determined by the dynamics of a number of partition and transport processes. Partition and transport processes are controlled mainly by physico-chemical properties of compounds as well as by the environmental setting. The present study was conducted to investigate the fate of low-molecular weight organic chemicals which are discharged into the river and partition between water and particulate phases. In order to improve the understanding of the relevant mechanisms and processes, the distribution of polycyclic synthetic musk substitutes was monitored in water and sediments of the Lippe river (a tributary to the Rhine river, Germany).

Since the 1950s, there is a widespread use of synthetic polycyclic musks as fragrances in cosmetics and detergents. After application, most of these materials are released to the sewer. Hence, polycyclic musks are present in influents of sewage treatment plants (STP) (Eschke et al., 1994; Simonich et al., 2000) as well as in sewage sludge (Herren & Berset, 2000). They are not completely removed during wastewater treatment (e.g. Eschke et al., 1994) and are therefore frequently found in river water (e.g. Franke et al., 1995; Müller et al., 1996; Heberer et al., 1999). Although polluted river water is highly diluted in the marine environment, PMF have still been detected in estuarine and marine waters from the German Bight (Bester et al., 1998). Concentrations were in the range of 0.09 – 4.8 ng/L for HHCB and of 0.08 – 2.6 ng/L for AHTN, respectively, with increasing concentrations between 1990-1995. Verbruggen et al. (2000) reported in a survey of all compartments of the Netherlands aquatic ecosystem SPE-extractable fractions of PMF in all samples, even in the North Sea.

Polycyclic musks are not only present in the aqueous phase. They also tend to be adsorbed to suspended particulate matter (Winkler et al., 1998) and accumulate in riverine sediments (Lach & Steffen, 1997). Additionally, Eschke et al. (1994, 1995a) reported the presence of synthetic musks in liver, muscle and fat tissues of different riverine fish species. The compounds were also found in marine invertebrates like North Sea shrimps (Rimkus & Brunn, 1996).

---

‡ Mainly adapted/reprinted from Dsikowitzky et al., 2002

The presence of HHCB and AHTN in fish and mussels was confirmed by enantioselective analysis (Franke et al., 1999). In the same study, the authors reported the occurrence of a transformation product of HHCB, the HHCB-lactone, found in Odra river and Elbe river water (Germany).

So far, the occurrence of seven polycyclic musks has been reported. Their chemical names, chemical structures and molecular formulae are described elsewhere (Rimkus, 1999). The structural feature of all polycyclic musks is an indane or tetraline skeleton, which is highly substituted mainly by methyl groups (Fig. 1). 7-Acetyl-1,1,3,4,4,6-hexamethyl-1,2,3,4-tetrahydronaphthalene (AHTN, trade name Tonalide) and 1,3,4,6,7,8-hexahydro-4,6,6,7,8,8-hexamethylcyclopenta[g]-2-benzopyrane (HHCB, trade name Galaxolide) are the most abundant ones. In this study, also 4-acetyl-1,1-dimethyl-6-*tert*-butylindane (ADBI, trade name Celestolide) and 6-acetyl-1,1,2,3,3,5-hexamethylindane (AHMI, AHDI, trade name Phantolide) are investigated (Fig. 1). Important physico-chemical properties of these compounds, which determine their environmental distribution and transport (e.g. solubility), are presented in chapter 4.1.1, in Ricking et al. (2003) as well as in Simonich et al. (2000).

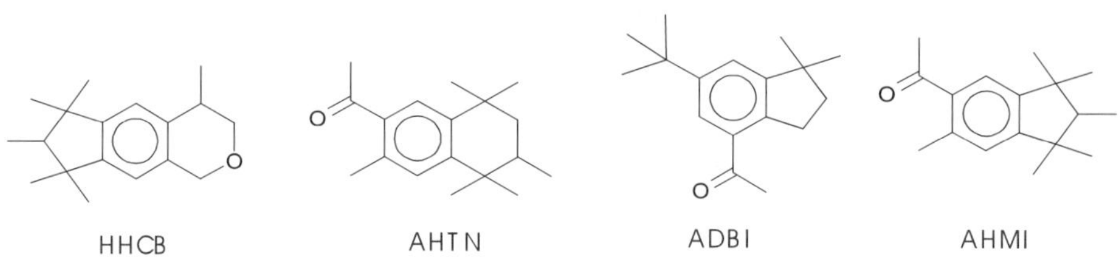


**Fig. 1:** Chemical structures of the investigated polycyclic musks HHCB ($C_{18}H_{26}O$), AHTN ($C_{18}H_{26}O$), ADBI ($C_{17}H_{24}O$) and AHMI ($C_{17}H_{24}O$).

The Lippe river is located in a densely populated, highly industrialised area with many point-sources of sewage effluents. Preliminary screening analyses of low-molecular weight organic compounds showed the presence of polycyclic musks in all water and nearly all surface sediment samples. As mentioned before, polycyclic musks are steadily discharged into the river by sewage effluents. Therefore, they are particularly suitable to observe the dynamic transport and partition processes within and between aqueous and particulate phases which sewage-derived organic chemicals undergo in a river system.

## Samples

Samples of river water and sediments were taken on 16 - 17 August 1999 at 19 sampling sites on a longitudinal section of the Lippe river, North-Rhine Westphalia, Germany (see Fig. 2). Water samples were scooped up from below the water surface at midstream and bottled in pre-cleaned glass flasks. Filled sample flasks were sealed free of air bubbles with glass stoppers. They were stored in the darkness at a temperature of approximately 4 °C. Relevant data of river runoff were supplied by the Environmental Agency of North-Rhine Westphalia (LUA NRW), the Wasserschiffahrtsamt Duisburg and the Emschergenossenschaft Lippeverband, Germany. The corresponding surface sediments of nearby river banks were sampled to a depth of approximately 5 cm. At sites 2 and 7 sediments were lacking due to river bank stabilization. Thus, sample collection was impossible at these locations.

On 22 August 1999 influent and effluent from a sewage treatment plant (STP) of the city of Hamm was sampled. Hamm is the largest city on the Lippe river (182,000 inhabitants) and marks the beginning of the densely populated lower reaches. Therefore, these samples represent a typical emission source for polycyclic musks into the Lippe river. The STP samples were extracted immediately after sampling.

Suspended particulate matter (SPM) from the river mouth was taken on 14 November 2000 by the LUA NRW. It was collected onboard the *R/V Max Prüss* using a centrifuge (Z61V4A, 16000 rpm, Carl Padberg Zentrifugenbau, Germany). All fresh, wet samples of particulate matter were kept in glass flasks with Teflon®-coated screw caps and stored in the darkness at a temperature of approximately 4 °C.

**Tab. 1:** Ions used for quantitation and recoveries of HHCB, AHTN, AHMI and ADBI from high-purity water (n = 4) and a pre-extracted sediment (n = 4). The data represent average recovery rates and relative standard deviations.

| Compound | Quantitation | % Recovery rate | |
|---|---|---|---|
| | Ions m/z | Water | Sediment |
| HHCB | 243; 258 | 94 ± 7 | 43±16 |
| AHTN | 243; 258 | 93 ± 7 | 46±17 |
| AHMI | 229; 244 | 88 ± 5 | 43±18 |
| ADBI | 229; 244 | 87 ± 3 | 49±16 |

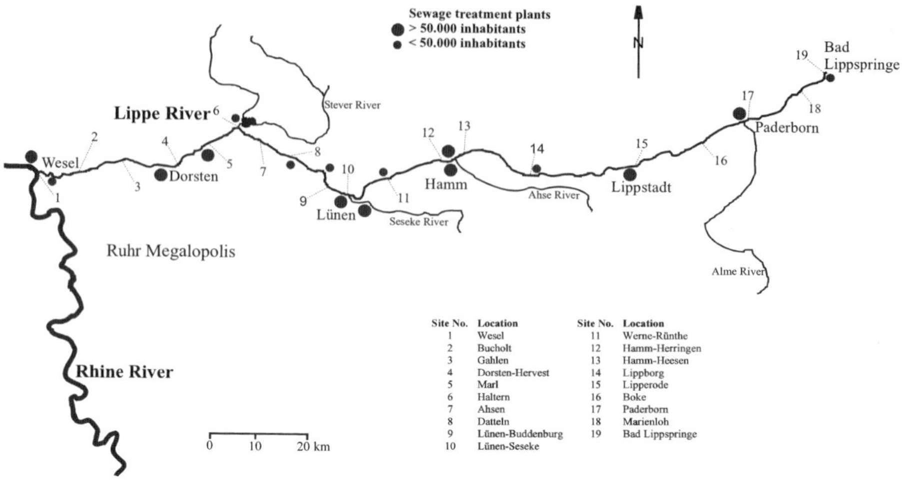


**Fig. 2:** Map showing the Lippe river system (North-Rhine Westphalia, Germany) with sampling locations (sites 1 to 19) and sewage treatment plants.

Identification of polycyclic musks was verified by comparison of the gaschromatographic retention times and mass spectra with those of the reference compounds. Quantitative data were obtained by integration of specific ion chromatograms extracted from the total ion current. The ions used for quantification were: m/z 243 and 258 for HHCB and AHTN and m/z 229 and 244 for AHMI and ADBI (see Table 1). The limit of quantitation of the applied method obtained from recovery experiments was in the range of 5 ng/L for water samples and in the range of 1 μg/kg for sediment samples. Considering the influence of varying matrices on the limit of quantitation, no attempts were made to quantify components at concentrations less than 10 ng/L in river water samples and, accordingly, less than 2 μg/kg in sediment samples.

## Results

### *Water samples*

GC-MS analyses of Lippe river water and sewage samples (STP Hamm) revealed the presence of four polycyclic musks (Table 2). HHCB and AHTN were detected in all river water samples with concentrations ranging from <10 to 180 ng/L and <10 to 70 ng/L, respectively. Generally, HHCB concentrations exceeded those of AHTN with HHCB/AHTN ratios varying from 1.8 to 5.0. Full-scan mass spectra of ADBI and AHMI were only obtained for five river water samples. However, their concentrations were always below the limit of quantitation (< 10 ng/L). In STP influent and effluent, 970 and 1400 ng/L HHCB were measured, respectively, whereas concentrations of AHTN were lower (320 and 360 ng/L). AHMI was also detected in both STP samples (20 and 60 ng/L). In contrast, ADBI was only found in the STP influent (20 ng/L). Influent and effluent samples were taken at the same time. During biological treatment sewage water is retained for several hours. Due to this time lag the effluent and influent sampled at the same time cannot be compared directly. The fact that musk concentrations in the influent were equal to or higher than in the effluent is attributed to fluctuations in the influent concentrations. In contrast to Simonich et al. (2000), no recovery experiments with the STP samples were performed so that matrix effects were not taken into account.

In Figure 3, concentrations of HHCB and AHTN along the river are shown. At the river source (site 19), the HHCB concentration was below the limit of quantitation (<10 ng/L). At site 18, which is only about two kilometres away from the river source, the concentration of HHCB was more than ten times higher (140 ng/L). In the remaining samples of the upper reaches (sites 17 to 14), HHCB concentrations were in the range of 50 to 90 ng/L. Site 13 marks the beginning of the densely populated lower reaches. From site 13 to site 9, HHCB concentrations were in the range of 90 to 120 ng/L. From site 8 to site 5, the highest concentrations were dctccted (about 170 ng/L). Downstream site 5, in the area near the river mouth, HHCB concentrations were lower and varied between 60 and 120 ng/L. AHTN concentrations were lower than HHCB concentrations, but they showed similar variations along the river.

**Tab. 2:** Concentrations of HHCB, AHTN, ADBI and AHMI and HHCB/AHTN ratios in water samples from the Lippe river (site numbers see Fig. 2) and in two samples from Hamm sewage treatment plant.

| Site No. | HHCB [ng/L] | AHTN [ng/L] | ADBI [ng/L] | AHMI [ng/L] | HHCB/AHTN ratio |
|---|---|---|---|---|---|
| 1 | 80 | 40 | <10 | <10 | 2.0 |
| 2 | 70 | 40 | <10 | <10 | 1.8 |
| 3 | 120 | 50 | <10 | <10 | 2.4 |
| 4 | 60 | 30 | <10 | <10 | 2.0 |
| 5 | 160 | 50 | <10 | <10 | 3.2 |
| 6 | 170 | 60 | <10 | <10 | 2.8 |
| 7 | 170 | 70 | <10 | <10 | 2.4 |
| 8 | 180 | 70 | <10 | <10 | 2.6 |
| 9 | 100 | 30 | <10 | <10 | 3.3 |
| 10 | 110 | 50 | <10 | <10 | 2.2 |
| 11 | 120 | 30 | <10 | <10 | 4.0 |
| 12 | 90 | 20 | <10 | <10 | 4.5 |
| 13 | 100 | 20 | <10 | <10 | 5.0 |
| 14 | 50 | 20 | <10 | <10 | 2.5 |
| 15 | 50 | 20 | <10 | <10 | 2.5 |
| 16 | 90 | 40 | <10 | <10 | 2.3 |
| 17 | 50 | 10 | <10 | <10 | 5.0 |
| 18 | 140 | 60 | <10 | <10 | 2.3 |
| 19 | <10 | <10 | <10 | <10 | - |
| STP influent | 970 | 320 | 20 | 20 | 3.0 |
| STP effluent | 1400 | 360 | <10 | 60 | 3.9 |

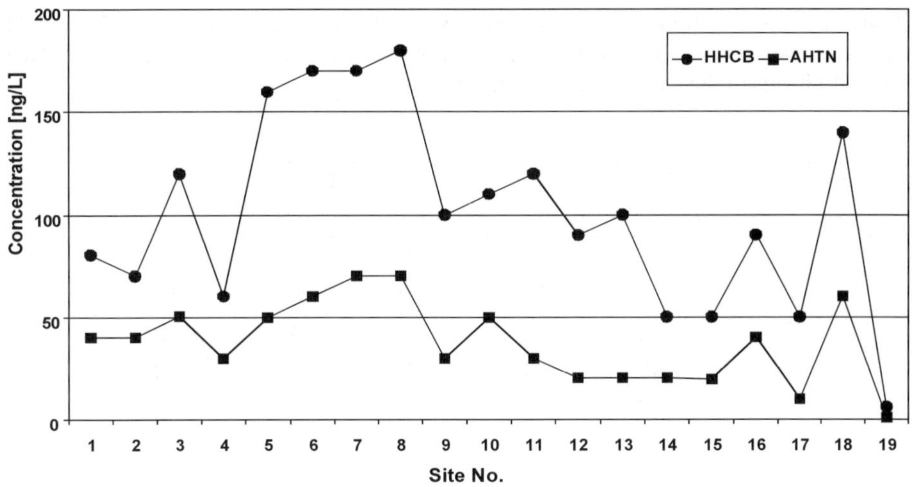


**Fig. 3:**. Concentration profile of HHCB and AHTN detected in water samples from the Lippe river source (site 19) to the river mouth (site 1).

### *Organic load of HHCB and AHTN*

Table 3 shows a compilation of river runoff data from several gauges (LUA NRW) along the Lippe river, which correspond to the accompanying sampling sites. In Hamm the Lippe river is connected to the western German channel system. Therefore, at site 12 (downstream weir) river runoff is lower than at site 13 (upstream weir).

The organic load ***L*** of dissolved HHCB and AHTN was determined on the basis of compound concentrations **C** (ng/L) in water and river runoff data **R** ($m^3/s$) and is here expressed in g/d:

$$\mathbf{L} = \mathbf{C} * \mathbf{R} * 0.0864.$$

The load of dissolved HHCB and AHTN along the river for the sampling period is plotted in Figure 4. It ranged from 3 to 293 g/d and from 1 to 108 g/d, respectively. Near the river source (sites 18 and 17), low HHCB loads were determined (5 and 3 g/d). Further downstream (sites 16 to 14), loads increased (34 to 47 g/d). From site 13, the beginning of the lower reaches, to site 9, the loads of HHCB were higher and range from 78 to 124 g/d. Maximum values resulted for the river between site 8 and site 5 (218 to 293 g/d). In the area near the river mouth, there was a strong variation between a high value at site 3 and low values (around 140 g/d at sites 4, 2 and 1). The variations of AHTN loads along the river showed a pattern similar to those of HHCB.

**Tab. 3:** River runoff data from several gauges along the Lippe river. The loads of dissolved HHCB and AHTN were calculated on the basis of river runoff and water concentrations at the corresponding sampling sites.

| Site No. | Date 1999 | Gauge | river runoff [$m^3/g$] | Load [g/d] HHCB | Load [g/d] AHTN |
|---|---|---|---|---|---|
| 1 | 16th Aug | Fusternberg [a)] | 23.0 | 159 | 79 |
| 2 | 16th Aug | Fusternberg [a)] | 23.0 | 139 | 79 |
| 3 | 16th Aug | Schermbeck [b)] | 24.8 | 258 | 108 |
| 4 | 16th Aug | Schermbeck [b)] | 24.8 | 129 | 64 |
| 5 | 16th Aug | Hüls [a)] | 19.0 | 263 | 82 |
| 6 | 16th Aug | Haltern [b)] | 19.9 | 293 | 103 |
| 7 | 16th Aug | Leven [b)] | 16.1 | 236 | 97 |
| 8 | 16th Aug | Vinnum [a)] | 14.0 | 218 | 85 |
| 9 | 16th Aug | Lünen [a)] | 13.0 | 112 | 34 |
| 10 | 16th Aug | Lünen [a)] | 13.0 | 124 | 56 |
| 11 | 17th Aug | Rünthe [a)] | 9.0 | 93 | 23 |
| 12 | 17th Aug | Hamm downstream weir [c)] | 10.0 | 78 | 17 |
| 13 | 17th Aug | Hamm upstream weir [c)] | 12.0 | 104 | 21 |
| 14 | 17th Aug | Kesseler [b)] | 10.9 | 47 | 19 |
| 15 | 17th Aug | Lippstadt Pegel 1 + Nordumflut [b)] | 7.9 | 34 | 13 |
| 16 | 17th Aug | Bentfeld [b)] | 5.2 | 41 | 18 |
| 17 | 17th Aug | Neuhaus [b)] | 0.8 | 3 | 1 |
| 18 | 17th Aug | Lippspringe [b)] | 0.4 | 5 | 2 |
| 19 | 17th Aug | river source | - | - | - |

a) Preliminary runoff data. verification pending; Emschergenossenschaft Lippeverband. Germany
b) Environmental Protection Agency of North-Rhine Westphalia. Germany (LUA NRW)
c) Wasserschiffahrtsamt Duisburg

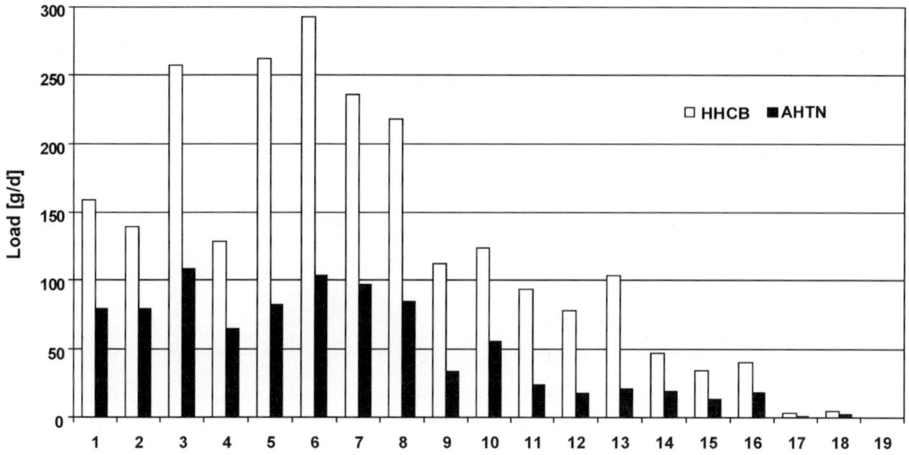


**Fig. 4:** Load of dissolved HHCB and AHTN along the Lippe river, which was calculated on the basis of compound concentrations in water and suitable river runoff data (Table 2).

*Sediment samples and suspended particulate matter*

Investigation of Lippe river sediments and suspended particulate matter showed the occurrence of only three polycyclic musk substitutes. Table 4 presents concentrations of HHCB, AHTN and AHMI. The raw data set was normalized to TOC content (µg/g TOC, Table 5), in order to take into account variable input of organic matter along the river. In Lippe river sediments, the detected concentrations of HHCB and AHTN varied from 1.3 to 15.2 µg/g and from 0.5 to 77.7 µg/g, respectively. In suspended particulate matter from the river mouth, 6.7 µg/g of HHCB and 11.0 µg/g of AHTN were measured. AHMI was present in suspended particulate matter (0.8 µg/g) and in two sediment samples (1.1 µg/g at site 3 and 3.3 µg/g at site 1).

**Tab. 4:** Concentrations of HHCB, AHTN and AHMI (normalized to dry sample weight) in Lippe river sediments and suspended particulate matter (SPM).

| Site No. | HHCB [µg/kg] | AHTN [µg/kg] | AHMI [µg/kg] |
|---|---|---|---|
| 1 | 191 | 1399 | 60 |
| 2 | - [a)] | - | - |
| 3 | 30 | 104 | 5 |
| 4 | 99 | 107 | <2 |
| 5 | 39 | 58 | <2 |
| 6 | 94 | 127 | <2 |
| 7 | - [a)] | - | - |
| 8 | 121 | 86 | <2 |
| 9 | 41 | 41 | <2 |
| 10 | 10 | 7 | <2 |
| 11 | <2 | <2 | <2 |
| 12 | 109 | 36 | <2 |
| 13 | <2 | <2 | <2 |
| 14 | 14 | 9 | <2 |
| 15 | 14 | 10 | <2 |
| 16 | <2 | <2 | <2 |
| 17 | 7 | 2 | <2 |
| 18 | 5 | 4 | <2 |
| 19 | <2 | <2 | <2 |
| SPM | 502 | 823 | 60 |

a) samples could not be collected

In Figure 5, a TOC-normalized concentration profile of HHCB and AHTN along the river is plotted. In this case, AHTN concentrations showed partly different trends than HHCB and are therefore also discussed in detail. In sediments of the upper reaches from site 19 to site 9 both compounds had relatively low concentrations (≤ 2.6 µg/g HHCB and ≤ 2.6 µg/g AHTN). At sites 16, 13, 11 and at the river source (site 19) they were not even detectable. An exception is site 18, where higher concentrations were observed (4.7 µg/g HHCB and 4.2 µg/g AHTN). However, the TOC content of this sediment was at the detection limit, so that these values should be regarded with caution. Concentrations of HHCB and AHTN in sediments of the lower reaches (downstream of site 9) were higher than in the upper reaches, although large variations occurred. The highest concentration of HHCB was measured at site 8 (15.2 µg/g). At the upper reaches, especially in the area near the river mouth, the highest concentrations of AHTN were observed (up to 77.7 µg/g). At sites 6, 5, 4, 3 and 1, AHTN concentrations even exceeded those of HHCB, which is reflected by a HHCB/AHTN ratio <1 (Table 5). HHCB/AHTN ratios in the other sediment samples varied between 1.0 and 3.1 and did not show a particular trend along the river. In suspended particulate matter from the river mouth the ratio was 0.6.

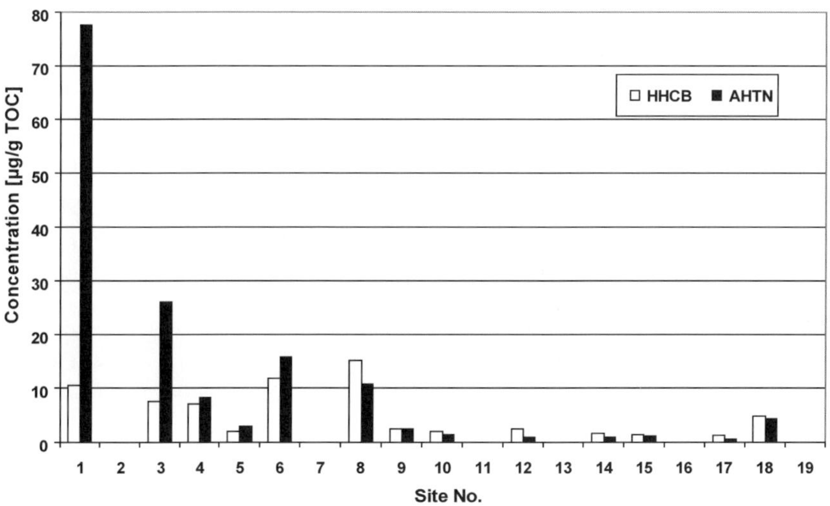


**Fig. 5:** Concentration profile of HHCB and AHTN detected in surface sediments from the Lippe river source (site 19) to the river mouth (site 1).

**Tab. 5:** Concentrations of HHCB, AHTN, ADBI and AHMI (normalized to TOC content) and HHCB/AHTN ratios in Lippe river sediments and suspended particulate matter (SPM).

| Site No. | TOC [%] | HHCB [µg/g TOC] | AHTN [µg/g TOC] | AHMI [µg/g TOC] | HHCB/AHTN ratio |
|---|---|---|---|---|---|
| 1 | 1.8 | 10.6 | 77.7 | 3.3 | 0.1 |
| 2 | -[a)] | - | - | - | - |
| 3 | 0.4 | 7.5 | 26.1 | 1.1 | 0.3 |
| 4 | 1.4 | 7.1 | 8.2 | <LOQ | 0.9 |
| 5 | 2.0 | 2.0 | 2.9 | <LOQ | 0.7 |
| 6 | 0.8 | 11.8 | 15.9 | <LOQ | 0.7 |
| 7 | -[a)] | - | - | - | - |
| 8 | 0.8 | 15.2 | 10.7 | <LOQ | 1.4 |
| 9 | 1.6 | 2.6 | 2.6 | <LOQ | 1.0 |
| 10 | 0.5 | 2.0 | 1.4 | <LOQ | 1.4 |
| 11 | 3.4 | <LOQ | <LOQ | <LOQ | <LOQ |
| 12 | 4.4 | 2.5 | 0.8 | <LOQ | 3.1 |
| 13 | 0.2 | <LOQ | <LOQ | <LOQ | <LOQ |
| 14 | 0.9 | 1.5 | 1.0 | <LOQ | 1.5 |
| 15 | 1.0 | 1.4 | 1.0 | <LOQ | 1.4 |
| 16 | 0.2 | <LOQ | <LOQ | <LOQ | <LOQ |
| 17 | 0.5 | 1.3 | 0.5 | <LOQ | 2.6 |
| 18 | 0.1 | 4.7 | 4.2 | <LOQ | 1.1 |
| 19 | 1.0 | <LOQ | <LOQ | <LOQ | <LOQ |
| SPM | 7.5 | 6.7 | 11.0 | 0.8 | 0.6 |

a) samples could not be collected, <LOQ: below limit of quantitation

## Discussion

*Water samples and organic load*

In general, polycyclic musks are discharged into rivers by effluents of sewage treatment plants (STP) (e.g. Eschke et al., 1994; Simonich et al., 2000). The permanent use of polycyclic musks in households causes their steady input via STP effluents into the rivers. This input is diffuse, because it takes place by numerous point sources along the river.

In the investigated STP samples HHCB dominated over AHTN. This reflects the higher usage rate of HHCB in Europe (HHCB 1482 tons/y,

AHTN 585 tons/y; Balk and Ford, 1999a). HHCB and AHTN concentrations were in the range of those determined by Eschke et al. (1994) who examined STP on the Ruhr river (500 to 2900 ng/L HHCB and 800 to 4400 ng/L AHTN). The occurrence of ADBI in sewage influent and effluent is also known (Eschke et al., 1994) whereas –to our knowledge- AHMI has not been detected prior to this study.

With respect to sewage samples, HHCB and AHTN concentrations in Lippe river water were in the same range as in other German rivers (<30-610 ng/L HHCB and <30-330 ng/L AHTN, data compilation by Rimkus, 1999). ADBI and AHMI concentrations were always below the limit of quantitation.

In the following section the changes of HHCB and AHTN concentrations along the river from the source to the mouth will be discussed (Fig. 3). Because river runoff affects compound concentrations in water, we additionally present a compilation of HHCB and AHTN loads along the river (Fig. 4).

The Lippe river source (site 19) is located in the city Bad Lippspringe (15,000 inhabitants). HHCB and AHTN were detected in the spring water but their concentrations were below the limit of quantitation. Lippe spring water is derived from a karst area. Thus, surface waters and rain are feeding the source. The karstic ground water flows through carbonate rocks which have a low filter capacity. Hence, contaminated surface water can find a way to the source, a process which might explain the occurrence of polycyclic musks therein. Only two kilometers further downstream (site 18), there is a significant increase of AHTN and HHCB concentrations. This is the result of sewage input by the STP of Bad Lippspringe. Due to low river runoff there is not much dilution. Hence, concentrations rise strongly and are among the highest along the river. The load also increases but is still low compared to the other sites further downstream.

West of the source region, there are four more sampling locations on the upper reaches of the Lippe river (sites 17 to 14). This area is not densely populated, but there are two major sewage treatment plants (Fig. 2; Paderborn and Lippstadt). The effluent discharge of the STP`s leads to significantly higher loads of HHCB and AHTN at sites 16 to 14. In contrast, concentrations are lower than at site 18. This is the result of Lippe water dilution by several small brooks (e.g. the Beke Brook) and by some larger tributaries (e.g. the Alme river).

The city of Hamm (site 13 and 12; 182,000 inhabitants) marks the beginning of the densely populated lower reaches of the Lippe river. There are two sewage treatment plants in Hamm, followed downstream by the plants of Werne, Lünen, Datteln and Haltern (Fig. 2). Accordingly, concentrations and organic loads of HHCB and AHTN rise. From site 8 to site 5 and also at site 3 the highest concentrations and loads of HHCB and AHTN were observed.

In the area near the river mouth (sites 4, 2 and 1), concentrations and loads of dissolved HHCB and AHTN decrease. This phenomenon is also visible further upstream at sites 9 and 12. Thus, at these sites the rate of removal is higher than the rate of input. Degradation and/or partitioning of compounds into other phases are processes which might explain this phenomenon and will be discussed below. Increasing loads can thus be interpreted as a compensation of transformation/partitioning processes by a larger input. The fact that loads are decreasing within a short flow distance (10 to 20 km; flow velocity in August 1999 was approximately 0.3 to 0.5 m/s, data from LUA NRW) implies that these processes take place within a short period (about 6 to 19 h).

In the following section, partitioning and transformation processes are discussed. Processes that may cause a disappearance of dissolved compounds in river systems are volatization, chemical degradation, biological degradation and adsorption to particulate matter.

The Henry`s law constants of HHCB and AHTN of 11.3 and 12.5 (Table 6), respectively, indicate that these compounds have very low vapour pressures. Therefore the tendency of these compounds to volatilize from the aqueous phase is not significant. On the other hand, chemical (e.g photochemical transformation, hydrolysis) and/or biological degradation may contribute significantly to their disappearance. Recent studies have shown that HHCB is not persistent in the atmosphere (Aschmann et al., 2001). Franke et al. (1999) identified an oxygenated transformation product of HHCB in river water. Balk and Ford (1999a) observed biotransformation of HHCB and AHTN by bluegill sunfish. These results suggest that HHCB and AHTN in river water and sediments undergo chemical and/or biological degradation, although no turnover rates were published so far. Adsorption to particulate matter may also explain the observed data due to the fact that HHCB and AHTN have relatively low water solubilities (Table 6). Apparent partition coefficients, Kp of HHCB and AHTN between water and suspended particulate matter in the Elbe river were calculated by Winkler et al. (1998). Both compounds have high Kp values (Table 6) and have therefore a high affinity to be adsorbed to particulate matter. Thus we postulate that adsorption and sedimentation processes as well as degradation processes are primarily responsible for the loss of dissolved HHCB and AHTN.

Ratios of dissolved HHCB/AHTN vary between 1.8 and 5.0 and do not show a clear trend. Variations of HHCB/AHTN ratios are possibly a result of ongoing supply of HHCB and AHTN with variable ratios. The median HHCB/AHTN ratio in Lippe river water is 2.9. This is close to the HHCB/AHTN ratio of 2.5 which is deduced from the 1995 European use volumes (Balk and Ford, 1999a). However, in water samples from the Elbe river, the Saale river and the Mulde river, Germany, lower median HHCB/AHTN ratios were determined (1.4 to 1.8, data compiled by Rimkus, 1999). Eschke (1996) reported median concentrations of 275 ng/L HHCB and 100 ng/L AHTN (HHCB/AHTN ratio 2.75), in Ruhr river water samples taken in 1995 and 1996. In Ruhr river water samples from 1994, the median HHCB/AHTN ratio was 1.9 (Eschke et al., 1994). Apparently, HHCB/AHTN ratios are subject to fluctuations. This is likely due to varying fragrance compositions in the products (personal communication, N. Rohde, Haarmann & Reimer GmbH, Germany), since Simonich et al. (2000) have shown that the percent removal of both compounds during wastewater treatment is very similar (AHTN removed at 86.2 %, HHCB removed at 87.4 %).

*Sediment samples and suspended particulate matter*

Concentrations of HHCB (5 to 191 μg/kg) and AHTN (2 to 1399 μg/kg) in Lippe river sediments were higher than those measured in a similar study by Lach and Steffen (1997), who investigated sediments of five German rivers (Elbe, Weser, Ems, Leine and Oker river). They reported concentrations of up to 54 μg/kg and up to 3.9 μg/kg for HHCB and AHTN, respectively. In contrast to Lach and Steffen (1997) ADBI was not found in this study, but AHMI could be quantified in two sediment samples collected near the river mouth.

In the following, concentrations of HHCB and AHTN in μg/g (normalized to TOC) along a river section from the source to the river mouth are discussed (Fig. 5). Sedimentation in river systems is a dynamic process with permanent settling and resuspension of particles depending on flow velocities and particle size. Thus, sediment samples do not necessarily represent the local pollution history of the river at the sampling sites. Additionally, sediments were not collected at representative cross sections, but only one hopefully representative sample at each site. This has to be considered in the following discussion.

Along the upper reaches of the river, from sites 19 to 13, concentrations of both compounds are low or not detectable. This is attributed to the low population density and expected low input of pollutants by sewage treatment plants effluents. Site 13 is located at the beginning of the densely populated lower reaches. Although there is a higher discharge of sewage effluents in this area, which is also reflected by higher loads of HHCB and AHTN in the aqueous phase, concentrations in the sediments remain low until site 9. Contaminated suspended particulate matter does not accumulate immediately but is transported further downstream. Accumulation in the sediment takes place in areas with low flow velocities. Due to this delay in sedimentation the higher input rate in this area is not necessarily evident in the sediments at site 13 to 9 but will be expected to manifest further downstream.

Indeed, at site 8 concentrations of HHCB and AHTN are significantly higher and are still high at site 6. Possibly, conditions for accumulation of suspended matter in the sediment are favourable due to low flow velocities there. Sediments at sites 5 and 4 contain less HHCB and AHTN. This could be the consequence of less favourable accumulation conditions or lower input in the area further upstream. Obviously, it is difficult to relate compound concentrations in riverine sediments to specific input situations, because other factors such as flow velocities and sediment composition are also of great importance.

By far the highest concentrations, especially of AHTN, were detected in sediments near the river mouth at sites 3 and 1. These are also the only sites where AHMI was detectable. In this area, the supply of polycyclic musks is relatively low because of a low population density (Fig. 2). Hence, particularly favourable conditions for the accumulation of suspended particulate matter from upstream areas must be the reason for this pattern. This could be due to three factors: i) low flow velocity, ii) high input of terrestrial detritus and, thus, large supply of suspended particulate matter for adsorption of contaminants, and iii) input of highly contaminated particles which were transported from the upper reaches of the river.

HHCB/AHTN ratios in Lippe river sediments are significantly lower (average 1.2) than in Lippe river water samples (average 2.9). Accordingly, AHTN is enriched in the sediments. Winkler et al. (1998) investigated musk concentrations in water and suspended particulate matter (SPM) from the Elbe river. They found a HHCB/AHTN ratio of about 1 in SPM, whereas river water analysis showed a HHCB/AHTN ratio of about 2 (which is also illustrated by the higher partition coefficient

Kp of AHTN, Table 6). This coincides with our results and suggests the preferential adsorption of AHTN to particulate matter in the riverine environment. Contrary, laboratory experiments revealed that sorptive behaviours log $K_{OC}$ of both compounds are very similar (Table 6). This discrepancy was observed by Balk and Ford (1999a). However, their median field derived log $K_{oc}$ calculated from the data of Winkler et al. (1998) was 4.8 for AHTN (range 4.5 to 5.2) and 4.7 for HHCB (range 4.3 to 5.1). It was therefore concluded that the sorption of both compounds apparently varies over time and with the composition and quantity of the suspended materials but that the behaviour of both compounds remains largely similar (Balk and Ford, 1999a). In our opinion, further research about sorptive behaviours of organic chemicals in the natural environment is necessary to enlighten this interesting aspect.

## Summary and conclusions

- Using polycyclic musks it was possible to monitor the distribution of sewage-derived organic chemicals in a river system. In the upper reaches of the Lippe river low loads of dissolved HHCB and AHTN were observed due to low population density and hence little input of sewage effluents. The central part of the Lippe river is densely populated, which was reflected by steadily increasing loads. On the contrary, loads decreased at some sites, especially along the less populated lower reaches.
- The data indicate that removal of musks was in some cases higher than input. Degradation and/or partitioning of compounds into other phases are processes that might explain this phenomenon. The fact that loads decrease within a short flow distance implies that these processes take place on a short time scale. Increasing loads further upstream can thus be interpreted as a compensation of degradation/partitioning processes by a larger input.
- On the basis of formerly reported properties of HHCB and AHTN it was concluded that their loss from the aqueous phase is due to degradation processes and/or adsorption to particulate matter. Consequently, high concentrations of HHCB and AHTN were detected in sediments and suspended particulate matter of the Lippe river. The highest concentrations were found in the area near the river mouth. Apparently in this area particularly favourable conditions for sedimentation of contaminated suspended particulate matter from the upper reaches prevailed. This was probably induced by low flow velocities which

enhanced particle sedimentation and availability of terrestrial suspended matter for adsorption of contaminants.

- HHCB/AHTN ratios in sediments and suspended particulate matter were lower than in water samples. The enrichment of AHTN in sediments may demonstrate its preferential adsorption to particulate matter. This is confirmed by field data of Winkler et al. (1998) but disagrees with the very similar sorptive behaviour of both compounds which were measured in laboratory tests (Balk and Ford, 1999a).

By observing the load of dissolved organic chemicals along a longitudinal river section we were able to demonstrate how degradation/partition processes induce the significant removal of these compounds from the aqueous phase. The data imply that these processes take place on a short time scale. Consequently, the investigated compounds were identified in high concentrations in riverine sediments and suspended particulate matter. The distribution processes which were identified in the present study are also relevant for other dissolved compounds with similar physico-chemical properties in comparable aquatic environments.

## 4.2 Compound specific stable carbon isotope analyses – a new tool for tracing the fate of organic riverine contaminants

### 4.2.1 Determination of $^{13}C/^{12}C$-ratios of anthropogenic organic contaminants in river water samples by GC-irmMS ‡

#### Introduction

First studies combining gas chromatography and mass spectrometry in order to determine variations in the stable carbon isotope composition of organic compounds were reported in 1976 and 1978 (Sano et al. 1976; Matthwes and Hayes 1978). Matthews and Hayes referred to their analytical approach as 'isotope ratio monitoring mass spectrometry' (irmMS) and approx. 10 years later this technique was commercially implemented. Subsequently the GC/irmMS technique and the corresponding compound specific carbon isotope analyses were applied to numerous fields of analytical research comprising e.g. food chemistry, organic geochemistry, archaeology, medicine and pharmacology (see also Brand 1996; Meier-Augstein 1999 and citations therein).

During the last years compound specific carbon isotope analyses were also performed in environmental studies. Investigations applied to field samples of the aquatic environment focused mainly on polycyclic aromatic hydrocarbons, monoaromatic compounds and halogenated aliphatics (e.g. Dempster et al., 1997; Sturchio et al., 1998; Mazeas et al., 2002; Stark et al.; 2003). In addition, several laboratory studies were reported investigating the isotopic fractionation as a result of environmental processes. Transport as well as transformation processes (e.g. vaporization, adsorption, abiotic and biotic degradation) of several contaminants comprising also polycyclic aromatic hydrocarbons as well as halogenated aliphatics and aromatics have been investigated (e.g. Harrington et al., 1999; Jackson and Pardue, 1999; Sherwood-Lollar et al., 1999; Mazeas and Budzinski, 2002; Slater et al., 2002; Yanick et al., 2003a). Carbon isotope ratios are not only useful data for evaluating the fate of organic pollutants in the environment but also for distinguishing between different emission sources discharging the same contaminants. Therefore isotopic characterization especially of technical mixtures of chlorinated aromatics were carried out (e.g. Jarman et al., 1998; Yanik et al., 2003b).

---

‡ Mainly adapted/reprinted from Schwarzbauer et al. 2005b

All of these isotopic studies were performed on systems with elevated contamination levels or high concentrations measured and mostly on simple compound mixtures. These applications avoid the two major limitations of compound-specific isotope analyses. These limitations lie in the lower sensitivity as compared to traditional GC/MS analyses and in the requirement of complete gas chromatographic separation of the contaminants. However, natural aquatic systems, especially riverine or groundwater systems, are mostly contaminated at a low or medium level and, therefore, concentrations of individual organic compounds are low. Consequently, information on the isotopic characterization of contaminants existing in these water systems is not available up to now.

In this context the present investigation focussed on the evaluation of a common analytical procedure adapted for the preparation of material utilizable for compound-specific stable carbon isotope analyses of riverine contaminants. Especially the sensitivity of the analytical method and the precision of the isotopic data obtained were studied. So far this analytical procedure has been applied to river water samples for qualitative and quantitative GC/MS analyses (Franke et al. 1995; Dsikowitzky et al. 2002).

## Samples

River water samples were taken at 8 sampling locations from the Rhine river (Germany) between Koblenz and Venlo in 2001 (see Figure 1) and at 19 sampling locations from the Lippe river (Germany) in 1999 (see chapter 3.1.1). 2 L - water samples were scooped up from below the water surface at a distance of approx. 2m from the river side and bottled in pre-cleaned glass flasks.

## Recovery experiments

Recovery experiments were performed by spiking high-purity water (Lichrosolv, Merck, Germany) with concentrations of 800 ng/L to 1500 ng/L of the respective reference compounds dissolved in methanol (see Table 1). After an equilibration time of 12 hours the extraction procedure was performed as described above and the extracts were analysed by GC/MS and GC/irmMS. Former investigations revealed recovery rates between 25 % and 95 % for the individual substances (Dsikowitzky et al., 2002, Dsikowitzky et al., 2004a/b).

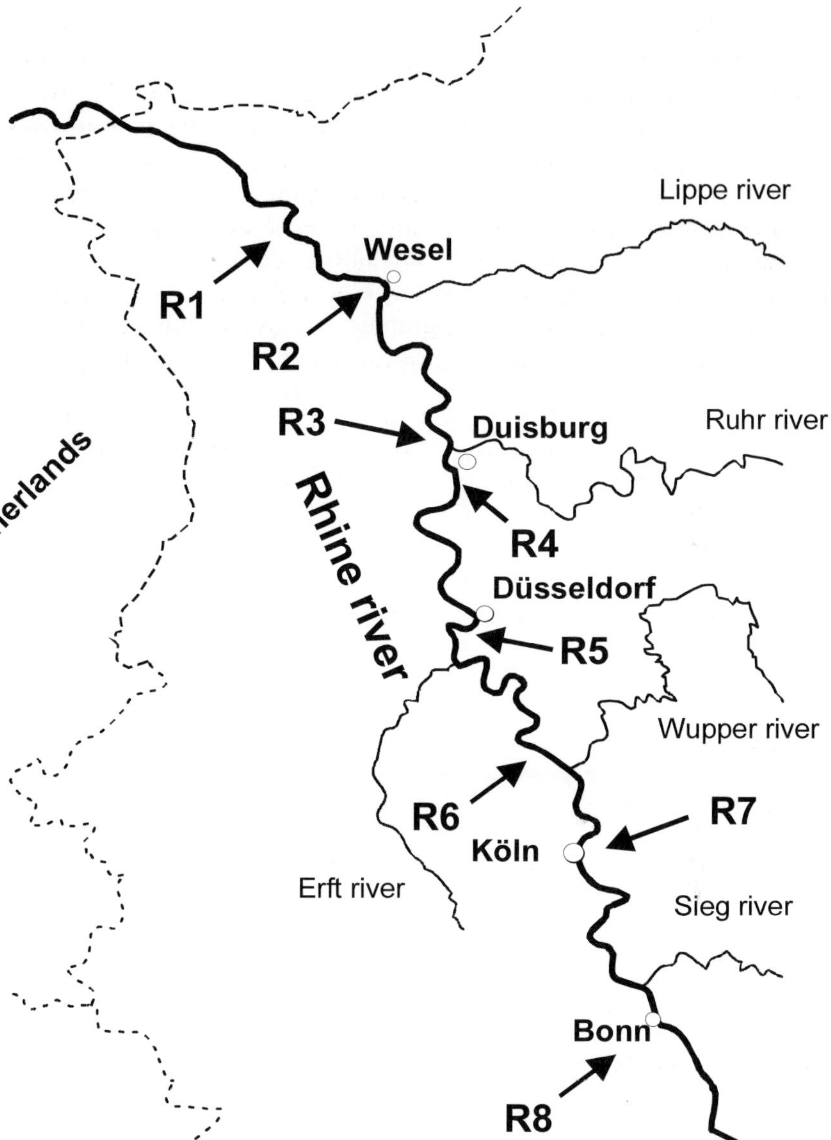


**Figure 1:** Sampling locations at the Rhine river.

## Results and Discussion

*Stable carbon isotope analyses of selected anthropogenic contaminants*

To evaluate the sensitivity and precision of compound specific carbon isotope analyses of river water contaminants a set of numerous anthropogenic and biogenic substances at different concentration levels were measured. Besides alkanes and fatty acid methyl esters xenobiotical compounds were considered including chlorinated aliphatics and aromatics, musk fragrances, phthalate based plasticizers and tetrabutyl tin. In order to avoid gas chromatographic coelutions the compounds were divided into appropriate subsets prior to the measurements. The resolution of the individual chromatograms is illustrated in Figure 2. All analyses were performed at four different concentration levels representing either a sufficient amount with respect to the technical conditions of the GC-irmMS system (approx. 25 ng, see Table 1a-c) and amounts prevalently detected during analyses of low contaminated river water extracts (1 to 5 ng, see Table 1a-c). Considering the analytical procedures applied to the river water samples (see 3.2.) the amounts selected represent concentration ranges between approx. 10 to 1000 ng/L in field water samples.

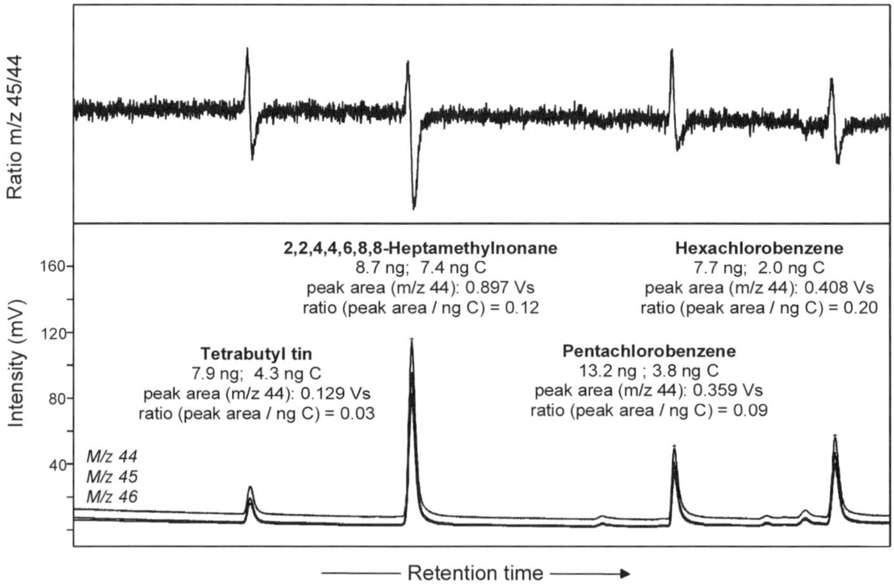


**Figure 2:** GC-irmMS chromatogram and corresponding m/z 44/45 ratio chromatogram obtained from a standard solution at the 10 ng level. Additionally, data of amounts (ng), corresponding amounts of carbon (ng C), the peak area derived from the m/z 44 trace (Vs) as well as the ratios of peak areas and carbon amounts are given for all individual compounds (Vs/ng C).

All $\delta^{13}C$-values determined are presented in Table 1a-c and range between approx. -20 ‰ and -30 ‰. The precision of the stable carbon isotope ratios detected is characterized by the standard deviation obtained by 10 repetitions of the measurements. These values covered the range from 0.2 ‰ and 3.2 ‰. Generally increasing standard deviation values were observed with decreasing amounts as a result of the technical limitations imposed by the irmMS system. Data obtained for the 25 ng- and 10 ng-levels are generally characterized by satisfactory deviations between 0.2 and 0.7 ‰. On the contrary at the 5 ng-level an increasing number of analyses revealed deviations of more than 0.7 ‰. Furthermore, at the 1 ng-level nearly all compounds were either not detected at all or their $\delta^{13}C$-values exhibited standard deviations higher than 1.0 ‰.

**Tab. 1a:** $\delta^{13}C$-values of selected organic compounds measured at different concentration levels. All substances are wellknown riverine contaminants. (nd= not detected)

| **Concentration levels (n = repetitions)** | | **n-Pentadecane** | **n-Docosane** | **n-Hexacosane** | **n- Tetradecanoic acid methylester** | **n- Pentadecanoic acid methylester** | **n- Hexadecanoic acid methylester** | **2,2,4,4,6,8,8-Hepta-methylnonane** |
|---|---|---|---|---|---|---|---|---|
| *rel. carbon fraction of compounds (%)* | | 85 | 85 | 85 | 74 | 75 | 76 | 85 |
| **1 ng-level** | amount (ng) | 0.8 | 0.9 | 0.9 | 1.3 | 1.2 | 1.0 | 0.9 |
| | $\delta^{13}C$-values (‰ VPDB) | **nd** | **nd** | **nd** | **-27.6** | **-29.9** | **-28.6** | **nd** |
| (n=10) | stand. dev. (‰ VPDB) | | | | *1.7* | *1.7* | *1.5* | |
| **5 ng level** | amount (ng) | 4.1 | 4.4 | 4.5 | 6.6 | 5.9 | 5.2 | 4.4 |
| | $\delta^{13}C$-values (‰ VPDB) | **-30.0** | **-24.5** | **-30.4** | **-28.9** | **-30.1** | **-29.9** | **-28.5** |
| (n=10) | stand. dev. (‰ VPDB) | *1.0* | *0.8* | *0.8* | *1.2* | *1.1* | *0.8* | *0.5* |
| **10 ng level** | amount (ng) | 8.2 | 8.7 | 8.9 | 13.1 | 11.7 | 10.5 | 8.7 |
| | $\delta^{13}C$-values (‰ VPDB) | **-30.00** | **-24.7** | **-29.4** | **-30.6** | **-29.8** | **-30.2** | **-29.7** |
| (n=10) | stand. dev. (‰ VPDB) | *0.3* | *0.3* | *0.5* | *0.6* | *0.5* | *0.6* | *0.6* |
| **25 ng level** | amount (ng) | 20.4 | 21.8 | 22.3 | 32.8 | 29.3 | 26.2 | 21.8 |
| | $\delta^{13}C$-values (‰ VPDB) | *-30.8* | *-25.1* | *-29.5* | *-30.6* | *-29.5* | *-30.1* | *-30.1* |
| (n=10) | stand. dev. (‰ VPDB) | *0.5* | *0.3* | *0.3* | *0.5* | *0.6* | *0.3* | *0.7* |

**Tab. 1b:** $\delta^{13}C$-values of selected organic compounds measured at different concentration levels. All substances are wellknown riverine contaminants. (nd= not detected)

| **Concentration levels (n = repetitions)** | | **Dimethylphthalate** | **Diethylphthalate** | **Di-iso-butylphthalate** | **Di-sec-butylphthalate** | **Di-n-butylphthalate** | **Bis-(2-ethylhexyl)-phthalate** |
|---|---|---|---|---|---|---|---|
| *rel. carbon fraction of compounds (%)* | | 62 | 65 | 69 | 69 | 69 | 74 |
| **1 ng-level** | amount (ng) | 0.8 | 0.9 | 1.1 | 0.9 | 0.9 | 1.0 |
| | $\delta^{13}C$-values (‰ VPDB) | **nd** | **nd** | **-24.8** | **-26.4** | **-25.6** | **-27.0** |
| (n=10) | stand. dev. (‰ VPDB) | | | *2.4* | *2.0* | *2.6* | *1.7* |
| **5 ng level** | amount (ng) | 3.9 | 4.7 | 5.3 | 4.6 | 4.4 | 5.0 |
| | $\delta^{13}C$-values (‰ VPDB) | **-24.6** | **-25.8** | **-26.8** | **-27.3** | **-27.7** | **-26.9** |
| (n=10) | stand. dev. (‰ VPDB) | *1.1* | *0.7* | *0.6* | *0.5* | *0.6* | *0.4* |
| **10 ng level** | amount (ng) | 7.8 | 9.5 | 10.6 | 9.1 | 8.7 | 10.0 |
| | $\delta^{13}C$-values (‰ VPDB) | **-25.9** | **-26.1** | **-27.2** | **-28.0** | **-27.8** | **-27.5** |
| (n=10) | stand. dev. (‰ VPDB) | *0.4* | *0.3* | *0.4* | *0.6* | *0.2* | *0.4* |
| **25 ng level** | amount (ng) | 19.4 | 23.7 | 26.6 | 22.8 | 21.8 | 25.1 |
| | $\delta^{13}C$-values (‰ VPDB) | **-26.1** | **-25.8** | **-27.0** | **-28.1** | **-27.9** | **-27.2** |
| (n=10) | stand. dev. (‰ VPDB) | *0.4* | *0.2* | *0.2* | *0.3* | *0.3* | *0.3* |

Additionally, significant variations of the standard deviation values were observed with respect to the individual compounds. The precision of $\delta^{13}C$-values of *n*-alkanes, 2,2,4,4,6,8,8-heptamethylnonane, fatty acid methyl esters, phthalates and musk fragrances was generally acceptable down to amounts of approx. 5 ng with standard deviation values below 1.0 ‰. On the contrary, already at the 25ng- and 10ng-levels the stable carbon isotope ratios of the chlorinated compounds ($Cl_5$- and $Cl_6$-benzene, $Cl_6$-butadiene) as well as of tetrabutyl tin were measured with high standard deviations between 0.8 ‰ and 3.2 ‰.

**Tab. 1c:** $\delta^{13}C$-values of selected organic compounds measured at different concentration levels. All substances are wellknown riverine contaminants. (nd= not detected)

| **Concentration levels (n = repetitions)** | | **Cashmerane** | **Celestolide** | **Pantolide** | **Tetrabutyl tin** | **Hexachlorobutadiene** | **Pentachlorobenzene** | **Hexachlorobenzene** |
|---|---|---|---|---|---|---|---|---|
| *rel. carbon fraction of compounds (%)* | | 82 | 84 | 84 | 55 | 16 | 29 | 26 |
| 1 ng-level | amount (ng) | 1.0 | 1.1 | 0.9 | 0.8 | 0.8 | 1.3 | 0.8 |
| | $\delta^{13}C$-values (‰ VPDB) | **-27.8** | **-24.8** | **-28.0** | **nd** | **nd** | **nd** | **nd** |
| (n=10) | stand. dev. (‰ VPDB) | *1.2* | *0.8* | *0.7* | | | | |
| 5 ng level | amount (ng) | 5.1 | 5.6 | 4.3 | 3.9 | 3.8 | 6.6 | 3.9 |
| | $\delta^{13}C$-values (‰ VPDB) | **-27.8** | **-24.8** | **-27.8** | **-24.0** | **-23.9** | **-20.8** | **-24.3** |
| (n=10) | stand. dev. (‰ VPDB) | *0.3* | *0.5* | *0.5* | *0.4* | *2.2* | *1.0* | *2.3* |
| 10 ng level | amount (ng) | 10.2 | 11.1 | 8.8 | 7.9 | 7.6 | 13.2 | 7.7 |
| | $\delta^{13}C$-values (‰ VPDB) | **-27.9** | **-25.6** | **-29.0** | **-25.2** | **-24.6** | **-21.5** | **-25.0** |
| (n=10) | stand. dev. (‰ VPDB) | *0.3* | *0.4* | *0.6* | *1.2* | *3.2* | *1.0* | *3.2* |
| 25 ng level | amount (ng) | 25.5 | 27.8 | 21.4 | 19.7 | 19.1 | 33.1 | 19.3 |
| | $\delta^{13}C$-values (‰ VPDB) | **-27.4** | **-24.1** | **-27.5** | **-25.1** | **-25.8** | **-22.2** | **-25.9** |
| (n=10) | stand. dev. (‰ VPDB) | *1.0* | *0.7* | *0.9* | *0.8* | *2.9* | *0.8* | *0.8* |

An influence on the precision of carbon isotope analyses by halogen and tin atoms became evident, possibly as a result of the low fraction of carbon in the polychlorinated substances (16 % to 29 %, see Table 1c) minimizing the effective amount of $CO_2$ measured in the irmMS. However, this assumption is contradicted by the varying standard deviations of analysis of indivdual compounds with comparable carbon amounts. E.g. the precision of pentachlorobenzene analysis at the 25ng-level (representing approx. 10 ng carbon) of 0.7 ‰ SD was significant higher as compared to analysis of cashmerane at the 10 ng-level (also representing approx. 10 ng carbon) with a standard deviation of 0.28 ‰. Therefore, a major influence on the oxidation processes due to interactions of halogens or tin with the catalysts (primarily with the copper) can be assumed. This assumption is partially supported by a quantitative comparison of combustion yields as illustrated in Fig. 2. Comparing the peak areas (m/z 44 trace) of pentachlorbenzene, hexachlorobenzene, tetrabutyl tin and 2,2,4,4,6,8,8-heptamethylnonane with the corresponding amounts of carbon (3.8 ng, 2.0 ng, 4.3 ng and 7.4 ng, respectively) significant lower ratios are obvious for the tin containing compound. This observation suggested an incomplete combustion of tin organic compounds as the result of poisoning effects on the catalyst. The same effect cannot be stated for halogenated compounds. Because penta- and hexachlorinated benzene generated a similar area to C-amount ratio as compared to the branched alkane representing very similar combustion yield for these compounds.

However, because many priority pollutants belong to the group of polyhalogenated substances the low degree of precision in the concentration ranges investigated affects the compound specific carbon isotope analysis of an important class of anthropogenic contaminants.

Considering both the compound and amount related effects compromising the precision of carbon isotope analyses, the sensitivity of the analytical method used can be appointed to an amount down to approx. 5 ng for numerous anthropogenic contaminants. However, it has to be noted that in comparison the precision of the analyses of halogenated and tin containing compounds is generally lower.

*Recovery experiments*

An important precondition for the successful determination of carbon isotope ratios is the prevention of isotopic shifts as a result of the analytical procedures applied. Therefore, five recovery experiments were performed in order to detect changes of the carbon isotope ratios during sample preparation and measurement. The compounds selected for these experiments are known riverine contaminants and comprise hexachlorobutadiene, several musk fragrances, phthalates and other plasticizers, a flame retardant and a pesticide. All recovery samples were spiked with concentrations between approx. 800 ng/L and 1500 ng/L for each compound representing a common abundance level in river systems.

The sequential extraction with *n*-pentane and dichloromethane resulted in an exclusive occurrence in one extract for most of the compounds selected. All musk fragrances investigated, most of the phthalates, N,N-diethyltoluamide (DEET), 2,2,4-trimethyl-1,3-pentanedioldi-*iso*-

butyrate, 2,6-di-*tert*-butyl-4-methylphenol and hexachlorobutadiene were detected in the pentane fractions. The more polar compounds 2,4,7,9-tetramethyl-5-decyne-4,7-diol, tri-*n*-butylphosphate and tris(2-chloroethyl)phosphate appeared exclusively in the dichloromethane fraction. 2-(2-Ethoxybutoxy)ethylacetate and dimethylphthalate were detected in both extracts with varying concentrations. All $\delta^{13}C$-values obtained after the recovery procedure are presented in Table 2 - 4 in the order of their occurrence in the separate extracts. Because the gas chromatographic separations of 2,6-di-*tert*-butyl-4-methylphenol and cashmeran as well as galaxolide and tonalide were insufficient regarding the strict chromatographic requirements of compound specific isotope analyses, the carbon isotope ratios were summarized for the unresolved peaks, respectively.

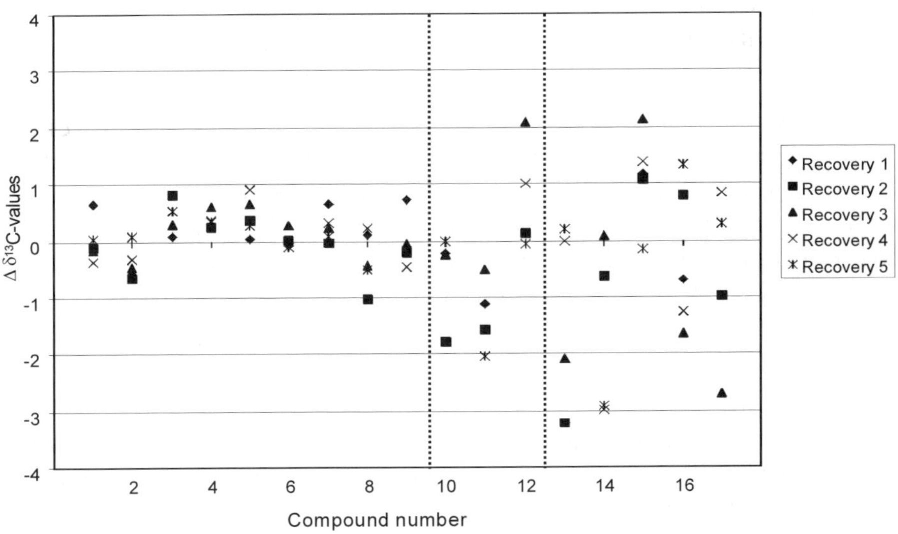


**Figure 3:** Differences of carbon isotope ratios measured after application of the recovery procedure. Compound numbers are: **1** - galaxolide/tonalide (coeluting), **2** - celestolide, **3** - 2-(2-ethoxybutoxy)ethylacetat, **4** - 2,2,4-trimethyl-1,3-pentandioldi-*iso*-butyrate, **5** - dimethylphthalate, **6** - diethylphthalate, **7** - di-*iso*-butylphthalate, **8** - di-*sec*-butylphthalate, **9** - di-*n*-butylphthalate, **10** - pantolide, **11** - tri-*n*-butylphosphate, **12** – bis(2-ethylhexyl)phthalate (DEHP), **13** – 2,4,7,9-tetramethyl-5-decyne-4,7-diol, **14** - hexachlorobutadiene, **15** - cashmerane/ionol, **16** – tris(2-chloroethyl)phosphate, **17** – N,N-diethyltoluamide (DEET).

**Tab. 2a:** $\delta^{13}C$-values of compounds exclusively detected in the *n*-pentane extracts of the recovery experiments.

| **Recovery experiments** | | **Hexachlorobutadiene** | **Cashmeran / 2,6-di-*tert*-butyl-4-methylphenol** | **Celestolide** | **Pantolide** | **Galaxolide/Tonalide** | **Diethylphthalate** |
|---|---|---|---|---|---|---|---|
| **Original** | $\delta^{13}C$-values (‰ VPDB) | **-32.1** | **-27.5** | **-26.0** | **-28.9** | **-27.8** | **-29.8** |
| | stand. dev. (‰ VPDB) | *1.3* | *0.4* | *0.5* | *0.3* | *0.2* | *0.2* |
| **1** | $\delta^{13}C$-values (‰ VPDB) | **n.d.** | **-28.7** | **-25.5** | **-28.7** | **-28.5** | **-29.7** |
| | stand. dev. (‰ VPDB) | | *0.5* | *0.1* | *0.2* | *0.1* | *0.3* |
| **2** | $\delta^{13}C$-values (‰ VPDB) | **-31.4** | **-28.6** | **-25.4** | **-27.1** | **-27.7** | **-29.8** |
| | stand. dev. (‰ VPDB) | *0.5* | *0.9* | *1.0* | *0.1* | *0.2* | *0.3* |
| **3** | $\delta^{13}C$-values (‰ VPDB) | **-32.1** | **-29.7** | **-25.6** | **-28.6** | **-27.7** | **-30.1** |
| | stand. dev. (‰ VPDB) | *1.1* | *0.5* | *0.3* | *0.3* | *0.1* | *0.1* |
| **4** | $\delta^{13}C$-values (‰ VPDB) | **-29.1** | **-28.9** | **-25.7** | **n.d.** | **-27.4** | **n.d.** |
| | stand. dev. (‰ VPDB) | *0.9* | *0.8* | *0.7* | | *0.7* | |
| **5** | $\delta^{13}C$-values (‰ VPDB) | **-29.1** | **-27.4** | **-26.1** | **-28.9** | **-27.8** | **-29.7** |
| | stand. dev. (‰ VPDB) | *1.5* | *0.3* | *1.4* | *0.8* | *0.7* | *0.1* |

In order to illustrate the variations of the carbon isotope ratios measured after execution of the five recovery experiments, Figure 3 presents the differences of the $\delta^{13}C$-values of the recovered substances as compared to the untreated compounds noted as $\Delta\ \delta^{13}C$-values.

A first group of contaminants including nearly all phthalates and musk fragrances, as well as 2-(2-ethoxybutoxy)ethylacetate and 2,2,4-trimethyl-1,3-pentanedioldi-*iso*-butyrate was recovered with only minor changes of the $\delta^{13}C$-values as compared to the untreated substances.

**Tab. 2b:** $\delta^{13}C$-values of compounds exclusively detected in the *n*-pentane extracts of the recovery experiments.

| **Recovery experiments** | | **Di-iso-butylphthalate** | **Di-sec.-butylphthalate** | **Di-n-butylphthalate** | **Bis(2ethylhexyl)phthalate** | **N,N-Diethyltoluamide (DEET)** | **2,2,4-trimethyl-1,3-pentandioldi-*iso*-butyrate** |
|---|---|---|---|---|---|---|---|
| **Original** | $\delta^{13}C$-values (‰ VPDB) | **-27.5** | **-27.5** | **-27.6** | **-26.9** | **-25.8** | **-27.4** |
| | stand. dev. (‰ VPDB) | *0.1* | *0.5* | *0.3* | *0.4* | *1.6* | *0.1* |
| **1** | $\delta^{13}C$-values (‰ VPDB) | **-28.2** | **-27.6** | **-28.3** | **-27.1** | **n.d.** | **n.d.** |
| | stand. dev. (‰ VPDB) | *0.2* | *0.1* | *0.3* | *0.4* | | |
| **2** | $\delta^{13}C$-values (‰ VPDB) | **-27.5** | **-26.5** | **-27.4** | **-27.0** | **-25.4** | **-27.7** |
| | stand. dev. (‰ VPDB) | *0.3* | *0.7* | *0.2* | *0.1* | *0.1* | *0.1* |
| **3** | $\delta^{13}C$-values (‰ VPDB) | **-27.7** | **-27.1** | **-27.5** | **-29.0** | **n.d.** | **-28.0** |
| | stand. dev. (‰ VPDB) | *0.2* | *0.1* | *0.1* | *1.5* | | *0.3* |
| **4** | $\delta^{13}C$-values (‰ VPDB) | **-27.9** | **-27.7** | **-27.1** | **-27.9** | **-30.6** | **-27.8** |
| | stand. dev. (‰ VPDB) | *1.3* | *0.5* | *1.6* | *1.2* | *0.3* | *1.6* |
| **5** | $\delta^{13}C$-values (‰ VPDB) | **-27.61** | **-27.0** | **-27.4** | **-26.8** | **n.d.** | **-27.8** |
| | stand. dev. (‰ VPDB) | *0.1* | *0.1* | *0.1* | *0.4* | | *0.3* |

Additionally, acceptable standard deviations were obtained for the analyses of these compounds (see Tables 2 to 4) and a slight tendency to lower $\delta^{13}C$-values was observed for almost all compounds of this first group. Pantolide, tri-*n*-butylphosphate, and bis(2-ethylhexyl)phthalate represent a second group of substances. This group is characterized by a distribution of the $\delta^{13}C$-values with differences up to 2 ‰ as compared to the original data of the untreated substances. General trends towards lower as well as to higher values were observed.

**Tab. 3:** $\delta^{13}C$-values of compounds exclusively detected in the dichloromethane extracts of the recovery experiments.

| Recovery experiments | | 2,4,7,9-Tetramethyl-5-decyne-4,7-diol | Tri-n-butyl-phosphatel | Tris(2-chloroethyl-phosphate) |
|---|---|---|---|---|
| **Original** | $\delta^{13}C$-values (‰ VPDB) | **-24.4** | **-28.1** | **-29.6** |
| | stand. dev. (‰ VPDB) | *0.4* | *0.4* | *0.4* |
| **1** | $\delta^{13}C$-values (‰ VPDB) | **-21.2** | **-26.9** | **-28.9** |
| | stand. dev. (‰ VPDB) | *0.7* | *1.8* | *0.3* |
| **2** | $\delta^{13}C$-values (‰ VPDB) | **-21.2** | **-26.5** | **-30.4** |
| | stand. dev. (‰ VPDB) | *0.3* | *1.1* | *0.4* |
| **3** | $\delta^{13}C$-values (‰ VPDB) | **-22.3** | **-27.6** | **-27.9** |
| | stand. dev. (‰ VPDB) | *0.8* | *0.7* | *0.4* |
| **4** | $\delta^{13}C$-values (‰ VPDB) | **-24.4** | **n.d.** | **-28.3** |
| | stand. dev. (‰ VPDB) | *0.7* | | *1.3* |
| **5** | $\delta^{13}C$-values (‰ VPDB) | **-24.6** | **-26.0** | **-30.9** |
| | stand. dev. (‰ VPDB) | *0.4* | *1.3* | *0.7* |

A third group is characterized by significant excursions of the carbon isotope ratios of up to 3 ‰ after application of the analytical procedures. It comprises hexachlorobutadiene, 2,4,7,9-tetramethyl-5-decyne-4,7-diol, cashmerane/2,6-di-*tert*-butyl-4-methylphenol, tris(2-chloroethyl)phosphate and N,N-diethyltoluamide (DEET). Interestingly, both halogenated compounds are included in this group reflecting also difficulties in precise carbon isotope analyses of chlorinated contaminants. The variations of the $\delta^{13}C$-values of the coeluting compounds cashmerane and ionol might be the effect of changing composition of the gas chromatographic peak as the result of different or fluctuating recovery rates. On the contrary the physico-chemical properties and, consequently, also the recovery rates of the insufficiently separated musk fragrances galaxolide and tonalide are very similar and, therefore, more accurate carbon isotope ratios were measured for these combined compounds as described above.

Two compounds were detected in both extracts. The $\delta^{13}C$-values of dimethylphthalate and 2-(2-ethoxybutoxy)ethylacetate analysed in the pentane and the dichloromethane extracts varied slightly. Generally, a slight enrichment of the heavier isotope was observed for the compounds in the more polar dichloromethane extracts. These observations suggest only a minor isotopic fractionation during the extraction procedures. However, this assumption is limited to a quite balanced distribution of the compounds.

With respect to the different shifts of $\delta^{13}$C-values in the recovery procedure it has to be noted, that a satisfactory determination of carbon isotope ratios depends not only on the analytical methods applied but also on the individual substances analysed. Therefore, in order to evaluate the quality of compound specific isotope analyses of riverine contaminants measurements of field samples have to be accompanied by recovery experiments of individual substances.

**Tab. 4:** $\delta^{13}$C-values of compounds detected in the pentane as well as in dichloromethane extracts of the recovery experiments.

| **Recovery experiments** | | **2-(2-Ethoxybutoxy)-ethylacteat** | | **Dimethylphthalate** | |
|---|---|---|---|---|---|
| **Original** | $\delta^{13}$C-values (‰ VPDB) | **-27.1** | | **-26.5** | |
| | stand. dev. (‰ VPDB) | *0.2* | | *0.3* | |
| | | **Pentane** | **DCM** | **Pentane** | **DCM** |
| **1** | $\delta^{13}$C-values (‰ VPDB) | **-27.2** | **-27.6** | **-26.5** | **-24.9** |
| | stand. dev. (‰ VPDB) | *0.5* | *0.7* | *0.3* | *0.5* |
| **2** | $\delta^{13}$C-values (‰ VPDB) | **-27.9** | **-26.1** | **-26.8** | **-26.2** |
| | stand. dev. (‰ VPDB) | *0.1* | *0.1* | *0.3* | *0.1* |
| **3** | $\delta^{13}$C-values (‰ VPDB) | **-27.4** | **-26.6** | **-27.1** | **-25.2** |
| | stand. dev. (‰ VPDB) | *0.1* | *0.2* | *0.7* | *0.4* |
| **4** | $\delta^{13}$C-values (‰ VPDB) | **n.d.** | **-27.5** | **-27.4** | **-26.5** |
| | stand. dev. (‰ VPDB) | | *2.0* | *0.8* | *0.4* |
| **5** | $\delta^{13}$C-values (‰ VPDB) | **-27.6** | **-25.8** | **-26.7** | **-25.6** |
| | stand. dev. (‰ VPDB) | *1.7* | *0.3* | *0.1* | *0.3* |

Further on, the matrix within the extracts might also affect the carbon stable isotope ratios. However, further purification or fractionation applied in order to minimize matrix effects is critical due to the low to very low amounts of the contaminants. A significant loss of numerous individual substances during further preparation steps and associated an isotopic fractionation has to be assumed. In particular during stable carbon isotope ratio analysis of a wide range of chemically different contaminants this adverse effects can appear. On the contrary, such approach might be helpful in analysing only a preselected and small set of pollutants of similar chemical and physical properties, for which the preparation can be optimized in terms of concentration and isotope shifts.

*Compound specific carbon isotope analyses of Rhine water samples*

Compound specific carbon isotope analyses were applied to 8 extracts of Rhine river water. Comparison of the chromatograms from GC/MS and GC/irmMS analyses shows a good correlation of the individual peaks as illustrated in Figure 4.

For the carbon isotope analyses a set of contaminants were selected considering the following criteria: (i) the substances were abundant in the river water extracts as revealed by GC/MS analyses, (ii) the determination of the stable carbon isotope ratios was expected to be successful as indicated by the results presented above and (iii) reference material had to be available in order to verifiy the identification of individual compounds in case of uncertain gas chromatographic correlations with GC/MS analyses. The following contaminants matched the described conditions and, consequently, their $\delta^{13}C$-values were determined in the Rhine water samples: Di-*n*-butylphthalate, bis(2-ethylhexyl)phthalate, galaxolide, *iso*-propylpalmitate, 2,4,7,9-tetramethyl-5-decyne-4,7-diol and 2,2,4-trimethyl-1,3-pentanedioldi-*iso*-butyrate. Furthermore, $d_{34}$-hexadecane was analysed as an internal standard. All results are presented in Figure 5. The attribution of the compounds analysed to the individual gas chromatographic peaks by GC-irmMS measurements based on the comparison of the retention times with those of reference substances. In addition the appearance or disappearance of all individual compounds in each sample were checked by GC-MS analyses.

Comparing the $\delta^{13}C$-values of the standard reference substances (see Tables 1 to 4) with the data analysed in the water samples consistent values were observed for di-*n*-butylphthalate and bis(2-ethylhexyl)phthalate. In addition the standard deviations of the water analyses ranging from 0.2 ‰ to 3.1 ‰ were very similar compared to the variations obtained from the recovery experiments (see Table 2).

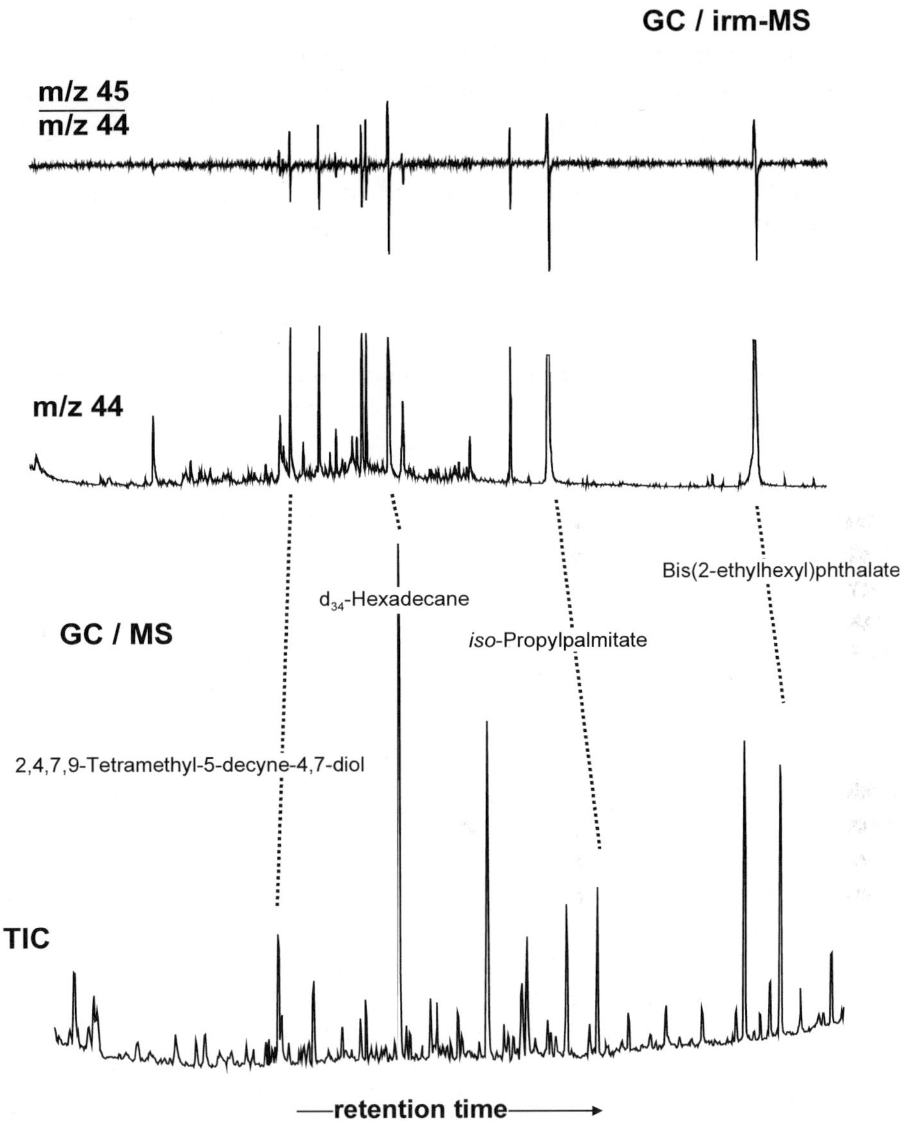


**Figure 4:** GC-irmMS chromatogram obtained from the pentane extract of water sample R2. Selected contaminants, which were identified and isotopically analysed, are marked.

This is in accordance with the quantitative data (unpublished results). Both phthalates occurred with concentrations between approx. 100 and 900 ng/L, which was considered to be sufficient for the determination of the carbon isotope ratios.

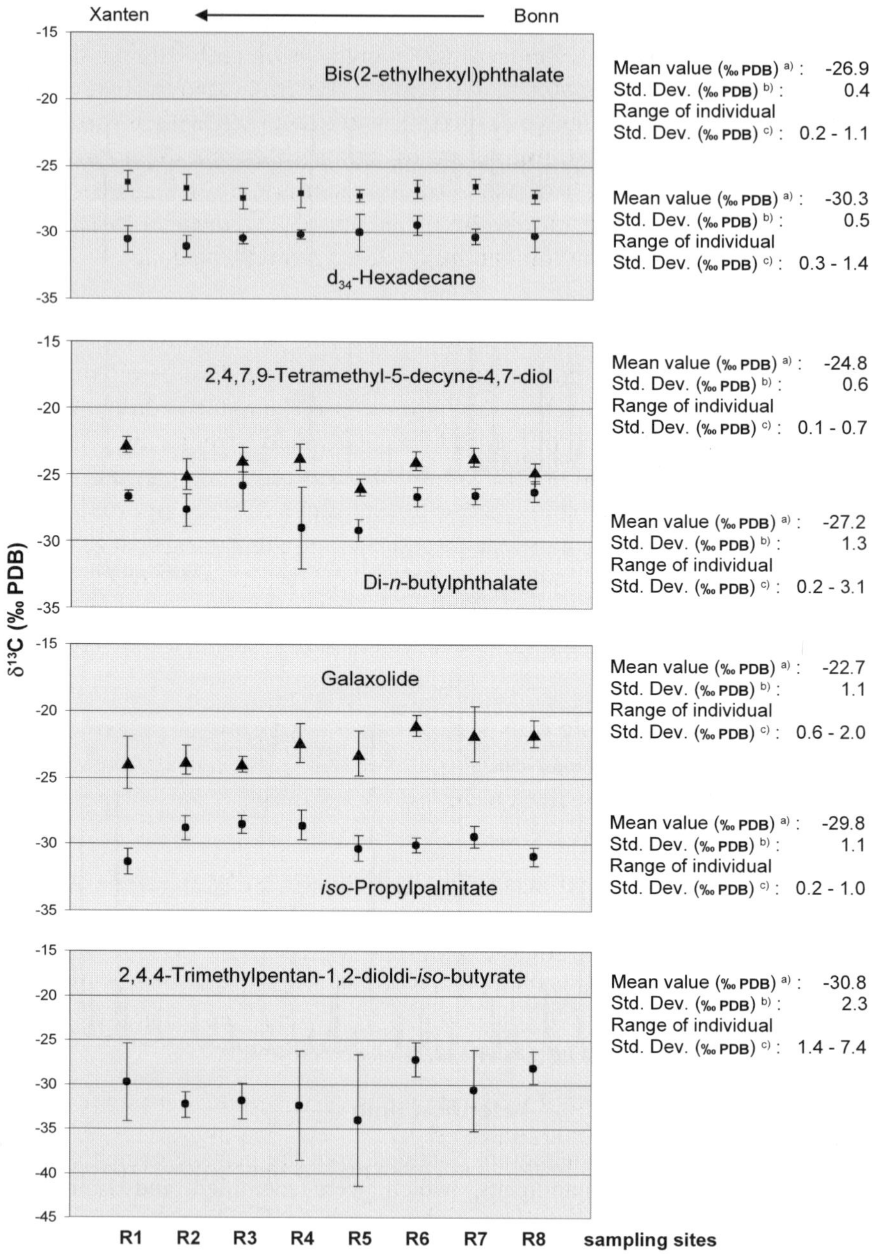


**Figure 5:** $\delta^{13}$C-values of selected contaminants in water samples from the Rhine river. On the right side the average value of the isotope ratios with the corresponding standard deviations obtained from the individual sampling locations is presented. (Calculation of the standard deviations was performed without consideration of the standard deviation of the individual data points).

On the contrary the $\delta^{13}$C-values of 2,2,4-trimethyl-1,3-pentanedioldi-*iso*-butyrate in the river water samples were significantly higher than the value of the reference material. However, the high standard deviations obtained from the analyses of the river water samples with up to 7.4 ‰ reflect an insufficient determination of the carbon isotope ratios. This might be mainly the result of very low concentrations of 2,2,4-trimethyl-1,3-pentanedioldi-*iso*-butyrate in the water samples (approx. 4 to 20 ng/L). Hence, no assured information can be obtained from these data.

For the interpretation of the distribution of the carbon isotope ratios within the riverine system investigated two different variations have to be considered. If the mean values summarized from all data of all sampling locations vary with a higher deviation than their individual standard deviations as well as the variations obtained from the recovery experiments a significant alteration of the isotopic composition of the contaminants can be inferred. These assumptions were supported by the $\delta^{13}$C-values of $d_{34}$-hexadecane. As expected for the internal standard the variation of the individual data points (standard deviation between 0.5 and 1.4 ‰) was in the same range as compared to the standard deviation derived from the data of all sample extracts analysed (standard deviation 0.5 ‰).

Significant variations are only obvious for the substances galaxolide and *iso*-propylpalmitate, because most or nearly all individual data points deviated less than the variation of the $\delta^{13}$C-values in the longitudinal profile (standard deviation 1.1 ‰).

On the contrary the variation of the individual $\delta^{13}$C-values of bis(2-ethylhexyl)phthalate in the range between 0.2 and 1.1 ‰ did not differ significantly as compared to the variation of the data derived from all samples (standard deviation 0.4 ‰). Hence no significant alteration of the composition of the stable carbon isotopes can be stated for this contaminant within the river water samples investigated.

With respect to di-*n*-butylphthalate most of the values analysed in the individual samples are characterized by a significantly lower variation (standard deviation 0.2 to 1.0 ‰ ) as compared to the summarized standard deviation derived from all samples (1.3 ‰). However, the $\delta^{13}$C-values analysed in extracts derived from sampling locations 5 and 6 exhibit higher standard deviations up to 3.1 ‰, hence the information obtained from these data are limited.

For 2,4,7,9-tetramethyl-5-decyne-4,7-diol the information derived from the stable carbon isotope ratios are also limited as the result of a non perfect base line separation in a couple of samples. This is illustrated in Fig. 4. In the TIC a minor peak at the right flank of 2,4,7,9-tetramethyl-5-decyne-4,7-diol was overlapped in the m/z 44 trace by a broad $CO_2$ peak. Hence, the variation of $\delta^{13}C$-values of 2,4,7,9-tetramethyl-5-decyne-4,7-diol can be induced by coeluting substances.

Generally, significant variations of carbon isotope ratios along the river section investigated has to be attributed either to an superimposition of several emissions sources discharging contaminants with different carbon isotope compositions or to processes within the river system modifying the $\delta^{13}C$-values of the affected substances. Those effects can be the result of degradation or transformation as well as of transfer processes within the aquatic environment.

*Compound specific carbon isotope analyses of Lippe water samples*

Additionally, 19 Lippe water samples collected in 1999 were investigated by compound specific isotope analysis. Also a sufficient detection of numerous anthropogenic contaminants was achieved in terms of adequate concentration levels and effectual gas chromatographic separation as illustrated in Fig. 6.

The isotopic analyses applied to the water samples derived from the Lippe river longitudinal profile comprised main contaminants as illustrated above and described intensively in chapter 3.1.1. In detail stable carbon isotope ratios of tri-*n*-butyl phosphate (TBP), tris(chloroethyl)phosphate (TCEP), 2,4,7,9-tetramethyl-5-decyne-4,7-diol (TPDB), di-*n*-butylphthalate, bis(2-ethylhexyl)phthalate (DEHP), galaxolide, tonalide, and squalene were determined. Further on, the internal standard $d_{34}$-hexadecane was analysed simultaneously. Noteworthy, several compounds presented were also determined in the Rhine river samples. All results are summarized in Figure 7.

The mean values of the summarized $\delta^{13}C$-values ranged between –24 ‰ and –30 ‰. with standard deviations along the longitudinal profile in the range between 0.8 ‰ and 6.2 ‰. These data supports the observations and conclusions drawn from the Rhine water analyses. Higher variations of isotopic data along the river sections reflect significant changes and alterations within the composition of individual contaminants as the result of transformation or partition processes as well as of superimposed emissions derived from different sources. The excessive deviation of the tonalide data has to be attributed to minor concentrations as discussed for the Rhine river contaminant 2,2,4-trimethyl-1,3-pentanedioldi-*iso*-butyrate (see chapter 4.2.1).

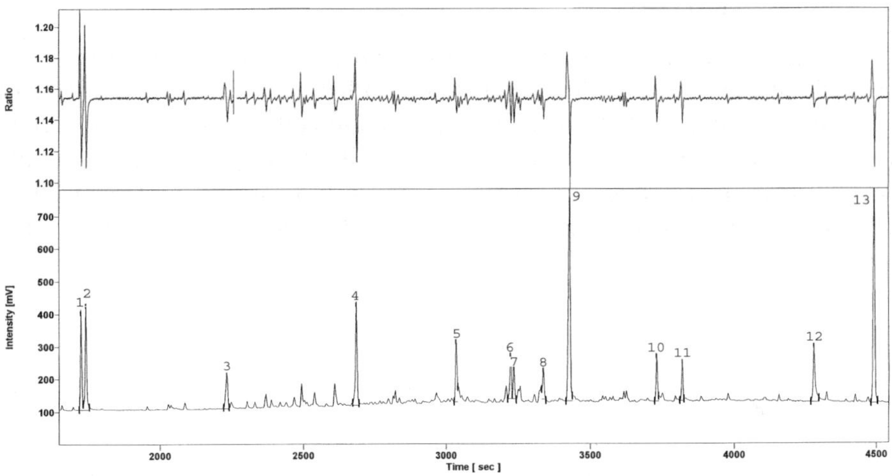


**Fig. 6:** GC/irmMS chromatgramm of a pentane extract of a Lippe water sample accomapnied by the m/z 45/44 ratio trace. Peak 1 and 2 = (hydroxypropoxy)propanols (-31.4‰ and –26.8‰) , 3 = 2,4,7,9 -tetramethyl-5-decine-4,7-diol (-24.4‰), 4 = $d_{34}$-hexadecane (- 29.4‰, internal standard), 5 = $d_{10}$-anthracene (-21.3‰, internal standard), 6 = Galaxolide (-25.1‰), 7 = tonalide (–29.1‰), 8 = 2,4-bis(1,1-dimethylpropyl)-phenol (insufficient peak separation), 9 = di-*n*-butylphthalate (-25.3‰), 12 = $d_{12}$-chrysene (-22.8‰, internal standard), 13 = DEHP (-27.5‰).

The average values of the substances analysed in both Rhine and Lippe water samples corresponded very well (see. Fig. 5 and Fig. 7). The averaged data of DEHP (-26.9 ‰ R (Rhine) / -26.6 ‰ L (Lippe)), DnBP (-27.2 ‰ R / -26.1 ‰ L), TPDB (-24.8 ‰ R / -24.4 ‰ L ) and of the internal standard (-30.3 ‰ R / -29.8 ‰ L) differed only by 0.3 ‰ to 1.1 ‰. Significant different $\delta^{13}$C-values were observed for galaxolide (-22.7 ‰ R / -26.7 ‰ L). Comparing the corresponding standard deviations derived from the fluctuations along the longitudinal profiles of both river systems a slightly higher variation was observed for the Lippe river samples. These differences might be the result of a more complex emission situation at the Lippe river system as formerly described (see chapter 3.1.1. and 3.1.2 or Dsikowitzky et al., 2004a,b)

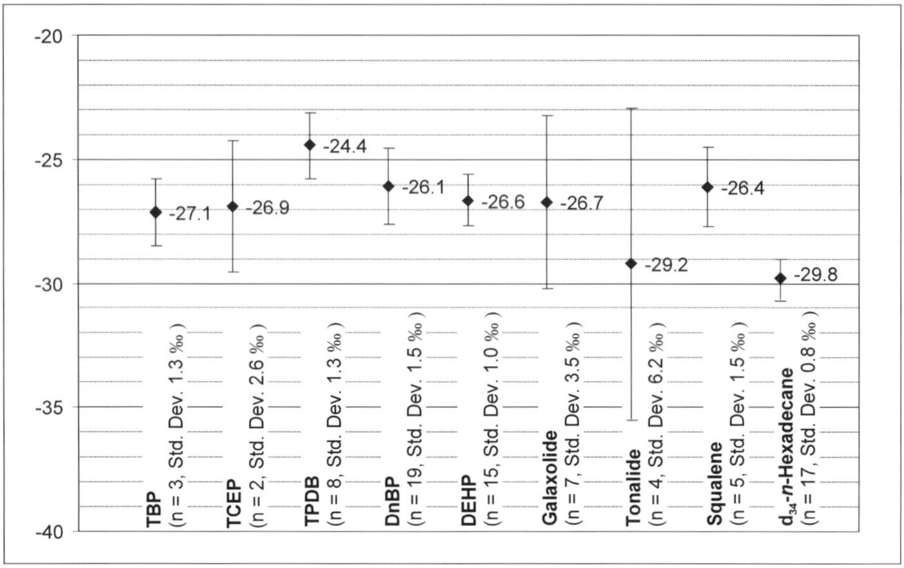


**Fig. 7:** Averaged stable carbon isotope ratios of main contaminants detected in 19 Lippe water samples derived from a longitudinal profile. Additionally the standard deviations and the number of detections are presented.

## Conclusions

The presented results obtained from compound specific carbon isotope analyses applied to standard solutions under various conditions support the following conclusions:

- For most of the anthropogenic contaminants investigated carbon isotope analyses were performed with an acceptable standard deviation down to amounts of approx. 5 ng absolutely applied to the gas chromatograph. These amounts correspond to concentrations in water samples at a natural abundance level of low to medium contaminated river systems. At lower amounts standard deviations of $\delta^{13}$C-values of individual compounds increase significantly.

- The precision as well as the sensitivity of the analytical methods depend partially on the chemical properties of the substances measured. Especially higher chlorinated compounds exhibited $\delta^{13}$C-values with elevated standard deviations. This fact is tentatively attributed to lower carbon contents of these compounds and to interactions of halogens or tin with the oxidation catalysts.

- Isotopic shifts or higher variations of the isotope ratios as a result of the analytical procedures applied were observed only for selected contaminants. Therefore, for accurate compound specific carbon isotope analyses of riverine contaminants it is recommended to conduct supplementary recovery experiments of the individual substances.

These conclusions were confirmed by stable carbon isotope analyses performed on river water extracts derived from the Rhine river. Using a sequential extraction procedure the determination of stable carbon isotope ratios of several riverine contaminants could be achieved by an appropriate gas chromatographic separation. Comparing the variation of the data of the individual compounds with the deviations obtained from the recovery experiments it was possible to differentiate contaminants with unaffected isotope ratios and substances with significant alterations of the $\delta^{13}C$-values. These significant variations reflect either multiple emission sources of different isotopic quality or environmental processes modifying the isotopic signature of the individual substances. With respect to the analyses of 2,2,4-trimethyl-1,3-pentanedioldi-*iso*-butyrate the low concentrations of riverine contaminants were pointed out as a major limitation of the analytical method.

## 4.3 The bound residues in riverine particulate matter

### 4.3.1 Anthropogenic organic contaminants incorporated into the non extractable particulate matter of riverine sediments from the Teltow Canal (Berlin)‡

#### Introduction

In urban and industrial regions the organic matter in riverine sediments is highly controlled by the anthropogenic input due to enhanced emissions of organic contaminants and pollutants. Not only the qualitative and quantitative composition of the extractable fraction but also of the non-extractable organic matter is affected by anthropogenic contributions. In contrast to the numerous investigations dealing with the occurrence and fate of low molecular weight pollutants in water and particulate matter of riverine systems only a few studies were carried out in order to analyse the alteration within the non-extractable fraction. This alteration via anthropogenic pollution was observed for both the macromolecular substances and the associated low molecular weight compounds.

The anthropogenic contribution to the macromolecular organic matter of riverine systems can be generally attributed to three different processes:

- Low molecular weight substances can be strongly incorporated into bio- or geopolymers (e.g. humic substances).
- Natural polymers can be alterated by technical processes (e.g. chlorination of drinking water or bleeching processes of paper) and released into the aquatic environment.
- Xenobiotic polymers can also be emitted into the aquatic environment (e.g. polysiloxanes).

Different kinds of anthropogenic compounds and their occurrence within the non-extractable matter are reported in a couple of studies. Most of the investigations are related to the occurrence and fate of associated low molecular contaminants, the so called "bound residues", and are published within the last 30 years (e.g. Li and Felbeck 1972; Kaufman et al. 1976; Liechtenstein et al. 1977; Wheeler et al. 1979; Liechtenstein 1980; Khan 1982, Boul et al. 1994; Lichtfouse 1997; Houot et al. 1997; Northcott and Jones 2000). Earlier investigations dealed especially with

‡ Mainly adapted/reprinted from Schwarzbauer et al. 2005a

the occurrence and fate of bound pesticides in soils. After introducing the pesticides and their metabolites into soils, sediments or waters a weak association of a significant proportion to geopolymers was pointed out leading to insufficient re-extraction rates of the observed compounds by means of regular solvent extraction procedures. Recent studies confirmed this environmental behaviour in soils for a couple of organic contaminants such as atrazine, 2,2′,5,5′-tetrachlorobiphenyl, 3,4-dichloroaniline, naphthalene and chlorinated phenols (Hsu and Bartha 1976; Palm and Lammi 1995; Barriuso and Houot 1996; Kan et al. 1997). The quota of the bound fraction ranged between 25 and 90 %. The linkage to the geopolymers covers a wide diversity of modes ranging from weaker interactions like adsorption or van der Waals forces up to strong ionic interactions and covalent bonds. The mechanism of incorporation depends on a variety of chemical and physico-chemical properties and conditions, e.g. functional groups within the molecules, pH and redox potential as well as charactersistics of the geopolymer (e.g. Ziechmann 1972; Parris 1980; Senesi 1992; Piccolo et al. 1992; Schulten and Leineweber 1996; Pignatello and Xing 1996; Luthy et al. 1997; Klaus et al. 1998a/b; Nanny 1999; Weber et al. 2001).

The occurrence of bound residues is not only restricted to soils but also to aquatic sediments and suspended matter. Similar associations and assimilations of low molecular weight compounds into aquatic geopolymers, aggregates or organo-mineral complexes were observed (e.g., Murphy et al. 1990; Chin and Gschwend 1992; Buffle and Leppard 1995; Klaus et al. 1998; Zwiener et al. 1999). An important aspect especially in riverine systems is the mobilisation of pollutants associated with colloids, suspended particulate matter or dissolved geopolymers and the subsequent enhanced spatial distribution (s.a. McCarthy and Zachara 1989; Johnson and Amy 1995). Modified transport processes in the presence of aquatic geopolymers are reported for pyrene and the pesticides amitrol, terbutylazine as well as pendimethaline (Huber et al. 1992; Herbert et al. 1993; Oesterreich et al. 1999).

Not only the mode of association but also the strength of bonding and the reversibility of the incorporation process are important for the microbial and abiotic degradation as well as toxicological aspects. Due to a limited bioavailability the toxicological effects of bound organic pollutants are generally reduced in comparison to the free substances (e.g. Lichtfouse 1997). Several investigations demonstrated a decrease of toxicity with progressive aging after application of pollutants to soils without a change of the absolute concentrations (Liechtenstein et al. 1977; Robertson and Alexander 1998). An increasing toxicity for substances (e.g. 2,4-dichlorophenol) in the presence of geopolymers was hardly reported (Steinberg et al. 1992).

In addition, the mode and rate of degradation processes of individual substances can be modified by the incorporation into geopolymers and the subsequent decrease of bioavailability. Frequently a higher stability or even persistence was observed for bound contaminants (e.g. Perdue and Wolfe 1982). Considering geochemical and biogeochemical cycles, the conservation of natural organic compounds in association with geopolymers is well investigated (Tissot and Welte 1984; Engel and Macko 1993). For bound residues not only a higher persistence but also modified degradation or transformation pathways were observed, e.g. for amitrole (Oesterreich et al. 1999). Jensen-Korte et al. (1987) reported an enhanced photolytic degradation of normally persistent pesticides in an aquatic environment induced by the addition of humic substances. Modified transformation processes were also observed in the case of an incorporation of metabolites into geopolmyers. Variations in the degradation pathway via the transfer of metabolites into the non-extractable matter has also to be assummed in the case of PCB and PAH. Richnow et al. (1994) and Michaelis et al. (1995) demonstrated an incorporation of chlorinated benzoic acids and polycyclic aromatic acids, wellknown metabolites of PCB and PAH, by means of covalent bonds. An association of mainly hydroxylated and/or dealkylated metabolites was formerly reported for atrazine in soils (Capriel et al. 1985).

Next to the alteration of non-extractable organic matter by bound residues, several anthropogenic activities also modify directly organic geopolymers. Following discharge these modified polymers affect mainly the aquatic environment. The most important technical modification of biopolymers and aquatic geopolymers are chlorination processes. Hence, in recent studies several chlorinated lignin-, humin- and cellulose-derived macromolecules were identified in the aquatic environment, mainly in Scandinavia (Dahlmann et al. 1993; Hyötyläinen et al. 1998a +1998b; Miikki et al. 1999), but also in Portugal, Germany and the Netherlands (Bultermann et al. 1997; Santos and Duarte 1998).

In addition to the input of modified natural macromolecules a further mode of anthropogenic alteration of the non-extractable organic matter is the emission of technical macromolecular products. Only very few investigations were reported concerning these emissions of xenobiotic polymers. Examples are the investigations by Fabbri et al. (1998a), as well as Requejo et al. (1985), characterising the input of polystyrene into the coastal and riverine aquatic environment via characteristic pyrolysis products as low molecular weight markers.

Considering the nature of the non-extractable organic matter the analytical methods used for the chemical characterisation and quantification can be divided generally into two different kinds of approaches. Non-destructive methods include IR-, UV/VIS-, NMR- and ESR-spectroscopy, also in combination with liquid chromatography or size exclusion chromatography (e.g. Burns et al. 1973; Senesi et al. 1987; Schlautmann and Morgan 1993; Fabbri et al. 1998b; Nanny 1999; Zwiener et al. 1999). Destructive analytical approaches transfer the bound or macromolecular fraction into low molecular weight compounds either by pyrolysis or by chemical degradation. Using pyrolysis different procedures were applied including on-line and off-line methods, with or without derivatisation of the products and subsequent gas chromatographic or gas chromatographic - mass spectrometric analysis (e.g. Horsfield et al. 1989; Schulten and Leineweber 1996; Stankiewicz et al. 1998; Asperger et al. 1999; Mongenot et al. 1999). Additionally, appropriate chemical degradation procedures allow the selective release of components depending on the mode of incorporation or binding. Common degradation reactions include acidic hydrolysis (e.g. Grasset and Ambles 1998), ether and ester cleavages using boron tribromide or boron trichloride (e.g. Richnow et al. 1994) or oxidation e.g. with CuO or $RuO_4$ (e.g. Hatcher et al. 1993; Hyötyläinen et al. 1998a). The degradation products were usually extracted, fractionated and analysed by traditional LC, GC or GC-MS. Using a sequential degradation approach with an increasing order of reactivity, a differentiation of individual modes of binding can be indicated for associated substances.

On the contrary, the modes of incorporation are also investigated on a laboratory scale by spiking appropriate samples with labelled ($^{14}C$, $^{13}C$, D) or non-labelled model compounds. Following, these artificial bound residues were released and characterized by selective analytical methods as described earlier (e.g. Hatcher et al. 1993; Richnow et al. 1998; Guthrie et al. 1999).

Recent reviews on the analytical topics are given by Northcott and Jones (2000), as well as Kögel-Knabner (2000).

In summary, anthropogenic activities cause significant alterations of the macromolecular organic matter (MOM) in riverine and lacustrine systems mainly by emission of pollutants and their subsequent incorporation into geopolymers (bound residues).

In the presented study we have characterized highly polluted MOM of riverine sediments (Spree River, Teltow Canal, Germany) in order to investigate the occurrence, alteration and distribution of several organic xenobiotics *in situ*. Thus, these investigations intend a comprehensive characterization of riverine MOM via combining different analytical techniques (pyrolytic analyses and chemical degradation techniques) in order to provide information concerning the incorporation mechanisms and the mode of binding for a variety of organic pollutants with different chemical properties.

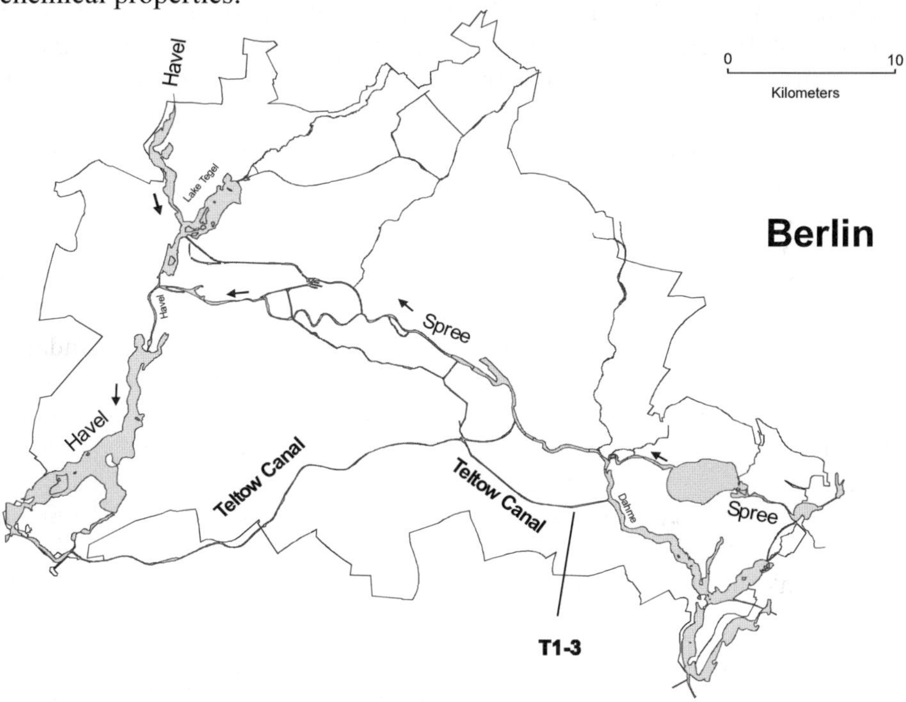


**Fig. 1:** Sampling location at the Teltow Canal, Berlin (Germany).

## Samples

Four sediment samples were taken in 1998 and 1999 from three locations at the Teltow Canal in Berlin, as indicated in Fig. 1. In addition to the surface sediment samples T1 and T3 taken by means of a 4L Ekman-Birge grab sampler a short sediment core T2 was obtained by using a tube coring device. The sediment core was subdivided into an upper part T2a (0-3 cm) and a lower part T2b (3-10 cm), that represents an older accumulation time between 1980 and 1990.

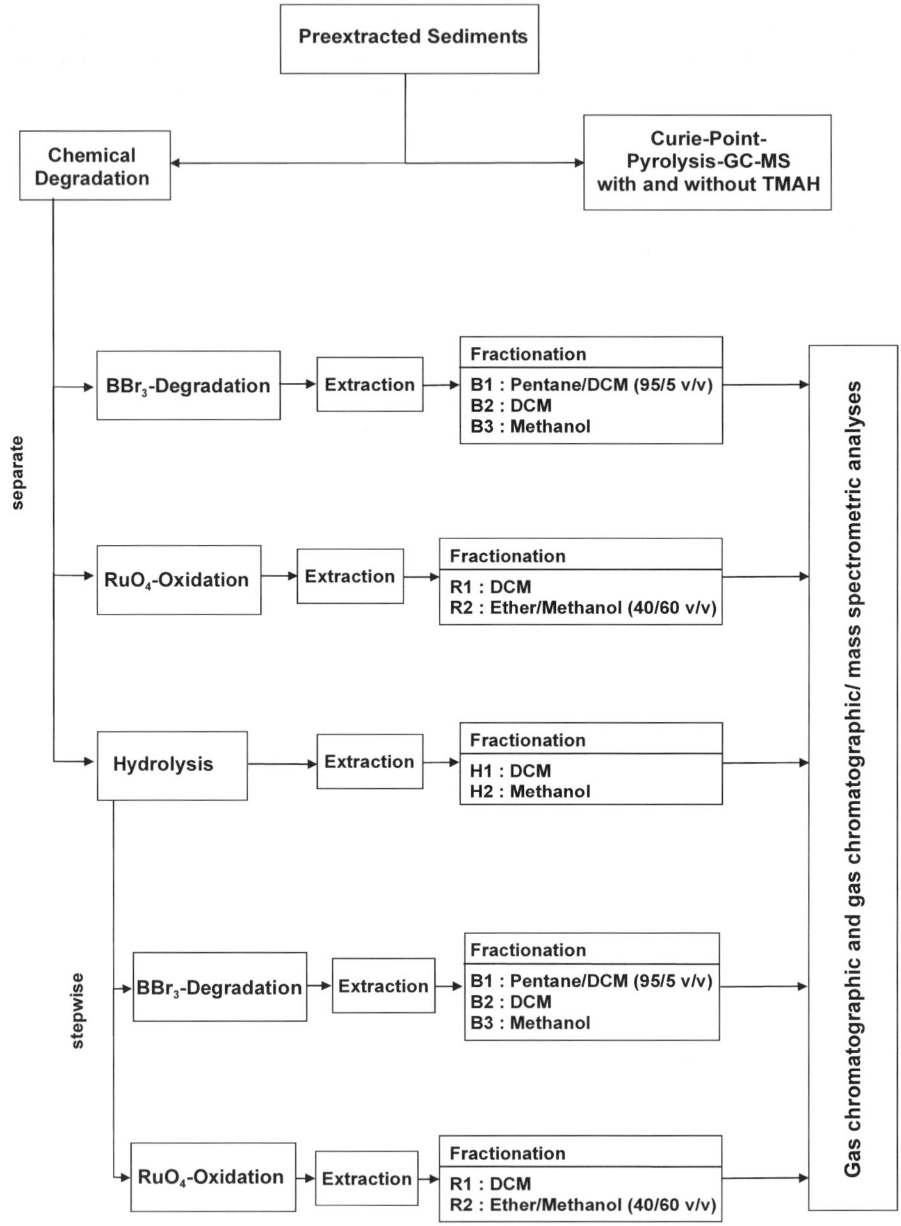


**Fig. 2:** Analytical flow scheme

All sediment samples were pre-extracted with methanol, butanol and solutions of hexane/acetone. The principal analytical flow scheme applied to the samples for the detection and determination of bound residues is given in Fig. 2.

## Results

### *Pyrolysis*

The main compounds yielded by flash pyrolysis are transformation or degradation products of biogenic precursors. Next to the amino acid glycine 1 most of the identified compounds are structurally related to carbohydrates, amino acids and condensed molecules of both components resulting from Maillard reactions. Examples include furfural 2, methylfurfural 3 and pyrrol-2-carboxaldehyde 4 (see Fig. 3).

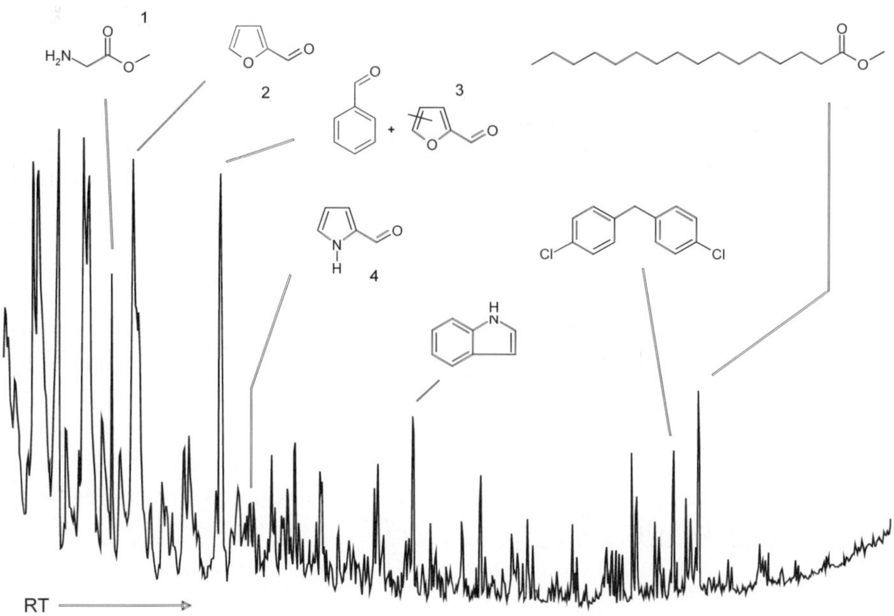


**Fig. 3:** Curie-point pyrolysis gas chromatogram of the non-extractable residue of a Teltow Canal sediment (Curie-point temperature 510° C).

Apart from these obviously biogenic compounds only very few definite anthropogenic substances were identified. Within the group of xenobiotics 2,4′- and 4,4′-dichlorophenylmethane, DDT-derived metabolites, were most abundant. A pyrolytic conversion of related DDT metabolites to DDM cannot be excluded. Thus the occurrence of DDM in the pyrogram is only suggestive for DDT group metabolites in general.

*Chemical degradation*

Apart from an alkaline hydrolysis affecting mainly ester and amide bonds a boron tribromide treatment and a ruthenium tetroxide oxidation were applied to the extracted residues. The $BBr_3$-treatment cleaves aromatic and aliphatic ether and ester bonds. The oxidation using $RuO_4$ attacks the aromatic carbon atoms and releases, therefore, mainly unaltered aliphatic moieties.

Depending on the degradation method a variety of biogenic compounds was identified reflecting the natural contribution to the MOM. Dominating groups of substances are illustrated in Fig. 4.

Hydrolysis revealed mainly fatty acids, fatty alcohols and long chain *n*-amides accompanied by unsaturated and branched isomers. In addition, steroid alcohols appeared in high amounts, but noteworthy the oxidized analogeous, the steroid ketones (e.g. cholestanone or coprostanone) were not present. Therefore a covalent association of the hydroxylated isomers by ester bondings is evident.

Boron tribromide cleavage released long chain carboxylic and dicarboxylic acids including their hydroxylated isomers as well as brominated alkanes, reflecting long chain aliphatic units of the naturally occuring macromolecules. The aromatic moieties were represented by brominated benzenes and alkylated homologues. Also most of the substances released after rutheniumtetroxid oxidation can be attributed to the aromatic proportion within the bio-/geopolymers. Examples including methoxylated and alkylated benzenes, methoxylated phenoles and phenylalkyl carboxylic acids.

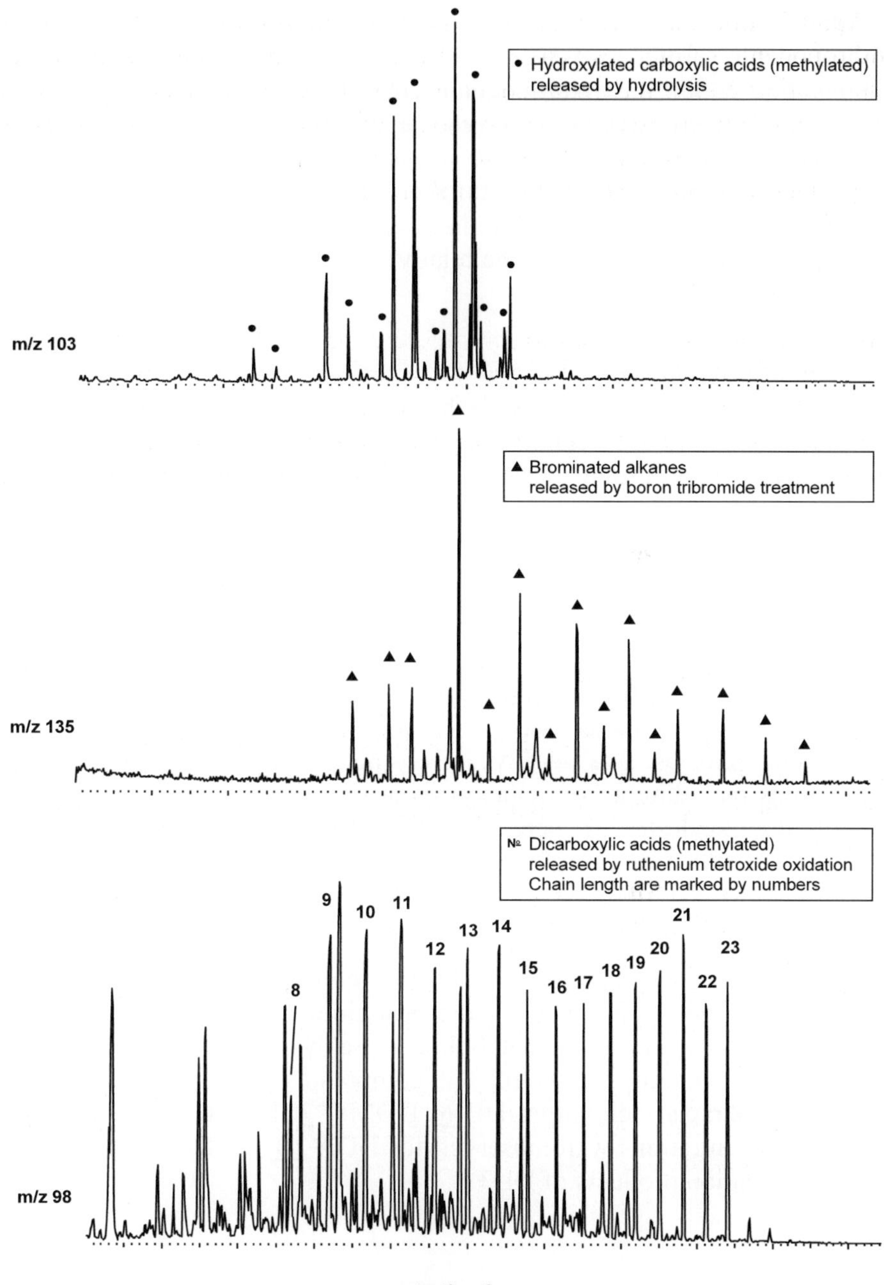


**Fig. 4:** Ion chromatogramms of selected substances identified in the non-extractable residue of Teltow Canal sediments reflecting the chemical composition of the natural macromolecular organic matter.

Additionally, in all degradation product mixtures numerous anthropogenic compounds were identified. Variations of the chemical composition within the anthropogenic proportion of the reaction mixtures were observed depending either on the degradation mode or on the sample. The following presentation of results, subdivided and arranged by the type of degradation reaction, focuses on the occurrence of the anthropogenic contaminants. All identified substances are summarized in Tab. 1 with a semi-quantitative estimation of the relative concentrations.

Hydrolysis

Hydrolysis revealed numerous anthropogenic compounds that can be attributed to different technical applications or widespread domestic usages. Within the group of plasticizers the phthalates, which are well-known and ubiquiteous pollutants, were most abundant. Also the plasticizers tributylphosphate and 2,4,4-trimethylpentane-1,3-dioldi-*iso*-butyrate occurred in minor concentrations mainly in sample T1. Due to their molecular structures a non-covalent association of these compounds to the geopolymers has to be assumed.

In addition a significant contribution of technical additives to the non-extractable organic matter was observed. Mainly the isopropyl ester of palmitinoic acid, used as additive in cosmetics and washing agents, and compounds including a 2-ethylhexyl moiety (2-ethylhexanol, hexanedioic acid 2-ethylhexylester 5) were identified. Both, the *iso*-propyl and the 2-ethylhexyl moieties (see Fig. 5), are very probably of anthropogenic origin due to the absence or rarely occurrence of similar molecular substructures within the biogenic compounds.

Furthermore as a result of industrial emissions the Teltow Canal sediments are highly polluted by the pesticides DDT and methoxychlor, accompanied by several metabolites (Schwarzbauer et al. 2001). Accordingly, numerous DDT-related compounds were identified in the hydrolysis extracts including DDE 6, DDMU 7, DDNU 8 and DDM 9. Highest concentrations were observed for DDA 10 and DBP 11, the more polar degradation products of DDT.

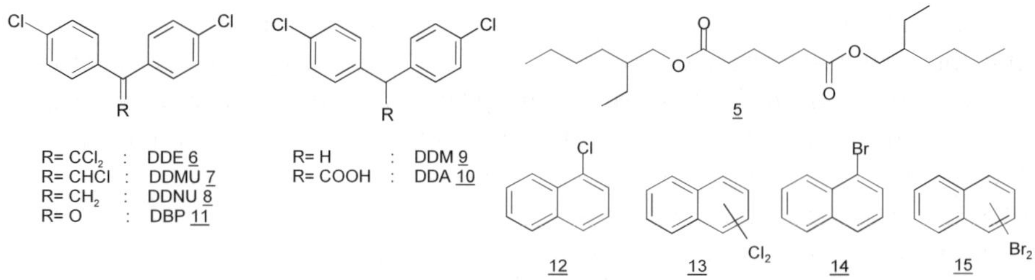


**Fig. 5:** Molecular structures of selected anthropogenic compounds released by hydrolysis

A second group of specific xenobiotics in Teltow Canal sediments are halogenated aromatics. Several chlorinated and brominated mono- and diaromatic hydrocarbons were detected in high amounts within the extractable organic matter as reported previously (Schwarzbauer et al. 2001). The halogenated arenes identified in the hydrolysis extracts included mono- and dichlorinated naphthalenes 12+13, mono- and dibrominated naphthalenes 14+15, tetra- to hexachlorinated biphenyls (PCB) and 2,4,6-tribomoaniline. The peak pattern of the chlorinated naphthalenes was similar to the congener distribution in technical mixtures e.g. Halowax 1000 (Falandysz 1998).

## $BBr_3$-treatment

The treatment of the extracted residues with the Lewis acid boron tribromide revealed numerous compounds that were also detected in hydrolysis extracts (see Fig. 6). Examples including hexanedioic acid 2-ethylhexylester 5, 2,4,6-tribromoaniline 16 as well as galaxolide 17 and tonalide 18, persistent synthetic musk substitutes widespread used as fragrances in soaps, perfumes, detergents and other household cleaning products. In addition, low amounts of bisphenol A 19, used as plasticizer, fungicide and intermediate in polymer syntheses, were detected. For bisphenol A estrogenic activities were evident (Safe and Gaido, 1998). Within the group of xenobiotics the DDT-related compounds became most abundant with DBP 11 as the main component.

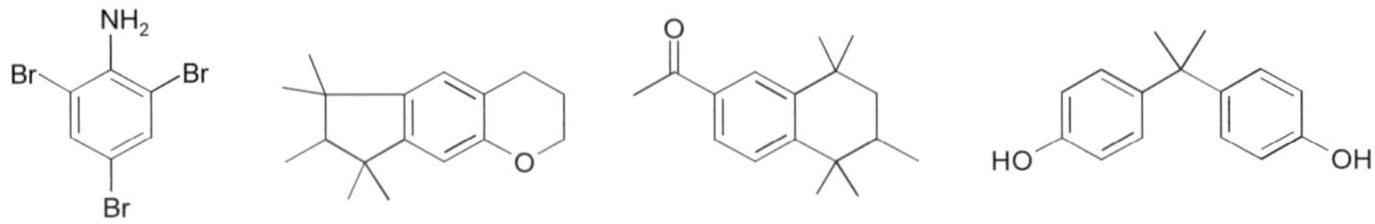

16 17 18 19

**Fig. 6:** Molecular structures of selected anthropogenic compounds released by boron tribromide teatment

Noteworthy, the boron tribromide treatment applied to saponified residues generated a slightly different pattern of compounds. As a result of the sequential degradation procedure additional compounds were identified and, furthermore, higher concentration of selected individual substances, detected in both $BBr_3$ extracts, were observed. In detail, DDA, hexanedioic acid di-*iso*-propyl ester and brominated naphthalenes were observed exclusively in the $BBr_3$ extract of the sequential procedure.

The origin of the brominated phenols is not obvious. As the most important result of boron tribromide treatment aliphatic and aromatic ethers are transformed to the corresponding alcohols and bromides. Hence, the occurrence of brominated phenols exclusively in the $BBr_3$ extracts indicated either a biogenic origin as a result of the cleavage of mono- to tetraalkoxylated aromatic substructures. However, the lack of further brominated aromatic alcohols (e.g. brominated catechols) contradicts this assumption. Alternatively, the existence of brominated phenoxy moieties linked by covalent ether bonds to the nonextractable particulate matter can be assumed, although not proven. It has to be noted, that a technical application and a subsequent emission of mono- and dibrominated phenols to the aquatic environment has not been reported so far. On the contrary, brominated phenols and anisols are well-known organohalogens derived from biogenic formation, but exclusively detected in the marine environment (Ballschmitter 2003). Thus, the origin of brominated phenols in the non-extracted matter of Teltow Canal sediments is still ambiguous.

## $RuO_4$-Oxidation

GC-MS analyses of the extracts obtained after rutheniumtetroxide oxidation revealed either compounds previously described as hydolysis or $BBr_3$ treatment products or novel compounds only occuring in $RuO_4$ extracts. Examples for the first group of anthropogenic contaminants are linear alkylbenzenes (LAB) with 11 to 13 carbon side chain length, bisphenol A 19, DDA, DBP, brominated naphthalenes and the pesticide chloropropylate 20 (see Fig. 7).

Numerous individual substances were detected only in $RuO_4$ extracts e.g. di- to pentachlorinated benzenes, 4-chlorobenzoic acid and 2,4-dichlorobenzoic acid 21, hexachlorocyclohexanes (α-,β-,γ- and δ-HCH) 22, a technical mixture obtained during the synthesis of lindane, and the plasticizers alkylsulfonic acid phenylesters 23. These plasticizers were recently identified in riverine sediments (Franke et al. 1998). Furthermore, nitro-substituted benzoic acid and alkylated phenols 24 were observed. The occurrence of aromatic nitro compounds as a result of the oxidation of anilines can be excluded due to the contemporary appearance of amino compounds, e.g. 4-aminobenzoic acid or N-ethylaniline. However, the origin as well as the emission pathway of these compounds is still unknown.

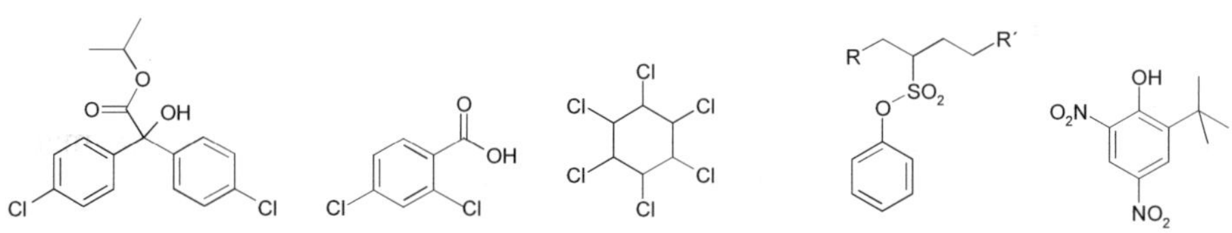

20 21 22 23 24

**Fig. 7:** Molecular structures of selected anthropogenic compounds released by ruthenium tetroxide oxidation

Due to the fact that $BBr_3$ degradation affects also ester linkages no phthalates were observed. Accordingly, an increasing amount of the degradation product phthalic acid was observed.

In contrast to the enhanced release of organic compounds by a sequential application of hydrolysis and $BBr_3$, the sequential procedure led in the case of $RuO_4$ to a minor portion of released organic compounds. Hence specific contaminants (e.g. halogenated arenes, nitro compounds) were not observed within the extracts of $RuO_4$ oxidation products applied to the saponified residues.

**Tab. 1a:** Organic contaminants identified in the non-extractable organic matter of Teltow Canal sediments after separate chemical degradation procedures.
o ; + ; ++ ; +++ = very low to high concentration
(me) = identified as methyl ethers; (m) = identified as methyl esters
( ) = not all isomers detected

| | T1 (u) | T1 (l) | T2 | T3 | T1 (u) | T1 (l) | T2 | T3 | T1 (u) | T1 (l) | T2 | T3 |
|---|---|---|---|---|---|---|---|---|---|---|---|---|
| **Compounds** | **$BBr_3$ treatment** | | | | **$RuO_4$ oxidation** | | | | **Hydrolysis** | | | |
| *Technical additives, solvents* | | | | | | | | | | | | |
| N-Methylpyrrolidone | | | | | | | | + | + | | | + |
| N-Ethylaniline | o | | | | | | | | | | | |
| Benzophenone | | | | | + | | | | | | | |
| Benzylbenzoat | | | | | + | | | | | | | |
| Dibutylmaleat | | | | | o | | | | | | | |
| Butylpalmitat | | | | | o | | | | | | | |
| Isopropyldodecanoate | | | | | o | | | | | | | |
| Isopropylpalmitate | | | | | + | | | + | ++ | ++ | + | + |
| Hexandioic acid di-*iso*-propyl ester | | + | | | + | | | | | | | |
| 2-Ethylhexanol | | | | | | | | | ++ | ++ | | |
| Hexandioic acid 2-ethylhexyl ester | ++ | + | | | + | + | + | | + | + | | |
| Bisphenol A (me) | + | o | o | o | | + | | o | | | | + |
| Ionol | | | | | + | | o | o | | o | o | o |
| Di-*tert*-butylchinone | | | | | + | | | | | | | |
| *Detergent residues, fragrances* | | | | | | | | | | | | |
| Linear alkylbenzenes | | | | | | o | o | o | | | o | o |
| Galaxolide | o | | | o | | | | + | | o | | |
| Tonalide | o | | | | | | | o | | o | | |
| *Plasticizers* | | | | | | | | | | | | |
| Alkylsulfonic acid phenyl esters | | | | | o | | o | o | | | | |
| Tri-*n*-butylphosphate | | | | | ++ | + | ++ | | o | o | o | |
| 2,4,4-Trimethylpentan-1,3-dioldi-*iso*-butyrate | | | | | + | o | | | + | + | | |
| Di-*n*-butylphthalate | o | | + | | ++ | | | | + | + | + | o |
| Di-*iso*-butyl-phthalate | | | + | | ++ | | | | + | | | o |
| Bis(2-ethylhexyl)-phthalate | + | | ++ | | ++ | + | | | + | + | + | + |
| Tris(2-ethylhexyl)-trimellitate | | | | | + | | | | | | | |

| | T1 (u) | T1 (l) | T2 | T3 | T1 (u) | T1 (l) | T2 | T3 | T1 (u) | T1 (l) | T2 | T3 |
|---|---|---|---|---|---|---|---|---|---|---|---|---|
| **Compounds** | **$BBr_3$ treatment** | | | | **$RuO_4$ oxidation** | | | | **Hydrolysis** | | | |
| Phthalic acid 2-ethylhexyl monoester (m) | | ++ | ++ | | + | | + | + | ++ | | | |
| Phthalic acid (m) | ++ | ++ | ++ | ++ | + | + | ++ | ++ | ++ | ++ | | + |
| *Nitro compounds* | | | | | | | | | | | | |
| 4-Nitrobenzoic acid (m) | | | | | ++ | ++ | | | | | | |
| 4-Aminobenzoic acid(m) | | | | | + | | | | | | | |
| N-Methyl-4-amino-benzoic acid (m) | | | | | o | | | | | | | |
| 2,4-Di-*tert*-butyl-6-nitrophenol | | | | | + | | | | | | | |
| 2-*tert*-Butyl-4,6-dinitrophenol | | | | | + | | | | | | | |
| *Pesticides and metabolites* | | | | | | | | | | | | |
| Hexachlorocyclohexanes (α-, β-, γ-, δ-) | | | | | | o | | | | | | |
| 2,4′-DDD | | | | | + | + | | | | | | |
| 4,4′-DDD | | + | | | + | + | + | | | | | |
| 4,4′-DDMS | | | | | o | | | | | | | |
| 2,4′-DDM | ++ | ++ | | | + | + | + | | + | + | + | + |
| 4,4′-DDM | ++ | ++ | | | ++ | + | ++ | ++ | + | + | + | + |
| 2,4′-DDE | | | o | o | | | | | | | o | o |
| 4,4′-DDE | | | + | o | | | | | o | o | o | + |
| 2,4′-DDMU | | | | | | | | | o | o | o | o |
| 4,4′-DDMU | | | | | | | | | o | + | + | + |
| 2,4′-DDNU | o | | | | | | | | o | + | o | o |
| 4,4′-DDNU | + | | | | | | | | + | + | + | + |
| 4,4′-DDEthane | | | | | | | | | o | o | o | |
| 2,4′-DDA (m) | | | | | ++ | ++ | | | + | + | | |
| 4,4′-DDA (m) | | | | | ++ | ++ | ++ | ++ | ++ | ++ | ++ | ++ |
| 4,4′-DDOH | + | | | | | | | | + | | + | |
| 2,4′-DBP | + | + | | | + | + | | ++ | | + | | |
| 4,4′-DBP | ++ | + | | | ++ | ++ | + | ++ | + | ++ | ++ | ++ |
| 2,4′-DDCN | + | | | | | | | | | | | |
| 4,4′-DDCN | + | | | | + | + | | | | | | |
| 4,4′-MDE | | | | | | | | | | + | | |
| 4,4′-Dimethoxy-benzophenone | | | | | | | | | + | + | + | + |
| 2,2-Bis(4 dimethoxy-phenyl)acetic acid (m) | | | | | | | | | o | + | o | ++ |
| Chloropropylate | ++ | | | | | ++ | | | | | | |

| | T1 (u) | T1 (l) | T2 | T3 | T1 (u) | T1 (l) | T2 | T3 | T1 (u) | T1 (l) | T2 | T3 |
|---|---|---|---|---|---|---|---|---|---|---|---|---|
| **Compounds** | **$BBr_3$ treatment** | | | | **$RuO_4$ oxidation** | | | | **Hydrolysis** | | | |
| *Halogenated aromatics* | | | | | | | | | | | | |
| Dichlorobenzene (1 isomer) | | o | | | | o | | | | | | |
| Trichlorobenzene (2 isomer) | | | | | (o) | o | | | | | (o) | |
| Tetrachlorobenzene (1 isomer) | | | | | o | o | | | | | | |
| Pentachlorobenzene | | | | | o | o | | | | | o | |
| 1-Chloronaphthalene | | | | | | | | | o | o | o | o |
| Dichloronaphthalene (3 isomers) | | | | | | | | | | (o) | (o) | + |
| 1-Bromonaphthalene | | | | | + | o | | | o | o | + | + |
| Dibromonaphthalenes (3 isomers) | | | | | (o) | | | | (o) | | + | + |
| PCB ($Cl_4$-$Cl_6$) | | | | | | | | | o | o | o | o |
| 4-Chlorobenzoic acid [(m)] | | | | | + | o | | | | | | |
| 2,4-Dichlorobenzoic acid [(m)] | | | | | + | | | | | | | |
| 2,4-Dibromoaniline | | | o | | | + | | | | | | |
| 2,4,6-Tribromoaniline | | o | o | o | | o | | | o | o | + | + |
| Bromophenol [(me)] | + | + | | | | | | | | o | | |
| Dibromophenol [(me)] | + | o | ++ | o | | | | | | | | |
| Tribromophenol [(me)] | o | | ++ | o | | | | | | | | |

**Tab. 1b:** Organic contaminants identified in the non-extractable organic matter of Teltow Canal sediments after sequential chemical degradation procedures.
o ; + ; ++ ; +++ = very low to high concentration
(me) = identified as methyl ethers; (m) = identified as methyl esters
( ) = not all isomers detected

| | T1 (u) | T1 (l) | T2 | T3 | T1 (u) | T1 (l) | T2 | T3 |
|---|---|---|---|---|---|---|---|---|
| **Compounds** | **$BBr_3$ treatment after hydrolysis** | | | | **$RuO_4$ oxidation after hydrolysis** | | | |
| *Technical additives, solvents* | | | | | | | | |
| N-Methylpyrrolidone | | | | | | + | | |
| N-Ethylaniline | ++ | | | | + | | o | + |
| Benzophenone | | | | | o | | | |
| Benzylbenzoat | | | | | | | | |
| Dibutylmaleat | | | | | | | | |
| Butylpalmitat | | | | | | | | |
| Isopropyldodecanoate | | | | | + | ++ | | |
| Isopropylpalmitate | | | | | | | o | |
| Hexandioic acid di-*iso*-propyl ester | | | ++ | | | | | |
| 2-Ethylhexanol | | | | | | | | |
| Hexandioic acid 2-ethylhexyl ester | + | | | | | + | o | + |
| Bisphenol A (me) | + | + | | o | | | | |
| Ionol | | | | | + | | o | |
| Di-*tert*-butylchinone | | | | | | | | |
| *Detergent residues, fragrances* | | | | | | | | |
| Linear alkylbenzenes | | | | o | | | | |
| Galaxolide | | | | | | | | |
| Tonalide | | | | | | | | |
| *Plasticizers* | | | | | | | | |
| Alkylsulfonic acid phenyl esters | | | | | | | | |
| Tri-n-butylphosphate | | | | | o | o | | |
| 2,4,4-Trimethylpentan-1,3-dioldi-*iso*-butyrate | | | | | | | | |
| Di-*n*-butylphthalate | + | | + | | + | ++ | ++ | +o |
| Di-*iso*-butyl-phthalate | + | | + | | + | ++ | + | ++ |
| Bis(2-ethylhexyl)phthalate | ++ | | ++ | | + | ++ | ++ | ++ |
| Tris(2-ethylhexyl)trimellitate | | | | | | | | |
| Phthalic acid 2-ethylhexyl monoester (m) | ++ | | ++ | | | | | |
| Phthalic acid (m) | ++ | ++ | ++ | ++ | ++ | ++ | ++ | ++ |
| *Nitro compounds* | | | | | | | | |
| 4-Nitrobenzoic acid (m) | | | | | | | | |
| 4-Aminobenzoic acid(m) | | | | | | | | |

| | T1 (u) | T1 (l) | T2 | T3 | T1 (u) | T1 (l) | T2 | T3 |
|---|---|---|---|---|---|---|---|---|
| **Compounds** | **$BBr_3$ treatment after hydrolysis** | | | | **$RuO_4$ oxidation after hydrolysis** | | | |
| N-Methyl-4-amino-benzoic acid (m) | | | | | | | | |
| 2,4-Di-*tert*-butyl-6-nitrophenol | | | | | | | | |
| 2-*tert*-Butyl-4,6-dinitrophenol | | | | | | | | |
| *Pesticides and metabolites* | | | | | | | | |
| Hexachlorocyclohexanes (α-, β-, γ-, δ-) | | | | | | | | |
| 2,4´-DDD | | | | | | | | |
| 4,4´-DDD | | | | | | | | |
| 4,4´-DDMS | | | | | | | | |
| 2,4´-DDM | o | o | | | + | + | | |
| 4,4´-DDM | + | o | | | + | + | | |
| 2,4´-DDE | | | | | | | | |
| 4,4´-DDE | | | | | | | | |
| 2,4´-DDMU | | | | | | | | |
| 4,4´-DDMU | | | | | | | | |
| 2,4´-DDNU | | | | | | | | |
| 4,4´-DDNU | | o | | | | | | |
| 4,4´-DDEthane | | | | | | | | |
| 2,4´-DDA (m) | + | + | | | | + | + | |
| 4,4´-DDA (m) | + | ++ | | | + | ++ | ++ | |
| 4,4´-DDOH | | | | | + | + | | |
| 2,4´-DBP | + | + | | | + | + | + | |
| 4,4´-DBP | + | ++ | | | + | ++ | ++ | |
| 2,4´-DDCN | | | | | | | | |
| 4,4´-DDCN | | | | | + | + | | |
| 4,4´-MDE | | | | | | | | |
| 4,4´-Dimethoxybenzophenone | | | | | | | | |
| 2,2-Bis(4 dimethoxyphenyl)acetic acid (m) | | | | | | | | |
| Chloropropylate | ++ | | | | | | | |
| *Halogenated aromatics* | | | | | | | | |
| Chlorinated benzenes ($Cl_2 – Cl_5$) | | | | | | | | |
| 1-Chloronaphthalene | | | | | | | | |
| Dichloronaphthalene (3 isomers) | | | | | | | | |
| 1-Bromonaphthalene | | o | | | | | | |
| Dibromonaphthalenes (3 isomers) | | o | o | o | | | | |
| PCB ($Cl_4$-$Cl_6$) | | | | | | | | |
| 4-Chlorobenzoic acid(m) | | | | | | | | |
| 2,4-Dichlorobenzoic acid (m) | | | | | | | | |
| 2,4-Dibromoaniline | | | | | | | | |
| 2,4,6-Tribromoaniline | | | | o | | | | |
| Bromophenol (me) | o | o | + | | | | | |
| Dibromophenol (me) | + | o | + | + | | | | |
| Tribromophenol (me) | + | | o | | | | | |

## Discussion

The occurrence of anthropogenic substances in the degradation reaction mixtures has to be discussed either by their mode of incorporation, their modification due to incorporation and their appearance in comparison to the substances obtained by traditional extraction techniques. Hence in Tab. 2 the anthropogenic compounds identified in the extracts are summarized as previously published (Schwarzbauer et al. 2001, Ricking et al. 2003).

**Tab. 2:** Summary of selected anthropogenic contaminants identified in the extracts of the Teltow Canal sediments as published previously (Schwarzbauer et al. 2001; Ricking et al. 2003). Compounds not identified in the non-extractable residue after application of chemical or pyrolytic degradation procedures are given in italics.

| | |
|---|---|
| **Technical additives, solvents** | **Detergents related compounds** |
| *Dibenzylether* | *Nonylphenols (10 isomers)* |
| Benzophenone | Linear alkylbenzenes (LAB) |
| *Benzylbenzoate* | |
| *Dimethyladipate* | **Halogenated aromatics** |
| | Chlorinated benzenes ($Cl_1$-$Cl_6$) |
| **Fragrances, UV-protectors** | *Chlorinated styrenes ($Cl_3$-$Cl_8$)* |
| *4-Methoxycinnamic acid* | Chlorinated naphthalenes ($Cl_1$-$Cl_7$) |
| *2-ethylhexyl ester* | Brominated naphthalenes ($Cl_1$-$Cl_2$) |
| Galaxolide | Polychlorinated Biphenyls ($Cl_4$-$Cl_7$) |
| Tonalide | *4,4′-Dibromobenzophenone* |
| *4-Oxoisophorone* | *Pentachloroanisole* |
| | *2,4-Dichlorobenzaldehyde* |
| **Plasticizers** | 2,4-Dichlorobenzoic acid (m) |
| Tributylphosphate | |
| *Tritolylphosphate* | **Pesticides and metabolites** |
| Alkylsulfonic acid phenyl/ | Hexachlorocyclohexanes ($\alpha$-,$\beta$-,$\gamma$-,$\delta$-) |
| cresyl esters | *DDT* |
| Dimethylphthalate | DDD |
| Di-*n*-butylphthalate | DDMS |
| Di-*iso*-butylphthalate | DDE |
| Bis(2-ethylhexyl)phthalate | DDMU |
| | DDCN |
| **Nitro compounds** | DDA methyl ester |
| *Nitrobenzene* | DBP |
| *4-Ethylnitrobenzene* | *Methoxychlor* |
| 4-Nitrobenzoic acid (m) | MDD |
| | MDE |
| | *Chlorfensone* |
| | *Bromopropylate* |

Most of the compounds identified in the degraded non-extractable residues were formerly reported as constituents of the extractable fraction (Schwarzbauer et al. 2001; Ricking et al. 2003). Hence these compounds known as pesticides, technical additives or industrial agents represent the unaltered bound substances and reflect the incorporated proportion of organic pollutants introduced into the aquatic environment by anthropogenic emissions.

Considering their molecular structures and their frequent but not systematical occurrence in extracts of different selective degradation steps it has to be stated that the major portion of these substances was not associated by covalent linkages but by weaker interactions like adsorption or van der Waals forces. Thus the majority of compounds was released by destruction of the macromolecular matrix and not by a selective bond breaking. Consequently no dramatically variation within the spectra of contaminants was observed with respect to the mode of chemical degradation. However, in detail minor differences of the contamination pattern obtained by the various degradation methods were noted. Substances released by hydrolysis and $RuO_4$ oxidation covered a wider range of substances as compared to the group of compounds revealed by the $BBr_3$ degradation.

With respect to the sequential degradation procedures we observed two slightly different trends. In the case of $RuO_4$ oxidation an decreasing quantity of contaminants can be stated within the already saponified residues as compared to the formerly untreated oxidation products, despite the different selectivity and reactivity of the degradation agents. On the contrary, the $BBr_3$ treatment released an pattern of compounds unaffected by a former hydrolysis.

Both observations, (i) the similar quality of released bound contaminants in case of hydrolysis and $RuO_4$ oxidation and (ii) the decreased quantity of compounds revealed by $RuO_4$ oxidation after hydrolysis indicate that organic contaminants comparablely associated to the macromolecular organic matter were affected by both degradation methods in a very similar mode. The $BBr_3$ treatment released organic contaminants which are incorporated in a different way.

Furthermore a correlation between the concentration analysed in the extractable fraction and the appearance of individual substances within the degradation extracts was not observed. The relative concentration of various abundant extractable compounds decreased in the degradation product mixtures and fell partly below the detection limit (e.g. alkylsulfonic acid phenylesters, tritolyl phosphates, hexachlorocyclohexanes). On the contrary, a few contaminants with very low concentration in the extractable fraction were also identified in the bound fraction (e.g. bisphenol A, chloropropylate). Additionally, several compounds occurred in both the bound and the extractable fraction at higher concentration levels (e.g. chlorinated and brominated naphthalenes, phthalates, DDT-group substances). With respect to the molecular structure and the relative concentrations these observations suggested no preference in the association of selected classes of compounds.

Most of the unaltered bound contaminants discussed were detected at a low to very low concentration level as compared to the degradation products of the natural organic components. Most abundant within the group of xenobiotics are the group of DDT-related compounds that were detected at elevated amounts. For the DDT metabolites a significant alteration has to be stated as compared to the DDT-related compounds detected in the extractable fraction. Main components in all degradation extracts were 4,4′- DBP, 4,4′-DDA and 4,4′-DDM. In addition, 4,4′-DDM was detected at rather high concentrations by pyrolytic analyses. The DDT metabolites DDMU, DDOH, DDMS occurred at minor concentrations, whereas DDD, DDE, DDCN and DDT itself were either not detected or at a significant lower level. This quantitative proportion of DDT metabolites was in contrast to the distribution observed in the extractable organic matter (Schwarzbauer et al. 2001). 4,4′-DDD was absolutely dominant in the extractable fraction according to the anaerobic degradation pathway of DDT (see: http://umbbd.ahc.umn.edu), whereas DDE, the main metabolite of aerobic microbial DDT-degradation, and the other metabolites occurred at significantly minor concentrations.

The alteration within the group of bound DDT-metabolites indicates either a different degradation pathway of incorporated DDT or the selective association of individual metabolites due to their different molecular structures. The significantly higher concentration of the DDT-related compounds as compared to unaltered bound contaminants suggests an enhanced incorporation of DDT or of its metabolites.

A very similar phenomenon was observed for methoxychlor related compounds. Also the more polar metabolites 4,4′-dimethoxy-benzophenone and 2,2-bis(4-methoxyphenyl)acetic acid became most abundant within the bound organic fraction.

## Summary

The aim of our investigation was to characterize the alteration of the non-extractable organic matter due to anthropogenic emissions and to elucidate the subsequent incorporation of the organic pollutants into riverine geopolymers. Hence we investigated the occurrence, alteration and distribution of several organic xenobiotics within the non-extractable organic matter of highly polluted riverine sediments (Spree River, Teltow Canal, Germany). We combined different analytical techniques (pyrolytic analyses and chemical degradation techniques) in order to provide information concerning the incorporation mechanisms and the mode of binding for a variety of organic pollutants with different chemical properties.

Briefly the following conclusions can be deduced from the results of the presented study considering the occurence, molecular structure and semi-quantitative amounts of the identified anthropogenic contaminants:

- Most of the compounds identified in the degraded non-extractable residues represent the unaltered bound substances and reflect the incorporated proportion of organic pollutants introduced into the aquatic environment by anthropogenic emissions. The major portion of these substances was not associated by covalent linkages but by weaker interactions.

- Hydrolysis and $RuO_4$ oxidation affected the interactions of the associated substances with the macromolecular organic matter and the alteration of the macromolecular matrix on a very similar mode, despite the different selectivity and reactivity of the degradation agents. The $BBr_3$ treatment affected the incorporation of organic contaminants in a rather different way.

- The appearance of individual substances within the degradation extracts did not correlate with the concentrations determined in the extractable fraction. Considering the molecular structures of the contaminants and the correspondig chemical and physico-chemical properties a favoured association of selected classes of compounds cannot be assumed.

- The altered distribution of the bound DDT-metabolites and the significantly higher concentrations as compared to unaltered bound contaminants indicates either a different degradation pathway of incorporated DDT or the selective and enhanced association of individual metabolites due to their different molecular structures.

### 4.3.2 DDT-related compounds bound to the non-extractable particulate matter in sediments of the Teltow Canal, Germany ‡

#### Introduction

Various degradation techniques have been applied to highly contaminated sediment samples from the Teltow Canal (Berlin, Germany) in order to reveal information about the incorporation of xenobiotics into the non-extractable particulate matter. Earlier reported studies have indicated a high contamination of the Teltow Canal by the pesticide 2,2-(4-chlorophenyl)-1,1,1-trichloroethane DDT and its metabolites due to industrial emissions (Heinisch et al. 1990, Heinisch 1992; Heberer and Dünnbier 1999; Schwarzbauer et al. 2001).

Although the globally distributed DDT is a very well investigated xenobiotic regarding the environmental occurrence and behaviour, detailed information about the fate of DDT in the bound residues fraction is very limited. Already in 1977 Lichtenstein et al. (1977) reported the formation of bound $^{14}$C-labelled DDT on agricultural soil accompanied by a drastically reduced insecticidal activity of the associated proportion. Also recent studies confirmed the decrease of DDT toxicity with time after application to soils as a result of less bioavailibility due the incorporation into the non-extractable particulate matter (Robertson and Alexander 1998). For a better understanding of the processes leading to these observations more information is required about the incorporation of DDT residues into the non-extractable particulate matter not only of soils but also of particulate matter within the aquatic environment.

Therefore the present investigation focussed on the occurrence and alteration of DDT and its metabolites within the non-extractable particulate matter (or matrix) of highly contaminated riverine sediments. It has to be noted, that this study is limited to extended analysis on only four sediment samples. Hence this report has to be characterized as a preliminary testing of the presented approach.

#### Samples

The sediment samples taken in 1998 and 1999 from three locations at the Teltow Canal in Berlin situated near a former pesticides producing

‡ Mainly adapted/reprinted from Schwarzbauer et al. 2003a

industrial plant are indicated in Fig. 1 and described in more detail in chapter 4.3.1.

Data on TOC and dry matter are listed in Tab. 1. The sampling locations were carefully chosen to represent the source, the accumulation a few hundred meters downflow the effluent and an additional potential source in former East-Berlin. All samples were taken in accumulation areas prior to the clean-up of former waste deposits in the Teltow Canal.

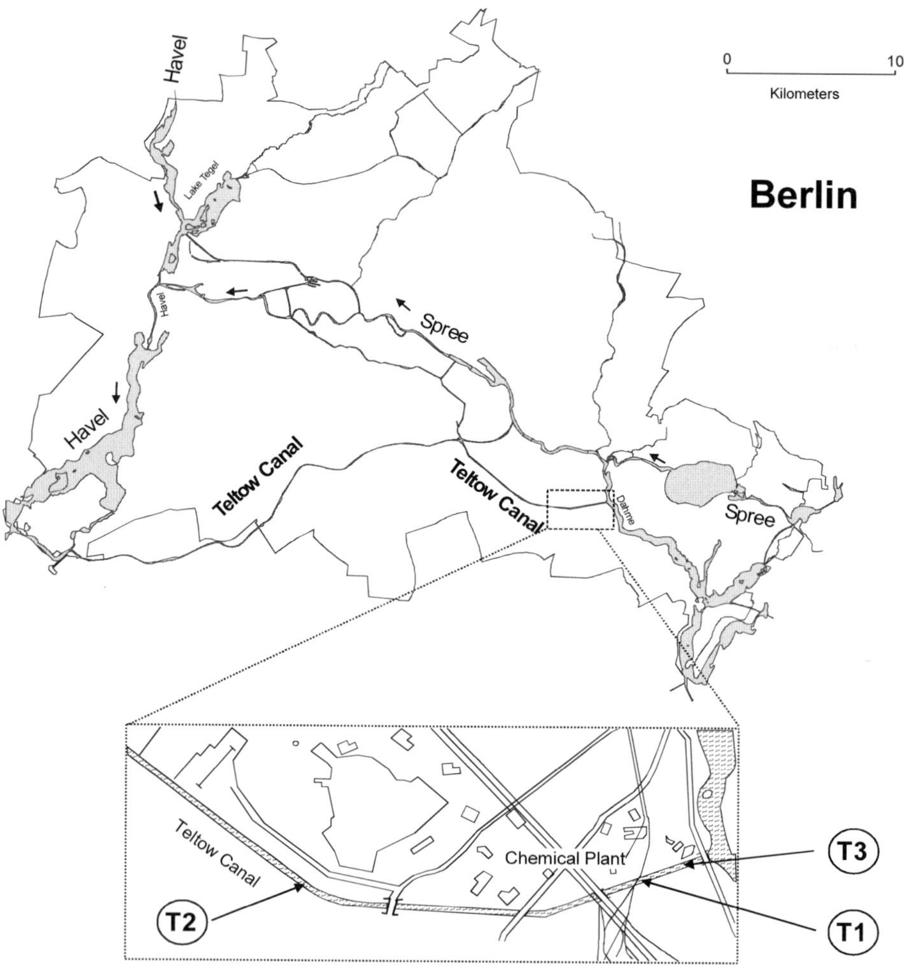


**Fig. 1:** Sampling locations of Teltow Canal sediments (T1–T3).

**Tab. 1:** Samples of Teltow Canal sediments with data on TOC and dry matter

| Abbreviation | Date of sampling | Description | Dry matter (%) | TOC (%) |
|---|---|---|---|---|
| T1a | 1999 | 0-3 cm of sediment core 500 m downflow of point source | 9.4 | 12.1 |
| T1b | 1999 | 3-10 cm of sediment core 500 m downflow of point source | 14.9 | 11.9 |
| T2 | 1998 | Grab sample directly at the effluent of the former pesticide manufacturer | 17.0 | 11.2 |
| T3 | 1999 | 0-15 cm grab sample directly at the western effluent of the sewage treatment plant Waßmannsdorf | 26.6 | 7.1 |

Quantitative data were obtained by integration of specific ion chromatograms extracted from the TIC. The ions used for quantification as well as the recoveries for the extraction and evaporating procedures are summarized in Tab. 2. All reference compounds (purity > 96-99 %) used as indicated in Tab. 2 were purchased from Promochem (Wesel, FRG) with exception of 4,4'-DDCN, which was synthesized according to Nyström and Berger (1958) and Beringer et al. (1959) The limit of quantitation was in the range of 5 µg/kg dry matter calculated from GC/MS-analyses of reference compounds (approx. 100 pg/µl of reference substances were determined with a signal to noise ratio of 10:1). Considering the influence of varying matrices and different degradation procedures on the limit of quantitation, no attempts were made to quantify components at concentrations of less than 20 µg/kg. All concentrations are given on a dry matter basis.

**Tab. 2:** DDT related compounds identified and quantified in Teltow Canal sediments

| **Compounds** | **Chemical structure** | **Ions used for quanti-fication (m/z)** | **Recovery rate (%)** | **Remarks** |
|---|---|---|---|---|
| 4,4'-DDT<br>2,2-Bis(4-chlorophenyl)-1,1,1-trichloroethane | 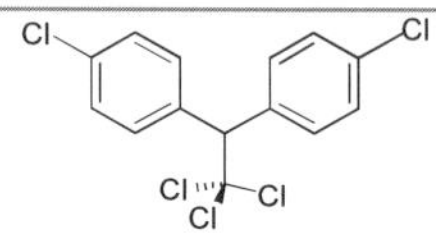 | 235, 237 | 64 | 2,4'-isomer was also quantified |
| 4,4'-DDD<br>2,2-Bis(4-chlorophenyl)-1,1-dichloroethane | 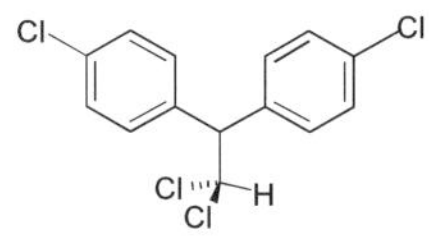 | 235,237 | 70 | 2,4'-isomer was also quantified |
| 4,4'-DDMS<br>2,2-Bis(4-chlorophenyl)-1-chloroethane | 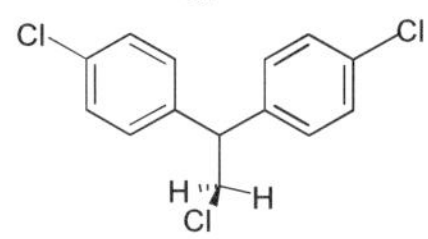 | 235,237 | | 4,4'- and 2,4'-isomer: recovery and calibration data were kept from 4,4'-DDD |
| 4,4'-DDEt<br>1,1-Bis(4-chlorophenyl)-ethane | 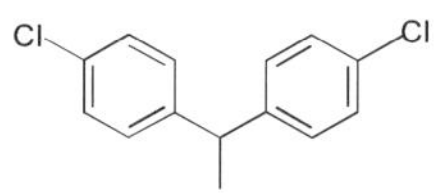 | 235,237 | | 4,4'- and 2,4'-isomer: recovery and calibration data were kept from 4,4'-DDD |
| 4,4'-DDE<br>2,2-Bis(4-chlorophenyl)-1,1-dichloroethene | 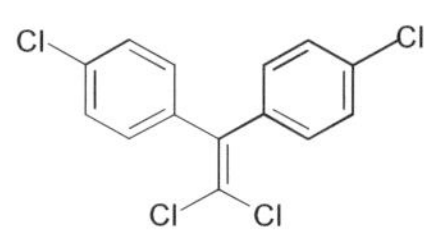 | 246,248 | 89 | 2,4'-isomer was also quantified |
| 4,4'-DDMU<br>2,2-Bis(4-chlorophenyl)-1-chloroethene | 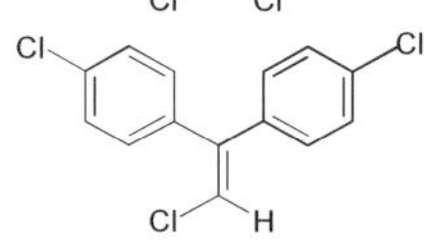 | 282,280 | 75 | 2,4'-isomer: recovery and calibration data were kept from 4,4'-DDMU |
| 4,4'-DDNU<br>1,1-Bis(4-chlorophenyl)-ethene | 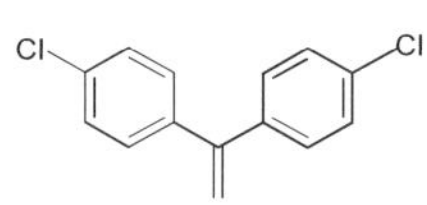 | 248,250 | | 4,4'- and 2,4'-isomer: recovery and calibration data were kept from 4,4'-DDMU |
| 4,4'-DDM<br>Bis(4-chlorophenyl)-methane | 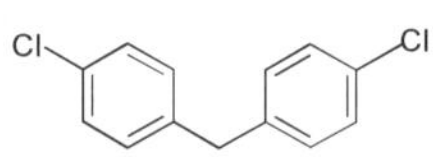 | 236,238 | 68 | 2,4'-isomer: recovery and calibration data were kept from 4,4'-DDM |

| Compounds | Chemical structure | Ions used for quanti-fication (m/z) | Recovery rate (%) | Remarks |
|---|---|---|---|---|
| 4,4'-DDCN Bis(4-chlorophenyl)-acetonitrile | 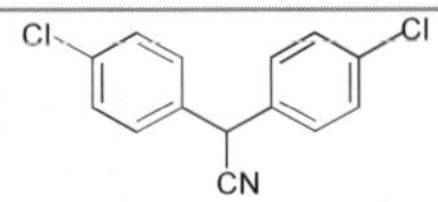 | 226,228 | 90 | 2,4'-isomer: recovery and calibration data were kept from 4,4'-DDCN |
| 4,4'-DBP 4,4'-Dichlorobenzo-phenone | 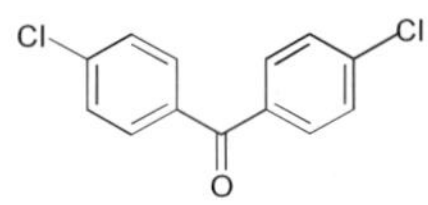 | 139,141 | 77 | 2,4'-isomer: recovery and calibration data were kept from 4,4'-DBP |
| 4,4'-DDOH 2,2-Bis(4-chlorophenyl)-ethanol | 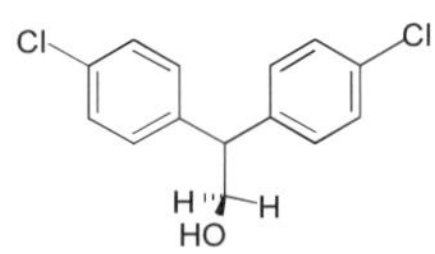 | 235,237 | | 4,4'- and 2,4'-isomer: recovery and calibration data were kept from 4,4'-DDA |
| 4,4'-DDA Bis(4-chlorophenyl)-acetic acid | 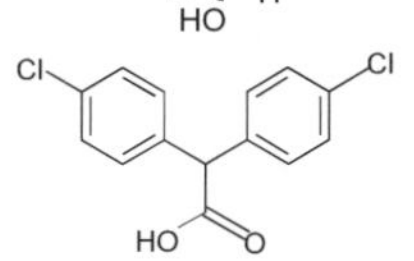 | 235,237 | 82 | Detected after derivatization as methyl ester. 2,4'-isomer: recovery and calibration data were kept from 4,4'-DDA |
| 4,4'-MDT 2,2-Bis(4-methoxy-phenyl)-1,1,1-trichloro-ethane | 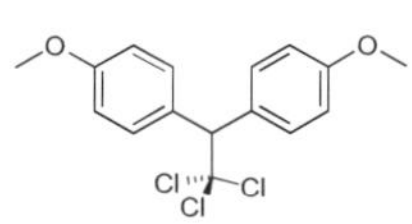 | 227, 344 | 91 | 2,4'-isomer: recovery and calibration data were kept from 4,4'-MDT |
| 4,4'-MDD 2,2-Bis(4-methoxy-phenyl)-1,1-dichloro-ethane | 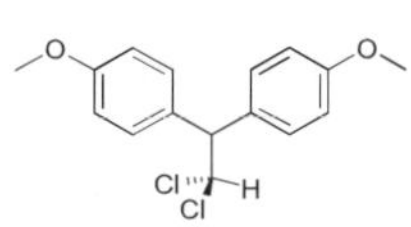 | 227, 310 | | 4,4'- and 2,4'-isomer: recovery and calibration data were kept from 4,4'-MDT |
| 4,4'-MDE 2,2-Bis(4-methoxy-phenyl)-1,1-dichloroethene | 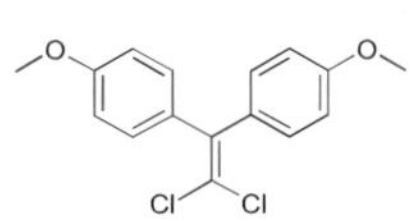 | 308,310 | | 4,4'- and 2,4'-isomer: recovery and calibration data were kept from 4,4'-MDT |

| Compounds | Chemical structure | Ions used for quanti-fication (m/z) | Recovery rate (%) | Remarks |
|---|---|---|---|---|
| 4,4'-MDB<br>4,4'-Dimethoxy-benzophenone | 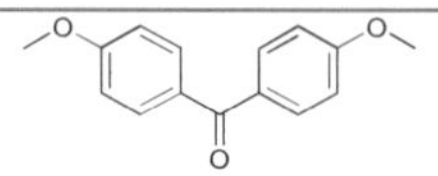 | 135, 242 | | 4,4'- and 2,4'-isomer: recovery and calibration data were kept from 4,4'-MDT |
| 4,4'-MDA<br>Bis(4-methoxyphenyl)-acetic acid | 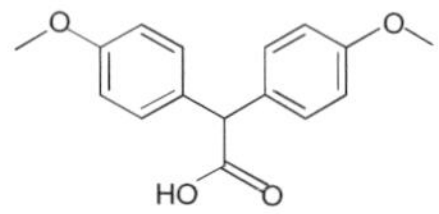 | 227, 286 | | 4,4'- and 2,4'-isomer: detected after methylation. Recovery and calibration data were kept from 4,4'-DDA. |

## Results and discussion

Four sediment samples (T1a, T1b, T2, T3) obtained from three different sampling locations (see Fig. 1) were investigated by analyses of extractable and non-extractable organic components. The sampling locations were situated in the Teltow Canal near a former pesticide producing chemical plant. Former analyses applied to sediment samples from the same area indicated a high contamination with halogenated compounds and pesticides as a result of industrial emissions (Schwarzbauer et al. 2001).

### *Extractable fraction*

Screening analyses by GC/MS applied to the extractable fraction of the samples investigated revealed a very high contamination with DDT and its metabolites DDD, DDE, DDCN, DDMU, 1,1-bis(chlorophenyl)ethene - DDNU, 2,2-bis(chlorophenyl)-1-chloroethane –DDMS, 1,1-bis-(chlorophenyl)ethane - DDEt, 2,2-bis(chlorophenyl)acetic acid - DDA and bis(chlorophenyl)methane DDM. All concentrations detected are presented in detail in Tab. 3.

**Tab. 3:** DDT related compounds quantified in extracts of Teltow Canal sediment samples (concentrations are given in µg/kg dry matter).

| Compounds | Extracts | | | |
|---|---|---|---|---|
| | T1a | T1b | T2 | T3 |
| 2,4'-DDT | 960 | <5 | 4200 | 1100 |
| 4,4'-DDT | 1900 | <5 | 9700 | 2500 |
| 2,4'-DDD | 5000 | 5200 | 1300 | 1200 |
| 4,4'-DDD | 1300 | 9500 | 3800 | 2000 |
| 2,4'-DDMS | 1100 | 2000 | 2800 | 1900 |
| 4,4'-DDMS | 5000 | 5100 | 1000 | 5000 |
| 2,4'-DDEt | 970 | 57 | 130 | 58 |
| 4,4'-DDEt | 2800 | 160 | 430 | 130 |
| 2,4'-DDE | 3700 | 1100 | 1300 | 360 |
| 4,4'-DDE | 1300 | 3000 | 1000 | 3500 |
| 2,4'-DDMU | 1500 | 190 | 530 | 51 |
| 4,4'-DDMU | 1300 | 2800 | 6100 | 1700 |
| 2,4'-DDNU | 1060 | 62 | 140 | 37 |
| 4,4'-DDNU | 3200 | 170 | 310 | 91 |
| 2,4'-DDM | 670 | 180 | 84 | 69 |
| 4,4'-DDM | 1900 | 470 | 290 | 160 |
| 2,4'-DBP | 2400 | 540 | 180 | 19 |
| 4,4'-DBP | 1900 | 1800 | 450 | 41 |
| 2,4'-DDCN | 6800 | 890 | 840 | 15 |
| 4,4'-DDCN | 3500 | 3100 | 4000 | 120 |
| 2,4'-DDOH | <5 | <5 | <5 | <5 |
| 4,4'-DDOH | <5 | <5 | <5 | <5 |
| 2,4'-DDA | <5 | 190 | 840 | <5 |
| 4,4'-DDA | <5 | 2300 | 1000 | <5 |
| Sum | 3038 | 3880 | 1133 | 4885 |

As expected, in all four sediment samples the DDT-related compounds are dominated by DDD (max. concentration 130000 µg/kg dry matter in T1a), the main metabolite of the anaerobic degradation pathway. A further group of metabolites including DDMS, DDE and DDMU were detected with constant concentrations not lower than 10 % of DDD. Beside this group several metabolites (DBP, DDCN and DDA) and DDT itself occurred with concentrations between 1 and 35 % relative to DDD, showing a higher variation of the relative amounts. All other metabolites were either not detected (DDOH) or at concentrations between the limit of quantification (LOQ) and approx. 5% as compared to DDD. The quantitative pattern of DDT and related compounds was in accordance with the results formerly published for sediment samples located nearby (Schwarzbauer et al., 2001).

*Non extractable residues*

The main objective of the present study was the investigation of organic substances incorporated into the non-extractable particulate matter. With respect to the results of the extractable fraction we consequently focussed on the identification and quantitation of associated DDT-related substances in order to obtain further information about the fate of DDT-derived compounds within the particulate matter of the aquatic environment. Using different kinds of chemical degradation techniques in a separate and a sequential application, the selective release of the compounds according to their mode of association was expected. The degradation procedures applied included alkaline hydrolysis, boron tribromide treatment and ruthenium tetroxide oxidation. These techniques are common organic reactions and widely used for organic geochemical analysis of macromolecular organic matter in soils, sediments, rocks, coals and kerogen (e.g. Palm and Lammi 1995; Gelin et al. 1997; Peng et al. 1997; Orem et al. 1997; Bajc et al. 2001). All procedures applied to the Teltow Canal sediments revealed specific degradation products as major constituents. Considering the specifity of both the reactions and the chemical structure of the products an abundant contribution of macromolecular organic matter to the non-extractable particulate matter as well as an intense modification or degradation due to the procedures applied has to be stated. This is in accordance with the elevated TOC values ranging between 7 and 12 %. A detailed view on the natural degradation products released from the Teltow Canal samples and subsequently of the biogenic macromolecular precursor is reported elsewhere (Schwarzbauer et al., 2005b).

Complementary to the chemical degradation techniques an on-line pyrolysis – GC/MS method was applied to the pre-extracted sediment residues.

*Pyrolysis:* Apart from obviously natural compounds only very few definitely anthropogenic substances were yielded by flash pyrolysis. Within the group of xenobiotics 2,4′- and 4,4′-DDM as DDT-related compounds were detected in high concentrations as compared to the main components within the pyrograms. A pyrolytic conversion of further DDT-metabolites to DDM cannot be excluded. Thus, the occurrence of DDM in the pyrogram is only indicative for DDT group metabolites in general. Furthermore, the analytical technique does not yield quantitative data due to unknown pyrolytic conversion rates of individual substances. Therefore, pyrolytic analyses were only useful to indicate semiquantitively DDT-related contamination.

*Chemical degradation:* In order to obtain more specific qualitative and quantitative information on DDT-related compounds within the non-extractable matter the extracts obtained after the chemical degradation procedures were investigated in more detail by quantitative target analyses considering all DDT-related compounds (2,4'- and 4,4'-isomers) previously identified in the sediment extracts. All results applied are given in Tab. 4 and 5 arranged according to the analytical methods.

In order to avoid misinterpretation of the occurrence and amount of bound DDT-residues due to the generation of artefacts, we applied all degradation procedures to selected reference substances including 2,4'-DDD, 4,4'-DDD, 4,4'-DDCN, 4,4'-DDE, and 4,4'-DDMU. We chose the most abundant metabolites as revealed by the screening analyses of the extracts. No alteration of the substances was observed after each chemical degradation step as illustrated in Fig. 2.

Alkaline hydrolysis, which attacks mainly ester linkages, applied to Teltow Canal sediment residues released a wide variety of DDT-related compounds including DBP, DDA, DDOH, DDM, DDNU, DDMU, DDE and DDEt (see Tab. 4).The total amount of DDT-related compounds released by hydrolysis ranged between 35000 and 130000 μg/kg. These results are in the same concentration range as the DDT–metabolites in the extractable fraction.

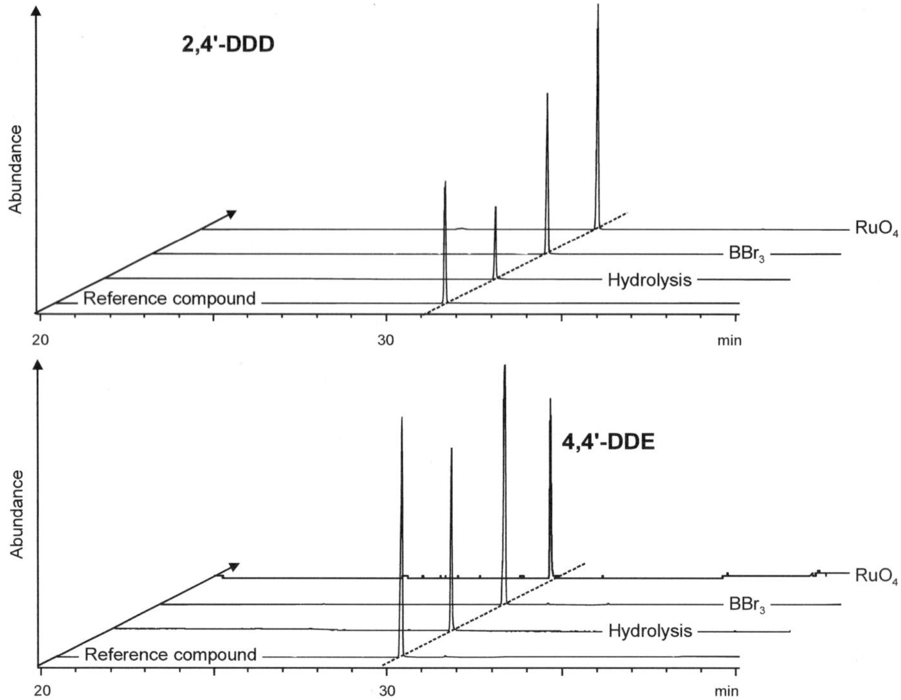


**Fig. 2:** Gas chromatograms of DDT-related compounds before (indicated as 'reference compound') and after application of the following chemical degradation procedures: hydrolysis (indicated as 'hydrolysis'), $BBr_3$-treatment (indicated as '$BBr_3$') and $RuO_4$-oxidation (indicated as '$RuO_4$'). No alteration of the educts was observed with respect to the applied reactions.

The most abundant compound was 4,4'-DDA with concentrations between 23000 and 90000 µg/kg. Also the 4,4'-isomers of DBP, DDM, DDMU and DDNU were detected at elevated concentrations in the range between 400 and 18000 µg/kg, whereas DDOH, DDEt and DDE were analysed at a low level. The concentrations of DDT, DDD, DDMS and DDCN fell below the detection limit. This quantitative distribution of DDT-metabolites was in contrast to the pattern detected in the corresponding extracts. With respect to their chemical structure only DDA and DDOH can be interpreted as products of ester bond breaking. Considering the high amounts of obviously non-covalently bound DDT-metabolites, like DBP or DDM, an exclusive origin of DDA and DDOH from former ester bindings has to be excluded and a significant contribution of DDA and DDOH molecules associated by weaker interactions is implied.

Boron tribromide treatment, frequently used in order to cleave ether and ester bonds was applied as a second degradation procedure to the pre-extracted residues as well as to the saponified samples. The results are presented in Tab. 4 and 5, respectively. In both cases only a few DDT-related substances were detected in the samples T1a and T1b, whereas in T2 and T3 no release of DDT-metabolites was observed.

Main components of the DDT-metabolites were 4,4'-DBP and 4,4'-DDM (3200 to 15000 μg/kg) in case of the separate as well as 4,4'-DBP and 4,4'-DDA (3100 to 31000 μg/kg) in case of the sequential application mode. In addition, high amounts of 4,4'-DDCN and 4,4'-DDOH (2300 and 4200 μg/kg, respectively) were observed after $BBr_3$-treatment of the pre-extracted sample T1a. The total amounts of DDT-related compounds ranged between 10000 and 75000 μg/kg and were in a similar concentration range as observed for the extractable fraction and the hydrolysis products. The individual degradation products cannot be attributed to specific covalent bond cleavages, because. the molecular structures of the main degradation products contained neither bromo or hydroxy substituents nor carboxylic groups. It should be noted, that high concentrations of DDA were observed after the sequential application of hydrolysis and $BBr_3$-treatment only, indicating an enhanced release of this DDT metabolite after prior hydrolysis. This finding may be interpreted as a hydrolysis of ester-bound DDA with a subsequent new association of a substantial proportion to the non-extractable particulate matter in a non-covalent mode of binding.

While the first proportion of DDA, formerly covalently bound or associated by weaker interactions, appeared within the hydrolysis product mixtures, the second proportion was not extractable until the subsequent $BBr_3$-degradation step alterated the non-extractable particulate matter more intensely. In contrast, the absence of DDA in the $BBr_3$-degradation products of the pre-extracted but not saponified residues suggests a high stability of ester bonded DDA versus this degradation agent.

**Tab. 4:** DDT related compounds quantified in the non-extractable organic matter of Teltow Canal sediment samples after application of different chemical degradation procedures (concentrations are given in µg/kg dry matter).

| | BBr₃ treatment | | | | RuO₄-Oxidation | | | |
|---|---|---|---|---|---|---|---|---|
| **Compounds** | **T1a** | **T1b** | **T2** | **T3** | **T1a** | **T1b** | **T2** | **T3** |
| 2,4'-DDT | <20 | <20 | <20 | <20 | <20 | <20 | <20 | <20 |
| 4,4'-DDT | <20 | <20 | <20 | <20 | <20 | <20 | <20 | <20 |
| 2,4'-DDD | <20 | 52 | <20 | <20 | 580 | 280 | 100 | <20 |
| 4,4'-DDD | <20 | 550 | <20 | <20 | 2900 | 3200 | 1100 | <20 |
| 2,4'-DDMS | <20 | <20 | <20 | <20 | <20 | <20 | <20 | <20 |
| 4,4'-DDMS | <20 | <20 | <20 | <20 | 760 | <20 | <20 | <20 |
| 2,4'-DDEt | <20 | <20 | <20 | <20 | <20 | <20 | <20 | <20 |
| 4,4'-DDEt | <20 | <20 | <20 | <20 | <20 | <20 | <20 | <20 |
| 2,4'-DDE | <20 | <20 | <20 | <20 | <20 | <20 | <20 | <20 |
| 4,4'-DDE | <20 | <20 | <20 | <20 | <20 | <20 | <20 | <20 |
| 2,4'-DDMU | <20 | <20 | <20 | <20 | <20 | <20 | <20 | <20 |
| 4,4'-DDMU | <20 | <20 | <20 | <20 | <20 | <20 | <20 | <20 |
| 2,4'-DDNU | 320 | <20 | <20 | <20 | <20 | <20 | <20 | <20 |
| 4,4'-DDNU | 3000 | <20 | <20 | <20 | <20 | <20 | <20 | <20 |
| 2,4'-DDM | 810 | 1500 | <20 | <20 | 5800 | 2700 | 5900 | 6400 |
| 4,4'-DDM | 4400 | 3900 | <20 | <20 | 11000 | 3600 | 39000 | 45000 |
| 2,4'-DBP | 2800 | 1500 | <20 | <20 | 2400 | 7900 | 1700 | 12000 |
| 4,4'-DBP | 15000 | 3200 | <20 | <20 | 27000 | 16000 | 5600 | 42000 |
| 2,4'-DDCN | 960 | <20 | <20 | <20 | 180 | 290 | <20 | <20 |
| 4,4'-DDCN | 2300 | <20 | <20 | <20 | 900 | 1100 | <20 | <20 |
| 2,4'-DDOH | 1100 | <20 | <20 | <20 | <20 | <20 | <20 | <20 |
| 4,4'-DDOH | 4200 | <20 | <20 | <20 | <20 | <20 | <20 | <20 |
| 2,4'-DDA | <20 | <20 | <20 | <20 | 16000 | 21000 | 9600 | 11000 |
| 4,4'-DDA | <20 | <20 | <20 | <20 | 91000 | 46000 | 24000 | 34000 |
| Sum | 34890 | 10702 | | | 158520 | 102070 | 87000 | 150400 |

**Tab. 4 continued**

| | Hydrolysis | | | |
|---|---|---|---|---|
| **Compounds** | **T1a** | **T1b** | **T2** | **T3** |
| 2,4'-DDT | <20 | <20 | <20 | <20 |
| 4,4'-DDT | <20 | <20 | <20 | <20 |
| 2,4'-DDD | <20 | <20 | <20 | <20 |
| 4,4'-DDD | <20 | <20 | <20 | <20 |
| 2,4'-DDMS | <20 | <20 | <20 | <20 |
| 4,4'-DDMS | <20 | <20 | <20 | <20 |
| 2,4'-DDEt | <20 | <20 | <20 | <20 |
| 4,4'-DDEt | 30 | 50 | 30 | <20 |
| 2,4'-DDE | <20 | 40 | 280 | 330 |
| 4,4'-DDE | 50 | 170 | 330 | 500 |
| 2,4'-DDMU | 40 | 120 | 60 | 370 |
| 4,4'-DDMU | 390 | 1200 | 2200 | 930 |
| 2,4'-DDNU | 220 | 520 | 240 | 240 |
| 4,4'-DDNU | 1520 | 2900 | 1900 | 640 |
| 2,4'-DDM | 1000 | 1200 | 980 | 380 |
| 4,4'-DDM | 3200 | 3700 | 2700 | 1100 |
| 2,4'-DBP | 120 | 2600 | 3200 | 2700 |
| 4,4'-DBP | 1400 | 13000 | 18000 | 17000 |
| 2,4'-DDCN | <20 | <20 | <20 | <20 |
| 4,4'-DDCN | <20 | <20 | <20 | <20 |
| 2,4'-DDOH | 180 | <20 | 420 | <20 |
| 4,4'-DDOH | 770 | <20 | 1400 | <20 |
| 2,4'-DDA | 3600 | 14000 | 12000 | 9800 |
| 4,4'-DDA | 23000 | 90000 | 82000 | 74000 |
| Sum | 35520 | 129500 | 125740 | 107990 |

As another more severe degradation technique $RuO_4$-oxidation was applied to the pre-extracted and saponified sediment residues. The compounds released by this procedure from Teltow Canal sediment residues were dominated by 4,4'-DBP and 4,4'-DDA. Apart from to the most abdundant substances minor concentrations of DDD, DDMS, DDOH and DDCN were detected. The total concentrations of DDT-related compounds ranged from 30000 to 150000 μg/kg with generally lower concentrations for the $RuO_4$-oxidation of the saponified residues. Note that also the relative concentrations of DDM as compared to DDA or DBP decreased significantly when $RuO_4$-oxidation was performed after saponification. Accordingly, most of the bound DDM was released by hydrolysis, whereas a following $RuO_4$-oxidation released an extended proportion of DBP and DDA most likely due to a more intense alteration or degradation of the non-extractable particulate matter. This observation corresponds to the results obtained for the $BBr_3$-treatment as discussed above.

Considering all observations presented, a distinct difference is evident in the distribution of abundances of the bound DDT-related compounds as compared to the substances within the extractable fraction. The main metabolite of the anaerobic degradation pathway (DDD), most abundant in the sediment extracts, had no relevance in the degradation products of all procedures applied.

Generally, the most abundant DDT-metabolites released by the degradation procedures used were 4,4'-DBP, 4,4'-DDA and 4,4'-DDM. The altered distribution of bound DDT-related compounds indicates either (i) a different biotic or abiotic degradation pathway of bound DDT and/or its main metabolite (DDD), or (ii) a selective association of individual metabolites due to their different molecular structures and physico-chemical properties without further alteration of the incorporated substances. Also a superimposition of both environmental fates suggested is conceivable, e.g., the association of selected metabolites followed by a modified transformation pathway as compared to the non-bound compounds.

Note that the main metabolites detected in the non-extractable particulate matter are either oxidized at the formerly trihalogenated methyl group (DDA) or formed by the loss of this moiety (DBP, DDM). These observations suggest on the one hand a better preservation or a preferred generation of oxidized DDT metabolites within the non-extractable particulate matter, likely due to enhanced interactions of this species either with the non-extractable particulate matter or in the bound state.

**Tab. 5:** DDT related compounds quantified in the non-extractable organic matter of Teltow Canal sediment samples after application of sequential chemical degradation prodcedures (concentrations are given in µg/kg dry matter).

| | **$BBr_3$ treatment after hydrolysis** | | | | **$RuO_4$ treatment after hydrolysis** | | | |
|---|---|---|---|---|---|---|---|---|
| **Compounds** | **T1a** | **T1b** | **T2** | **T3** | **T1a** | **T1b** | **T2** | **T3** |
| 2,4'-DDT | <20 | <20 | <20 | <20 | <20 | <20 | <20 | <20 |
| 4,4'-DDT | <20 | <20 | <20 | <20 | <20 | <20 | <20 | <20 |
| 2,4'-DDD | <20 | <20 | <20 | <20 | <20 | <20 | <20 | <20 |
| 4,4'-DDD | <20 | <20 | <20 | <20 | <20 | <20 | <20 | <20 |
| 2,4'-DDMS | <20 | <20 | <20 | <20 | <20 | <20 | <20 | <20 |
| 4,4'-DDMS | <20 | <20 | <20 | <20 | <20 | <20 | <20 | <20 |
| 2,4'-DDEt | <20 | <20 | <20 | <20 | <20 | <20 | <20 | <20 |
| 4,4'-DDEt | <20 | <20 | <20 | <20 | <20 | <20 | <20 | <20 |
| 2,4'-DDE | <20 | <20 | <20 | <20 | <20 | <20 | <20 | <20 |
| 4,4'-DDE | <20 | <20 | <20 | <20 | <20 | <20 | <20 | <20 |
| 2,4'-DDMU | <20 | <20 | <20 | <20 | <20 | <20 | <20 | <20 |
| 4,4'-DDMU | 60 | <20 | <20 | <20 | <20 | <20 | <20 | <20 |
| 2,4'-DDNU | <20 | <20 | <20 | <20 | <20 | <20 | <20 | <20 |
| 4,4'-DDNU | <20 | 160 | <20 | <20 | <20 | <20 | <20 | <20 |
| 2,4'-DDM | 290 | 150 | <20 | <20 | 590 | 800 | <20 | <20 |
| 4,4'-DDM | 740 | 490 | <20 | <20 | 1600 | 1600 | <20 | <20 |
| 2,4'-DBP | 1400 | 9800 | <20 | <20 | 1800 | 7900 | 5900 | <20 |
| 4,4'-DBP | 3100 | 31000 | <20 | <20 | 9100 | 45000 | 18000 | <20 |
| 2,4'-DDCN | <20 | <20 | <20 | <20 | 1200 | 590 | <20 | <20 |
| 4,4'-DDCN | <20 | <20 | <20 | <20 | 8600 | 3200 | <20 | <20 |
| 2,4'-DDOH | <20 | <20 | <20 | <20 | 280 | 310 | <20 | <20 |
| 4,4'-DDOH | <20 | <20 | <20 | <20 | 1100 | 1400 | <20 | <20 |
| 2,4'-DDA | 2400 | 5700 | <20 | <20 | 2100 | 9100 | 8300 | <20 |
| 4,4'-DDA | 9000 | 29000 | <20 | <20 | 5300 | 15000 | 14000 | <20 |
| Sum | 16990 | 76300 | | | 31670 | 84900 | 46200 | |

On the other hand the abundance of the non-polar DDM accompanied by the more polar compound DBP might characterize a facilitated biotic or abiotic attack on the aliphatic C-C bond of slightly degraded DDT metabolites (e.g., DDD) in bound conditions. The spectra of DDT related contaminants covered generally a wide range irrespective of the degradation procedure applied. In addition, the structures of the released compounds cannot be attributed to the selective degradation mechanism. Therefore, a weak association mainly based on non-covalent interactions is implied for the observed DDT-related contaminants. Regarding the degradation techniques characterized as usual organic reactions, a modification and degradation, especially of the organic fraction consisting mainly of macromolecular organic matter has to be suggested. This is in accordance with the most abundant proportion of organic degradation products which has to be clearly characterized as formerly biogenic compounds (Schwarzbauer et al., 2005b). Only the enhanced release of DDA after alkaline hydrolysis indicates that a significant proportion of DDA is attached to the macromolecular organic matter by ester bonds.

*Methoxychlor related compounds*

Further investigations revealed not only the presence of DDT and its metabolites in Teltow Canal sediments but also the occurrence of 2,4' and 4,4'-methoxychlor (MDT) at elevated concentrations (up to 1100 µg/kg) (22). As methoxychlor is structurally related to DDT, MDT-related compounds were also included in the quantitative analyses (see Tab. 6). We detected MDT, MDD, MDE, MDB and MDA in the extracts and partly in the hydrolysis product mixtures of all four sediment samples. The total amounts ranged between 600 and 8000 µg/kg in the extracts and between 1200 and 6000 µg/kg after application of the hydrolysis procedure. In degradation products after $BBr_3$-treatment and $RuO_4$-oxidation no MDT-related substances were analysed likely due to the lower concentration level as compared to the DDT-related compounds.

**Tab. 6:** Methoxychlor (MDT) related compounds quantified in extracts and hydrolysed residues of Teltow Canal sediment samples (concentrations are given in µg/kg dry matter).

| | **Extracts** | | | | **Hydrolysis** | | | |
|---|---|---|---|---|---|---|---|---|
| **Compounds** | **T1a** | **T1b** | **T2** | **T3** | **T1a** | **T1b** | **T2** | **T3** |
| 2,4'-MDT | <5 | <5 | 110 | <5 | <20 | <20 | <20 | <20 |
| 4,4'-MDT | 120 | 47 | 430 | 61 | <20 | <20 | <20 | <20 |
| 2,4'-MDD | 120 | 12 | <5 | 62 | <20 | <20 | <20 | <20 |
| 4,4'-MDD | 2800 | 230 | 1000 | 2200 | <20 | <20 | <20 | <20 |
| 4,4'-MDE | 61 | <5 | 90 | 37 | <20 | 310 | <20 | <20 |
| 4,4'-MBP | 2500 | 170 | 180 | 1700 | 500 | 1100 | 190 | 340 |
| 4,4'-MDA | 2300 | 130 | 260 | 930 | 5500 | 1000 | 1200 | 880 |
| Sum | 7901 | 589 | 2070 | 4995 | 6000 | 2510 | 1390 | 1220 |

Similarly to the DDT-related compounds the MDT-metabolites in the extractable fraction are dominated by MDD. Furthermore, MDA was very abundant, in contrast to low concentrations of the corresponding DDA in the same extracts. The pattern of MDT-related contaminants released by alkaline hydrolysis differed significantly in the same way as observed for the DDT-related compounds. Mainly 4,4'-MDA and 4,4'-MDB were enriched with concentrations up to 5500 and 1100 µg/kg, respectively.

The results presented support the conclusions drawn from the observations of bound DDT-related compounds.

## 4.3.3 Quantitation of nonextractable anthropogenic contaminants released from Teltow Canal sediments after chemical degradation ‡

### Introduction

The association of DDT-related pollutants to the non extractable matter of sediments from the Teltow Canal has been reported recently (Chapter 4.3.2 and Schwarzbauer et al. 2003a). This study was primarily initiated to quantify concentrations of bound 2,2-bis(chlorophenyl)-1,1,1-trichloroethane (DDT) residues in order to obtain further information about the fate of DDT-derived compounds within the particulate matter of the aquatic environment. Generally, the distribution of the bound DDT-related compounds was found to differ distinctly from the distribution within the extractable fraction (Schwarzbauer et al. 2003a).

However, Teltow Canal sediments are not only contaminated by DDT and its metabolites (2,2-bis(chlorophenyl)-1,1-dichloroethane - DDD, 2,2-bis(chlorophenyl)-1-chloroethane DDMS, 2,2-bis(chlorophenyl)-1,1-dichloroethene - DDE, 2,2-bis(chlorophenyl)-1-chloroethene DDMU, 1,1-bis(chlorophenyl)ethene DDNU, 1,1-bis(chlorophenyl)ethane - DDEt, bis(chlorophenyl)methane – DDM, 2,2-bis(chlorophenyl)acetonitrile DDCN) but also by halogenated aromatics and additional pesticides (e.g. bromonaphthalenes, tribromoanilines, the acaricide 2,2-bis(4-bromophenyl)-2-hydroxyacetic acid *iso*-propylester - bromoproylate and the insecticide 2,2-bis(methoxyphenyl)-1,1,1-trichloroethane - methoxychlor). Apart from these contaminants derived from huge local emissions of an industrial point source, numerous anthropogenic organic substances were additionally detected, representing a diffuse contamination mainly due to sewage effluents (Schwarzbauer et al. 2001). Considering this wide spectrum of free solvent extractable organic contaminants, the incorporation of DDT residues and of several supplementary anthropogenic substances has to be assumed. Therefore, this study will also present a quantification of concentrations of these numerous organic contaminants associated with the non extractable matter of Teltow Canal sediments as a contribution to a comprehensive view on the state of pollution.

‡ Mainly adapted/reprinted from Schwarzbauer et al. 2003c

## Samples

Details of the samples are given in chapter 4.3.1 and 4.3.2. The chemical degradation steps applied to the samples were carried out in two different modes. In a first set aliquots of the preextracted samples were treated separately with KOH/MeOH, $BBr_3$ and $RuO_4$. In a second step the preextracted and saponified residues were treated once more with $BBr_3$ or $RuO_4$ (according to Schwarzbauer et al. 2005a). Quantitative data were obtained by integration of specific ion chromatograms extracted from the TIC. The ions used for quantification as well as the recovery rates for the extraction and evaporating procedures are summarized in Table 1.

**Tab. 1:** Selected contaminants and ion fragments used for quantification as well as determined recoveries.

| Compounds | Ion fragments m/z | Recovery % |
|---|---|---|
| *Halogenated aromatics* | | |
| 1-Chloronaphthalene | 162, 164 | 35 |
| Dichoronaphthalenes | 196, 198 | 53 |
| 1-Bromonaphthalene | 206, 208 | 37 |
| 2,4,6-Tribromoaniline | 329, 331 | 63 |
| Bromophenols | 186, 188 | 40 |
| Dibromophenols | 264, 266 | 40 |
| *Nitro compounds* | | |
| 4-Nitrobenzoic acid (methylester) | 150, 181 | 44 |
| 2,4-Di-*tert*-butyl-6-nitrophenol | 236, 251 | 41 |
| 2-*tert*-Butyl-4,6-dinitrophenol | 225, 240 | 41 |
| *Plasticizers* | | |
| Di-*iso*-butylphthalate | 149 | 45 |
| Di-*n*-butylphthalate | 149 | 45 |
| Bis(2-ethylhexyl)phthalate | 149 | 45 |
| 2,4,4-Trimethylpentane-1,3-dioldi-*iso*-buytrate | 71, 111 | 41 |
| Tributylphosphate | 99, 155 | 78 |
| Bisphenol A | 215, 230 | 39 |
| *Fragrances* | | |
| Galaxolide | 243, 258 | 43 |
| Tonalide | 243, 258 | 46 |

## Results and discussion

Using common extraction procedures and chemical degradation techniques as described above the extractable and nonextractable fraction of four sediment samples of the Teltow Canal were investigated. The extracts as well as the degradation products were analysed by means of gas chromatographic - mass spectrometric analyses.

The GC-MS screening analyses applied to the extracts of sediment samples T1 – T3 revealed numerous anthropogenic organic contaminants including halogenated aromatics, pesticides, technical additives, fragrances, plasticizers, polycyclic aromatic compounds, organotin compounds and detergent related substances. This spectrum of compounds was in accordance with analytical results obtained from Teltow Canal sediments situated nearby as previously published (Schwarzbauer et al. 2001; Ricking et al. 2003).

With respect to the nonextractable fraction also numerous obviously anthropogenic compounds were identified, but they were superimposed by huge amounts of degradation products derived from natural macromolecules (e.g. brominated alkanes and cycloalkanes, fatty acids, hydroxylated carboxylic acids, fatty alcohols). Table 2 summarizes significant contaminants of the nonextractable fraction as recently reported (Schwarzbauer et al. 2005a). These nonextractable contaminants can be attributed to two different groups considering the concentrations as well as their spatial distribution within the riverine systems of the urban area of Berlin. On the one hand a distinct local emission was identified representing industrial emissions of a chemical plant, which formerly produced pesticides and halogenated technical additives. Therefore, this emission can be characterized mainly by the occurrence of DDT and its metabolites as well as of brominated aromatics (e.g. brominated naphthalenes) in the nonextractable organic fraction of the affected sediment samples. This spectrum of compounds was in accordance with analytical results published for the extractable fraction of comparable sediment samples of the Teltow Canal (Schwarzbauer et al. 2001).

In addition to the local contamination numerous substances were detected within the nonextractable fraction, that are common riverine contaminants including e.g. phosphates, nitro compounds, UV-protectors, pesticides, fragrances, organotin compounds and halogenated aromatics (see Table 2).

**Tab. 2:**Organic contaminants identified in the nonextractable organic matter of Teltow Canal sediments after chemical degradation procedures

| | |
|---|---|
| *Technical additives, solvents* | *Pesticides and metabolites* |
| N-Methylpyrrolidone | Hexachlorocyclohexanes (α-, β-, γ-, δ-) |
| N-Ethylaniline | 2,4′-DDD |
| Benzophenone | 4,4′-DDD |
| Benzylbenzoat | 4,4′-DDMS |
| Dibutylmaleat | 2,4′-DDM |
| Butylpalmitat | 4,4′-DDM |
| Isopropyldodecanoate | 2,4′-DDE |
| Isopropylpalmitate | 4,4′-DDE |
| Hexandioic acid di-*iso*-propyl ester | 2,4′-DDMU |
| 2-Ethylhexanol | 4,4′-DDMU |
| Hexandioic acid 2-ethylhexyl ester | 2,4′-DDNU |
| Bisphenol A $^{(me)}$ | 4,4′-DDNU |
| Ionol | 4,4′-DDEthane |
| Di-*tert*-butylchinone | 2,4′-DDA $^{(m)}$ |
| | 4,4′-DDA $^{(m)}$ |
| *Detergent residues, fragrances* | 4,4′-DDOH |
| Linear alkylbenzenes | 2,4′-DBP |
| Galaxolide | 4,4′-DBP |
| Tonalide | 2,4′-DDCN |
| | 4,4′-DDCN |
| *Plasticizers* | 4,4′-MDE |
| Alkylsulfonic acid phenyl esters | 4,4′-Dimethoxybenzophenone |
| Tri-n-butylphosphate | 2,2-Bis(4-dimethoxyphenyl)acetic acid $^{(m)}$ |
| 2,4,4-Trimethylpentane-1,3-dioldi-*iso*-butyrate | Chloropropylate |
| Di-*n*-butylphthalate | *Halogenated aromatics* |
| Di-*iso*-butylphthalate | Dichlorobenzene (1 isomer) |
| Bis(2-ethylhexyl)phthalate | Trichlorobenzene (2 isomers) |
| Tris(2-ethylhexyl)trimellitate | Tetrachlorobenzene (1 isomer) |
| Phthalic acid 2-ethyl-hexylmonoester $^{(m)}$ | Pentachlorobenzene |
| Phthalic acid $^{(m)}$ | 1-Chloronaphthalene |
| | Dichloronaphthalene (3 isomers) |
| *Nitrogen containing compounds* | 1-Bromonaphthalene |
| 4-Nitrobenzoic acid $^{(m)}$ | Dibromonaphthalenes (3 isomers) |
| 4-Aminobenzoic acid$^{(m)}$ | PCB ($Cl_4$-$Cl_6$) |
| N-Methyl-4-amino-benzoic acid $^{(m)}$ | 4-Chlorobenzoic acid$^{(m)}$ |
| 2,4-Di-*tert*-butyl-6-nitrophenol | 2,4-Dichlorobenzoic acid $^{(m)}$ |
| 2-*tert*-Butyl-4,6-dinitrophenol | 2,4-Dibromoaniline |
| | 2,4,6-Tribromoaniline |
| | Bromophenol $^{(me)}$ |
| | Dibromophenol $^{(me)}$ |
| | Tribromophenol $^{(me)}$ |

$^{(me)}$ = identified as methyl ethers; $^{(m)}$ = identified as methyl esters; (underlined) = quantified and presented in this study

These compounds were frequently detected in Havel and Spree river sediments at lower concentrations compared to the elevated amounts of the point source contaminants. Thus, a widespread distribution of these compounds in minor concentrations was stated (Ricking et al. 2003).

Apart from these qualitative results also quantitative data are of great importance for the environmental assessment of the bound organic contaminants. Quantitative analyses of DDT and its metabolites revealed not only an accumulation of these compounds within the bound fraction but also significant differences in the quantitative distribution as compared to the corresponding extractable substances (Schwarzbauer et al. 2003a).

Accordingly, the investigations were extended to the specific organic contaminants reflecting not only the industrial point source emission but also the widespread contamination of riverine particulate matter by non-point sources. Selective degradation techniques were used to release the bound contaminants and to allow a differentiation of the mode of incorporation into the nonextractable particulate matter. Apart from an alkaline hydrolysis affecting mainly ester and amide bonds a boron tribromide treatment and a ruthenium tetroxide oxidation were applied to the extracted residues. The $BBr_3$-treatment cleaves aromatic and aliphatic ether and ester bonds. The oxidation using attacks the aromatic carbon atoms and releases, therefore, mainly unaltered aliphatic moieties.

In the following sections the quantified occurrence of specific riverine contaminants including their mode of association is discussed in detail for the Teltow Canal sediments investigated. The contaminants determined were attributed to halogenated aromatics, nitro compounds, plasticizers, technical additives and fragrances. The chemical structures are illustrated in Figure 1 and all quantitative results are presented in Tables 3 to 5.

Halogenated naphthalenes in Teltow Canal sediments are specific pollutants reflecting the industrial point source emission. Chlorinated naphthalenes are known environmental pollutants, which are emitted due to their usage as technical additives, as a result of pyrolytic processes or as an impurity associated with PCB products (Haglund et al. 1993; Järnberg et al. 1993). In the extractable fraction mono- and dichlorinated naphthalenes were determined with total concentrations between 120 and 493 µg/kg . The peak pattern was similar to the congener distribution of technical mixtures (Halowax 1000, Halowax 1001) (Falandysz 1998). The concentration of 1-bromonaphthalene ranged between the limit of quantification (LOQ) and 250 µg/kg (see Table 1).

**Tab. 3:** Specific contaminants quantified in the extractable fraction of Teltow Canal sediment samples (contents are given in µg/kg dry matter).

| | Extracts | | | |
|---|---|---|---|---|
| **Compounds / samples** | **1a** | **1b** | **2** | **3** |
| *Halogenated aromatics* | | | | |
| 1-Chloronaphthalene | 34 | 62 | 5 | 11 |
| 1,3-Dichloronaphthalene | 98 | 44 | 21 | 33 |
| 1,4-Dichloronaphthalene | 190 | 140 | 51 | 84 |
| 1,5-/1,6-Dichloronaphthalene | 120 | 68 | 27 | 47 |
| 1,7-/2,6-/2,7-Dichloronaphthalene | 42 | 38 | 16 | 31 |
| 1,2-Dichloronaphthalene | 9 | 5 | nd | 5 |
| 1-Bromonaphthalene | 250 | 35 | 5 | nd |
| 2,4,6-Tribromoaniline | nd | 28 | 35 | 16 |
| Bromophenols | nd | nd | nd | nd |
| Dibromophenols | nd | nd | nd | nd |
| *Nitro compounds* | | | | |
| 4-Nitrobenzoic acid | nd | nd | nd | nd |
| 2,4-Di-*tert*-butyl-6-nitrophenol | nd | nd | nd | nd |
| 2-*tert*-Butyl-4,6-dinitrophenol | nd | nd | nd | nd |
| *Plasticizers* | | | | |
| DiBP | 340 | nd | 48 | 160 |
| DnBP | 540 | nd | 120 | 436 |
| DEHP | 4400 | nd | 2210 | 2131 |
| 2,4,4-Trimethylpentane1,3-dioldi-*iso*-butyrate | 9 | 92 | nd | nd |
| TBP | nd | nd | nd | nd |
| Bisphenol A | nd | nd | nd | nd |
| *Fragrances* | | | | |
| Galaxolide | nd | nd | nd | 410 |
| Tonalide | nd | nd | nd | 340 |

Samples T1a and T1b were higher contaminated by halogenated naphthalenes as compared to samples T2 and T3. With respect to the nonextractable particulate matter the occurrence of halogenated naphthalenes was limited to the extracts obtained after alkaline hydrolysis.

Although the concentration range was in the same order of magnitude, it has to be noted, that the observed concentrations differed significantly for the individual samples as compared to the extractable fraction. The highest concentrations were determined in sample T1b and T3 with 1380 and 1600 µg/kg of chlorinated and brominated naphthalenes, respectively (see Table 4). In both samples the concentrations were clearly higher compared to the extractable fraction. Although the highest concentration in the extractable fraction was detected in sample T1a, this sample revealed only minor concentration of 1-chloro- and 1-bromonaphthalene after hydrolysis. In sample T2 nearly the same concentrations in the extractable as well as as in the nonextractable fraction were determined. Thus, a distinct preference of enrichment cannot be stated for the halogenated naphthalenes with respect to the extractable or nonextractable fraction.

Within the group of halogenated aromatics also 2,4,6-tribromoaniline was determined, which was previously characterized as a specific pollutant derived from the local industrial emissions (Schwarzbauer et al. 2001). Brominated anilines are used as coupling components in azo dye production Hunger et al. 1998). In contrast to the halogenated naphthalenes this compound was not only detected in all hydrolysis extracts but also in extracts obtained after $BBr_3$-treatment and, exclusively, after $RuO_4$-oxidation. The concentrations ranged between the LOQ and 2400 µg/kg (see Table 4). Interestingly, the higher concentrations were observed in the samples T2 and T3, whereas the concentration of the extractable 2,4,6-tribromoaniline peaked at samples T1b and T2 (see Table 3). In addition, the concentrations in the bound fractions of samples T1b, T2 and T3 were higher than the non-bound brominated aniline.

As a third group of brominated aromatics mono- and dibrominated phenols were determined, which appeared both after a separate and a sequential application of the $BBr_3$-treatment procedure. Concentrations ranged between the LOQ and 45000 µg/kg (see Table 4 and 5). In contrast to the halogenated naphthalenes and tribromoaniline the brominated phenols have not been previously reported as extractable contaminants of Teltow Canal sediments. Considering the occurrence of these compounds, which are only observed in extracts obtained after $BBr_3$-treatment, a natural origin cannot be excluded. $BBr_3$-treatment affects heavily aliphatic and aromatic ether linkages and generates corresponding alcohols and bromides. Therefore, the appearance of brominated phenols can be the result of the cleavage of mono- to tetraalkoxylated aromatic structures within the organic macromolecular matter.

**Tab. 4:** Specific contaminants quantified in the nonextractable fraction of Teltow Canal sediment samples after a separate application of different chemical degradation procedures (contents are given in µg/kg dry matter, nd = not determined).

| | **$BBr_3$ treatment** | | | | **$RuO_4$-Oxidation** | | | |
|---|---|---|---|---|---|---|---|---|
| **Compounds / samples** | **T1a** | **T1b** | **T2** | **T3** | **T1a** | **T1b** | **T2** | **T3** |
| *Halogenated aromatics* | | | | | | | | |
| 1-Chloronaphthalene | nd | nd | nd | nd | nd | nd | nd | nd |
| 1,3-Dichloronaphthalene | nd | nd | nd | nd | nd | nd | nd | nd |
| 1,4-Dichloronaphthalene | nd | nd | nd | nd | nd | nd | nd | nd |
| 1,5/1,6-Dichloronaphthalene | nd | nd | nd | nd | nd | nd | nd | nd |
| 1,7-/2,6-/2,7-Dichloro-naphthalene | nd | nd | nd | nd | nd | nd | nd | nd |
| 1,2-Dichloronaphthalene | nd | nd | nd | nd | nd | nd | nd | nd |
| 1-Bromonaphthalene | nd | nd | nd | nd | 1000 | 18 | nd | nd |
| 2,4,6-Tribromoaniline | nd | 220 | 480 | 650 | nd | 181 | nd | nd |
| Bromophenols | 2700 | 1500 | nd | nd | nd | nd | nd | nd |
| Dibromophenols | 1600 | 620 | 43000 | 4500 | nd | nd | nd | nd |
| Tribromophenols | 47 | nd | 25000 | 400 | nd | nd | nd | nd |
| *Nitro compounds* | | | | | | | | |
| 4-Nitrobenzoic acid | nd | nd | nd | nd | 38000 | nd | nd | nd |
| 2,4-Di-*tert*-butyl-6-nitrophenol | nd | nd | nd | nd | 2200 | nd | nd | nd |
| 2-*tert*-Butyl-4,6-dinitrophenol | nd | nd | nd | nd | 2000 | nd | nd | nd |
| *Plasticizers, Additives* | | | | | | | | |
| DiBP | nd | nd | 3500 | nd | 47000 | nd | nd | nd |
| DnBP | 930 | nd | 7700 | nd | 65000 | nd | nd | nd |
| DEHP | 3500 | nd | 20000 | nd | 160000 | 2970 | nd | nd |
| 2,4,4-Trimethylpentane-1,3-dioldi-*iso*-butyrate | nd | nd | nd | nd | 8000 | 160 | nd | nd |
| TBP | nd | nd | nd | nd | 15000 | 2400 | 17000 | nd |
| Bisphenol A | 1100 | 110 | 110 | 120 | nd | 1100 | nd | 140 |
| *Fragrances* | | | | | | | | |
| Galaxolide | 410 | nd | 21 | nd | nd | nd | nd | 2500 |
| Tonalide | 200 | nd | nd | nd | nd | nd | nd | 950 |

**Tab. 4** continued

| | Hydrolysis | | | |
|---|---|---|---|---|
| **Compounds / samples** | **T1a** | **T1b** | **T2** | **T3** |
| *Halogenated aromatics* | | | | |
| 1-Chloronaphthalene | 33 | 150 | 110 | 720 |
| 1,3-Dichloronaphthalene | nd | 160 | nd | 22 |
| 1,4-Dichloronaphthalene | nd | 460 | 54 | 380 |
| 1,5-/1,6-Dichloronaphthalene | nd | 200 | 44 | 180 |
| 1,7-/2,6-/2,7-Dichloro-naphthalene | nd | nd | nd | 58 |
| 1,2-Dichloronaphthalene | nd | nd | nd | 20 |
| 1-Bromonaphthalene | 130 | 76 | 1600 | 1600 |
| 2,4,6-Tribromoaniline | 260 | 400 | 2400 | 2100 |
| Bromophenols | nd | 230 | nd | nd |
| Dibromophenols | nd | nd | nd | nd |
| Tribromophenols | nd | nd | nd | nd |
| *Nitro compounds* | | | | |
| 4-Nitrobenzoic acid | nd | nd | nd | nd |
| 2,4-Di-*tert*-butyl-6-nitrophenol | nd | nd | nd | nd |
| 2-*tert*-Butyl-4,6-dinitrophenol | nd | nd | nd | nd |
| *Plasticizers, Additives* | | | | |
| DiBP | 2000 | nd | nd | 180 |
| DnBP | 1400 | 1000 | 7200 | 940 |
| DEHP | 5300 | 9400 | 15000 | 1400 |
| 2,4,4-Trimethylpentane-1,3-dioldi-*iso*-butyrate | 2800 | 1400 | nd | nd |
| TBP | nd | nd | nd | nd |
| Bisphenol A | nd | nd | nd | 2400 |
| *Fragrances* | | | | |
| Galaxolide | nd | 49 | nd | nd |
| Tonalide | nd | 21 | nd | nd |

Alternatively, with respect to the missing corresponding brominated catechols the chemical degradation released already brominated phenoxy moieties linked by covalent ether bondings to the geopolymers.

Apart from the halogenated substances a few nitro compounds were analysed exclusively in extracts obtained after the separate application of the $RuO_4$-oxidation procedure (see Table 4). Furthermore, the appearance of these compounds was limited to sample T1a. The concentration of the alkylated nitrophenols ranged between 2000 and 2200 µg/kg, whereas the concentration of 4-nitrobenzoic acid reached 38000 µg/kg. It has to be noted, that these nitrogen containing compounds were also not detected in the extractable fraction as well as in the extracts obtained from the $RuO_4$-oxidation applied to the saponified residues. The occurrence of the nitro compounds as a result of the oxidation of amino substituents can be excluded due to the contemporary appearance of anilines in the same extracts as published elsewhere (Schwarzbauer et al. 2005a).

A third group of anthropogenic contaminants included plasticizers, additives and fragrances. These compounds mainly appeared in the riverine particulate matter as the result of sewage effluents. Apart from phthalic acid esters, a common and widespread group of plasticizers frequently detected in the hydrosphere, the aditives tributylphosphate, 2,4,4-trimethylpentane-1,3-dioldi-*iso*-butyrate and bisphenol A were evaluated with concentrations between the LOQ and 17000 µg/kg (see Tables 4 and 5). Higher concentrations were observed for the phthalates in extracts obtained after $RuO_4$-oxidation, hydrolysis and $BBr_3$-treatment, but not in general in all samples. Therefore, a distinct preference of a degradation procedure for an elevated release of phthalates or an accumulation in an individual sample cannot be stated. In contrast, tributylphosphate was only detected in extracts obtained after $RuO_4$-oxidation with concentrations of up to 17000 µg/kg. The sequential application of hydrolysis and $RuO_4$-oxidation released TBP only from the nonextractable residues of sample T1a and T1b with low concentrations between 130 and 820 µg/kg. Bisphenol A, used as plasticizer, fungicide and intermediate in polymer syntheses, was analysed in extracts obtained after a $BBr_3$-treatment and $RuO_4$-oxidation with concentrations between LOQ and 1100 µg/kg. After application of the alkaline hydrolysis procedure this additive was only determined in sample T3 with a concentration of 2400 µg/kg. $BBr_3$-treatment of the saponified residues of samples T1a and T2 released bisphenol A with higher concentrations of 2800 and 5300 µg/kg, respectively.

Finally the two synthetic musk fragrances galaxolide and tonalide were analysed with concentrations between the LOQ and 2500 µg/kg (see Table 4). The occurrence of these fragrances was also not limited to individual samples or degradation procedures. In the extractable fraction galaxolide and tonalide were detected only in sample T3 with concentrations of 410 and 340 µg/kg, respectively.

**Tab. 5:** Specific contaminants quantified in the nonextractable fraction of Teltow Canal sediment samples after a sequential application of different chemical degradation procedures (contents are given in µg/kg dry matter, nd = not determined).

| | **$BBr_3$ treatment after hydrolysis** | | | | **$RuO_4$-Oxidation after hydrolysis** | | | |
|---|---|---|---|---|---|---|---|---|
| **Compounds / samples** | **T1a** | **T1b** | **T2** | **T3** | **T1a** | **T1b** | **T2** | **T3** |
| Halogenated aromatics | | | | | | | | |
| 1-Chloronaphthalene | nd | nd | nd | nd | nd | nd | nd | nd |
| 1,3-Dichloronaphthalene | nd | nd | nd | nd | nd | nd | nd | nd |
| 1,4-Dichloronaphthalene | nd | nd | nd | nd | nd | nd | nd | nd |
| 1,5-/1,6-Dichloro-naphthalene | nd | nd | nd | nd | nd | nd | nd | nd |
| 1,7-/2,6-/2,7-Dichloro-naphthalene | nd | nd | nd | nd | nd | nd | nd | nd |
| 1,2-Dichloronaphthalene | nd | nd | nd | nd | nd | nd | nd | nd |
| 1-Bromonaphthalene | nd | 86 | nd | nd | nd | nd | nd | nd |
| 2,4,6-Tribromoaniline | nd | nd | nd | 217 | nd | nd | nd | nd |
| Bromophenols | 710 | 210 | 1500 | nd | nd | nd | nd | nd |
| Dibromophenols | 5900 | 660 | 5800 | 4500 | nd | nd | nd | nd |
| Tribromophenols | 2600 | nd | 74 | nd | nd | nd | nd | nd |
| *Nitro compounds* | | | | | | | | |
| 4-Nitrobenzoic acid | nd | nd | nd | nd | nd | nd | nd | nd |
| 2,4-Di-*tert*-butyl-6-nitro-phenol | nd | nd | nd | nd | nd | nd | nd | nd |
| 2-*tert*-Butyl-4,6-dinitro-phenol | nd | nd | nd | nd | nd | nd | nd | nd |
| *Plasticizers, Additives* | | | | | | | | |
| DiBP | 1300 | nd | 7200 | nd | 4300 | 16000 | 39000 | 11000 |
| DnBP | 7600 | nd | 8400 | nd | 4600 | 15000 | 55000 | 7100 |
| DEHP | 27000 | nd | 28000 | nd | 6300 | 21000 | 150000 | 43000 |
| 2,4,4-Trimethylpentane-1,3-dioldi-*iso*-butyrate | nd | nd | nd | nd | nd | nd | nd | nd |
| TBP | nd | nd | nd | nd | 130 | 820 | nd | nd |
| Bisphenol A | 2800 | nd | 5300 | 810 | nd | nd | nd | nd |
| *Fragrances* | | | | | | | | |
| Galaxolide | 410 | nd | 21 | nd | nd | nd | nd | nd |
| Tonalide | 200 | nd | nd | nd | nd | nd | nd | nd |

**Halogenated aromatics**

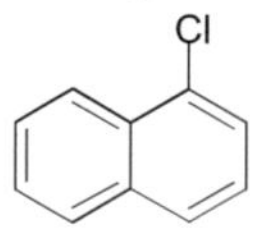


1

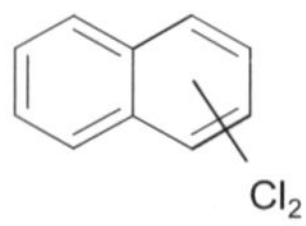


2

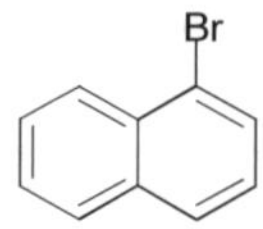


3

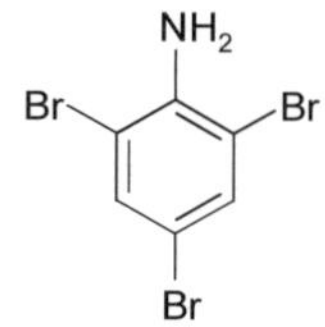


4

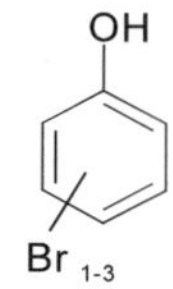


5

**Nitro compounds**

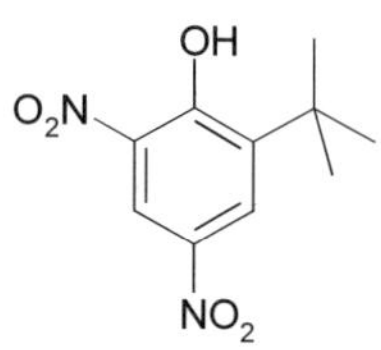


6

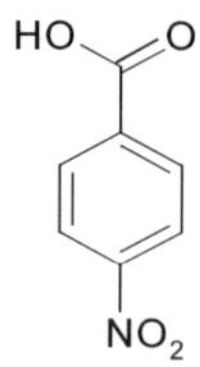


7

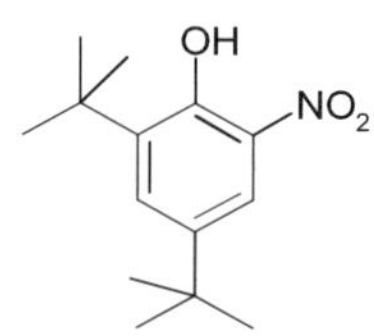


8

**Plasticizers, additives and fragrances**

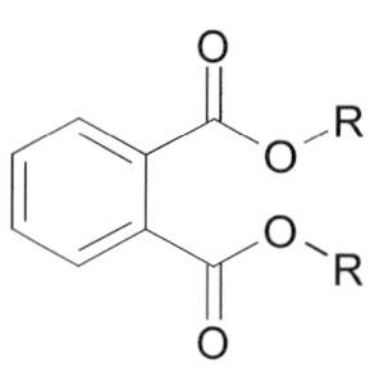


9

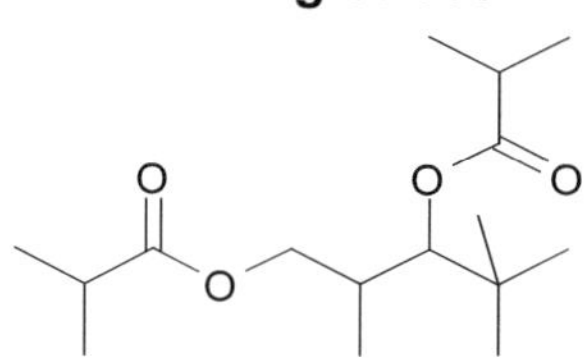


10

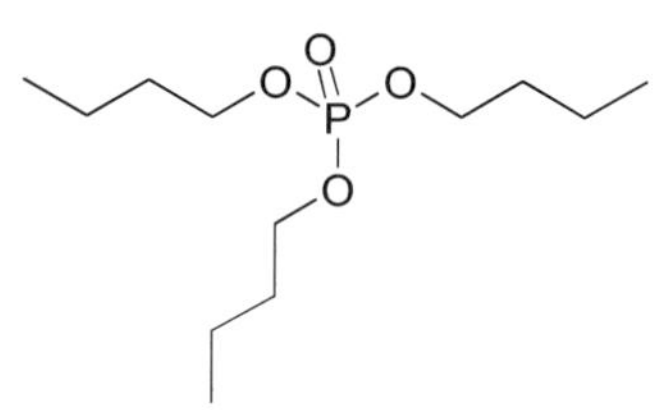


11

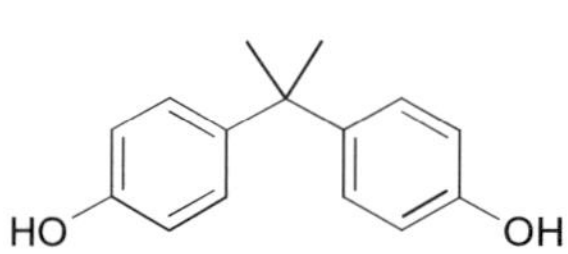


12

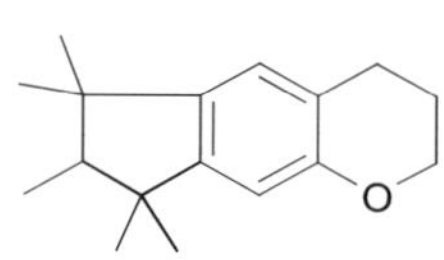


13

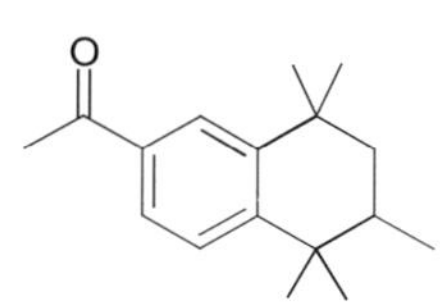


14

**Fig. 1:** Molecular structures of the contaminants investigated in this study.

Comparing the results obtained from the sequential and the separate application of the degradation procedures (see Tables 4 and 5), only minor variations can be pointed out. Generally, no significant alteration of the quantitative distribution can be observed. But the concentrations found in extracts obtained after the sequential application of $BBr_3$-treatment and $RuO_4$-oxidation were significantly lower than in extracts after separate application of the degradation procedures. Only the concentrations of phthalates peaked in the extracts obtained after $RuO_4$-oxidation of the saponified residues.

With respect to the concentrations of the individual compounds in the free and bound fractions (see Tables 3 to 5) different kinds of contamination can be distinguished. The major portion of the solvent extractable anthropogenic compounds (as formerly reported, Schwarzbauer et al. 2001, Ricking et al. 2003) was not detected in the extracts obtained after degradation procedures (see Table 2). A second group of contaminants including chlorinated and brominated naphthalenes, 2,4,6-tribromoaniline, phthalates, 2,4,4-trimethylpentane-1,3-dioldi-*iso*-butyrate, galaxolide and tonalide appeared in the extractable as well as in the nonextractable fraction. For these compounds, it has to be noted, that the concentrations within the bound fraction were generally higher than in the solvent extracts. Especially the brominated and chlorinated aromatics as well as the phthalates appeared with very high concentrations, whereas the concentrations of galaxolide and tonalide were only slightly higher than in solvent extracts.

Brominated phenols, nitrobenzoic acid, butylated nitrophenols, tributylphosphate and bisphenol A formed a third group of contaminants that occurred only in the nonextractable fraction.

Besides the quantitative differentiation of the free and bound contaminants the application of different but selective degradation methods was also used to obtain information about the mode and strength of incorporation. Extended non-target screening analyses of all extracts derived from the different degradation procedures revealed only unaltered contaminants but no further compounds structurally related to these compounds (Schwarzbauer et al. 2005a). Thus, incorporation and subsequent remobilisation processes, which modifies the moleculare structure of the associated substances, can be excluded for the observed substances.

Furthermore, a covalent linkage to the nonextractable organic matter can be excluded for numerous components according to the data on the release of different degradation techniques (see also Tab. 6). Consequently, the release of the bound substances has to be attributed to the modification and degradation of the nonextractable matter. Regarding the mode of degradation procedures a transformation mainly of the organic macromolecular matter has to be assumed.

**Tab. 6:** Summarized characteristics of the incorporation of individual contaminants into sediment samples of the Teltow Canal with respect to the frequency of detection and the release after different degradation procedures.

| | **Frequency of detection** | | **Release after application of different degradation procedures** | |
|---|---|---|---|---|
| **Compounds** | **1 sample** | **2-4 samples** | **selective** | **nonselective** |
| *Halogenated aromatics* | | | | |
| 1-Chloronaphthalene | | + | Hydrolysis | |
| Dichloronaphthalenes | | + | Hydrolysis | |
| 1-Bromonaphthalene | | + | | + |
| 2,4,6-Tribromoaniline | | + | | + |
| Bromophenols | | + | $BBr_3$-treatment | |
| Dibromophenols | | + | $BBr_3$-treatment | |
| *Nitro compounds* | | | | |
| 4-Nitrobenzoic acid | T1a | | $RuO_4$-oxidation | |
| 2,4-Di-*tert*-butyl-6-nitrophenol | T1a | | $RuO_4$-oxidation | |
| 2-*tert*-Butyl-4,6-dinitro-phenol | T1a | | $RuO_4$-oxidation | |
| *Plasticizers* | | | | |
| DiBP | | + | | + |
| DnBP | | + | | + |
| DEHP | | + | | + |
| 2,4,4-Trimethylpentane-1,3-dioldi-*iso*-butyrate | T1a | | | + |
| TBP | | + | $RuO_4$-oxidation | |
| Bisphenol A | | + | | + |
| *Fragrances* | | | | |
| Galaxolide | | + | | + |
| Tonalide | | + | | + |

Interestingly, for several compounds including chlorinated naphthalenes, nitro compounds, brominated phenols and tributylphosphate a restricted occurrence after the application of one selective degradation method can be observed (see Tab. 6). These observations lead to the assumption, that these contaminants interact more specifically with the macromolecular organic matter of the nonextractable fraction. As an example, alkylated nitrophenols were released from the preextracted sample T1a only after application of the $RuO_4$-oxidation. This reaction mainly affects the aromatic moieties of the macromolecular organic matter. Hence, we suggest a stronger association of the nitro groups with the aromatic systems. Brominated phenoles were also released selectively after the application of $BBr_3$-treatment. As mentioned above a covalent ether linkage can neither be confirmed nor excluded for this group of compounds. But a stronger and more specific interaction of these compounds with the macromolecular organic matter can be stated than for the compounds discussed above.

However, in summary we conclude, that the mode and strength of incorporation is less influenced by the physico-chemical or chemical properties of the substances investigated. But a dominant influence of the non-extractable matter including the inorganic (e.g. minerals) as well as the organic composition (e.g. humic substances) on the mode of association of low molecular weight contaminants has to be assumed.

## Conclusions

Based on their distribution and concentration data two groups of bound contaminants can be differentiated. A major portion of the contaminants determined appeared in the extractable as well as in the nonextractable fraction with concentrations generally higher in the latter. A second group occurred only in the nonextractable fraction with substantial concentrations but not in the extractable fraction.

With respect to the spatial distribution and the different selectivity of the degradation methods a second differentiation was observed. Most contaminants appeared frequently and independently of the kind of degradation procedure applied. Also a preferred occurrence at individual sampling locations was not observed for most of the substances. Considering additionally the specifity of the degradation reactions as well as the molecular structures of the contaminants investigated a weaker noncovalent association to the nonextractable particulate matter has to be stated for these compounds.

In contrast the occurrence of only a few substances was spatially restricted. Also only few compounds were determined exclusively after application of one degradation method. For these contaminants a more specific interaction with the macromolecular organic matter of the nonextractable fraction has to be assumed.

The mode and strength of the association of the contaminants investigated seems to be mainly influenced by the composition of the non-extractable matter including the inorganic and the organic components. However, the release of all bound substances investigated has to be attributed mainly to the modification and degradation of the nonextractable organic matter.

# 5 Particle associated output of organic contaminants from riverine systems

## 5.1 The contribution of riverine contamination to an estuarine area of the North Sea - the German Bight

### 5.1.1 Identification of specific organic contaminants for estimating the contribution of the Elbe river to the pollution of the German Bight‡

**Introduction**

The marine environment of the North Sea is highly influenced by anthropogenic input as a result of intense navigation, petroleum and gas production, atmospheric deposition as well as riverine contribution of terrestrial pollutants. The qualitative and quantitative composition of the organic material in the sediment and the water column reflects the discharge of anthropogenic contaminants. Hence, natural biogenic substances are accompanied by a wide variety of anthropogenic compounds.

In the German Bight a significant proportion of the anthropogenic organic matter is contributed by contaminated riverine systems discharging into the North Sea. Next to the Ems and Weser rivers, the Elbe river is the most polluted riverine system due to well known industrial emissions and partly deficient sewage treatment in the catchment area. In order to estimate the fluvial input to the organic contamination of marine sediments, analyses of carbon fluxes by bulk parameter such as dissolved organic carbon (DOC), particulate organic matter (POM), terrestrial organic matter (TOM) are useful (Ittekkot, 1988; Keil et al., 1997; Gupta et al., 1997; Alberts and Takács, 1999). More detailed information about the terrestrial contribution to the marine organic matter could be obtained

‡ Mainly adapted/reprinted from Schwarzbauer et al., 2000

by identification and quantification of several specific molecular markers (Hedges et al., 1997). For example, analyses of carbohydrates, amino carbohydrates and amino acids in the suspended particulate matter of the river Indus indicated a significant terrigenous input to the marine environment (Ittekkot and Arain, 1986). Hedges and Parker (1976) characterized the terrestrial organic matter in surface sediments from the Gulf of Mexico using lignin oxidation products as markers. The contribution of the Mackenzie River to the Beaufort Sea coastal sediments was assessed by examination of specific aliphatic and aromatic hydrocarbons (Yunker et al., 1991; Yunker et al., 1993). With a similar aim, Zegouagh et al. (1996; 1998) studied the molecular and isotopic properties both of hydrocarbons and acids in sediments of the Lena River delta and the Laptev Sea.

Information about the fluvial discharge into the marine environment can also be achieved with anthropogenic substances (Eganhouse, 1997; Takada et al., 1997). Well known anthropogenic markers include e.g. tetrapropylene-based alkylbenzenes (TAB), linear alkylbenzenes (LAB) and the sulfonated analogues (linear alkylbenzenesulfonates, LAS). The LAS are widely used surfactants and the LAB are synthetic raw material. Their occurrence in the aquatic environment reflects the emission of municipal waste water effluents (Takada and Eganhouse, 1998, and references cited therein). Fecal steroids derived from sewage contaminated rivers have also been used to document the discharge of fluvial anthropogenic matter to coastal sediments (Takada and Eganhouse, 1998, and references cited therein).

Natural and anthropogenic marker compounds are useful to distinguish between marine and terrigenous organic matter in coastal sediments and water. But in order to point out the contributions made by different estuaries situated close together, such as in the area of the German Bight, more specific molecular marker information is needed. The emission of specific anthropogenic compounds from point sources or the common usage of characteristic technical formulations results in quite different patterns of organic substances. These can be used as river-specific molecular markers.

Detailed analyses of the organic matter are necessary to identify such specific marker compounds, but only a few investigations have been carried out on organic contaminants in coastal water and sediments of the North Sea. Previous detailed analyses of lipophilic and low molecular compounds in Elbe river water have indicated a high abundance of river-specific organic substances (Franke et al., 1995; Theobald et al., 1995).

For preselected compounds (e.g. thiophosphates, benzothiazoles, nitrobenzenes and polycyclic musk fragrances) the contribution of the Elbe river to the contamination of coastal waters has been demonstrated quantitatively (Gatermann et al., 1995; Gatermann et al., 1996; Bester et al., 1997; Bester et al., 1998).

In order to document the state of organic pollution in the sediments, GC/MS screening analysis has been applied to samples from the German Bight. Based on full scan electron impact mass spectra, gas chromatographic retention times and synthetic reference compounds, several classes of biogenic and anthropogenic compounds were identified. The main focus of our study was to isolate Elbe-specific molecular markers that are appropriate to estimate the discharge of riverine anthropogenic compounds to the sediments of the German Bight. These compounds can only be potential markers because little information is available about the organic matter of the Ems and Weser rivers, which are situated close to the Elbe and also influence the sediments of the German Bight.

## Samples

Sediment samples (Tab. 1) were taken in 1998 by the German Federal Maritime and Hydrographic Agency (Hamburg) using a van Veen grab. This yielded material from the sediment surface to a depth of approximately 15 cm. All sampling locations are shown in Fig. 1. The wet sediments were stored in glass flasks with teflon lined screw caps at 4°C in the dark.

**Tab. 1:** Sediment samples collected from the German Bight (see Fig. 1 for locations).

| Sample | Altitude | Latitude | Fraction of grain size < 63 mm | Dry weight (%) |
|---|---|---|---|---|
| A | 54°02' | 8°12.5' | > 50 % | 73.4 |
| B | 54°04' | 8°07.5' | > 50 % | 58.3 |
| C | 54°22.5' | 7°38.75' | < 5 % | 79.9 |
| D | 54°30' | 6°30' | 5 – 10 % | 78.8 |
| E | 54°50' | 6°35' | 11 – 20 % | 72.1 |
| F | 55°00' | 6°30' | 21 – 50 % | 69.5 |
| G | 53°49' | 6°23' | < 5 % | 82.6 |

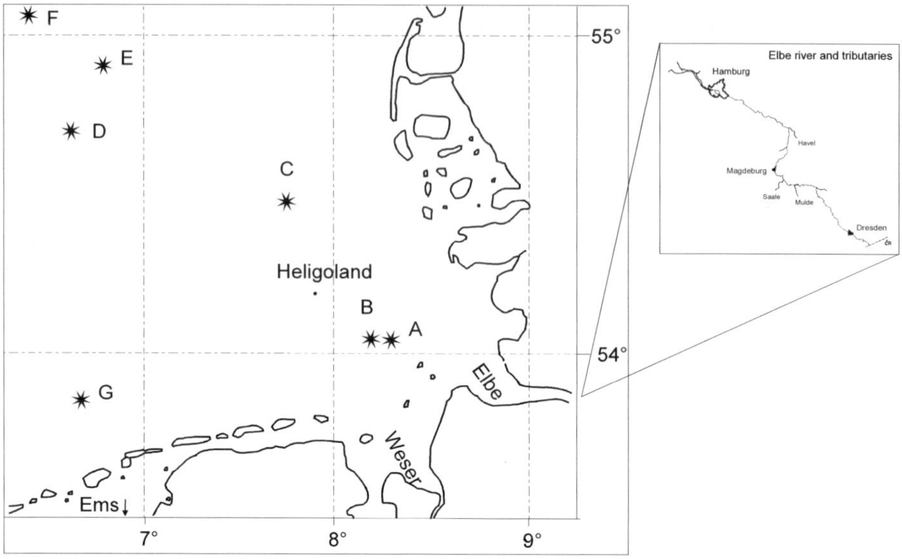


**Fig. 1:** Sampling locations of sediments in the German Bight.

Because of the intense analytical approach (Fig. 2) only the samples of these seven locations were investigated. Three of the samples seemed to be directly influenced by the Elbe river (A,B) and the Weser and Ems rivers (G), whereas the contributions of the riverine systems to the organic matter of the sediments situated farer from the coastal area (C,D,E,F) are still ambiguous.

## Results and discussion

The non-target screening analyses revealed a large number of individual organic compounds occurring in sediments of the German Bight. All identified contaminants are listed in Tab. 2. These are subdivided and arranged either by their structural properties or technical applications in case of some anthropogenic substances.

**Tab. 2:** Organic compounds identified in sediments of the German Bight.

| **Compounds** | **A** | **B** | **C** | **D** | **E** | **F** | **G** |
|---|---|---|---|---|---|---|---|
| *Alkanes, cycloalkanes and alkenes* | | | | | | | |
| Homologous series of n-alkanes ($C_9$ to $C_{30}$) [a)] | + | + | + | + | + | + | + |
| 2,2,4,4,6,8,8-Heptamethylnonane [a)] | + | + | + | + | + | + | + |
| Homologous series of n-alkylcyclohexanes ($C_4$- to $C_{20}$ side chain) [c)] | + | (+) | (+) | + | + | + | (+) |
| Homologous series of n-alkenes ($C_{12}$ to $C_{20}$) [a)] | | + | | | | | |
| *Terpenoids and degradation products* | | | | | | | |
| Limonene [a)] | + | + | + | + | + | + | + |
| Dihydroactinidiolide [c)] | | + | + | + | + | + | + |
| α-Ionone [c)] | | + | + | + | + | + | + |
| β-Ionone [c)] | | + | + | + | + | + | + |
| β-Cyclocitral [a)] | | | | + | + | | + |
| Cadalene [c)] | | + | | | | | |
| Calamenene [c)] | + | + | + | + | + | + | + |
| α-Cedrene [c)] | | + | | | | | |
| Longicyclene [c)] | + | + | | | | | |
| Junipene [c)] | + | + | | | | | |
| 7-Isopropyl-1-methylphenanthrene [e)] | + | + | + | + | + | + | + |
| 1,2,3,4-Tetrahydro-7-isopropyl-1-methylphenanthrene [e)] | + | + | + | + | + | + | + |
| Dehydroabietin [e)] | | + | | | | | |
| Dehydroabietane [e)] | | + | | | | | |
| 2,6-Dimethylundecane [e)] | + | + | + | + | + | + | + |
| Pristane [a)] | + | + | + | + | + | + | + |
| Phytane [a)] | + | + | + | + | + | + | + |
| Phytene (1 isomer) [d)] | + | + | | + | + | + | + |
| Phytadienes (3 isomers) [d)] | + | + | | + | + | + | + |
| 6,10-Dimethylundecan-2-one [e)] | + | + | + | + | + | + | + |
| 6,10-Dimethyl-5,9-undecadien-2-one [e)] | + | + | + | + | + | + | + |
| 6,10,14-Trimethylpentadecan-2-one [e)] | + | + | + | + | + | + | + |
| Phytol [a)] | | | | | | | |
| Phytolic acid [e)] | + | | + | + | | + | + |
| 4,8,12-Trimethyltridecanoic acid [e)] | + | + | + | + | | + | + |
| 4,8,12-Trimethyltetradecanoic acid [c)] | + | | + | + | | + | + |
| Squalene [a)] | + | + | + | + | + | + | + |
| *Hopanoids* | | | | | | | |
| 18α(H)-Trisnorneohopane, $T_s$ [c)] | + | + | + | + | + | + | + |
| 17α(H)-Trisnorhopane, $T_m$ [c)] | + | + | + | + | + | + | + |

| Compounds | A | B | C | D | E | F | G |
|---|---|---|---|---|---|---|---|
| 18α-Norneohopane [c] | + | | + | + | + | + | + |
| 17α(H),21β(H)-Norhopane [c] | + | + | + | + | + | + | + |
| 17β(H),21α(H)-Norhopane [c] | + | + | + | + | + | + | + |
| 17β(H),21β(H)-Norhopane [c] | + | + | + | + | + | + | + |
| 17α(H),21β(H)-Hopane [c] | + | + | + | + | + | + | + |
| 17β(H),21β(H)-Hopane [c] | + | + | + | + | + | + | + |
| (22S)-17α(H),21β(H)-Homohopane [c] | + | + | + | + | + | + | + |
| (22R)-17α(H),21β(H)-Homohopane [c] | + | + | + | + | + | + | + |
| (22S)-17α(H),21β(H)-Bishomohopane [c] | + | + | + | + | + | + | + |
| (22R)-17α(H),21β(H)-Bishomohopane [c)] | + | + | + | + | + | + | + |
| (22S)-17α(H),21β(H)-Trishomohopane [c] | + | + | + | + | + | + | + |
| (22R)-17α(H),21β(H)-Trishomohopane [d] | + | + | + | + | + | + | + |
| *Steranes and steroids* | | | | | | | |
| $C_{20}$-5 α(H),14α(H),14α(H)-Sterane [c] | + | + | + | + | + | + | + |
| $C_{21}$-5α(H),14β(H),17β(H)-Sterane [c] | + | + | + | + | + | + | + |
| $C_{22}$-5α(H),14β(H),17β(H)-Sterane [c] | + | + | + | + | + | + | + |
| $C_{27}$-20S-13β(H),17α(H)-Diasterane [c] | + | + | + | + | + | + | + |
| $C_{27}$-20R-13β(H),17α(H)-Diasterane [c] | + | + | + | + | + | + | + |
| $C_{27}$-20S-13α (H),17β(H)-Diasterane [c] | + | + | + | + | + | + | + |
| $C_{27}$-20R-13α(H),17β(H)-Diasterane [c] | + | + | + | + | + | + | + |
| $C_{28}$-20S-13β(H),17α(H)-Diasterane [c] | + | + | + | + | + | + | + |
| $C_{29}$-20S-13β(H),17α(H)-Diasterane [c] | + | + | + | + | + | + | + |
| 20S-5α(H),14α(H),17α(H)-Cholestane [c] | + | + | + | + | + | + | + |
| 20R-5α(H),14β(H),17β(H)-Cholestane [c] | + | + | + | + | + | + | + |
| 20S-5α(H),14β(H),17β(H)-Cholestane [c] | + | + | + | + | + | + | + |
| 20R-5α(H),14α(H),17α(H)-Cholestane [c] | + | + | + | + | + | + | + |
| Cholestenes (1 isomer) [d] | + | + | + | + | + | + | + |
| Cholestadienes (3 isomers) [d] | (+) | (+) | + | (+) | + | + | + |
| Cholestatrienes (3 isomers) [d] | + | + | + | + | + | + | + |
| 5α(H)-Cholestan-3-one [a] | + | + | | + | + | + | + |
| 5β(H)-Cholestan-3-one [a] | + | + | | + | + | + | + |
| 20S-5a(H),14α(H),17α(H)-Ergostane [c] | + | + | + | + | + | + | + |
| 20R-5a(H),14b(H),17b(H)-Ergostane [c] | + | + | + | + | + | + | + |
| 20S-5a(H),14b(H),17b(H)-Ergostane [c] | + | + | + | + | + | + | + |
| 20R-5a(H),14a(H),17a(H)-Ergostane [c] | + | + | + | + | + | + | + |
| Ergostanones (2 isomers) [d] | | + | | | + | + | |
| 20S-5a(H), 14a(H),17a(H)-Stigmastane [c] | + | + | + | + | + | + | + |
| 20R-5a (H),14b(H),17b(H)-Stigmastane [c] | + | + | + | + | + | + | + |
| 20S-5a(H),14b(H),17b(H)-Stigmastane [c] | + | + | + | + | + | + | + |

| **Compounds** | **A** | **B** | **C** | **D** | **E** | **F** | **G** |
|---|---|---|---|---|---|---|---|
| 20R-5a(H),14a(H),17a(H)-Stigmastane [c] | + | + | + | + | + | + | + |
| Stigmastene [d] | + | + | + | + | + | + | + |
| Stigmastadienes (1 isomer) [d] | + | + | + | + | + | + | + |
| Stigmastatrienes (2 isomers) [d] | + | + | + | + | + | + | + |
| Stigmastanones (2 isomers) [d] | | + | | | + | + | + |
| *Alkylbenzenes* | | | | | | | |
| Ethylbenzene [a] | + | + | + | + | + | + | + |
| m-/p-Xylene [a] | + | + | + | + | + | + | + |
| o-Xylene [a] | + | + | + | + | + | + | + |
| $C_3$-Benzenes (8 isomers) [d] | (+) | + | + | (+) | + | (+) | + |
| $C_4$-Benzenes (15 isomers) [d] | (+) | (+) | + | + | + | + | + |
| $C_5$-Benzenes (10 isomers) [d] | (+) | + | + | + | + | + | + |
| $C_6$-Benzenes (6 isomers) [d] | (+) | + | (+) | (+) | + | + | + |
| $C_7$-Benzenes (7 isomers) [d] | (+) | + | (+) | + | (+) | + | + |
| $C_8$-Benzenes (9 isomers) [d] | (+) | + | (+) | (+) | (+) | + | (+) |
| $C_9$-Benzenes (11 isomers) [d] | (+) | + | (+) | (+) | + | + | (+) |
| $C_{10}$-Benzenes (8 isomers) [d] | (+) | + | (+) | (+) | + | + | (+) |
| $C_{11}$-Benzenes (3 isomers) [d] | (+) | + | (+) | (+) | + | + | (+) |
| $C_{12}$-Benzenes (5 isomers) [d] | (+) | + | (+) | + | + | + | (+) |
| $C_{13}$-Benzenes (5 isomers) [d] | (+) | + | (+) | + | + | + | (+) |
| $C_{14}$-Benzenes (11 isomers) [d] | (+) | + | (+) | + | | + | (+) |
| | | | | | | | |
| 5-Phenyldecane [a] | + | + | + | + | + | + | + |
| 4-Phenyldecane [a] | + | + | + | + | + | + | + |
| 3-Phenyldecane [a] | | + | | | | + | |
| 6-Phenylundecane [a] | + | + | + | + | + | + | + |
| 5-Phenylundecane [a] | + | + | + | + | + | + | + |
| 4-Phenylundecane [a] | + | + | + | + | + | + | + |
| 3-Phenylundecane [a] | + | + | + | + | + | + | + |
| 2-Phenylundecane [a] | + | + | + | + | + | + | + |
| 6-Phenyldodecane [a] | + | + | + | + | + | + | + |
| 5-Phenyldodecane [a] | + | + | + | + | + | + | + |
| 4-Phenyldodecane [a] | + | + | + | + | + | + | + |
| 3-Phenyldodecane [a] | + | + | + | + | + | + | + |
| 2-Phenyldodecane [a] | + | + | + | + | + | + | + |
| 7-/6-Phenyltridecane [a] | + | + | + | + | + | + | + |
| 5-Phenyltridecane [a] | + | + | + | + | + | + | + |
| 4-Penyltridecane [a] | + | + | + | + | + | + | + |
| 3-Phenyltridecane [a] | + | + | + | | + | + | + |
| 2-Phenyltridecane [a] | + | + | + | | + | | + |

| **Compounds** | **A** | **B** | **C** | **D** | **E** | **F** | **G** |
|---|---|---|---|---|---|---|---|
| *Polycyclic aromatic compounds, PACs* | | | | | | | |
| Naphthalene [a)] | + | + | + | + | + | + | + |
| Ethylnaphthalene (1 isomer) [d)] | + | + | + | + | + | + | + |
| 1-Methylnaphthalene [a)] | + | + | + | + | + | + | + |
| 2-Methylnaphthalene [a)] | + | + | + | + | + | + | + |
| C2-Naphthalenes (6 isomers) [d)] | + | + | + | + | + | + | + |
| C3-Naphthalene (10 isomers) [d)] | + | + | + | + | + | + | + |
| C4-Naphthalene (8 isomers) [d)] | + | + | (+) | (+) | + | + | (+) |
| Biphenyl [a)] | + | + | + | + | + | + | + |
| 3-Methylbiphenyl [b)] | + | + | + | + | + | + | + |
| 4-Methylbiphenyl [b)] | + | + | + | + | + | + | + |
| $C_2$-Biphenyls (3 isomers) [d)] | + | + | + | + | + | + | + |
| $C_3$-Biphenyls (3 isomers) [d)] | (+) | + | + | + | (+) | (+) | + |
| 1,1-Diphenylethane [b)] | | | | | | + | |
| Acenaphtylene [a)] | + | + | + | + | + | + | + |
| Acenaphtene [a)] | + | + | + | + | + | + | + |
| Fluorene [a)] | + | + | + | + | + | + | + |
| 9-Methylfluorene [a)] | + | + | + | + | + | + | + |
| 2-Methylfluorene [a)] | + | + | + | + | + | + | + |
| 1-Methylfluorene [a)] | + | + | + | + | + | + | + |
| Methylfluorenes (1 isomer) [d)] | + | + | + | + | + | + | + |
| $C_2$-Fluorenes (6 isomers) [d)] | (+) | + | (+) | (+) | (+) | (+) | (+) |
| 1-Phenylnaphthalene [a)] | + | + | + | + | + | + | + |
| 2-Phenylnaphthalene [a)] | + | + | + | + | + | + | + |
| Phenanthrene [a)] | + | + | + | + | + | + | + |
| Anthracene [a)] | + | + | + | + | + | + | + |
| 3-Methylphenanthrene [b)] | + | + | + | + | + | + | + |
| 2-Methylphenanthrene [b)] | + | + | + | + | + | + | + |
| 2-Methylanthracene [b)] | + | + | | + | + | + | + |
| 4-/9-Methylphenanthrene [b)] | + | + | + | + | + | + | + |
| 1-Methylphenanthrene [b)] | + | + | + | + | + | + | + |
| Dimethylphenanthrene (1 isomer) [d)] | + | + | + | + | + | + | + |
| 3,5-Dimethylphenanthrene [b)] | + | + | + | + | + | + | + |
| 2,7-Dimethylphenanthrene [b)] | + | + | + | + | + | + | + |
| 1,3-3,10-Dimethylphenanthrene [b)] | + | + | + | + | + | + | + |
| 1,6-/2,9-Dimethylphenanthrene [b)] | + | + | + | + | + | + | + |
| 1,7-Dimethylphenanthrene [b)] | + | + | + | + | + | + | + |
| 1,9-/4,9-Dimethylphenanthrene [b)] | + | + | + | + | + | + | + |
| 1,8-Dimethylphenanthrene [b)] | + | + | + | + | + | + | + |
| 1,2-Dimethylphenanthrene [b)] | + | + | + | + | + | + | + |

| **Compounds** | **A** | **B** | **C** | **D** | **E** | **F** | **G** |
|---|---|---|---|---|---|---|---|
| $C_3$-Phenanthrenes/-anthracenes (7 isomers) [d)] | + | + | + | + | + | + | + |
| 4H-Cyclopenta(def)phenanthrene [b)] | + | + | + | + | + | + | + |
| Fluoranthene [a)] | + | + | + | + | + | + | + |
| Acephenanthrylene [b)] | + | + | | + | + | + | + |
| Pyrene [a)] | + | + | + | + | + | + | + |
| Methylfluoranthenes/-pyrenes (6 isomers) [d)] | + | + | + | + | + | + | + |
| Dimethylfluoranthenes/-pyrenes (9 isomers) [d)] | + | + | + | + | + | + | + |
| Ethylfluoranthenes/-pyrenes (2 isomers) [d)] | | + | | | | | |
| o-Terphenyl [b)] | + | + | + | + | + | + | + |
| m-Terphenyl [b)] | + | + | | + | + | + | |
| p-Terphenyl [b)] | + | + | + | + | + | + | + |
| Benz(ghi)fluoranthene [a)] | + | + | + | + | + | + | + |
| Benz(c)phenanthrene [a)] | + | + | + | + | + | + | + |
| Benz(a)anthracene [a)] | + | + | + | + | + | + | + |
| Cyclopenta(cd)pyrene [a)] | | | + | | | + | |
| Chrysene/Triphenylene [a)] | + | + | + | + | + | + | + |
| Naphthacene [b)] | + | + | + | + | + | + | + |
| Methyl-228 (5 isomers) [§, d)] | + | + | + | + | + | + | + |
| $C_2$-228 (7 isomers) [§, d)] | + | + | + | + | + | + | + |
| 1,2´-Binaphthyl [b)] | + | + | + | + | + | + | + |
| 2,2´-Binaphthyl [b)] | + | + | + | + | + | + | + |
| 9-Phenylphenanthrene [b)] | + | + | + | + | + | + | + |
| Phenylphenanthrene/-anthracene (3 isomers) [d)] | + | + | + | + | + | + | + |
| Benz(x)fluoranthene (x=j,b,k) [a)] | + | + | + | + | + | + | + |
| Benz(e)pyrene [a)] | + | + | + | + | + | + | + |
| Benz(a)pyrene [a)] | + | + | + | + | + | + | + |
| Perylene [a)] | + | + | + | + | + | + | + |
| Methyl-252 (5 isomers) [§, d)] | + | + | + | + | + | + | + |
| Indeno(1,2,3-cd)pyrene [a)] | + | + | + | + | + | + | + |
| Benzo(ghi)perylene [a)] | + | + | + | + | + | + | + |
| Dibenzo(a,h)anthracene [b)] | + | + | + | + | + | + | + |
| Dibenzo(a,c)anthracene [b)] | + | + | + | + | + | + | + |
| Benzo(b)chrysene [a)] | + | + | + | + | + | + | + |
| Picene [b)] | + | + | + | + | + | + | + |
| Anthanthrene [b)] | | + | | | | | |
| *Hydrogenated aromatic compounds* | | | | | | | |
| Decalin [a)] | + | + | | + | + | + | + |
| Tetralin [a)] | + | + | | | + | + | |
| 1,1,6-Trimethyletralin [e)] | + | + | | + | + | + | |
| 3,3,7-Trimethyl-1,2,3,4-tetrahydrochrysene [e)] | + | + | | | | + | |

| **Compounds** | **A** | **B** | **C** | **D** | **E** | **F** | **G** |
|---|---|---|---|---|---|---|---|
| Cyclohexylbenzene [e)] | + | + | + | + | + | + | |
| Cyclohexylcyclohexane [e)] | + | | | + | + | + | |
| *Sulphur containing PACs* | | | | | | | |
| Benzo(b)thiophene [a)] | + | + | | + | + | + | |
| Dibenzothiophene [a)] | + | + | | + | + | + | |
| Benzo(b)naphtho(2,1-d)thiophene [b)] | + | + | + | + | + | + | + |
| Benzo(b)naphtho(1,2-d)thiophene [b)] | + | + | + | + | + | + | + |
| Phenanthro(9,10-b)thiophene [b)] | + | + | + | + | + | + | + |
| Benzo(b)naphtho(2,3-d)thiophene [b)] | + | + | + | + | + | + | + |
| Methylbenzonaphthothiophenes (4 isomers) [d)] | + | + | + | + | + | + | + |
| *Oxygen containing PACs* | | | | | | | |
| Benzofuran [a)] | + | + | + | + | + | + | + |
| Methylbenzofuran (2 Isomers) [d)] | + | + | + | + | + | + | + |
| Dibenzofuran [a)] | + | | + | + | + | | + |
| Methyldibenzofuran (2 isomers) [d)] | + | | + | + | + | + | + |
| $C_2$-Dibenzofuran (4 isomers) [d)] | + | + | + | + | + | + | + |
| Benzonaphthofuran (6 isomers) [d)] | + | + | + | (+) | + | + | + |
| Methylbenzonaphthofuran (4 isomers) [d)] | + | + | | | + | + | + |
| *Nitrogen containing PACs* | | | | | | | |
| Carbazol [a)] | + | + | + | + | + | + | + |
| Methylcarbazoles (2 isomers) [d)] | + | + | + | + | + | + | + |
| Benzcarbazoles (2 isomers) [d)] | + | + | + | + | (+) | + | |
| *Oxygenated aromatic compounds* | | | | | | | |
| Cyclpenta(def)phenanthren-4-one [b)] | + | + | | | + | | |
| 9,10-Anthraquinone [a)] | | + | | | + | | |
| Benzanthrone [a)] | + | + | | + | + | + | |
| *Sulfides* | | | | | | | |
| Di-*iso*-propyldisulfide [a)] | + | + | + | + | + | + | + |
| Di-*iso*-propyltrisulfide [e)] | + | + | + | | + | + | + |
| *Alcohols and phenols* | | | | | | | |
| 2,9-Dihydroxy-6-methyl-4,7-dioxadecane [e)] | + | + | + | + | + | + | + |
| Phenol [a)] | + | + | + | + | | | + |
| 4-Cresol [a)] | | + | | + | | | + |
| 2-/3-Cresol [a)] | | + | | + | | | + |

| **Compounds** | **A** | **B** | **C** | **D** | **E** | **F** | **G** |
|---|---|---|---|---|---|---|---|
| *Aldehydes and ketones* | | | | | | | |
| Homologous series of *n*-aldehydes ($C_9$ to $C_{32}$) [a] | + | + | | | + | | + |
| Homologous series of 2-alkanones ($C_{23}$ to $C_{29}$) [b] | + | + | | | (+) | | (+) |
| Benzaldehyde [a] | + | + | | | + | + | + |
| Methylacetophenone [d] | + | + | + | + | + | + | + |
| 1-(2,3-Dihydro-1,1-dimethyl-1H-inden-4-yl)-ethanone [e] | + | + | | + | | | + |
| *Alkanoic acids* | | | | | | | |
| Homologous series of *n*-alkanoic acids ($C_9$ to $C_{26}$) [a] | | | | (+) | (+) | (+) | + |
| Homologous series of *iso*-alkanoic acids ($C_{12}$ to $C_{20}$) | | | | + | (+) | (+) | + |
| Homologous series of *anteiso*-alkanoic acids ($C_{12}$ to $C_{20}$) | | | | + | (+) | (+) | + |
| 9-Hexadecenoic acid, palmitoleic acid [a] | | | | + | | | + |
| 9,12,15-Octatrienoic acid, linolenic acid [a] | | | | + | | | + |
| 9,12-Octadienoic acid, linoleic acid [a] | | | | | | | |
| 9-Octadecenoic acid, oleic acid [a] | | | | + | + | + | + |
| Eicosatetraenoic acid, arachidonic acid [a] | | | | | | | + |
| Eicosapentaenoic acid [d] | | | | | | | + |
| Phenylacetic acid [a] | + | | + | + | | + | + |
| *Esters* | | | | | | | |
| Complex mixture of wax esters ($C_{27}$ to $C_{32}$) [d] | (+) | + | | | + | + | + |
| Methyldodecanoate [a] | | + | | | + | | + |
| Methyltetradecanoate [a] | | + | | | + | | + |
| Methyl-*iso*-pentadecanoate [e] | | + | | | + | | + |
| Methyl-*anteiso*-pentadecanoate [e] | | + | | | + | | + |
| Methylpentadecanoate [a] | | + | | | + | | + |
| Methyl-*iso*-hexadecanoate [e] | | + | | | + | | + |
| Methylpalmitoleate [a] | | + | | | | | + |
| Methylhexadecanoate [a] | | + | | | + | | + |
| Methyl-*iso*-heptadecanoate [e] | | + | | | | | + |
| Methyl-*anteiso*-heptadecanoate [e] | | | | | + | | + |
| Methylheptadecanoate [a] | | + | | | + | | + |
| Methyloleate [a] | | + | | | | | + |
| Methyloctadecanoate [a] | | + | | | + | | + |
| Methyleicosanoate [a] | | + | | | | | + |
| Isopropyldodecanoate [e] | | | | | + | | + |
| Isopropyltetradecanoate [e] | + | + | | + | + | + | + |

| **Compounds** | **A** | **B** | **C** | **D** | **E** | **F** | **G** |
|---|---|---|---|---|---|---|---|
| Isopropylhexadecanoate [e] | | + | | | + | | + |
| Dibutyl-2-butenoate [e] | | + | | | + | | + |
| *Amides* | | | | | | | |
| Pentadecanamide [e] | | + | | | + | + | + |
| Hexadecanamide [a] | + | + | + | + | + | + | + |
| Heptadecanamide [e] | + | + | + | + | + | + | + |
| Octadecanamide [a] | | + | | | + | + | + |
| Ethylmethylmaleimide [e] | + | + | + | + | | + | |
| *Tocopheros and degradation products* | | | | | | | |
| α-Tocopherol [a] | + | + | + | + | + | + | + |
| 3,4-Dimethyl-2,5-furandione [a] | | + | | | + | + | |
| 4,8,12,16-Tetramethylheptadecan-4-olide [a] | + | + | + | + | + | + | + |
| *Plasticizers* | | | | | | | |
| Dimethylphthalate [a] | + | + | + | + | + | + | + |
| Dimethylterephthalate [a] | + | + | + | + | + | + | + |
| Diethylphthalate [a] | + | + | + | + | + | + | + |
| Di-*iso*-butylphthalate [a] | + | + | + | + | + | + | + |
| Di-*n*-butylphthalate [a] | + | + | + | + | + | + | + |
| Bis(2-ethylhexyl)phthalate [a] | + | + | + | + | + | + | + |
| 2,4,4-Trimethylpentan-1,3-diol-diisobutyrate [a] | + | + | + | | + | | + |
| Tributylphosphate [a] | + | + | + | + | + | | + |
| Complex mixture of alkylsulfonic acid phenylesters ($C_{12}$ to $C_{18}$ side chains) [a] | (+) | + | | | | | |
| *Fragrances and UV-protectors* | | | | | | | |
| Galaxolide [a] | + | + | + | + | + | + | + |
| Tonalide [a] | + | + | + | + | + | + | + |
| 2,6,6-Trimethyl-2-cyclohexen-1,4-dione [a] | + | + | + | + | + | | + |
| 1,2,3,6,7,8-Hexahydro-1,1,6,6-tetramethyl-4-isopropyl-as-indacene | | + | | | | | |
| 4-Methoxycinnamic acid 2-ethylhexyl ester [a] | + | + | | + | + | + | + |
| *Halogenated compounds* | | | | | | | |
| Complex mixture of tetra- to heptachlorinated biphenyls [a), d)] | + | + | (+) | (+) | + | + | + |
| Hexachlorobutadiene [a] | + | + | | | | | |
| 1,3-Dichlorobenzene [a] | + | + | + | + | + | + | + |
| 1,4-Dichlorobenzene [a] | + | + | + | + | + | + | + |
| 1,2-Dichlorobenzene [a] | + | + | + | + | + | + | + |
| 1,3,5-Trichlorobenzene [a] | + | + | + | + | + | + | + |

| Compounds | A | B | C | D | E | F | G |
|---|---|---|---|---|---|---|---|
| 1,2,4-Trichlorobenzene [a)] | + | + | + | + | + | + | + |
| 1,2,3,5-/1,2,4,5-Tetrachlorobenzene [a)] | + | + | | | + | + | |
| 1,2,3,4-Tetrachlorobenzene [a)] | + | + | | | + | + | |
| Pentachlorobenzene [a)] | + | + | + | | + | + | |
| Hexachlorobenzene [a)] | + | + | + | | + | + | |
| 4,4`-Dichlorodiphenylsulfide [a)] | + | + | + | | + | + | |
| 1-Chloronaphthalene [a)] | + | + | + | | | | |
| 1,4-Dichloronaphthalene [a)] | + | + | + | | | | |
| 1,5-/1,6-Dichloronaphthalene [a)] | + | + | | | | | |
| 1,7-/2,6-/2,7-Dichloronaphthalene [a)] | + | + | | | | | |
| Bromophenol [d)] | | + | | | | | |
| *Pesticides and degradation products* | | | | | | | |
| 4,4`-DDMU [a)] | | + | | | | | |
| 4,4`-DDE [a)] | + | + | + | | | | |
| 4,4`-DDD [a)] | + | + | + | + | + | + | |
| *Organometallic compounds* | | | | | | | |
| Tetrabutyltin [a)] | + | + | + | | + | + | |

(+)= not all isomers were detected
§ = only molecular masses of the parent PAH are given
a) = identified by comparision of gas chromatographic and mass spectral data with those of reference compounds
b) = identified by comparison of gas chromatographic and mass spectral data with those of mass spectral data bases and published retention indices.
c) = identified by comparison of gas chromatographic and mass spectral data with those of mass spectral data bases and published gas chromatographic elution patterns.
d) = molecular structure is not more specified.
e) = identified by comparison of mass spectral data with those of mass spectral data bases

## *Commonly observed organic compounds in coastal sediments of the German Bight*

Biological activity in the German Bight produces several groups of low molecular lipophilic compounds deposited in the surface sediments together with land-derived organic matter. Carotenoid degradation products (e.g. ionenes, dihydroactinidiolide ) and compounds structurally related to phytol (e.g. pristane, phytane, phytene, phytadiene, 6,10,14-trimethylpentadecan-2-one, 4,8,12-trimethyltetradecanoic acid) were observed. Also fatty acids from $C_9$ to $C_{26}$ were identified as main components in the fifth liquid chromatographic fraction. Next to the *n*-alkanoic acids, several methyl branched acids with *anteiso* and *iso* substitution (chain length from $C_{12}$ to $C_{17}$) as well as the unsaturated components, palmitoleic acid, linolenic acid, linoleic acid, oleic acid, arachidonic acid and an eicosapentaenoic acid, were detectable at most of the sampling locations. It is well known that these classes of biogenic compounds detected in lacustrine sediments are derived from both autochthonous and allochthonous emission and are, therefore, of marine and terrigenious origin (Cranwell, 1981a; Cranwell et al., 1987; Riley et al., 1991). In addition, complex mixtures of wax esters with chain lengths from $C_{27}$ to $C_{32}$ could be detected in sediment samples of sites B, F,G,and E. The mass spectral data and gas chromatographic properties (Fig. 2) suggest an isomer distribution similar to that described by Cranwell (1981b), who assumed that the wax esters were produced by microbial activity.

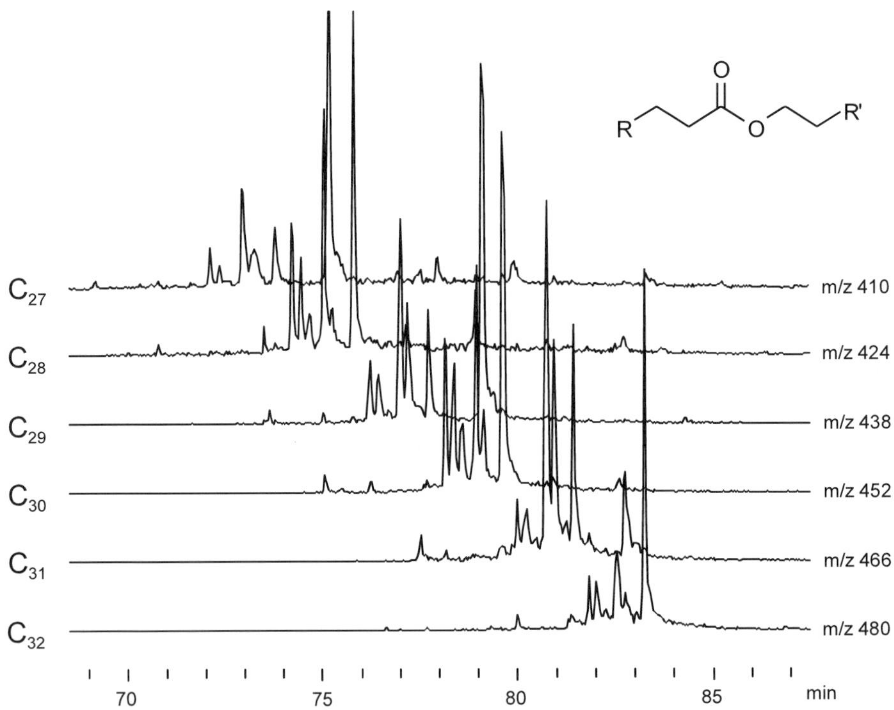


**Fig. 2:** Gas chromatographic elution of wax esters in the range from $C_{27}$ to $C_{32}$ illustrated by their molecular ion chromatograms. The chain lengths of the acidic and alcoholic components (R and R′) range between $C_7$ and $C_{21}$.

Terpenoids are not always diagnostic of biogenic contributions to the organic matter in sediments. Widespread use of monoterpenoic compounds in products such as perfumes and odour agents (e.g. mixtures of α- and β-ionenes, limonene), in organic solvents (oil of turpentine) and as plasticizers (e.g. campher, fenchone) and the resulting discharge into riverine systems prevents an absolute association of such compounds with natural sources. Also α-tocopherol, detected in all sediment extracts, could not be associated only with biogenic sources because of its widespread use as an antioxidant and vitamin in food as well as food supplements (Eganhouse and Kaplan, 1985). The occurrence of 3,4-dimethyl-2,5-furandione and 4,8,12,16-tetramethylheptadecan-4-olide, previously described as oxidation products for the structure elucidation of tocopherols (Fernholz, 1938), suggests a possible oxidative degradation of α-tocopherol in the aquatic environment.

Long chain *n*-aldehydes ($C_9$ to $C_{32}$) found at sampling locations A,B and G, situated near to the estuaries of the Elbe and Ems rivers, could be indicative of the contribution of terrestrial organic matter. Such long-chain *n*-aldehydes ($C_{20}$ to $C_{32}$) with even-to-odd carbon chain length predominance have been attributed to the input of terrestrial plant waxes, whereas the origin of *n*-aldehydes with shorter side chains is not known (Prahl and Pinto, 1987; Stephanou, 1989). In the studied sediments the homologous series of the *n*-aldehydes show a distribution in the range from $C_{20}$ to $C_{32}$ according to previously published results (Fig. 3). *N*-aldehydes of lower molecular weight ($C_9$ to $C_{19}$) maximizing at $C_{10}$ and $C_{15}$ occured in the same concentration range as the higher homologous, but a carbon chain length predominance is not observable. In sample A, mainly influenced by the Elbe river, the composition of *n*-aldehydes was different to that of sample G, which is situated near the estuary of the Ems Rivers. In both samples distributions maximized at $C_{26}$, but the even-over-odd predominance is higher at sample location G.

Also the amount of short-chain *n*-aldehydes ($C_9$ to $C_{18}$) relative to the long-chain components ($C_{19}$ to $C_{32}$) in sample G is greater than in sediments from sample site A. The presence of *n*-aldehydes in sediments from sampling location E indicates a terrigenious contribution at this site too. The homologous pattern in this sample could be constructed by mixing of the *n*-aldehydes at sampling location A and G. Accordingly, the land-derived organic matter at site E seemed to originate from the Elbe river as well as from the Ems and Weser rivers.

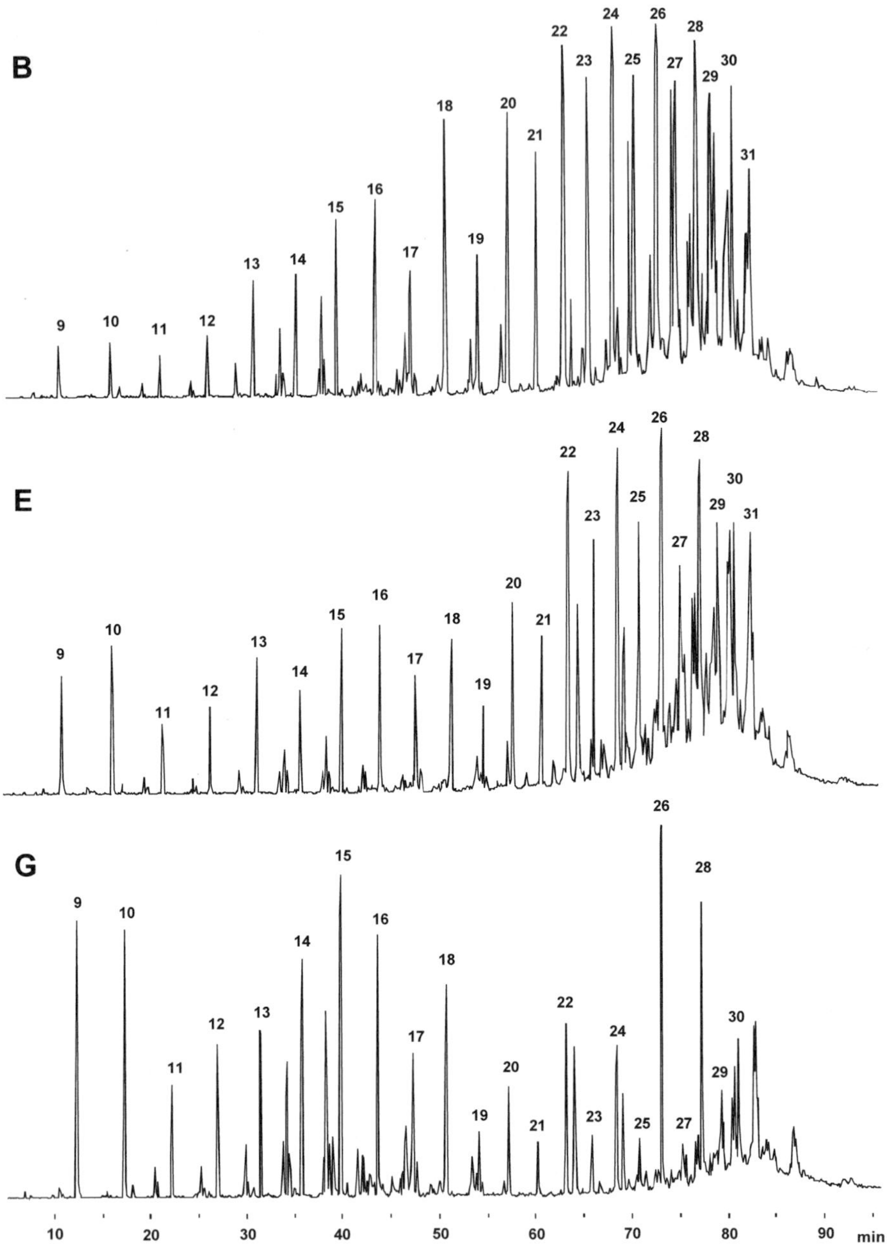


**Fig. 3:** Comparison of gas chromatographic elution pattern of *n*-aldehydes at sample sites B,E and G. Chain lengths are marked by numbers.

In addition, the occurrence of *n*-alkan-2-ones ($C_{23}$ to $C_{29}$) at sampling locations A,B,E and G suggested terrigenous contributions to the coastal sediments due to the proposed formation by microbial β-oxidation of corresponding higher plant derived *n*-alkanes (Allen et al., 1971; Cranwell, 1981a; Riley et al., 1991).

A second group of lipophilic compounds in sediments of the German Bight were petrogenic substances. The distributions of hopanoids and steranes identified in the sediment samples were indicative of petrogenic input due to their unique structures, persistence and well known diagenetic and catagenetic transformations (Volkman et al., 1997). For instance the occurrence of $C_{31}$- to $C_{33}$-hopanes with similar concentrations of 22R- and 22S-isomers as well as the ratio of trisnorhopanes ($T_s/T_s$-$T_m$) is characteristic of thermally mature organic material that could only be introduced to recent sediments by contamination from fossil fuels (see Fig. 4). Unresolved complex mixtures (UCM´s) are generally observed in the aliphatic fractions. These UCMs and high amounts of alkylated cyclohexanes and benzenes also reflect the contribution of petrogenic emissions, whereas the petrogenic *n*-alkanes were superposed by biogenic contributions indicated by an odd-over-even carbon chain length predominance. A successful differentation of aliphatic hydrocarbons from both biogenic and petrogenic sources in coastal sediments is rather difficult (Fernandes et al., 1997; Tran et al., 1997).

Polycyclic aromatic compounds (PACs) are ubiquitous contaminants in several environmental compartments. In coastal sediments, direct input of PAC by atmospheric deposition or discharge of petroleum products is supplemented by riverine contribution of urban and industrial effluents as well as urban runoff containing PACs from asphalt and car exhaust particles (Mattiasson et al., 1977; Wakeham et al., 1980; Wang et al., 1995; Bence et al., 1996; Aboul–Kassim and Simoneit, 1996; Burns et al., 1997; Dachs et al., 1997). In sediments of the German Bight we identified di- to hexacyclic aromatic compounds accompanied by $C_1$- to $C_3$-substituted isomers in the same concentration range. Pyrogenic PACs originating from incomplete combustion processes are represented by a predominance of fluoranthene and pyrene as well as five- and six-membered ring compounds in the fourth chromatographic fraction of each sediment extract. PACs of fossil origin are indicated by high amounts of naphthalene, phenanthrene and alkylated isomers of several parent polycyclic aromatic hydrocarbons that were main components in the third chromatographic fraction.

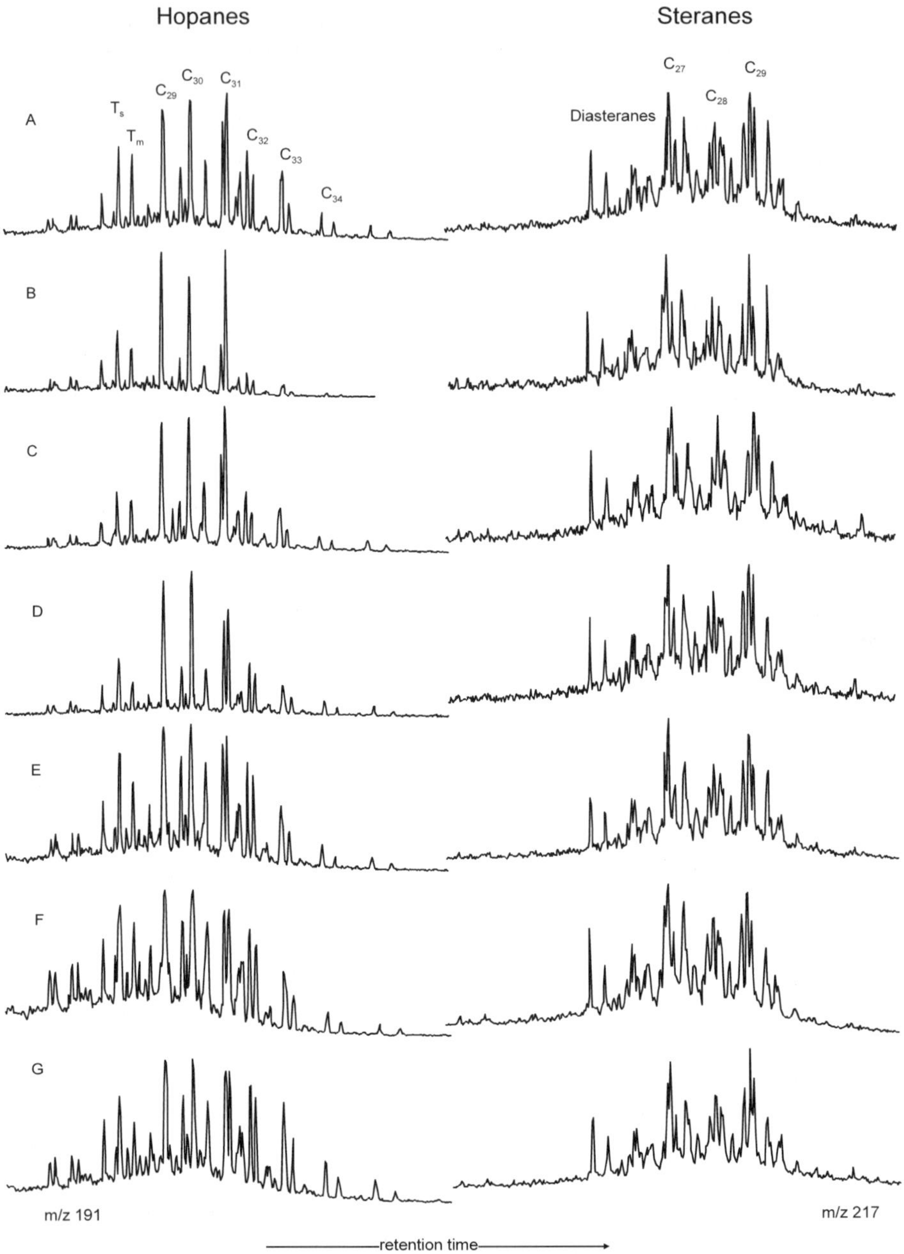


**Fig. 4**: Gas chromatographic elution profiles of hopanes (m/z 191) and steranes (m/z 217) reflecting petrogenic input into coastal sediments.

Hydrogenated PACs were identified in minor concentrations such as tetralin, decalin, 1,1,6-trimethyltetralin and hydrogenated biphenyls. Nitrogen-, sulphur- and oxygen-containing PAC with alkylated isomers occurring in higher amounts were indicative for organic matter of fossil origin. Sulphur containing PACs are well documented constituents of crude oils and petrogenic products (e.g. Later et al., 1981; Glinzer et al., 1983; Grimmer et al., 1981a; Grimmer et al., 1981b; Arpino et al., 1987; Wang and Fingas, 1995; Chakhmakhchev et al., 1997). The identified oxygenated PACs, cyclopenta(def)phenanthren-4-one, 9,10-anthraquinone and benzanthrone, probably originate from oxidation of polycyclic aromatic hydrocarbons during incomplete combustion or atmospheric transport (Schuetzle et al., 1981; König et al., 1983; Tong and Karasek, 1984).

In addition to biogenic and petrogenic compounds, a wide variety of anthropogenic substances were identified. Several low molecular weight organic compounds representing domestic source contamination were introduced to the coastal sediments by riverine discharge. Methyl and isopropyl esters of fatty acids originating from soaps and washing agents are main components of household effluents (Paxeus, 1996). Their occurrence is associated with anthropogenic emissions and furthermore riverine contribution to coastal sediments. Also the higher concentrations of coprostanone in comparison to cholestanone suggest contributions from sewage effluents and are therefore indicative for organic matter of riverine origin (Chalaux et al., 1995; Takada et al., 1997; Chan et al., 1998). Major components in the fifth liquid chromatographic fraction of each analysed sediment extract were a group of plasticizers, including phthalates, 2,4,4-trimethylpentane-1,3-dioldi-*iso*-butyrate and tributylphosphate. These compounds are well known ubiquitous pollutants and were also detected in Elbe river water (Franke et al., 1995). Fragrances and ordorants also indicate municipal and, therefore, land derived contamination in coastal sediments. In the sediment samples from the German Bight we identified in most samples the musk fragrances galaxolide and tonalide used as odorants in detergents as well as 2,6,6-trimethyl-2-cyclohexen-1,4-dione (4-oxoisophorone). This compound is industrially produced via oxidation of isophorone and mainly used in perfumes (Papa and Sherman, 1981). In addition a 2-ethylhexyl ester of 4-methoxycinnamic acid was frequently identified which is commonly used as a UV-protector in cosmetics. Furthermore a group of organic contaminants could be detected at all sample locations reflecting the widespread use of synthetic detergents. Linear alkylbenzenes (LABs), with side chain length from $C_{10}$ to $C_{13}$, are common detergent residues due to their use as a raw material for the synthesis of the alkylbenzenesulfonate surfactants and are therefore useful anthropogenic markers for domestic waste emissions (Takada and Eganhouse, 1998, and references cited therein). The isomeric composition of LABs (Fig. 5) differs from the pattern of technical formulations because of the more rapid degradation of external isomers (phenyl group attached near the end of the alkyl chain, e.g. 2- and 3-substitution) in comparison to internally substituted isomers (phenyl group attached near the middle of the alkyl chain, e.g. 5- and 6-substitution). The relatively decreasing abundance of external isomers described as increasing I/E-ratio has been used as indicator for the state of LAB degradation in the aquatic environment (Takada and Ishiwatari, 1990).

Also alkylated benzenes ($C_{10-14}$-benzenes) with molecular masses corresponding to those of LABS were detected. But shifted gas chromatographic retention times and different mass spectral data of these compounds in comparison to those of LABs and TABs indicated a multiple substitution at the benzene ring.

Well known halogenated organic pollutants of anthropogenic origin include the polychlorinated biphenyls (PCBs) frequently used in a wide range of technical applications. Low aqeuous solubility, low vapor pressure and a high resistance to chemical and biological degradation result in the ubiquity of PCBs in aquatic sediments (Lang, 1992). In each sample investigated in this study a complex mixture of tetra- to hepatchlorinated congeners could be identified by gas chromatographic analyses with simultanous electron capture and flame ionisation detection (ECD/FID). The individual congeneric composition corresponding to technical mixtures with a similar content of chlorine is illustrated in Fig. 5.

All anthropogenic land-derived contaminants described above are mainly introduced to the coastal sediments by atmospheric deposition and in higher quantity by riverine contributions. But these compounds can only be linked to multiple sources or widespread technical use and are therefore inappropriate to assess the discharge of the Elbe river to the German Bight.

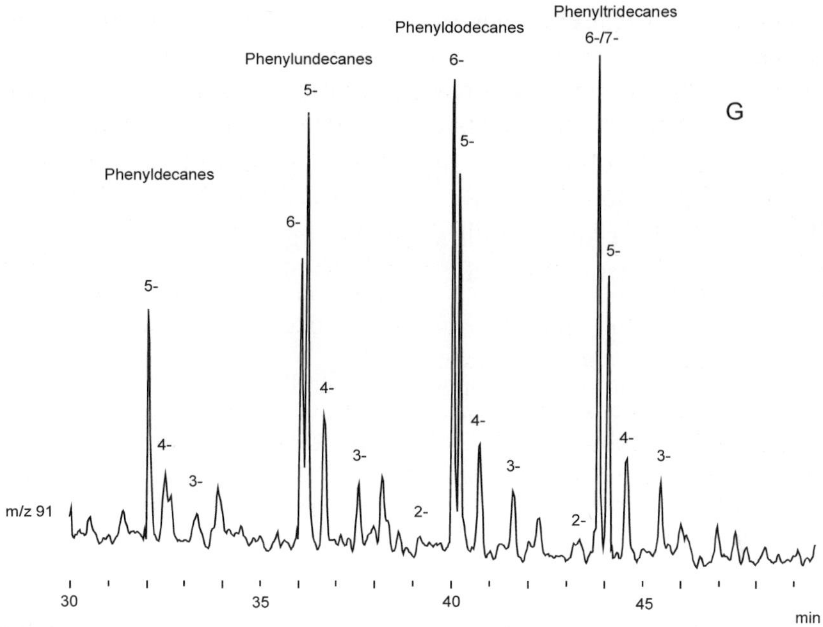


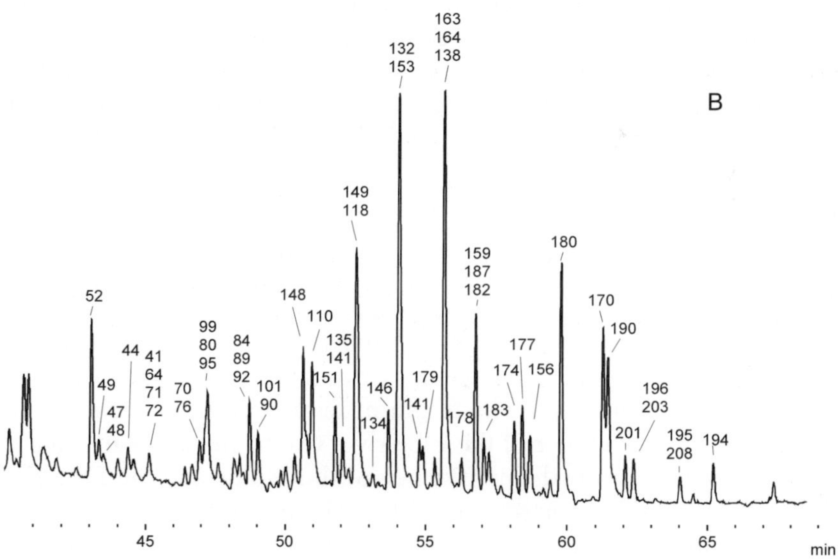


**Fig. 5:** Ion chromatogram of LABs (m/z 91) and ECD chromatogram of PCBs indicating anthropogenic inputs from non-point sources. The superscript on each peak of the LABs indicates the substitution position. Congeners of PCBs are numbered according to Ballschmitter et al. (1987).

*Potential organic molecular marker compounds for estimating the contribution of the Elbe river*

Beside the components described above several compounds could be identified that seemed to be indicative of the contribution of the Elbe river to the pollution of the sediments in the German Bight (Tab. 3). The use of these compounds as potential Elbe-specific markers is, however, limited because of the limited information about the organic matter derived from other rivers also discharging into the German Bight.

**Tab. 3:** Potential organic marker compounds of the Elbe river.

| Halogenated compounds | Nonhalogenated compounds |
|---|---|
| 4,4′-Dichlorodiphenylsulfide | Complex mixture of alkylsulfonic phenylesters |
| Tetrachlorobenzenes | 1,2,3,6,7,8-Hexahydro-1,1,6,6-tetramethyl-4-isopropyl-*as*- indacene |
| Pentachlorobenzene | Tetrabutyl tin |
| Hexachlorobenzene | |
| 1-Chloronaphthalene | |
| Dichloronaphthalenes | |
| Hexachlorobutadiene | |

Two classes of substances are appropriate as Elbe-specific marker compounds: (i) common compounds only detected in high amounts at the locations close to the Elbe river (A,B), and (ii) compounds with specific molecular structures that are not ubiquitous contaminants or have been described formerly as contaminants of the Elbe river.

A well known Elbe-specific compound is tetrabutyl tin, the parent substance for the synthesis of mono- to tributyl tin compounds widely used as antifoulants, stabilizers in poly(vinyl chloride)s (PVC) and industrial as well as agricultural biocides. The origin of tetrabutyl tin in sediments and suspended particulate matter of the Elbe river can be linked to an industrial point source situated near the confluence of the Mulde and the Elbe rivers (Wilken et al. 1994, Schwarzbauer, 1997). The occurence of tetrabutyl tin not only at sample sites mainly influenced by the Elbe river (sample sites A,B and C), but also in sediments situated farer from the Elbe estuary (sample sites E,F) indicates a wide spatial distribution of Elbe-derived organic matter in the German Bight.

Also 1,2,3,6,7,8-hexahydro-1,1,6,6-tetramethyl-4-isopropyl-*as*-indacene, attributed to a group of synthetic fragrances, can be used as Elbe-specific tracer molecule. It was detectable only at sampling location B, whereas 4-oxoisophorone, 2,4,4-trimethylpentan-1,3-diol-diisobutyrate and the musk fragrances, galaxolide and tonalide, are ubiquitous contaminants of both the Elbe river (Franke et al., 1995) and the sediments investigated in this study. The indacene compund is part of the large emission of an industrial plant situated near the Mulde river (Schwarzbauer, 1997).

As described above, plasticizers were main components in all gas chromatograms of the semi-polar and polar fractions obtained from the sediment samples. Examples include phthalates, tributyl phosphate and in addition a technical mixture of alkylsulfonic acid phenylesters. These arylesters were detected recently in high amounts in sediments as well as in the particulate matter of the Elbe system (Franke et al., 1998). We detected these plasticizers only at sampling locations A and B with a pattern similar to the isomeric distribution found in the Elbe river (Franke et al., 1998). The sample sites A and B are most influenced by the Elbe river and, therefore, we suggest that alkylsulfonic acid phenylesters are Elbe-specific.

With respect to halogenated compounds, several substances were identified that are also frequently found in sediments of the Elbe river (Schwarzbauer, 1997). Chlorinated benzenes with 2 to 6 chlorine substituents were identified in sediments of the Elbe river as well as in the samples of the German Bight. Dichloro- and trichlorobenzenes are used as synthetic raw material for many technical products such as antiseptic agents, solvents and additives (Bryant, 1993). These compunds were detected in all samples and are therefore not Elbe-specific. On the contrary higher substituted congeneres appeared at sample sites A,B,C,E and F, but not at sample site D and G, which are mainly influenced by input of the Ems river. The occurence of tetra- and pentachlorbenzenes in the environment is not attributed to specific sources due to industrial processes or technical applications. Only hexachlorobenzene is well known as a formerly used herbicide and synthetic by-product in a number of organic syntheses (Bryant, 1993, Schlör, 1970). Therefore, the appearance of tetrachloro- to hexachlorobenzenes in the sample sites described above could be suggestive for Elbe-specific contribution of organic matter.

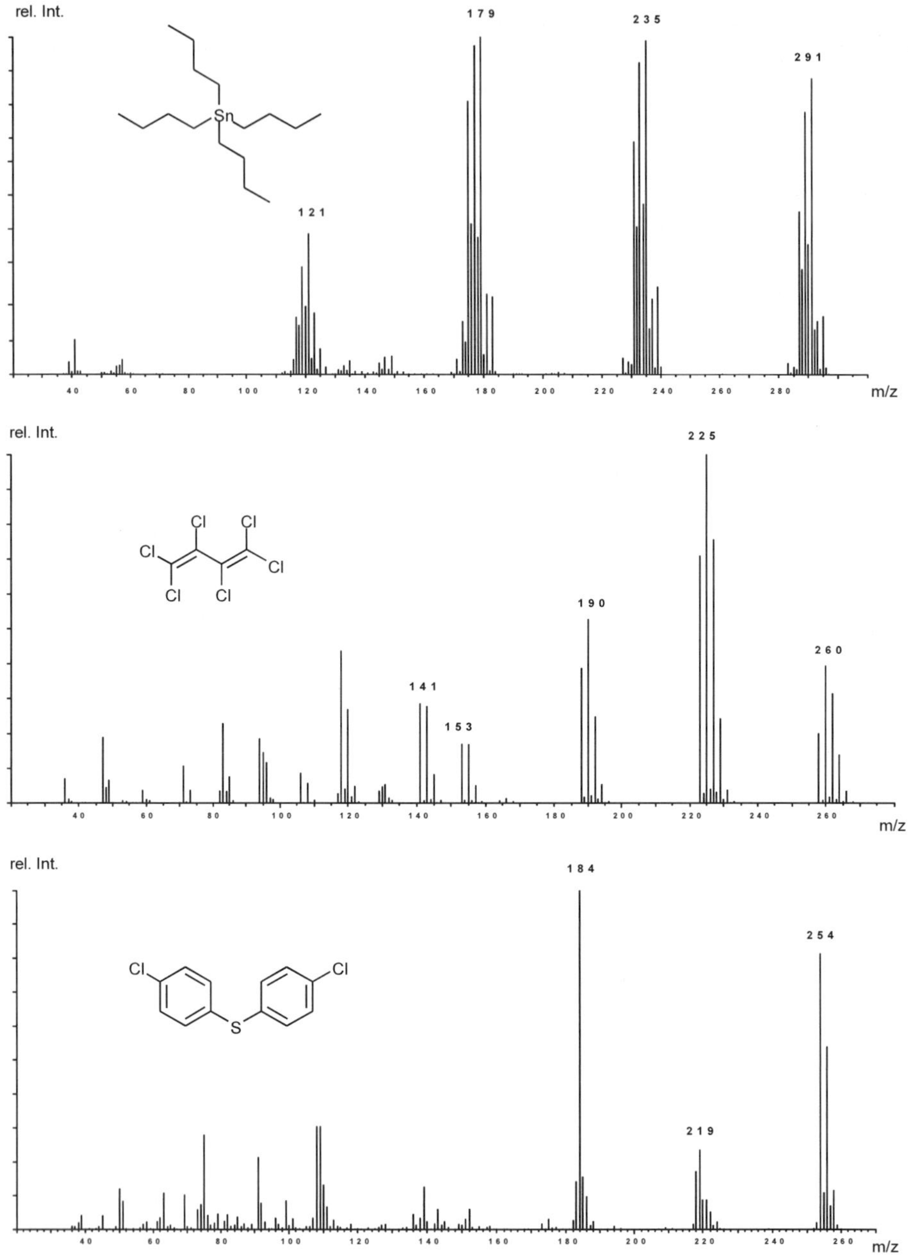


**Fig. 6:** Mass spectra and molecular structures of selected Elbe specific compounds

Mono- and disubstituted chloronaphthalenes with patterns related to those of technical agents (e.g. Halowax 1000) may be appropriate as Elbe-specific molecular markers due to their presence only at locations A,B and C. Although the occcurence of only low chlorinated naphthalenes is rarely reported (Falandysz, 1998), these substances were identified in sediments of the Elbe river (Schwarzbauer, 1997) as well as in the sediment samples described above. In contrast higher chlorinated naphthalenes with up to 8 chlorine atoms were detected frequently in several environmental compartments (Falandysz, 1998).

Beside the congeneric groups of chlorinated aromatic substances, individual halogenated compounds such as 4,4′-dichlorodiphenylsulfide and hexachlorobutadiene, a known organic agent used as solvent and in hydraulic fluids (Koch, 1995), could only be detected at sample sites A,B and C. The origin of 4,4′-dichlorodiphenylsulfide in sediments of the Elbe river and in the investigated samples of the German Bight is still unknown.

To a minor degree degradation products of the pesticide DDT might be useful Elbe-specific marker compounds, because of the longer period of DDT application in the catchment area of the Elbe river in contrast to the Weser and Ems rivers and the high concentration of DDD and DDE found in water as well as suspended particulate matter of the Elbe river (Goetz et al., 1994, lit). The para-substituted isomer of DDE could be detected only in sediment samples mainly influenced by the Elbe river (sites A,B,C), whereas 4,4′-DDD occurred additonally at sites D, E and F, but not at site G that is influenced by the Weser river.

## Summary and conclusions

Detailed screening analyses revealed a wide variety of organic lipophilic compounds of biogenic, petrogenic and anthropogenic origin in sediments of the German Bight. The biological activity in the marine environment is be reflected by the occurence of several compounds, e.g. carotenoids, fatty acids and wax esters. Contamination of petrogenic origin are indicated by an isomeric distribution of saturated steranes and hopanes, characteristic fossil markers occurring in high amounts in the examined sediments. Substances indicating the terrestrial input of low molecular organic matter to the coastal sediments might be both biogenic long chain *n*-aldehydes and isomers with shorter side chains of unknown origin. Also several anthropogenic compounds characterized the land-derived contribution to the organic matter in the coastal sediments. Sewage-specific marker compounds include coprostanone, linear alkylbenzenes, plasticizers and fragrances.

Specific organic marker compounds indicating the contribution of the Elbe river to the pollution in sediments of the German Bight were attributed mainly to chlorinated aromatic contaminants. Specifically, tetra- to hexachlorbenzenes, mono- and dichloronaphthalenes, hexachlorobutadiene and 4,4′-dichlorodiphenylsulfide occurring only in sediments influenced by the Elbe river were useful in describing the spatial distribution of Elbe-derived organic matter.

Also, an isomeric mixture of alkylsulfonic acid phenylesters, the individual contaminants tetrabutyl tin and 1,2,3,6,7,8-hexahydro-1,1,6,6-tetramethyl-4-isopropyl-*as*-indacene as well as degradated compounds of DDT could be attributed to the group of Elbe-specific marker compounds.

All identified Elbe-specific substances are only potential molecular markers at present, because marker compounds reflecting the input of the Weser and the Ems river to the sediments of the German Bight are not known.

## 5.1.2 Quantitative evaluation of Elbe river derived organic marker compounds in sediment samples from the German Bight[‡]

### Introduction

A large proportion of pollution in the coastal zone derived from anthropogenic contamination is a result of riverine contributions. Accordingly, the pollution of North Sea estuaries and coastal areas (e.g. Wash and Humber estuaries, Dutch Delta area, Wadden Sea) by organic xenobiotics is dominantly characterized by land-derived contamination discharged by the major riverine systems of the Rhine, Thames, Weser and Elbe (Ducrotoy et al. 2000). Numerous investigations have described the state of pollution of the North Sea water especially by organic contaminants (e.g., Dewulf et al. 1998, Buser et al. 1998, Weigel et al. 2002). In addition quantitative data on organic pollutants analysed in North Sea biota have been reported (e.g., Shawky and Emons 1998, Roose and Brinckmann 2000, Birchenough et al. 2002). Although a high proportion of organic contaminants is associated with particulate matter and therefore transported with these particles, only a few studies have been published regarding contaminations in North Sea sediments (e.g., Scrimshaw et al., 1996, Sündermann and Radach, 1997). Primary sinks of particulate matter are concentrated along the North Sea margins. Thus, elevated concentrations of particulate associated pollutants of both marine and terrestrial origin can be expected for these coastal regions.

The German Bight is an estuarine system situated in the south-eastern part of the North Sea. It is mainly influenced by fluvial input of the rivers Elbe, Weser and Ems. Numerous recently published investigations describe the organic contamination of sea water by substances such as hexachlorocyclohexanes (HCH), polychlorinated biphenyls (PCB), pesticides, organo tin compounds, volatile organic compounds (VOC), oxygenated aromatics, nitroaromatics, benzothiazoles, synthetic fragrances, pharmaceuticals as well as chlorinated alkylethers and -formals. Several of these contaminants were characterized as Elbe pollutants derived from the Elbe river (Gatermann et al. 1995, Bester et al. 1997, Hühnerfuss et al. 1997, Franke et al. 1998a, Bester et al. 1998a, Bester et al. 1998b, Bester et al. 1998c, Bester and Theobald 2000). Due to the lack of chemical analyses of North Sea sediments investigations on the pollution of sedimentary particulate matter in the German Bight are also

---

‡ Mainly adapted/reprinted from Schwarzbauer and Littke 2003

still limited (e.g., Sündermann and Radach 1997, Hühnerfuss et al. 1997). Many of the investigations mentioned above revealed a major impact of the Elbe river on the pollution of the German bight. This includes also the pollution of the sediments by heavy metals as reported e.g. for cadmium and lead (Puls et al. 1997, Radach and Heyer 1997, ).

Although the German bight is a system influenced by the inflow of various rivers, information on the individual contribution of each river system is very limited. Especially for the differentiation of superimposed contributions the application of anthropogenic organic marker compounds is very useful. Source specific organic substances, that are characterized by a high environmental stability and lipophilicity, are appropriate to trace the pollution of particulate matter in coastal areas back to the individual riverine contributions. In addition contributions of several rivers discharging close together into the marine environment can be differentiated. Compared to a direct measurement of the pollution in the discharge of the respective rivers, the analyses of organic marker compounds in coastal samples additionally provide very specific information on the spatial and quantitative distribution of the river derived contamination within the marine ecosystem. Furthermore, the application of inorganic compounds to discriminate diverse riverine contributions is impeded by a minor source specificity as compared to organic substances. However, to our knowlegde the anthropogenic organic marker approach has been applied very rarely for environmental investigations on coastal areas so far.

In a previous study we have identified a high number of organic contaminants in sediments of the German Bight emitted by various sources of pollution (Schwarzbauer et al. 2000). Apart from biogenic and petrogenic substances numerous compounds were attributed to industrial and municipal emissions, primarily as a result of riverine contributions. In addition, a group of contaminants was identified representing Elbe specific marker compounds. The characterization of the specific marker compounds was based on (i) their spatial distribution within the German Bight, (ii) a formerly reported occurrence in the Elbe river system and (iii) on published information regarding their origin or emission. However, quantitative analyses are more adequate to describe especially the spatial distribution as compared to qualitative analyses. Hence, in the present investigation determination of various anthropogenic contaminants has been performed in order to evaluate the proposed Elbe specific marker compounds. These new analyses were conducted using the original sample set and are discussed in the following. With respect to the limited number of samples investigated these results have to be considered preliminary.

## Samples

Sediment samples (see Tab. 1 and Fig. 1) were taken in 1998 by the German Federal Maritime and Hydrographic Agency (Hamburg) using a van Veen grab. This yielded material from the sediment surface to a depth of approximately 15 cm.

**Tab. 1:** Sampling locations, fraction of grain size, as well as data on TOC and dry weight of sediments from the German Bight.

| Sample | altitude | latitude | fraction of grain size < 63 mm | TOC (%) | dry weight (%) |
|---|---|---|---|---|---|
| A | 54°02' | 8°12.5' | > 50 % | 0.65 | 73 |
| B | 54°04' | 8°07.5' | > 50 % | 1.07 | 58 |
| C | 54°22.5' | 7°38.75' | < 5 % | 0.19 | 80 |
| D | 54°30' | 6°30' | 5 – 10 % | 1.10 | 79 |
| E | 54°50' | 6°35' | 11 – 20 % | 0.29 | 72 |
| F | 55°00' | 6°30' | 21 – 50 % | 0.43 | 70 |
| G | 53°49' | 6°23' | < 5 % | 0.46 | 83 |

The sampling locations were selected considering two criteria: (i) the main flow circle as suggested by Ducrotoy et al. (2000) and (ii) significant sedimentation properties (grain size and TOC values) characterizing main sedimentation areas. Furthermore, besides definitely Elbe influenced samples (A,B,C) sampling sites, that were definitely not affected (G) or influenced to a minor extent by Elbe derived organic matter (D,E.F), have been chosen.

Data on total organic carbon (TOC) and grain size fractions as presented in Tab. 1 were also provided by the German Federal Maritime and Hydrographic Agency (Hamburg).

The ions used for quantification and the signal accumulation time as well as the recoveries for the extraction and evaporating procedures are summarized in Tab. 2. An external four-point-calibration generated from a mixture of authentic reference compounds was used for quantification. The limit of quantitation (LOQ) was 0.01 µg/kg dry matter calculated from GC/MS-analyses of reference compounds (approx. 100 pg/µL of reference substances were determined with a signal to noise ratio of 10:1).

**Tab. 2:** Selected contaminants (partly with abbreviations applied in Fig.1 and 2) and ion fragments used for quantification as well as determined recoveries.

| **Compounds** | **Abbreviation (used in Fig. 1 and 2)** | **Ions (m/z)** | **Signal accumulation time (ms)** | **Recovery rates (%)** |
|---|---|---|---|---|
| *Alkanes* | | | | |
| 2,2,4,4,6,8,8-Heptamethyl-nonane | HMN | 99, 113, 155 | 200 | 88 |
| *Chlorinated compounds* | | | | |
| Hexachlorobutadiene | HCBu | 225, 227, 260, 262 | 200 | 41 |
| Dichlorobenzenes | DCB | 146, 148 | 400 | 43 |
| Trichlorobenzenes | | 180, 182 | 200 | 53 |
| Tetrachlorobenzenes | | 214, 216 | 300 | 55 |
| Pentachlorobenzene | | 250, 252 | 400 | 66 |
| Hexachlorobenzene | HCB | 284, 286 | 400 | 81 |
| Chloronaphthalenes | CN | 162, 164 | 200 | 35 |
| Dichloronaphthalenes | DCN | 196, 198 | 300 | 53 |
| 4,4'-Dichlorodiphenylsulfide | DDS | 254, 256 | 500 | 84 |
| *DDT-metabolites* | | | | |
| 4,4'-DDE | | 246, 248 | 400 | 89 |
| 2,4'-DDD | | 235, 237 | 400 | 70 |
| 4,4'-DDD | DDD | 235, 237 | 400 | 70 |
| 4,4'-DDCN | | 226, 228, 261, 263 | 400 | 90 |
| *Plasticizers* | | | | |
| Alkylsulfonic acid phenylester | ASPE | 94 | 900 | 80 |
| Tri-*n*-butyl phosphate | | 99, 155 | 400 | 78 |
| 2,4,4-Trimethylpentane-1,3-diol-di-*iso*-butyrate | TPDB | 159, 243 | 400 | 61 |
| *Fragrances* | | | | |
| Galaxolide | Gal | 243, 258 | 400 | 43 |
| Tonalide | | 243, 258 | 400 | 43 |
| *Tinorganic compounds* | | | | |
| Tetrabutyl tin | TeBT | 235, 237 | 400 | 84 |

## Results and discussion

Numerous anthropogenic contaminants were quantified in seven surface sediment samples from the German Bight (A – F, see Tab. 1 as well as Fig.1 and 2) formerly analysed by means of a non target screening. The selection of the analytes is based on a qualitative evaluation of Elbe specific marker compounds (Schwarzbauer et al. 2000). Besides a set of proposed marker compounds including 4,4'-dichlorodiphenylsulfide, tetra- to hexachlorinated benzenes, chloro- and dichloronaphthalenes, hexachlorobutadiene, alkylsulfonic acid phenylesters and tetrabutyl tin, further contaminants (heptamethylnonane, DDT group substances, di- and trichlorinated benzenes, tri-*n*-butylphosphate, 2,4,4-trimethylpentane-1,3-dioldi-*iso*-butyrate, galaxolide and tonalide) were determined in order to evaluate the source specifity of the Elbe-derived compounds. All quantitative results are presented in Tab. 3 given in µg/kg dry matter. In general, very low to low concentrations were detected ranging between the LOQ ( 0.01 µg/kg dry matter) and 6.85 µg/kg dry matter.

As reported earlier in detail three of the samples were assumed to be directly influenced by the Elbe river (A,B) and the Weser and Ems rivers (G), whereas the contributions of the riverine systems to the organic matter in the sediments situated at greater distance from the coastal area (C,D,E,F) are still ambiguous (Schwarzbauer et al. 2000).

Hence, for the evaluation of the proposed Elbe marker compounds the quantitative and spatial distribution as illustrated for selected substances in Fig.1 and 2 are the most important criteria. The quantitative evaluation of the molecular markers has to consider the accumulation rates of organic matter at the different sampling sites. Thus, in order to avoid misinterpretation all quantitative data presented in Fig.1 and 2 are normalized to TOC content (given in µg/kg TOC).

A first group of contaminants clearly has a limited extension appearing only at sampling locations A and B. These contaminants unambiguously reflect Elbe river derived contamination. This first group of specific marker compounds includes hexachlorobutadiene and a complex mixture of alkylsulfonic acid phenyl esters, which were detected in samples A and B with concentrations in the range of 280 to 520 mg/kg and 23 to 750 µg/kg TOC, respectively. Alkylsulfonic acid aryl esters are common PVC plasticizers and are known as Elbe river contaminants, which occur in high concentrations in sediments and suspended particulate matter (Franke et al. 1998b). Hexachlorobutadiene is used as industrial solvent as well as in hydraulic fluids (Koch 1995) and has previously been detected in Elbe river sediments (Schwarzbauer 1997).

Mono- and dichlorinated naphthalenes, tetra- to hexachlorinated benzenes, 4,4'-dichlorodiphenylsulfide and tetrabutyl tin represent a second group of Elbe marker compounds. The distribution of these compounds is characterized by higher concentrations at sampling locations A and B and much lower concentrations along the sampling locations C,D,E, and F, which are situated at greater distance from the coastal area. Considering the principal flow direction of the North Sea water according to Ducrotoy et al. (2000) (see Fig.1 and 2) a contribution of Elbe derived sediments has to be assumed for sampling locations C, E and F due to the transport of suspended particulate matter. This contribution is reflected by the quantitative distribution of the substances of the second group.

**Tab. 3:** Concentrations of organic contaminants in sediment samples of the German Bight (all data are given in µg/kg dry matter).

| **Compounds / Sampling sites** | **A** | **B** | **C** | **D** | **E** | **F** | **G** |
|---|---|---|---|---|---|---|---|
| *Alkanes* | | | | | | | |
| 2,2,4,4,6,8,8-Heptamethylnonane | 0.68 | 0.83 | 0.69 | 0.92 | 0.51 | 0.73 | 0.31 |
| *Chlorinated compounds* | | | | | | | |
| Hexachlorobutadiene | 1.83 | 5.53 | | | | | |
| 1,2-Dichlorobenzene | 0.23 | 1.16 | 0.16 | 0.23 | 0.19 | 0.58 | 0.07 |
| 1,3-Dichlorobenzene | 0.12 | 0.65 | 0.05 | 0.02 | 0.04 | 0.28 | 0.02 |
| 1,4-Dichlorobenzene | 0.28 | 1.6 | 0.23 | 0.12 | 0.33 | 1.02 | 0.12 |
| 1,2,3-Trichlorobenzene | | 0.17 | | | | 0.04 | |
| 1,2,4-Trichlorobenzene | 0.06 | 1.82 | 0.05 | 0.02 | 0.05 | 0.43 | 0.08 |
| 1,3,5-Trichlorobenzene | 0.09 | 0.15 | 0.02 | 0.01 | 0.02 | 0.04 | 0.01 |
| 1,2,3,4-Tetrachlorobenzene | 0.08 | 0.24 | | | 0.03 | 0.07 | |
| 1,2,3,5-Tetrachlorobenzene | 0.14 | 0.1 | | | 0.04 | 0.12 | |
| Pentachlorobenzene | 0.60 | 0.23 | 0.05 | | 0.04 | 0.12 | |
| Hexachlorobenzene | 0.14 | 0.50 | 0.02 | | 0.04 | 0.19 | |
| 1-Chloronaphthalene | 0.16 | 0.75 | 0.08 | | | | |
| 1,3-Dichloronaphthalene | | 0.12 | | | | | |
| 1,4-Dichloronaphthalene | 0.06 | 0.45 | 0.01 | | | | |
| 1,5-/1,6-Dichloronaphthalene | 0.04 | 0.24 | | | | | |
| 1,7-/2,6-/2,7-Dichloronaphthalene | 0.01 | 0.04 | | | | | |
| 1,2-Dichloronaphthalene | | 0.03 | | | | | |
| 1,8-Dichloronaphthalene | | 0.23 | | | | | |
| 4,4'-Dichlorodiphenylsulfide | 0.15 | 0.48 | 0.02 | | 0.02 | 0.06 | |
| *DDT-metabolites* | | | | | | | |
| 4,4'-DDE | 0.17 | 0.34 | 0.06 | | | | |
| 2,4'-DDD | | 0.11 | | | | | |
| 4,4'-DDD | 0.16 | 0.30 | 0.09 | 0.01 | 0.02 | 0.02 | |
| 4,4'-DDCN | | 0.17 | | | | | |

| Compounds / Sampling sites | A | B | C | D | E | F | G |
|---|---|---|---|---|---|---|---|
| *Plasticizers* | | | | | | | |
| 7-Tridecylsulfonic acid phenylester | 0.02 | 0.76 | | | | | |
| 6-Tridecylsulfonic acid phenylester | | 0.45 | | | | | |
| 5-Tridecylsulfonic acid phenylester | | 0.24 | | | | | |
| 4-Tridecylsulfonic acid phenylester | | 0.18 | | | | | |
| 3-Tridecylsulfonic acid phenylester | | 0.30 | | | | | |
| 2-Tridecylsulfonic acid phenylester | | 0.20 | | | | | |
| 7-Tetradecylsulfonic acid phenylester | 0.05 | 1.12 | | | | | |
| 6-Tetradecylsulfonic acid phenyester | 0.02 | 0.40 | | | | | |
| 5-Tetradecylsulfonic acid phenylester | | 0.22 | | | | | |
| 4-Tetradecylsulfonic acid phenylester | | 0.14 | | | | | |
| 3-Tetradecylsulfonic acid phenylester | | 0.35 | | | | | |
| 2-Tetradecylsulfonic acid phenylester | | 0.18 | | | | | |
| 7-/8-Pentadecylsulfonic acid phenylester | 0.04 | 1.11 | | | | | |
| 6-Pentadecylsulfonic acid phenylester | 0.01 | 0.59 | | | | | |
| 5-Pentadecylsulfonic acid phenylester | | 0.30 | | | | | |
| 4-Pentadecylsulfonic acid phenylester | | 0.31 | | | | | |
| 3-Pentadecylsulfonic acid phenylester | | 0.28 | | | | | |
| 2-Pentadecylsulfonic acid phenylester | | 0.06 | | | | | |
| 7-/8-Hexadecylsulfonic acid phenylester | 0.01 | 0.38 | | | | | |
| 6-Hexadecylsulfonic acid phenylester | | 0.3 | | | | | |
| 5-Hexadecylsulfonic acid phenylester | | 0.12 | | | | | |
| Tri-*n*-butyl phosphate | 6.85 | 5.86 | 0.37 | 2.11 | 1.30 | | 0.93 |
| 2,4,4-Trimethylpentane-1,3-dioldi-*iso*-butyrate | 0.16 | 0.26 | 0.52 | | 1.72 | | 1.07 |
| *Fragrances* | | | | | | | |
| Galaxolide | 0.87 | 0.65 | 0.91 | 2.86 | 0.93 | 7.06 | 1.70 |
| Tonalide | 0.38 | 0.32 | 0.70 | 1.55 | 0.74 | 4.13 | 1.32 |
| *Tinorganic compounds* | | | | | | | |
| Tetrabutyl tin | 0.55 | 1.39 | 0.05 | | 0.04 | 0.06 | 0.02 |

Hence, the decreasing concentration profiles of the substances of the second group indicate an Elbe-derived contamination but with an extended spatial distribution as compared to the substances previously discussed. Also the slightly higher concentration at sampling site F as compared to sampling sites C and E as well as a sharp decrease at sampling site D can be explained by the flow circle of the water body turning westward to the sampling location F after passing the German west coast at sampling site C. Regarding the presented concentration profiles and using these marker compounds a significant contribution of Elbe river derived contamination can be stated at sites A, B, C, E and F but not for samples D and G.

The concentrations of tetrabutyl tin range from 130 µg/kg TOC at sample site B to 14 µg/kg TOC at sample sites E and F and is below the LOQ at sample site D. The emission of this well known Elbe-specific compound was formerly linked to an industrial point source situated near the confluence of the Mulde and the Elbe rivers (Wilken et al., 1994, Schwarzbauer, 1997). Mono- and disubstituted chloronaphthalenes were also detected with concentrations between the LOQ at sample sites not influenced by Elbe derived contaminations and 70 and 100 µg/kg TOC at sample site B. These low chlorinated naphthalenes occur in the sediments with patterns similar to those of technical agents (e.g. Halowax 1000) and were formerly identified with comparable isomer distributions in sediments of the Elbe river and its tributaries (Schwarzbauer 1997, Schwarzbauer et al. 2001). On the contrary the origin of the Elbe specific contaminant 4,4′-dichlorodiphenylsulfide detected at sample sites A,B,C,E and F with concentrations ranging from 7 to 45 µg/kg TOC is still unknown.

Di- and trichlorinated benzenes were detected not only at sample sites mainly influenced by the Elbe river but also at sampling sites G and D, which are mainly influenced by input of the Ems river, with similar concentrations as in samples E and F (see Tab. 2). Hence these lower substituted compounds, which are used as synthetic raw material for many technical products such as antiseptic agents, solvents and additives (Bryant, 1993), are not appropriate marker compounds for tracing Elbe river derived contaminations. In contrast the tetra- to hexachlorinated benzenes, which have formerly been identified as contaminants in particulate matter of the Elbe river system (Schwarzbauer 1997, Schwarzbauer et al. 2001), appeared in the North Sea sediments only at sampling sites A,B,C,E and F (see Tab. 2). Fig. 1 illustrates the spatial distribution of hexachlorobenzene with concentrations between 2 and 50 µg/kg TOC in samples influenced by the Elbe river.

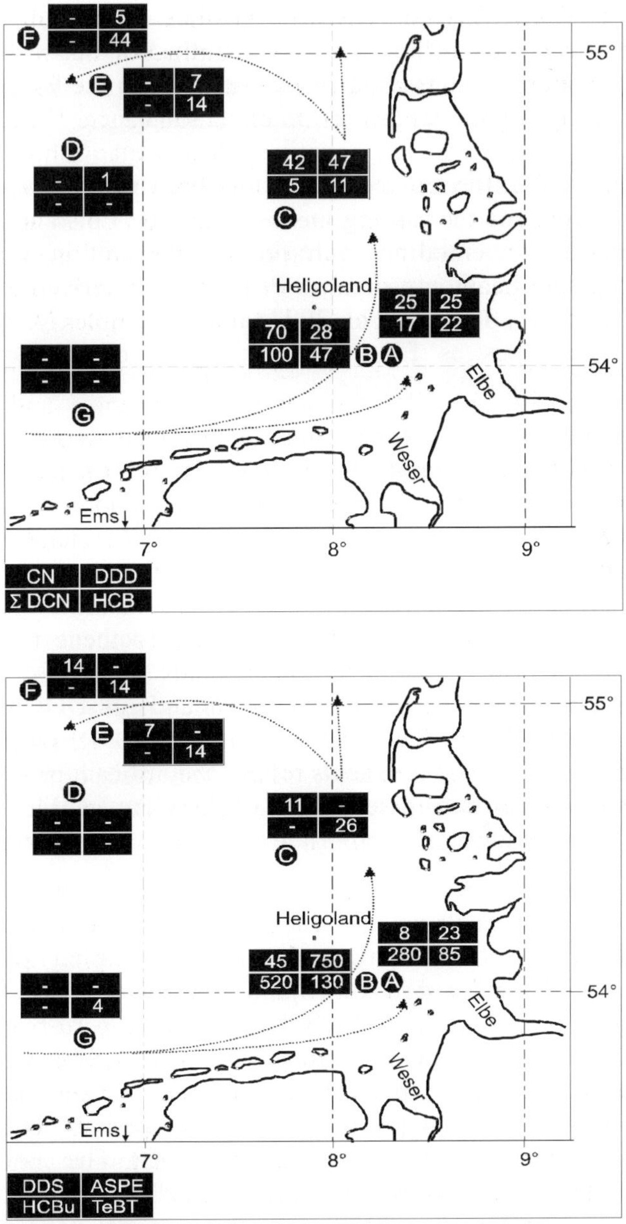


**Fig. 1:** Concentration and spatial distribution of selected Elbe specific marker compounds in sediments of the German Bight (all data are given in µg/kg dry matter and normalized to TOC). All abbreviations of the substances are explained in Tab. 2. The dotted lines indicate the main flow circle of the water body according to Ducrotoy et al. (2000).

2,2-Bis(4-chlorophenyl)-1,1,1-trichlorethane (DDT) and related substances were also detected in the North Sea sediments and were previously characterized as abundant compounds in the Elbe river system. They are only of minor use as marker substances because of their widespread occurrence in the aquatic environment (Schwarzbauer et al. 2000). However, DDT was applied in the Elbe river catchment area for a longer period of time than in regions affecting the Ems and Weser rivers. Hence, elevated concentrations were recently detected in Elbe river water and sediments (e.g., Goetz et al. 1994, Dannenberger and Lerz 1998, Ricking et al. 2003, Schulze et al. 2003) and, consequently, the application of DDT metabolites as Elbe specific marker compounds has to be reevaluated by quantitative analyses.

As illustrated in Fig. 1 for 2,2-bis(4-chlorophenyl)-1,1-dichloroethane (4,4'-DDD) higher concentrations (25 to 47 µg/kg TOC) were observed at sampling sites A, B and C, whereas the concentrations decreased at sites D, E and F (1, 7 and 5 µg/kg TOC, respectively). At sampling location G, which is not influenced by Elbe river derived sedimentary matter, the concentration of 4,4'-DDD fell below the LOQ. A similar distribution was observed for 2,2-bis(4-chlorophenyl)-1,1-dichloroethene (4,4'-DDE) (see Tab. 3). Low concentrations of DDT metabolites bis(4-chlorophenyl)acetonitrile (4,4'-DDCN) and 2,4'-DDD were detected only at sampling site B. The concentrations of DDD, DDE and DDCN analysed in the North Sea sediments reflect a significant proportion of Elbe derived DDT metabolites in sediments of the German Bight and confirm DDT related compounds as appropriate Elbe marker compounds.

In addition to the discussed Elbe marker compounds further contaminants were analysed in order to illustrate the quantitative and spatial distribution of non-marker substances. These compounds include the common fragrances galaxolide and tonalide, the plasticizers tri-*n*-butylphosphate and 2,4,4-trimethylpentane-1,3-dioldi-*iso*-butyrate (TPDB) as well as the highly branched alkane 2,2,4,4,6,8,8-heptamethylnonane (HMN). Galaxolide occurred with concentrations between 61 µg/kg TOC at sampling site B and 34 to 79 µg/kg TOC (sites A,C,D,E,G) with the highest concentration at site F (1600 µg/kg TOC). The concentrations of TPDB ranged between LOQ and 270 µg/kg TOC with a maximum at sampling site E (590 µg/kg TOC). Also the concentrations of HMN were very similar at sites A to G (67 to 180 µg/kg TOC) and peaked at sampling location C with 370 µg/kg TOC. In summary, the quantitative distribution of theses compounds in the sediments was characterized by no significant accumulation at sampling locations mainly influenced by either the Elbe, the Ems or the Weser rivers. Mainly low concentrations were observed at sampling sites A and B, whereas higher concentrations were detected at locations situated more distant from the coastal area or near the Weser and Ems river estuaries. Examples are illustrated in Fig. 2.

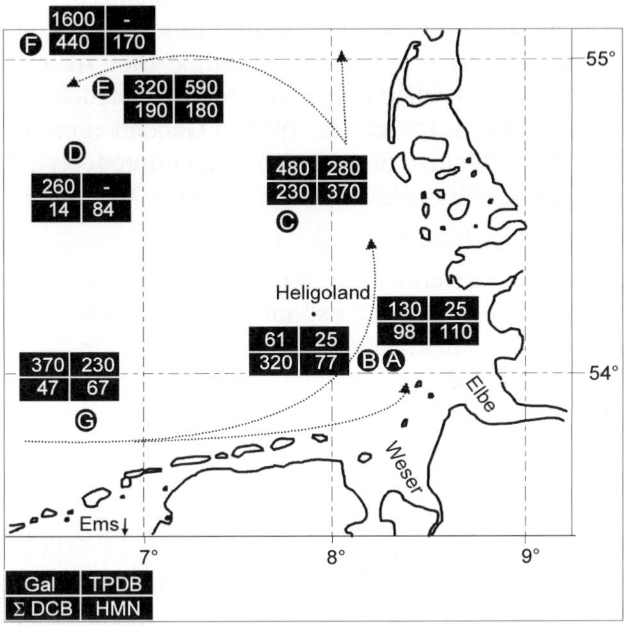


**Fig. 2:** Concentration and spatial distribution of nonspecific contaminants in sediments of the German Bight (all data are given in μg/kg dry matter and normalized to TOC). All abbreviations of the substances are explained in Tab. 2. The dotted lines indicate the main flow circle of the water body according to Ducrotoy et al. (2000).


## Summary

Selected contaminants were quantified in seven sediment samples of the German Bight. The compounds analysed include substances formerly proposed as Elbe specific marker compounds and contaminants reflecting a diffuse contribution to the pollution of North Sea sediments. For the evaluation of the proposed Elbe marker compounds the quantitative and spatial distribution was investigated and discussed as the most important criterion to define their source specifity.

Two groups of Elbe derived contaminants were differentiated. Substances of the first group include hexachlorobutadiene and alkylsulfonic acid phenyl esters and appear only at sampling locations A and B directly influenced by the Elbe river. The second group consists of compounds, which occur not only at sampling locations A and B but also in minor concentrations at sites situated at greater distance from the coastal area, which are therefore less influenced by the Elbe river. However, their spatial and quantitative distribution characterizes them clearly as Elbe river derived components. Compounds of the second group are tetra- to hexachlorinated benzenes, mono- and disubstituted chloronaphthalenes, tetrabutyl tin, 4,4′-dichlorodiphenylsulfide and DDT related compounds.

In order to discern the discussed Elbe marker compounds from non-marker substances several nonspecific contaminants were analysed additionally. These contaminants include galaxolide, tonalide, tri-*n*-butylphosphate, 2,4,4-trimethylpentane-1,3-dioldi-*iso*-butyrate (TPDB) and 2,2,4,4,6,8,8-heptamethylnonane (HMN). No significant accumulation of these compounds was observed at sampling locations mainly influenced by either the Elbe river or the Ems and Weser rivers.

These results illustrate how quantitative data on source specific organic contaminants and their spatial distribution can be used to trace individual riverine contributions to the pollution of coastal areas. This approach might be especially useful to differentiate the contribution of several rivers discharging close together into the marine environment. Nevertheless, these results are limited due to the restricted number of samples. Hence, more comprehensive analyses on numerous coastal and estuarine areas have to be done further on to support this approach.

## 5.2 Geochronology of terrestrial or subaquatic deposited riverine particulate matter

### 5.2.1 Organic pollutants in riparian wetlands of the Lippe river (Germany)‡

#### Introduction

Aquatic particulate matter is an important sink for numerous lipophilic anthropogenic contaminants as the result of adsorption and accumulation processes. In the adsorbed state organic contaminants can be subject to transformation as well as to transportation processes within the river system. The anthropogenic contribution to the organic load of river systems can be attributed mainly to sewage effluents derived from industrial and domestic point sources, to agricultural emissions or to shipping activities. Numerous specific compounds reflect the different sources of pollution and, therefore, are frequently detected in river systems.

During flood events, the particulate matter can be deposited on flood plains and riparian wetlands. Accordingly these areas can act as ecological archives in case of undisturbed and periodical sedimentation. The dating of these sediment deposits can be performed by gamma spectrometric analysis of the radio-nuclides $^{210}Pb$ and $^{137}Cs$. Organic geochemical investigations on anthropogenic contaminants in these dated layers allow a detailed description of the emission of organic contaminants into the corresponding riverine system. However, this approach is restricted to substances, which are characterized by a sufficient environmental stability and immobility in the particulate phase.

In 1993 Klös and Schoch published a study characterizing the organic contamination of a riparian wetland situated at the confluence of the Lippe and Rhine rivers, Germany. Based on $^{210}Pb$ and $^{137}Cs$ radiodating the geochronological distribution of a limited number of selected priority pollutants (heavy metals, dioxines, hexachlorobenzene, hexachlorocyclohexane, hexachlorobiphenyle) has been presented.

In the present, prelimenary study (see also chapter 5.2.2) we have extended these investigations to a wide variety of organic contaminants reflecting the state of pollution in a more comprehensive way. Detailed

‡ Mainly adapted/reprinted from Heim et al., 2003

and extensive GC/MS-screening analyses applied on two selected sediment horizons of the same sampling material revealed a diversified and specific spectrum of anthropogenic pollutants. Quantitative analyses of selected marker compounds correlated with the time periods of sedimentation reflected not only the history of pollution, but also information on the effectiveness of measures for water protection.

## Samples

A sediment core from a floodplain of the Lippe river, a tributary of the Rhine river (Germany) was taken in 1989/90 by the North-Rhine Westphalian Environment Agency.

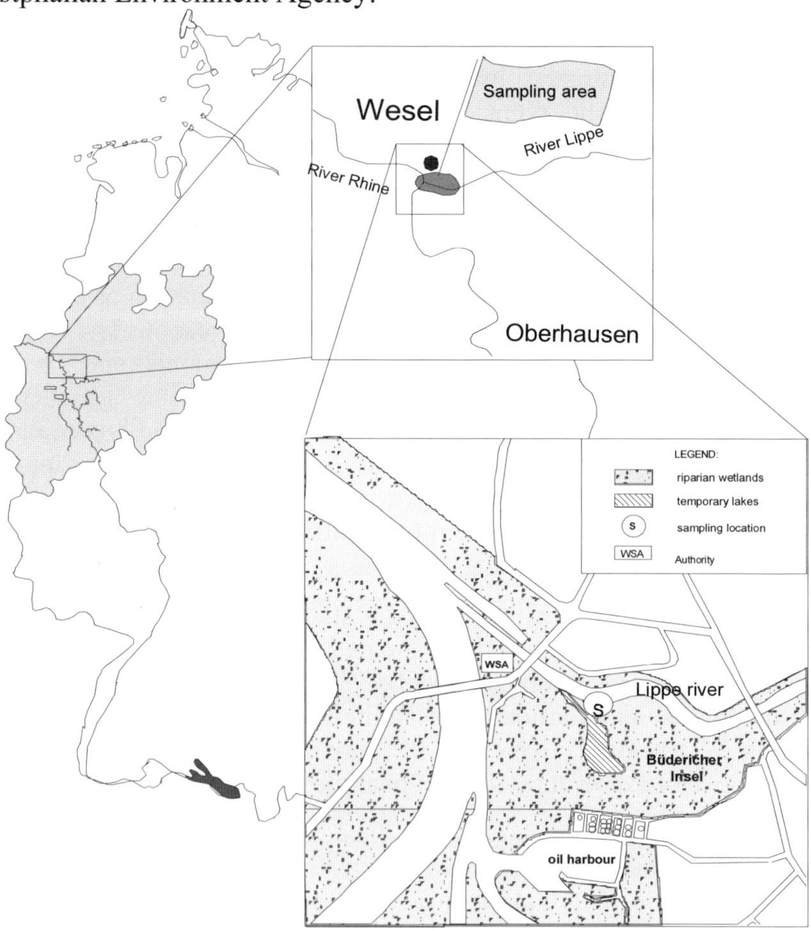


**Fig. 1:** Map of sampling location

The core of 180 cm length was divided to 4 and 8 cm segments. γ-Spectrometry dated the bottom layer to an age of ± 100 years (approx. 1900). The accumulation rate was fixed to about 2 cm/year (Klös and Schoch, 1993). For the present study sediment samples of two separate horizons of this core were investigated. The upper sector included 7 sediment layers (4 cm segments) at a depth-range between 8-36 cm correlated to a sedimentation time period of approx. 1985 to 1972. The second horizon consisted of two sediment layers (8 cm segments) at a depth between 76 and 92 cm (sedimentation time period approx. 1952 to 1944). The sediment core was characterized by a distinct bedding with evident colour variations in single layers.

## Results and Discussion

### *Non-target screening*

Detailed GC/MS-screening analyses were applied to samples selected from two separate stratigraphical horizons of a sediment core derived from a riparian wetland of the Lippe river (Germany). The analyses revealed a wide variety of lipophilic organic compounds of both biogenic and anthropogenic origin. Characteristic anthropogenic compounds identified are presented in Table 1.

### *Quantification of selected marker compounds*

An assortment of the anthropogenic contaminants has been selected for quantitative analyses (Tab. 1). For the quantification numerous marker compounds were considered, which were specific for different anthropogenic emission sources comprising e.g. agricultural activities, municipal and industrial effluents. Additionally, the criteria for the selection included a high environmental stability and immobility within the sediment layers in order to avoid misinterpretation due to transport or degradation processes. The quantitative data measured permitted the establishment of depth-correlated concentration profiles. Using the radiodating by Klös and Schoch (1993) a time-correlated distribution was obtained as illustrated in Figure 2. In the following section the time depending occurrence as well as the source specificity of the marker compounds analysed are discussed in detail.

**Tab. 1:** Selected organic contaminants identified in the sediment core of a riparian wetland of the Lippe river (Germany). The compounds selected for quantitative analyses are underlined.

| | |
|---|---|
| *Aromatics* | *Chlorinated aromatics* |
| Di-*iso*-propylnaphthalenes | $Cl_2$- to $Cl_5$-Benzenes |
| 2,2-Diphenylpropane | Polychlorinated biphenyles (PCB) |
| Carbazole | Tetrachlorobenzyltoluenes (TCBT) |
| | Dichloroanisoles |
| *Benzenes* | Trichloroanisoles |
| *p*-Xylene | Octachlorostyrene (OCS) |
| $C_3$- to $C_5$-Benzenes | |
| | *Chlorinated aliphatics* |
| *Alcohols and ethers* | Tetrachlorobutadiene |
| | Pentachlorobutadiene |
| 2,6-Di-*tert*-butylphenol | Hexachlorobutadiene |
| Phenyl-*o*-tolylether | Bis(chloropropyl)ether |
| *Synthetic musks* | *Vitaminoids* |
| Galoxolide | Phthiocol |
| Tonalide | |
| | *Pesticides and bacteriocides* |
| *Metallorganic compounds* | Dichlorobenzonitrile |
| Tetrabutyl tin | Methyltriclosan |

Contamination of riverine particulate matter by emissions of sewage treatment plants orginated either from municipal or industrial effluents. As markers for municipal effluents the vitaminoide phthiocol and the methylated metabolite of the bactericide triclosan (5-chloro-2(2,4-dichlorodiphenyl) ether), commonly used e.g. in tooth paste, shampoo or detergents, were analysed. In the horizons investigated phthiocol and methyltriclosan were not detected in the lower layers (Fig. 2 A, B) but showed a significant increase in concentration ranges from 2 and 3 ng/g up to 20 and 130 ng/g in the upper layers, respectively. It has to be noted, that the samples contained only methyltriclosan but not triclosan. We interpret this as a result of environmental transformation of triclosan to methyltriclosan by biomethylation as also suggested by Lindström et al. (2002).

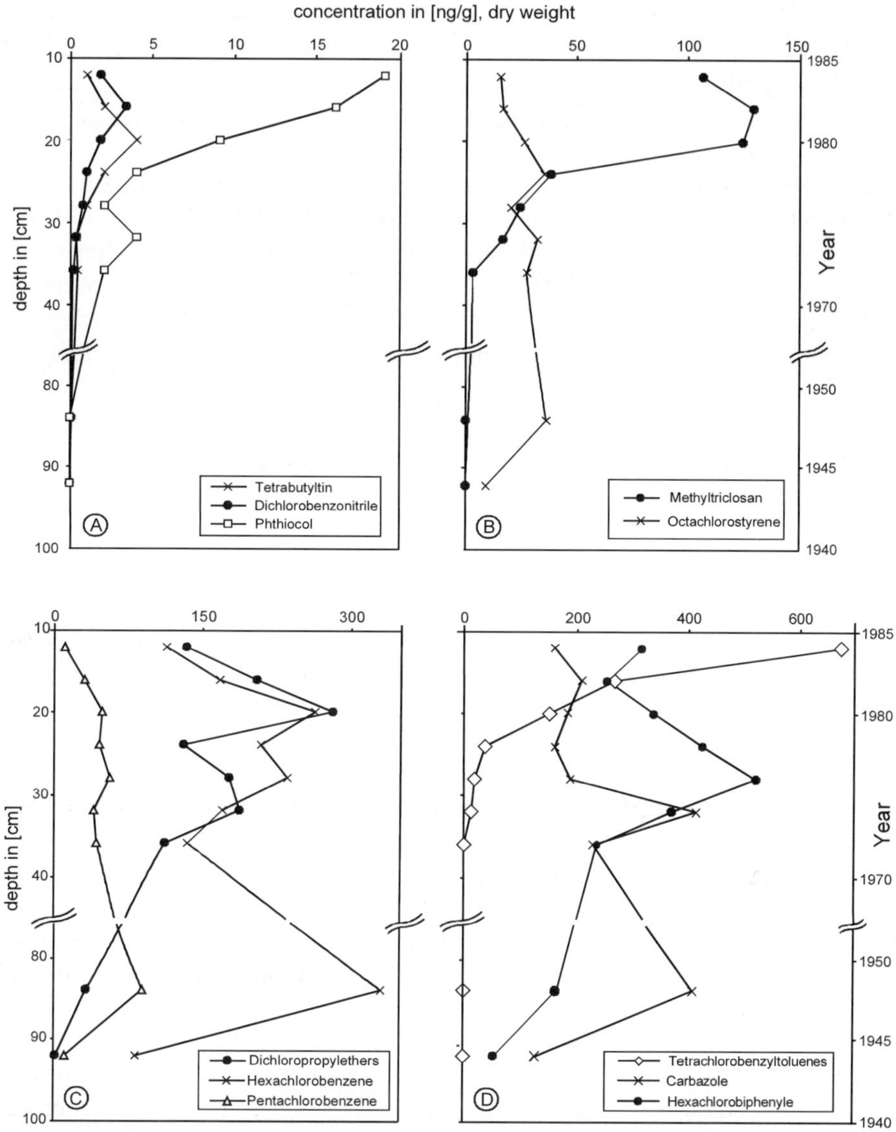


**Fig. 2 A-D:** Depth and time correlated concentration ranges of selected anthropogenic contaminants (listed from right to left by increasing concentrations); dichloropropylethers means bis-(1-chloro-2-propyl)ether and (1-chloro-2-propyl-2-chloro-1-propyl)ether

Besides the municipal markers numerous compounds were quantified representing the industrial contribution to the contamination of the Lippe river. Hexachlorobiphenyl, a representative of the congeneric group of polychlorinated biphenyls and tetrachlorobenzyltoluenes (Ugilec®) were appointed as markers for the mining industry. Their distribution within the sediment core showed opposing trends (Fig 2 D). Hexachlorobiphenyl was encountered with concentrations increasing from the deepest layer (50 ng/g) up to 520 ng/g at a depth of 28 cm (approx. 1976). Afterwards slightly decreasing concentrations towards the upper layers were observed. On the contrary, tetrachlorobenzyltoluenes were only detected with concentrations increasing in the upper layers representing accumulation times younger than 1974.

A second set of industrial markers characterized emissions of the chemical industry. Tetrabutyltin, key chemical material of butylated organotin compounds, was detected only in sediment layers deposited since 1970 with low concentrations ranging between 0.3 and 4 ng/g (Fig. 2A). Octachlorostyrene, e.g. specific by-product of the magnesium production, was detected with concentrations from 9 to 35 ng/g in all samples investigated without significant maxima (Fig. 2 B). Dichloropropylethers (Fig 2 C), representing by-products of the propylenoxide chemistry were detected continuously from the bottom to the top layers with a concentration maximum of 280 ng/g at a depth of 20 cm (approx. 1980). Also pentachlorobenzene represents industrial emissions and occurred with comparatively low but steady concentrations between 10 and 90 ng/g (Fig. 2 C).

Emissions derived from agricultural activities contained the herbicide dichlorobenzonitrile, and to a minor extend hexachlorobenzene and carbazol. The concentrations of dichlorobenzonitrile (Fig. 2 A) ranged between 0.1 and 3.5 ng/g, with higher values in the upper layers. The emission of hexachlorobenzene and carbazol can not be definitely attributed to agricultural sources. Although these compounds were used as herbicides or insecticides other technical applications as well as petrogenic sources might be responsible for the distribution of these compounds in the riverine environment. However, the concentration profiles of both substances in the sediment layers investigated are very similar. A high variation of the values at a high level of contamination (up to 330 and 40 ng/g, respectively) without a preferred accumulation neither in the upper nor in the lower layers was observed. Besides the source specific interpretation of the compounds investigated the results have also to be discussed with respect to the general concentration trends as follows.

A first group of substances comprising phthiocol, methyltriclosan, dichloropropylethers and tetrachlorobenzyltoluenes occur in decreasing concentrations from the top to the bottom layers. This occurrence characterizes an increasing pollution of the river system by these substances as a result of increasing production, usage and, consequently, emission rates. In a second group all substances with increasing concentrations from the top to the bottom layers are included (octachlorostyrene, pentachlorobenzene). This distribution indicates formerly higher emission rates with a subsequent decrease of contamination e.g. due to reduced production and usage rates or an improved elimination from effluents. The third group of contaminants (tetrabutyl tin, hexachlorobenzene, carbazol, dichlorobenzonitrile and hexachlorinated biphenyles) is characterized by significant variations of the concentrations and several concentration maxima representing an unsteady emission rate for the time period investigated. With respect to chlorinated biphenyl a causal connection between using period and emission trends was obvious. A first appearance of tetrachlorobenzyltoluenes can be linked with the decrease of polychlorinated biphenyl concentrations in the sediments as a results of the substitution of polychlorinated biphenyl by tetrachlorobenzyltoluenes in mining devices during the early 70´ies.

## Conclusion

GC/MS analyses revealed a significant anthropogenic contamination in sediment layers of a riparian wetland of the Lippe river (Germany). From a variety of organic contaminants we selected specific anthropogenic marker compounds assigned to different emission sources. Different concentration profiles with distinctive maxima indicated an individual and compound-specific pollution of the investigated sediment layers. Due to a periodical and undisturbed sedimentation this study allowed the determination of specific load histories associated to the time period considered. Generally, this kind of investigation can be used as a long term evaluation of the anthropogenic contribution to the pollution of riverine environments.

## 5.2.2 Geochronology of a load history: anthropogenic pollutants in riparian wetland sediments of the Lippe river (Germany)[‡]

### Introduction

Anthropogenic contaminants reach the fluvial environment by different emission pathways. Besides atmospheric deposition, the major proportion of contamination can be attributed to direct or indirect anthropogenic discharge of noxious substances by municipal, industrial and agricultural sources. The municipal input includes particulate matter derived from sewage sludge as well as waste water from household effluents and surface and street runoff (Kolpin et al., 2002; Takada et al., 1997). Numerous organic constituents of municipal effluents occur as the result of careless or incorrect usage of chemical products, (excessive) usage of detergents, personal care products, paints, varnish and further technical compounds as well as the application and disposal of pharmaceuticals in our daily routine. Besides municipal effluents, industrial discharge also represents an important pollution source in riverine systems (Di Corcia et al., 1999, Kjeller and Rappe, 1995, Sanders et al., 1992). Industrial sewage and waste water contains a wide spectrum of noxious substances including synthesis products, corresponding by-products and production wastes, as well as contaminations as a result of incomplete combustion (Bruns et al., 1997, Kaplan et al., 1997, Volkman et al., 1997).

As a third principal emission source, intensive agricultural activities discharge numerous pollutants e.g. organic and inorganic fertilizers, herbicides and insecticides (Schwarzbauer et al., 2001, Venkatesan et al., 1999; Zhang et al., 1999). The agricultural input has to be characterized as a diffuse source emission reaching the aquatic environment either by surface runoff or by the interaction of the riverine system with corresponding contaminated groundwater. Further non point source emissions can be attributed to shipping activities discharging, for example, fuel contaminants or constituents of anti-fouling paint like tributyl tin (Bence et al., 1996, Poppe et al., 1988, 1991). Besides the permanent emissions described, also short term, catastrophic contamination occurs as the result of single incidents including e.g. industrial accidents and disasters during careless transport and storage of chemicals. In summary, a wide spectrum of different specific anthropogenic pollutants and man

[‡] Mainly adapted/reprinted from Heim et al. 2004

made chemicals reach the riverine environment by various different sources.

The environmental fate of anthropogenic pollutants depends on their chemical and physical behaviour in the aquatic system. In the river system, water-soluble (hydrophilic) components predominantly accumulate in the aqueous phase and are consequently transported rapidly as a mobile fraction, whereas more hydrophobic (lipophilic) contaminants tend to be enriched in the solid phase. The particulate matter is then deposited according to hydrological and sedimentological conditions. Therefore, associated lipophilic compounds are transferred into the sediments, as a more immobile fraction. Because of more anaerobic conditions, the sedimentary compartment shows an enhanced environmental stability for many lipophilic contaminants of geogenic, biogenic and anthropogenic origin, accumulated in and incorporated into riverine sediment deposits.

Based on these processes, environmental investigations on fluvial systems focus on the particle-bound contaminants. Besides the lateral transport by temporary deposition and subsequent erosion, the long-term storage of contaminated particulate matter in undisturbed sediments is of major interest (Takada et al., 1998 and 1994). In case of undisturbed periodical or aperiodical sedimentation, for example during flood events, the accumulated sediment deposits (flood plains and wetlands) can act as ecological archives. Dating of these sediment layers by radiological analyses in combination with quantitative chemical analyses reveals a detailed record of the riverine load histories for preserved particle bound contaminants (Catallo et al., 1995, Kähkönen et al., 1998). Hence, fluvial sediment deposits in particular on flood plains and riparian wetlands, can act as the "memory of the river" (Klös and Schoch, 1993a).

In the past, numerous studies have been performed on dated sediment systems in order to reconstruct historical inputs of organic and inorganic contaminants. Most of these investigations have analysed contaminated sediment cores derived from two different depositional areas: coastal zones as the final sedimentation area of riverine particulate matter (e.g. Fox et al., 2001, Venkatesan et al., 1999, Winkels et al., 1998, Latimer and Quinn, 1996, Gustafsson et al., 2001), and lacustrine systems (e.g. Reiser et al., 1997, Van Metre et al., 1997, Sanders et al., 1992, Eisenreich et al., 1989).

In contrast, geochronological studies on riverine or riparian sediments have been rarely reported (Winkels et al., 1998). Furthermore, it must be noted that the major proportion of these geochemical studies dominantly focused on pre-selected organic and inorganic contaminants such as heavy metals or priority organic pollutions (Keith and Telliard, 1979). Examples include PAHs and PCBs (Gevao et al., 2000, Latimer and Quinn, 1996; Van Metre et al.,1997, Sanders et al., 1993,), synthetic detergents (Reiser et al., 1997) and chlorinated hydrocarbons (Winkels et al., 1998, Eisenreich et al., 1989). However, the documentation of a wide range of river-specific contaminants of various emission sources in dated riverine sediment cores is quite rare.

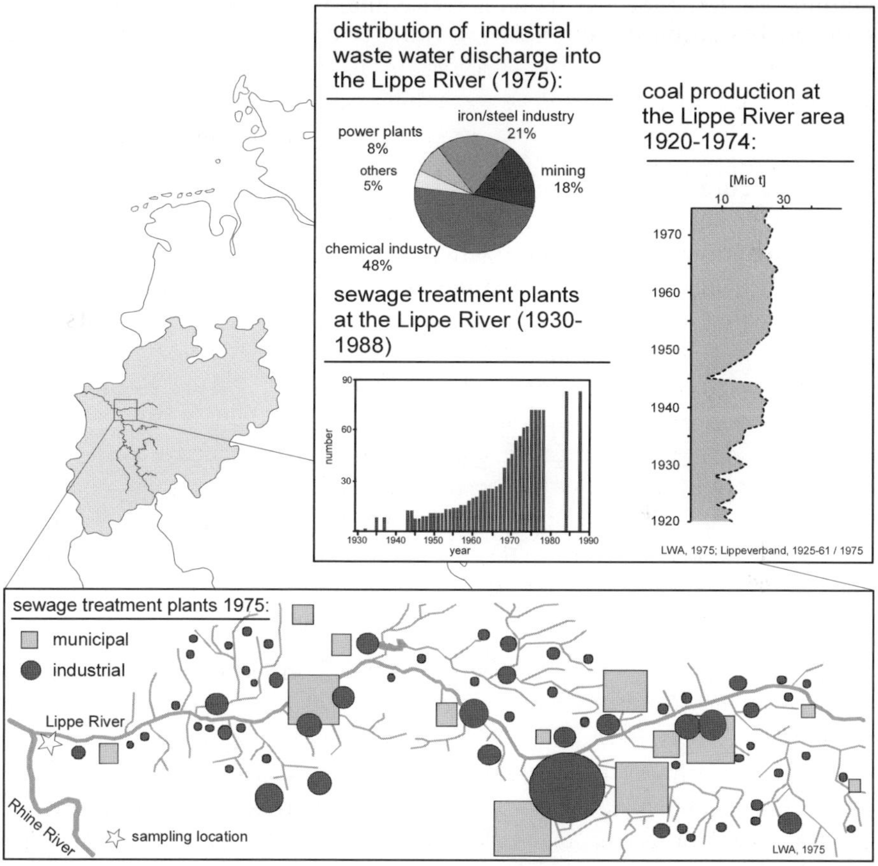


**Fig. 1:** Map of sampling location at the Lippe river (Ruhr District, Germany) (A) exemplary distribution of waste water supply (LWA 1975), coal production at the time period of 1920-1974, and number of sewage treatment plants at the Lippe river (1930-1988); (B) industrial and municipal sewage treatment plants at the Lippe river, indicated by the recorded waste water in proportionate symbol size (Lippeverband, 1975).

Therefore, this study was performed to produce a more comprehensive reconstruction of the pollution history of a riverine system. Based on an earlier, preliminary investigation (Heim et al., 2003), a sediment core derived from a riparian wetland of the Lippe river (Germany) was examined by means of radiological and organic geochemical analyses.

Additionally, several inorganic contaminants were analysed. In the present work, time-resolved concentration profiles of a variety of source specific substances are discussed in detail in order to gain a deeper insight into contamination vs. time trends and, consequently, the pollution history of the Lippe river system.

## Samples

A sediment core from a floodplain of the Lippe river, a tributary of the Rhine river (Germany) was taken in October 2001 (see Fig. 1). The sampling location was characterized as an undisturbed sediment deposit with periodical/aperiodical sedimentation (Klös and Schoch, 1993a).

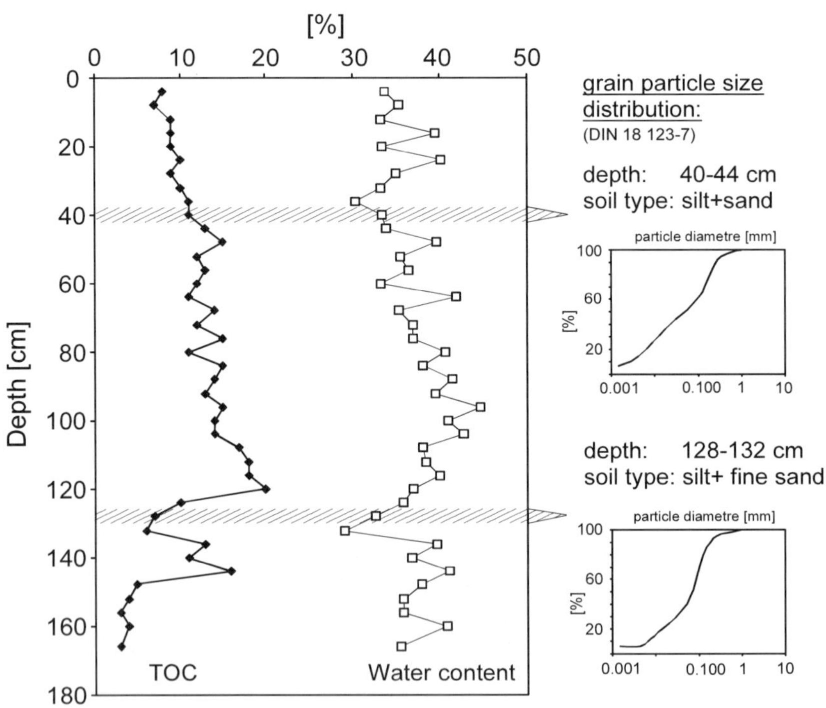


**Fig. 2:** Petrogenic data of the investigated sediment core of the Lippe river wetland; (TOC = Total Organic Carbon)

The core of 166 cm length was divided into 4 cm segments. The fresh material was placed in solvent-cleaned glass flasks and stored at 4°C in the dark. Prior to the organic, inorganic and radiometric analyses, the layer samples were subdivided into several aliquots. The sediment was characterized as loamy fine sand (Fig. 2) with distinct bedding and with evident colour variations among the layers.

## Radiological dating methods

The chronology of the Lippe sediment core was determined using $^{210}$Pb and $^{137}$Cs dating methodologies (Bollhöfer et al. 1994). 40 samples representing the top 166 cm were analysed using a Canberra n-type $\gamma$-spectrometer. Since considerable amounts of excess $^{226}$Ra were observed (see below), the equilibrium activity of $^{210}$Pb and $^{226}$Ra produced by detritic materials was determined via the $^{234}$Th activity. Within the limits of error, a constant value for $^{234}$Th is observed for the core, the weighted average being (1.25 ± 0.03) dpm/g.

## Results

To study the historical input of anthropogenic contaminants into the Lippe river system, organic and inorganic geochemical and radiometric investigations were applied to a sediment core of a corresponding riparian wetland. The sampling location was situated on a frequently flooded area at the downstream end of the Lippe river near the confluence with the Rhine river. Therefore, the particle-bound contaminants deposited on this area represent the accumulated load of the emissions discharged to the overall riverine system.

### *Radiological dating*

Cs-137

From the $^{137}$Cs profile of the core (Fig. 3 A), three time markers can be derived: The beginning of the nuclear weapon tests (1952) in the depth interval from 92-96 cm, the culmination of the nuclear weapon tests (1963-64) in the depth interval from 52-56 cm and the Tschernobyl reactor accident (1986) in the top four cm. This implies, however, that the last 16 years are missing in the core. This may be due to either partial erosion of the sediment or infrequent occurrence or absence of flooding events during the past decade. From the three time markers a mean sedimentation rate of approx. 2.4 cm/a for the lower 54 cm and approx. 3.6 cm/a for the interval of 54-94 cm can be calculated.

Pb-210/Ra-226

The $^{210}Pb$ and $^{226}Ra$-profile shown in Fig. 3B displays a significant excess of $^{226}Ra$. A similar profile was observed by Klös and Schoch (1993a) on a sediment deposit from a nearby location of the same river. The authors explain the excess of $^{226}Ra$ in the sediments by discharge of radium-enriched mine water into the Lippe river, a common feature in mining areas. In comparison, the annual $^{210}Pb$ deposition from the atmosphere is negligible. This assumption is based on the following estimation: Taking a mean sedimentation rate of 3 cm/a, derived from the $^{137}Cs$ profile, an atmospheric flux of $^{210}Pb$ of 0.6 dpm/cm$^2$/a (Dominik et al., 1981, Hemmerich, 1980) and an average dry bulk density of 0.93 g/cm$^3$, we arrive at a specific activity of atmospheric (unsupported) $^{210}Pb$ of approx. 0.2 dpm/g, a value ca. 60 times smaller than the measured value of $^{210}Pb$ in the top layer. We can thus assume a pure radium excess model for the present study:

$$\frac{^{226}Ra(z)-{}^{210}Pb(z)}{^{226}Ra(z)-A(z)} = \exp\left[-\lambda_{210}\left(t_0+\frac{m}{r}\right)\right]$$

where $^{226}Ra$ and $^{210}Pb$ are the measured specific activities at depth $z$, $A$ the specific activity of the equilibrium component of both $^{210}Pb$ and $^{226}Ra$, produced by detrital materials, $\lambda_{210}$ the decay constant of $^{210}Pb$, $t_0$ the age of the top layer, $m$ the cumulative dry mass per unit area and $r$ the mean accumulation rate (Bollhöfer et al., 1994). For $A(z)$ the weighted average of (1.25±0.03) *dpm/g* was inserted. The plot of the ratio on the left hand side of the equation versus $m$ is shown in Fig. 3C. From an exponential fit to the data points, a mean accumulation rate of (3.57±0.27) g/cm$^2$/a and an age of the top layer of $t_0$ = (21±1) *a* is determined. Using the average dry bulk density of 0.93 g/cm$^3$, a mean sedimentation rate of (3.8±0.3) *cm/a* is calculated. Since compaction effects are almost negligible for this core, accumulation and sedimentation rate yield identical ages within limits of error.

The sedimentation rate compares well with the caesium-derived sedimentation rate for the depth interval from 54-94 cm, but is significantly larger than the caesium-derived rate for the upper 50 cm.

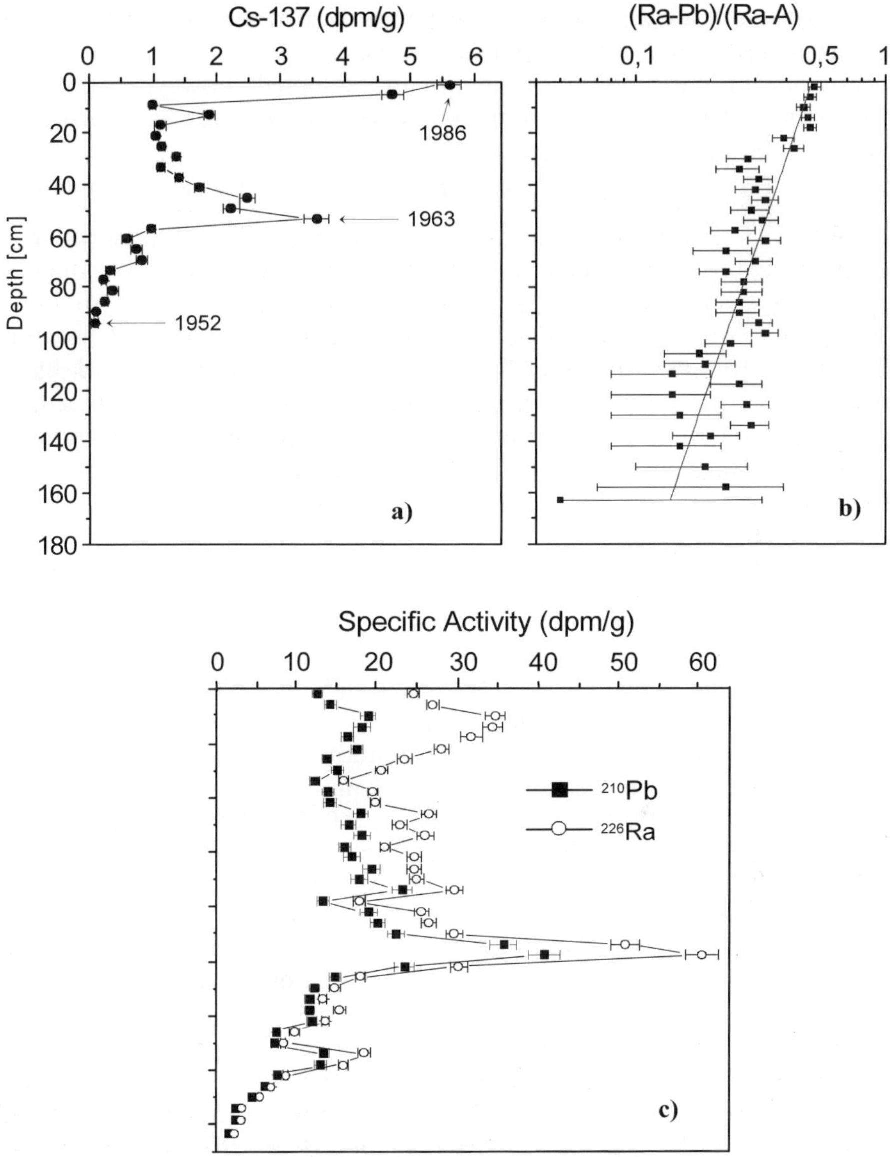


**Fig. 3:** (a) $^{137}$Cs-profile of the sediment core of a riparian wetland of the Lippe river (Germany), (b) $^{210}$Pb/$^{226}$Ra-Profile and (c) the $^{210}$Pb/$^{226}$Ra resulting plot

In addition, the $^{210}$Pb age of the top layer is somewhat higher than the one indicated by the $^{137}$Cs profile. It is possible that the discharged mine water is not only enriched in radium but also transports particle-bound

$^{210}$Pb in variable concentration. This would then be an additional, hardly quantifiable source of $^{210}$Pb in the sediments, which results in a systematic uncertainty in the $^{210}$Pb ages. It should be noted, however, that age discrepancies between the two clocks never exceed five years. We thus conclude that the $^{137}$Cs time markers should be more reliable, but that the derived $^{210}$Pb/$^{226}$Ra chronology can be useful for coarse age estimations below 54 cm.

*Inorganic geochemical analysis*

The inorganic contamination of the sediment core was assessed by a determination of selected heavy metals. According to the screening programme of the ICPR (International Commission for the Protection of the Rhine), the elements cadmium, nickel, chromium, lead, copper and zinc were analysed. The vertical concentration profiles are shown in Fig. 4 including a corresponding time scale.

The concentration profiles of all elements analysed exhibited similar trends, but on different concentration levels. Lead, copper and zinc occurred in significantly higher concentrations as compared to cadmium, nickel and chromium with values ranging from 95 to 400 µg/g for Pb, from 56 to 430 µg/g for Cu and from 517 to 2373 µg/g for Zn. These data correspond to values measured for other study areas on the Lippe river, which are highly influenced by industrial emissions (Poppe et al., 1991). Lower concentrations were determined for Cd (2.7 to 15.7 µg/g), Ni (28 to 62 µg/g) and Cr (34 to 199 µg/g). With respect to the concentration trends, two significant maxima at a depth of approx. 25 cm and between 88 and 115 cm were observed for most of the metals analysed as indicated by the hatched areas in Fig. 4. These horizons are correspond to sedimentation periods from 1975 to 1980 and from 1945 to 1955, respectively.

In detail, most of the concentration profiles reveal increasing values from the bottom of the core up to a depth of 100 cm, reaching a first maximum there. Subsequently, the values decrease to a fairly constant level at a depth between ±80 cm and ±30 cm. The averaged values in this zone are: Cd ±6 µg/g, Ni ±50 µg/g, Cr ±85 µg/g, Cu ±180 µg/g, Pb ±240 µg/g and Zn ±1100 µg/g. A second concentration maximum was observed at a depth of approx. 25 cm. In the top layers of the sediment core the concentrations decrease down to values similar to those in the 30-80 cm zone. These trends are best developed in the concentration profiles of lead, copper, zinc and cadmium (Fig. 4).

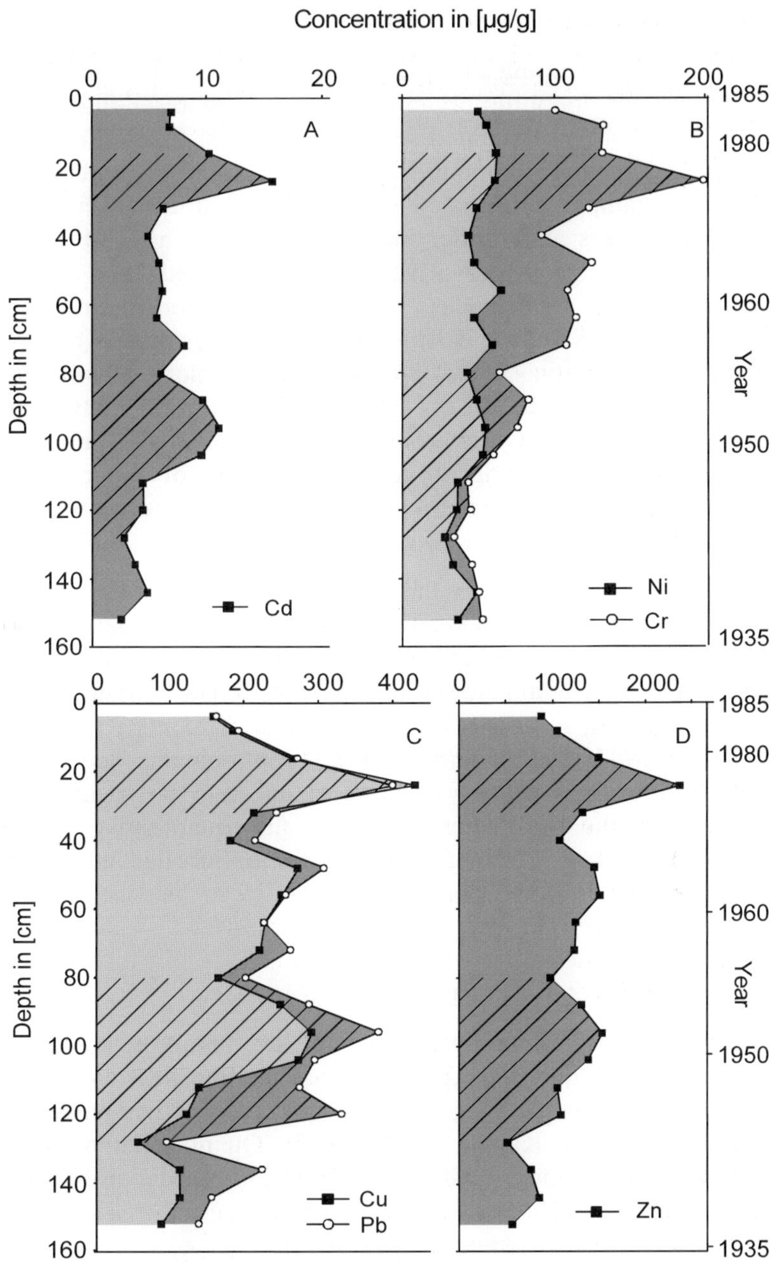


**Fig. 4:** Depth and time correlated concentration range of detected heavy metals cadmium, nickel, chorme, copper, lead and zinc, detected in a undisturbed sediment core of the Lippe river (Germany).

*Organic geochemical analyses*

Detailed GC/MS screening analyses were performed to identify a wide range of persistent lipophilic organic compounds representing the organic contamination of the investigated sediment samples. In addition to a large proportion of biogenic compounds, numerous anthropogenic substances including, for example, chlorinated aromatics, constituents of personal care products, plasticizers and further industrial additives as well as artificial food ingredients (e.g. preservatives) were identified. For quantitative analyses, numerous substances were selected considering different selection criteria: (i) source specifity, (ii) fluvial origin and (iii) appropriate environmental stability. Only a minor proportion of the organic contaminants fulfilled all three criteria. Furthermore, only lipophilic (hydrophobic) compounds were considered in this study in order to avoid misinterpretation as a result of vertical transport of the contaminants associated with pore water. Such vertical transport processes are impeded in particular by strong interactions of compounds with particulate matter.

For most of the substances the riverine origin was confirmed by a recently reported investigation on surface sediments derived from a longitudinal profile of the Lippe river (Kronimus et al., 2004). Also, the attribution of the individual contaminants to different emission sources in particular municipal, industrial and agricultural effluents was established (LWA, 1982/1987/1991). Furthermore, contaminants derived from superposed emission sources were considered, e.g. polycyclic aromatic compounds. In the following sections, the quantitative results on contaminants and the corresponding concentration profiles obtained for the dated sediment layers are presented in detail.

Polycyclic aromatic compounds, PACs

Polycyclic aromatic compounds (PACs) are common environmental pollutants derived either from fossil fuels or as atmospheric contaminants as a result of incomplete combustion processes (pyrogenic sources; Volkman et al., 1997; Kaplan et al., 1997). Quantitative results of the PACs investigated are presented as concentration profiles in Fig. 5A-C or are listed in Tab. 1.

The concentration of the representative pyrogenic PAC benzo[a]pyrene ranged between 81 and 11061 ng/g, with only a few significant concentration maxima (Fig. 5A). The profile shows slowly increasing concentrations from the bottom of the core up to a depth of 120 cm. Subsequently, a significant maximum at the depth of 102 cm and a second maximum value at 86 cm are recorded. Above this level the concentrations decrease to an average of ± 1400 ng/g.

Benz[a]anthracene and benzo[x]fluoranthene (x= b,k] were detected at elevated concentrations, 280 - 19920 ng/g and 417 - 36116 ng/g, respectively (Fig. 5B and 5A). The concentration profile of benzo[x]fluoranthene (x = b,k) is very similar to the one of benzo[a]pyrene. In contrast, the concentration profile of benz[a]anthracene exhibited a broad maximum between 120 cm and 60 cm depth rather than two distinct but narrow maxima.

With respect to further PACs analysed in the sediment core (see Tab. 1), concentration profiles very similar to the one of benz[a]anthracene were detected for most of the substances. This group of polycyclic aromatic hydrocarbons (PAHs) include acenaphthylene, anthracene, chrysene, fluoranthene, phenanthrene, fluorene and pyrene. Only for perylene the distribution within the sediment core resembled the ones of benzo[a]pyrene and benzo[x]fluoranthene (x = b,k).

As common sulphur-containing aromatic compounds, dibenzothiophene and benzo[b]naphtho[2,1-d]thiophenes were determined as illustrated in Fig. 5 C and 5B. Interestingly, the concentration profiles of both contaminants differ significantly. Dibenzothiophene was detected with concentration values between 66 and 2939 ng/g. The first occurrence at a depth of 152 cm can be attributed to a period around 1940. The concentration profile shows relatively consistent values from the bottom to the top of the core with two local maxima at a depth of 142 cm and at a depth range between 62 and 38 cm. In contrast, the concentrations of benzo[b]naphtho[2,1-d]thiophenes are very low in the bottom sediment layers (170 to 135 cm), but increase rapidly towards the depth of 110 cm with values up to approx. 4400 ng/g. This high concentration level persists to a depth of 80 cm followed by a slight decrease towards the top layers with concentrations around 2000 ng/g. Interestingly, also acenaphtylene (Tab. 1) was detected with a similar concentration profile.

In addition, nitrogen- and oxygen-containing PACs were analysed comprising dibenzofurane, carbazole and benzocarbazole. Dibenzofurane was found in high concentrations of up to 6000 ng/g, whereas the nitrogen containing compounds occur at a lower concentration level between 10 to 170 ng/g. The concentration profiles of the nitrogen- and oxygen-containing PACs match very well the distribution pattern observed for most of the PAHs, e.g. benz[a]anthracene.

**Tab. 1:** Concentration values of quantified PACs, determined in the dated sediment core of the Lippe river, Germany (n.q. = not quantified).

| **Depth** | Acenaphthylene | Ace-naphthene | Fluorene | Phenanthrene | Anthracene | Fluoranthrene | Pyrene |
|---|---|---|---|---|---|---|---|
| [cm] | [ng/g] | [ng/g] | [ng/g] | [ng/g] | [ng/g] | [ng/g] | [ng/g] |
| 2 | 65 | 1 | 438 | 742 | 387 | 878 | 558 |
| 6 | 92 | 1 | 573 | 1125 | 515 | 1245 | 773 |
| 14 | 235 | 20 | 514 | 895 | 569 | 1209 | 640 |
| 22 | 255 | 7 | 681 | 943 | 767 | 1221 | 716 |
| 30 | 143 | 3 | 613 | 886 | 450 | 1021 | 498 |
| 38 | 127 | 112 | 766 | 1533 | 890 | 1048 | 778 |
| 46 | 480 | 1462 | 1447 | 3256 | 1881 | 1806 | 1467 |
| 54 | 241 | 195 | 1159 | 2664 | 1122 | 2697 | 1490 |
| 62 | 251 | 24 | 585 | 953 | 628 | 1082 | 637 |
| 70 | 178 | 2 | 790 | 1191 | 757 | 1682 | 993 |
| 78 | 334 | 10 | 626 | 1014 | 868 | 1743 | 902 |
| 86 | 230 | 31 | 878 | 1401 | 1197 | 2129 | 1192 |
| 94 | 362 | 10 | 1253 | 1490 | 916 | 1938 | 1091 |
| 102 | 397 | 10 | 817 | 1304 | 1135 | 2536 | 1115 |
| 110 | n.q. | n.q. | n.q. | n.q. | n.q. | n.q. | n.q. |
| 118 | 467 | 45 | 1015 | 1237 | 1018 | 2570 | 828 |
| 126 | 203 | 5 | 473 | 612 | 486 | 1382 | 552 |
| 134 | 274 | 95 | 732 | 861 | 882 | 905 | 589 |
| 142 | 14 | | 108 | 283 | 59 | 325 | 174 |
| 150 | 20 | 1 | 100 | 282 | 56 | 353 | 159 |
| 158 | 11 | 1 | 41 | 121 | 19 | 219 | 58 |
| 166 | 3 | | 8 | 19 | | 60 | 10 |

**Tab. 1** continued

| **Depth** | Chrysene | Benzo[e]pyrene | Perylene | Dibenzofuran | Dibenzothiophen | Carbazole | Benzocarbazole |
|---|---|---|---|---|---|---|---|
| [cm] | [ng/g] | [ng/g] | [ng/g] | [ng/g] | [ng/g] | [ng/g] | [ng/g] |
| 2 | 6022 | 3260 | 984 | 618 | 181 | 54 | 38 |
| 6 | 6332 | 2034 | 502 | 1624 | 71 | 76 | 94 |
| 14 | 3078 | 1484 | 608 | 1423 | 271 | 90 | 77 |
| 22 | 7414 | 3374 | 947 | 2050 | 385 | n.q. | n.q. |
| 30 | 5385 | 1205 | 298 | 1917 | 223 | 95 | 48 |
| 38 | 4116 | 795 | 284 | 2551 | 674 | 156 | 55 |
| 46 | 5136 | 883 | 206 | 6315 | n.q. | 160 | 50 |
| 54 | 6595 | 2381 | 542 | 4096 | 2939 | 171 | 88 |
| 62 | 5119 | 2120 | 380 | 2346 | 1392 | 87 | 53 |
| 70 | 11110 | 2210 | 395 | 3514 | n.q. | 86 | 60 |
| 78 | 7497 | 7362 | 922 | 2452 | 414 | 59 | 94 |
| 86 | 10574 | 13156 | 2223 | 3880 | n.q. | 130 | 92 |
| 94 | 10570 | 2363 | 657 | 5070 | 505 | 103 | 107 |
| 102 | 8592 | 12119 | 3000 | 3259 | 66 | 93 | 90 |
| 110 | n.q. | n.q. | n.q. | n.q. | 349 | 143 | 146 |
| 118 | 7376 | 3754 | 837 | 3401 | 343 | 113 | 130 |
| 126 | 4120 | 1971 | 479 | 1522 | 200 | 59 | n.q. |
| 134 | 4156 | 824 | 202 | 2634 | 201 | 107 | 97 |
| 142 | 1284 | 321 | 80 | 667 | 1978 | 19 | 16 |
| 150 | 1483 | 401 | 85 | 480 | | | n.q. |
| 158 | 888 | 246 | 47 | 176 | | 11 | 7 |
| 166 | 350 | 117 | 28 | 17 | | 6 | 8 |

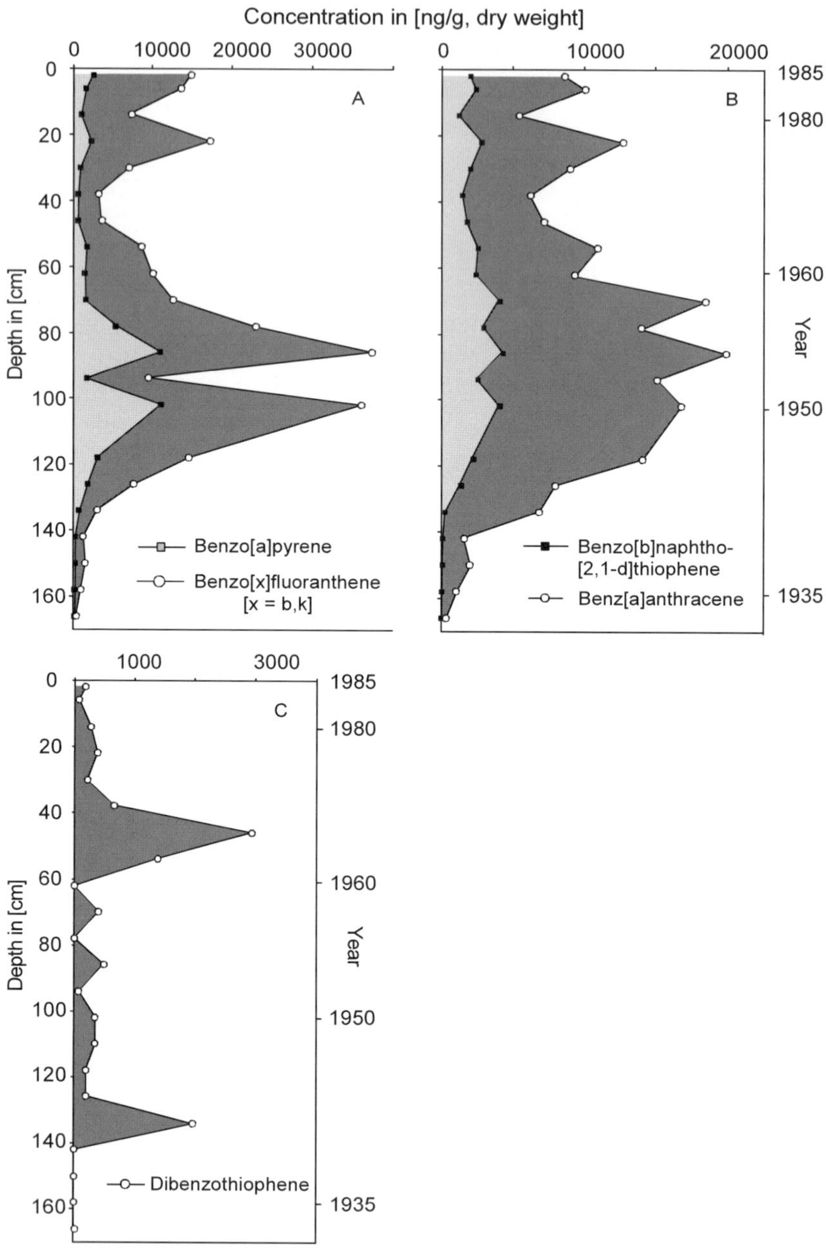


**Fig. 5 A-C:** Depth- and time-correlated concentration profiles of detected PACs benzo[a]pyrene, benz[a]anthracene, benzo[x]fluoranthene (with x=b,k), dibenzothiophene and benzo[b]naphtho[2,1-d]thiophene; determined in dated sediments of a Lippe river wetland (Germany).

PCBs, Tetrachlorobenzyltoluenes and polychlorinated aromatic compounds

Polychlorinated biphenyls (PCBs) are well known persistent pollutants ubiquitously distributed in the aquatic environment (Takada and Eganhouse, 1998). Therefore, PCBs as well as their industrial substitutes tetrachlorobenzyltoluenes (TCBT, commercial product: Ugilec® 141) were both formerly identified in the Lippe river sediments (Kronimus et al. 2004, Poppe et al., 1991; LUA, 2000).

In this study, seven PCB congeners (PCB 8, 28, 52, 101, 153 and 180) were quantified. Fig. 6A illustrates the concentration profiles of the PCBs in sum (ΣPCB), Fig. 6B shows the data for the PCB substitute TCBT. The concentrations of ΣPCB range between 1 and 2632 ng/g. The first occurrence of PCBs was detected at a depth of 150 cm with very low values. Subsequently, the concentrations increase slowly up to values between 100 and 300 ng/g (38 cm layer). A significant increase of the concentration level up to a maximum of 2632 ng/g was observed between the depth of 30 and 6 cm, followed by a distinct decrease towards the surface. The concentration ranges of the individual PCB congeners are different. PCB 8, 28 and 180 occur in concentrations up to 100 ng/g, whereas the congeners PCB 52, 101 and 153 predominate with concentrations up to 1580 ng/g per individual compound. Nevertheless, all corresponding concentration profiles exhibit similar trends. In contrast to the occurrence of the PCBs, the concentration profile of TCBT shows a continuous increase from its first occurrence at a depth of 70 cm up to the top of the core with a concentration maximum of 13421 ng/g. These results confirm former data on Lippe river sediments analysed in 1987 and 1989 (Poppe et al., 1991).

Hexachlorobenzene represents both, industrially derived contaminations and agricultural emissions. The concentration profile shows two distinct maxima (Fig. 6C) with values of 10 and 6500 ng/g. At the bottom of the core the concentrations are very low, but a significant increase towards a depth of 118 cm was observed. Subsequently, a second local maximum (94 cm) occurs, followed by rapidly decreasing values (±1743 ng/g). Up to the top of the core the concentrations stagnate at values of around 1500 ng/g.

Octachlorostyrene is a very specific industrial pollutant. The emission trend, as revealed by the concentration profile, is similar to hexachlorobenzene. There are two concentration maxima at a depth of 118 and 94 cm, respectively (Tab. 2). In the shallower section there is a decrease down to average values of ±280 ng/g between 54 und 14 cm and up to values of ± 100 towards the surface layer.

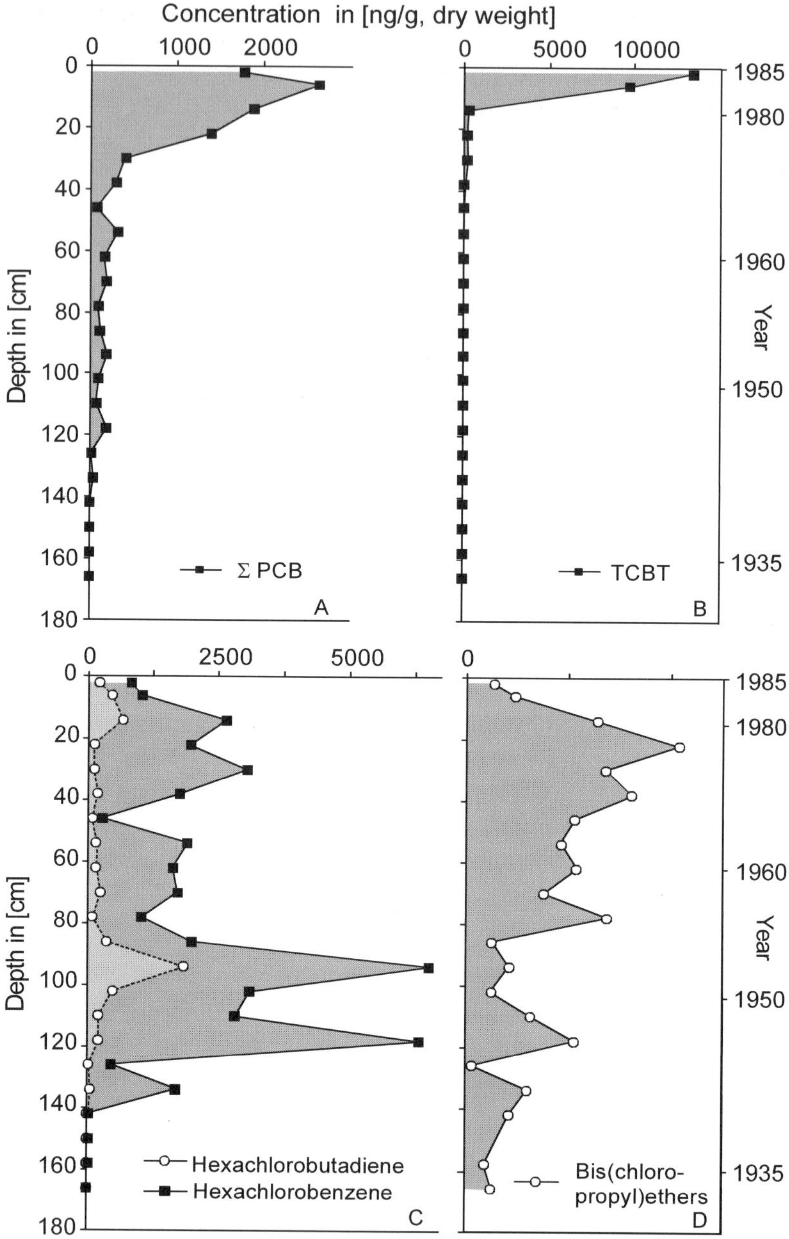


**Fig. 6 A-D:** Depth- and time-correlated concentration ranges of detected industrial markers ΣPCB, TCBT, hexachlorobenzene, hexachloro-1,3-butadiene and bischlorobutylethers (2 isomers) in dated sediments of a Lippe river wetland.

**Tab. 2:** Concentration values of quantified compounds with different emission sources: linear alkylbenzenes (LAB), bis(-4-octylphenylamine) (Vanlube 81 ®), organochlorine compounds and limonene, determined in dated sediments of the Lippe river wetland (Germany). (n.q. = not quantified)

| Depth | $\Sigma$ $C_{10}$-LAB | $\Sigma$ $C_{11}$-LAB | $\Sigma$ $C_{12}$-LAB | $\Sigma$ $C_{13}$-LAB | Bis(4-octyl-phenylamine) | Octachloro-styrene | Limonene |
|---|---|---|---|---|---|---|---|
| [cm] | [ng/g] | [ng/g] | [ng/g] | [ng/g] | [ng/g] | [ng/g] | [ng/g] |
| 2 | 17 | 77 | 85 | 55 | 149 | 89 | 8 |
| 6 | 23 | 94 | 99 | 67 | 9 | 106 | 11 |
| 14 | 37 | 154 | 158 | 103 | 1 | 294 | 14 |
| 22 | 38 | 132 | 115 | 63 | | 259 | 13 |
| 30 | 24 | 56 | 65 | 35 | 1 | 342 | 7 |
| 38 | 31 | 112 | 82 | 45 | | 213 | 11 |
| 46 | n.q. | n.q. | n.q. | n.q. | n.q. | n.q. | 1 |
| 54 | 19 | 28 | 18 | 10 | | 208 | 19 |
| 62 | n.q. | n.q. | n.q. | n.q. | | 97 | 12 |
| 70 | 22 | 22 | 33 | 26 | | 93 | 7 |
| 78 | 12 | 29 | 67 | 47 | | 50 | 10 |
| 86 | 49 | 66 | 126 | 83 | | 131 | 12 |
| 94 | 54 | 87 | 203 | 154 | | 776 | 11 |
| 102 | 32 | 113 | 192 | 109 | | 443 | 6 |
| 110 | 40 | 33 | 45 | 19 | | 202 | 7 |
| 118 | 19 | 12 | 17 | 10 | | 639 | 5 |
| 126 | 6 | 2 | 1 | 1 | | 31 | 2 |
| 134 | 21 | 13 | 5 | 1 | | 81 | 9 |
| 142 | 1 | 1 | 1 | | | 2 | 2 |
| 150 | 2 | 1 | 1 | | | 2 | 3 |
| 158 | 1 | 1 | 1 | | | 1 | 2 |
| 166 | | | | | | 1 | 1 |

**Tab. 2**. continued

| Depth | Dichloro-carbazole | Pentachloro-butadiene | Dichloro-benzene | Trichloro-benzene | Tetrachloro-benzene | Pentachloro-benzene |
|---|---|---|---|---|---|---|
| [cm] | [ng/g] | [ng/g] | [ng/g] | [ng/g] | [ng/g] | [ng/g] |
| 2 | 16 | 47 | 88 | 21 | 11 | 125 |
| 6 | 32 | 81 | 175 | 28 | 18 | 202 |
| 14 | 8 | 262 | 282 | 64 | 56 | 493 |
| 22 | n.q. | 181 | 248 | 40 | 47 | 588 |
| 30 | | 396 | 95 | 26 | 48 | 903 |
| 38 | | 584 | 211 | 47 | 72 | 919 |
| 46 | | 406 | n.q. | n.q. | n.q. | n.q. |
| 54 | | 454 | 153 | 43 | 46 | 492 |
| 62 | | 82 | 113 | 43 | 24 | 475 |
| 70 | | 27 | 47 | 37 | 18 | 553 |
| 78 | | 83 | 110 | 42 | 20 | 463 |
| 86 | | 307 | 174 | 146 | 47 | 964 |
| 94 | | 218 | 211 | 294 | 199 | 5663 |
| 102 | | 98 | 76 | 116 | 55 | 1636 |
| 110 | | 106 | 83 | 18 | 13 | 671 |
| 118 | | 14 | 18 | 5 | 11 | 1083 |
| 126 | | 17 | 6 | 2 | 2 | 66 |
| 134 | | 7 | 49 | 11 | 6 | 210 |
| 142 | | 9 | 3 | 2 | 1 | 7 |
| 150 | | 2 | 5 | 2 | 2 | 14 |
| 158 | | 1 | | 1 | 1 | 8 |
| 166 | | | | | | 2 |

Hexachloro-1,3-butadiene is another representative of emission of industrial origin (Fig. 6C). The concentrations determined range between 3 and 1900 ng/g with a maximum value at a depth of 94 cm. Below and above this layer, similar average values of ±170 ng/g were detected. Furthermore, a second concentration maximum at a depth of 18 cm is followed by slowly decreasing values towards the top layer.

(1-Chloro-2-propyl-2-chloro-1-propyl)ether was the last industrial marker compound analysed in 2 isomers. The concentration profile (Fig. 6D) exhibits a generally increasing trend from the bottom up to a depth of 22cm with concentrations between 7 and 210 ng/g. Subsequently, a significant decrease towards the top of the core was detected.

Linear alkylbenzenes, LABs

Linear alkylbenzenes (LABs) are alkylated benzenes with linear $C_{10}$ – $C_{14}$ alkyl chains. As a residue of detergents, LABs reach the river system mainly by municipal sewage and waste water. The systematic change in their isomeric composition is used as an indicator for the extent of microbial degradation. For this purpose, two types of isomers are considered: (i) the external isomers (benzene ring attached to the end of the alkyl chain or at the 2- or 3-position) and (ii) the internal isomers (benzene ring attached in the middle of the alkyl chain). The ratio of internal to external isomers (I/E-ratio) has been proposed as an index of the degree of selective microbiological degradation. Thus, higher I/E ratios indicate a higher microbial degradation (Takada and Eganhouse, 1998). The concentration profiles of several $C_{10}$ – $C_{13}$ linear alkylated benzenes were determined as presented in Tab. 2. As an example, the concentration profiles of the external $C_{12}$-LABs are represented in Fig. 7A, together with the calculated I/E-ratios. The concentrations of the isomers detected (Σ 5-$C_{12}$- and 6-$C_{12}$-LABs) range between 1 and 62 ng/g. The concentration profile reveals two distinct maxima at a depth of 78 cm and 14 cm, respectively. Between both maxima concentrations are in the range of approx. 25 ng/g. Much lower concentrations - between the quantitation limit and approx. 5 ng/g – are encountered in the lower part of the profile. The I/E-ratios (black dots in Fig. 7A) within the sediment profile range between 0.4 and 0.8 with higher ratios in the deeper layers.

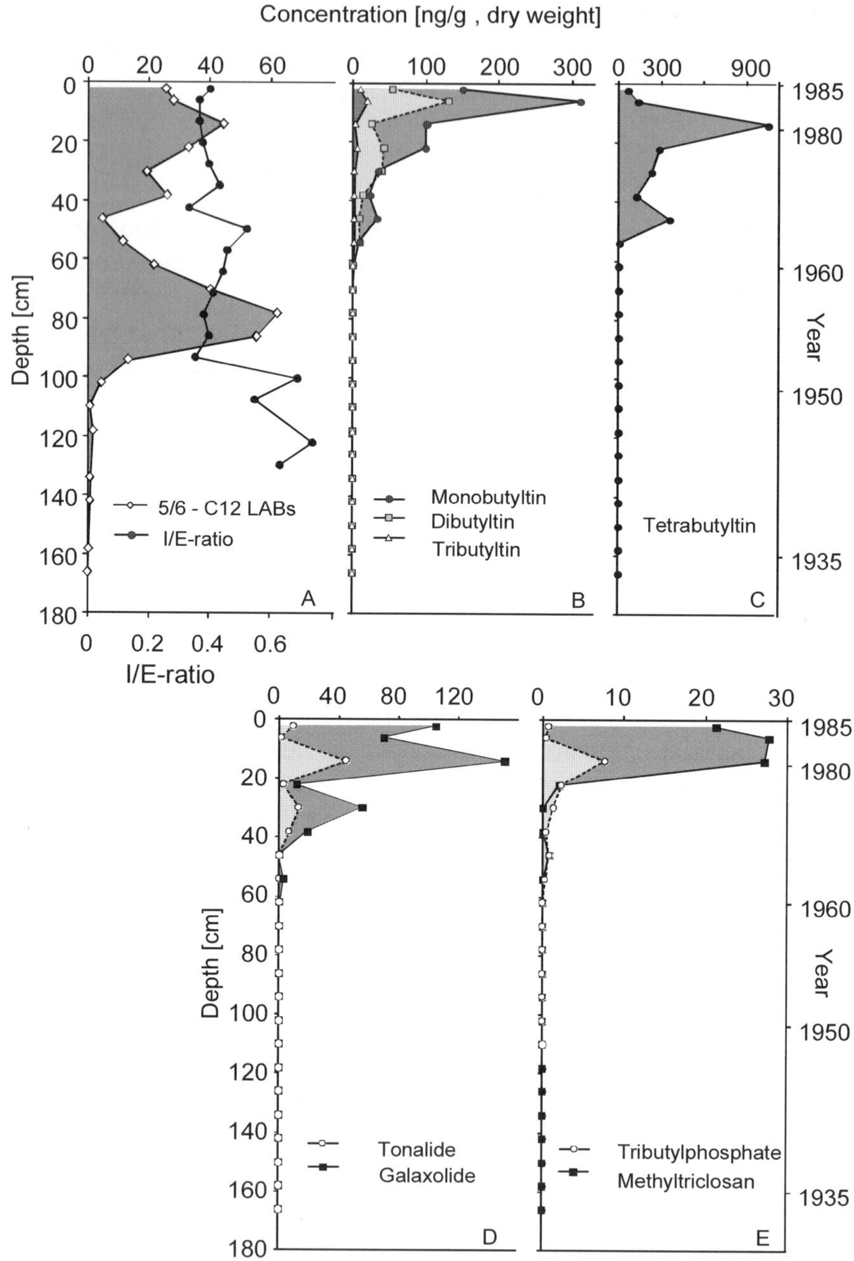


**Fig. 7 A-E:** Depth- and time-correlated concentration ranges of LABs, organotin compounds, synthetic musks, tributylphosphates and methyltriclosan, determined in dated sediments of a Lippe river wetland (Germany).

Organotin compounds

In addition to tetrabutyl tin, the key chemical material of butylated organotin compounds, also mono-, di- and tributyl tin (TBT) were detected. TBT reaches the Lippe river system by industrial waste water as well as emissions from anti-fouling agents used in marine paints for shipping and fish cages. The concentration profiles generally reveal an onset of contamination in younger sediment levels above a depth of 62 cm. Increasing values up to a maximum peak at a depth of 6 and 14 cm (Fig. 7B and 7C) are followed by significantly lower concentrations in the top layer. The concentrations were found to range between 10 - 1051 ng/g for tetrabutyl tin, 1 - 19 ng/g for tributyl tin, 9 - 130 ng/g for dibutyl tin and 2 - 310 ng/g for monobutyl tin.

Galaxolide / tonalide

The musk fragrances galaxolide (1,3,4,6,7,8-hexahydro-4,6,6,7,8,8-hexamethyl-cyclo-penta-(g)-2benzopyrane, HHCB) and tonalide (7-acetyl-1,1,3,4,4,6-hexamethyltetraline, AHTN) enter fluvial systems mainly by municipal emissions. The concentration profiles of both synthetic musks (Fig. 7D) exhibit very similar concentration trends with only little variation. Galaxolide and tonalide were not detected in the lower horizons. The first occurrences of both compounds were encountered at a depth of 54 cm with concentration levels increasing towards the top. The maximum values were determined in the 14 cm layer with 151 ng/g and 44 ng/g, respectively.

Methyltriclosan

Methyltriclosan is a rarely reported contaminant in riverine sediments. It enters the fluvial environment via municipal effluents. The concentrations detected in the sediment core investigated range between 1 and 28 ng/g. The first occurrence of methyltriclosan was detected at a depth of 46 cm with very low values (Fig. 7E). A concentration maximum is reached at a depth between 14 and 6 cm.

Tributylphosphate

The last anthropogenic marker compound analysed in this study was tributylphosphate (TBP). TBP enters the aquatic environment by municipal as well as industrial sewage effluents . The concentration trend illustrated in Fig. 7E shows a high similarity with the one of methyltriclosan. The first occurrence of tributylphosphate was detected in the 54 cm layer, followed by low concentrations up to 1 ng/g without any significant variation. Towards the top layers (14 cm), rapidly increasing concentrations, up to 7.8 ng/g, were observed.

## Discussion

A combination of radiological and geochemical analyses of a sediment core derived from a riparian wetland of the Lippe river was used to determine the deposition and emission periods of individual contaminants. The layers of the investigated sediment core cover a sedimentation period from 1935 to 1986. Given an appropriate environmental stability, the concentration profiles detected for inorganic and in particular organic compounds thus reflect the emission of the contaminants to the corresponding riverine system. Consequently, variations in riverine contamination loads were reflected in the sediment core by varying concentrations.

The Lippe river contains an extensive catchment area with numerous municipal, industrial and agricultural emission sources. With the beginning of mining activities in the Lippe river area around 1900, the waste water discharge increased significantly. Because of obvious damage caused to the environment, sewage treatment plants were installed since 1930. Altogether, 83 municipal and industrial sewage treatment plants were put into service until 1988 (Fig. 2, LWA, 1989). Nevertheless, extensive investigations into the organic and inorganic content of the waste water discharge into the river did not start until the 1980s. Documentations and specific environmental data are very rare prior to 1980. Therefore, the interpretation dates determined in this study are based on more or less coherent contemporary documents such as reports on water quality, published by several water authorities (e.g. Lippeverband) and the Environmental Agency of North-Rhine Westphalia (LUA-NRW). Nevertheless, determined concentration values of the investigated components are comparable with data published earlier (e.g. Klös and Schoch, 1993a/b).

In the following sections, the time-correlated concentration profiles are discussed with respect to the main pollution sources (industrial, municipal and agricultural emissions) in order to provide a comprehensive view on the pollution history of the Lippe river.

*Mining activities*

The pollution discharge of numerous heavy metals and specific organic compounds in this area could be attributed to mine-water drainage and sewage water from coking plants and mining industry, as documented already by numerous former studies (e.g. LWA, 1982/1987/1991; LUA, 2000, Lippeverband 1975; Klös and Schoch, 1993a/b; see Fig. 2).

Heavy metals are biologically and chemically non-degradable and therefore their persistence reveals not only their ecological and ecotoxicological relevance but characterizes them as stable marker substances. Significant maximum concentrations of all heavy metals investigated were correlated to an emission time between 1945 and 1955. This pollution trend can be associated with increasing mining activities after the Second World War and related mine water drainage. The results and evaluations are consistent with results published by Klös and Schoch (1993a). Decreasing heavy metal contaminations and a reduction of pollution levels in the 30-80 cm layers corresponding to 1955 and 1975 were correlated with: (i) the stepwise closure of Pb/Zn-mines, (ii) the reduction of mine water contributions derived from the coal mining industry (Klös and Schoch, 1993a), and (iii) the increased river development and new sewage treatment plants (LWA, 1991). Furthermore, high concentrations for the sedimentation period from 1975 to 1980 were associated with renewed increase of mining waste water and industrial sewage discharge, as well as increasing car traffic and rising discharge by heat generation and production of electricity (Klös and Schoch, 1993a).

However, anthropogenic emissions can also be traced to other sources than mining. This can be reflected in more detail by specific organic marker compounds. Most of these compounds have entered the list of priority pollutants due to their biological and chemical persistence (Takada and Eganhouse, 1998). This reveals not only their ecological and ecotoxicological relevance but characterizes them as stable marker substances.

Polycyclic aromatic compounds are common constituents of fossil fuels. Due to the intensive usage of petroleum, coal and oil related products; these compounds are released in high amounts into the aquatic environment. Additionally, PACs enter the aquatic system and corresponding flood plains after formation during incomplete combustion processes and subsequent atmospheric deposition (Sanders et al., 1993). Perylene and benzo[a]pyrene were analysed as representative PAHs of pyrogenic origin (Volkman et al., 1997). But, there was no clear assignment, based on fingerprints of significant PAHs, to make an obvious statement regarding their origin (fossil fuel or atmospheric input).

Generally, all PACs exhibit concentration profiles with only minor variations. A significant accumulation was observed within the time period between 1947 and 1960. This contamination trend exhibits a high similarity to those of the mining-derived heavy metals, as described above. Thus, a close correlation of the PAC emission with mining activities in particular with the coal mining industry can be inferred. Another contamination source of PACs in the investigated sediments is a refinery, situated close to the sampling location. Decreasing contamination trends in PAC-contaminations are attributed to improved sewage treatment plants (Lippeverband, 1975).

Further indicators of mining activities are polychlorinated biphenyls (PCBs), due to their extended usage in hydraulic oil in the mining industry (Poppe et al., 1988). The variety of their technical applications (e.g. in capacitors, transformer oils and high-pressure lubricants) and resulting emission sources associate a distinct correlation of PCB contamination with local mining activities. In the 1980'ies the mining industry substituted PCBs gradually by tetrachlorobenzyltoluenes (TCBT, commercial product: Ugilec® 141). But the ecotoxicological behaviour of TCBTs is very similar to PCBs (Poppe et al., 1991). Because of their toxicity, persistence and bioaccumulation TCBTs have been included into the list of priority pollutants and their use was also restricted since 1990.

The substitution of PCBs by TCBTs is clearly reflected in the sediment core investigated (Fig. 6A and 6B). The first appearance of TCBT coincides with the decrease of PCB concentrations since the mid 1970´s. However, the contamination history of PCB i.e. their distribution within the sediment core differs significantly from that of the mining markers discussed above. In particular, the sharp increase of concentrations since 1975 is interpreted as a superimposition of further emission sources.

*Industrial emissions*

Further halogenated compounds identified in the sediment core specifically reflect the influence of industrial activities on the pollution level of the Lippe river in time. In detail, $Cl_2$- to $Cl_6$-benzenes, octachlorostyrene, hexachloro-1,3-butadiene and bis(chloropropyl)ethers were quantified. A distinct association to emissions derived from industrial areas situated at the lower river section has already been proposed by Kronimus et al. (2004).

Chlorinated benzenes are key precursors for the industrial production of colours, pesticides, rubber products and disinfectants. Additionally, hexachlorobenzene (HCB) was used for a long time as a pesticide. Octachlorostyrene is a specific by-product of the production of magnesium, as well as a by-product of the technical production of chlorine (Kaminski and Hites, 1984). Hexachloro-1,3-butadiene is used in the rubber production, as hydraulic liquid, solvent and as a by-product of the technical synthesis of chlorinated compounds like tetrachloroethene (Booker and Pavlostathis, 2000). Bis(chloropropyl)ethers represent by-products of the technical synthesis of epichlorhydrin, used for the production of epoxy resins and disinfectants. The concentration profile of the individual chlorinated compounds reveals different emission histories.

For the chlorinated benzenes, a very similar distribution within the sediment core is observed as for some PAHs, e.g. benzo[a]pyrene. An elevated large-scale industrial activity related to these compounds can be deduced for the time between 1947 and 1955. We attribute the decrease in contamination towards the top layers to a reduction of emissions as a result of more efficient sewage treatment plants (Fig. 1A,B) as well as a modified array of products. The concentration profile of HCB (Fig. 6C) and all lower chlorinated benzenes (Tab. 2) suggests the dominance of industrial sources responsible for the contamination as contrasted to agricultural emission derived from pesticide usage. It should be noted that the contamination level of 1,4-dichlorobenzene was elevated in the time period between 1975 and 1980, comparable with concentration levels determined in Rhine river sediments 1982/83. The extensive use of 1,4-dichlorobenzene as an odorous ingredient of toilet cleaners contributed additionally to the contamination via sewage effluents (LWA, 1987/1989).

The vertical distribution of octachlorostyrene agrees very well with that of the chlorinated benezenes, reflecting a similar industrial source and usage. With respect to hexachlorobutadiene and its de-chlorinated congener penatchlorobutadiene, slight differences within the contamination history were observed. Concentrations of hexachlorobutadiene maximise in a narrow time interval around 1955. A second maximum was detected near the top of the core. The concentration and correlated contamination profile of bis(chloropropyl)ethers reflect a different load history. The emissions increased continuously from 1945 to 1978, but were significantly reduced thereafter. The recently decreasing pollution level is probably due to improved sewage treatment processes.

Apart from chlorinated compounds, also butyl tin compounds reflect the contribution of industrial effluents to the contamination of the Lippe river system. Tributylated tin compounds enter the aquatic environment due to its migration from antifouling paints. Sources of di- and monobutylated tin compounds are municipal and industrial waste water, sewage sludge, and landfill leaks (Fent and Hunn, 1991). The production and usage in particular of tributyl tin (TBT) was restricted in Germany since 2001. The major contribution to the contamination of the Lippe river system can be attributed to an organotin-consuming and -producing industrial plant (Kronimus et al., accepted). All butylated tin compounds are dominated by tetrabutyl tin, the synthetic precursor for the preparation of all lower butylated substances. This pattern was confirmed as a typical industrial signature of riverine contamination e.g. by Poppe et al. (1991).

Butyl tin compounds were detected in the core layers representing an input time since approx. 1970. This observation agrees very well with the initiation of a large scale production in the area investigated. The concentration profile reveals a significant increase to the upper layers until approx. 1980. However, thereafter a sharp decrease in the top layer was observed corresponding to a similar contamination trend in all other industrial markers.

*Municipal effluents*

As one of the most important pollution sources of riverine systems, municipal effluents have been considered in this study by analyses of linear alkylbenzenes (LABs), limonene, methyltriclosan, synthetic musk fragrances, tributylphosphate and phosphate-based flame retardant. All compounds are known constituents of effluents of sewage treatment plants (Takada and Eganhouse, 1998; and references therein) and were already identified in surface sediments of the Lippe river (Kronimus et al., 2004). Interestingly, some of these compounds occur exclusively since 1970.

LABs are raw materials for the industrial production of linear alkylbenzenesulfonates (LASs), the most widely used anionic surfactants. Large scale production and usage of LAS type detergent started in the early 1960s as biodegradable substitutes of the non-biodegradable tetrapropylen-based alkylbenzenesulfonate (ABS). In the sediment core of the Lippe river floodplain, alkylbenzenes with linear $C_{10}$ - $C_{13}$-alkyl chains were detected since 1950 (Fig. 7A), with rapidly increasing concentrations until 1960, representing the ABS/LAS-substitution (Reiser et al., 1997). The subsequent decrease in contamination has to be attributed to improved sewage treatment techniques in the time period until 1960 in this region. In order to avoid misinterpretations due to degradation processes affecting these more biodegradable compounds, the I/E-ratios approach (see chapter 3.1) was applied. The calculated ratios indicate a slowly increasing degradation towards the bottom of the sediment core (Fig. 7A). Hence, using the more stable internally substituted isomers for concentration profiles, a realistic view on the contamination history was achieved.

As further municipal contaminants, synthetic musk fragrances, galaxolide (HHCB) and tonalide (AHTN), were determined. They are used widely in soaps, cosmetics and laundry detergents (Rimkus et al., 1997). Also methyltriclosan, the methylated metabolite of triclosan, commonly used as an antibiotic or disinfectant in personal care products, is regarded as a common environmentally stable sewage effluent indicator (Lindström et al., 2002, Kronimus et al., 2004). In addition, tris(chloropropyl)phosphates (TCPP), belonging to the group of technical flame retardants and mainly used in polyurethane foams and discharged by municipal effluents, were determined. Tributylphosphate (TBP) can be assigned to both municipal and industrial effluents, because of its usage as plasticizer and additive in cellulose-varnish as well as constituent during cellulose production, and, to a minor extent, as solvent for rare earths (Fries and Püttmann, 2001).

In contrast to common contaminations like heavy metals, PACs and LABs, we classify butylated tin compounds, synthetic musk fragrances, methyltriclosan, TBP, TCPP and some chlorinated compounds as "modern contaminants" (Fig. 7B-E, Tab. 2). They first occur in the sediment layers in 1975. Subsequently, sharply increasing concentrations levels were determined with compound-specific variations towards the top layer. With respect to their increasing production rates and usage, their reduced concentration in the top layer implies an enhanced elimination as the result of improved sewage treatment processes.

*Miscellaneous*

Further contaminants detected in the sediment core could not be definitely attributed to individual emission sources, because the knowledge about their environmental occurrence and behaviour is very limited. Examples include the high-viscosity lubricant bis(-4-octylphenyl)amine (Vanlube 81 ®), 3,6-dichlorocarbazole and limonene. The quantification of limonene revealed a very uniform concentration profile, which cannot be attributed to any other contamination history discussed so far. In contrast, the occurrence of 3,6-dichlorocarbazole, formerly not described as an environmental contaminant, exhibits high similarities to more modern sewage marker compounds (see Tab. 2).

## Conclusions

The TCBTs are very specific marker compounds for mining activities. However, their usage as indicator substances is restricted to their application time since approx. 1970. In contrast, PCBs which were formerly used in the same field of technical application are no specific mining marker compounds due to multiple emission sources.

Besides the mining industry, two further important industrial fields were characterized. Halogenated aromatic and aliphatic compounds indicated a huge discharge derived from the chlororganic chemical industry. Maximum emissions were attributed to the time period between 1947 and 1955. These emissions are paralleled by decreasing emissions related to mining activities after 1955. In accordance with other publications we suggest that both types of contaminations were reduced mainly by improved waste water treatment techniques. In addition, the concentration profiles of organotin compounds as another group of industrial pollutants, indicate a pollution history corresponding to their individual usage time period since 1970.

In addition to the main classification system which relates the described contaminants to mining and industrial activities, municipal sewage and agricultural effluents, we suggest, as a second set of criteria regarding pollution history, to classify contaminants as either common or modern. Common contaminants occurred in the sedimentary record over a long period of time, whereas more modern contaminants occurred only in sediment layers representing a time period since 1970. However, for most of the contaminants described, a significant decrease of concentration and, consequently, of corresponding emissions into the Lippe river was observed in the top layers. These observations indicate the beginning of an efficient cleaning process covering a broad spectrum of contaminants, probably as the result of modern sewage treatment techniques.

Interestingly, indications for a significant contribution of agricultural contaminants were not observed. However, several obviously anthropogenic compounds (e.g. 3,6-dichlorocarbazole) remain uncorrelated with distinct emission sources due to limited information concerning their technical application and usage or their emission pathway. In consequence, they could not be considered for the comprehensive reconstruction of the pollution history of the Lippe river for the time period between 1935 and 1986.

## 5.2.3 Geochronology and specific anthropogenic markers in a sediment core from the Teltow Canal in Berlin‡

### Introduction

The Teltow Canal situated in the urban area of Berlin (Germany) is a very slow flowing canal with a high sedimentation rate as the result of a barrier near the former GDR-border. As formerly reported numerous halogenated and non-halogenated compounds were detected in Teltow canal sediments nearby an former industrial point source (Schwarzbauer et al. 2001, Ricking et al. 2003). Further investigations on Teltow Canal sediments considering the solvent extractable as well as the non extractable organic fraction revealed concentrations of extractable DDT related compounds up to 300000 ng/g (d.w.). More than 125000 ng/g (d.w.) of these contaminants were determined in the non-extractable fraction (bound residues) as recently published (Schwarzbauer et al. 2003a). In addition, the bound residues fraction, as analysed by flash pyrolysis and chemical degradation, indicated either transformation or degradation processes affecting the DDT pesticides, that differed significantly as compared to the well known processes affecting the extractable compounds.

To prolong the previously published analysis results and to study the geochronology of anthropogenic markers in the past decades, two additional sediment cores were taken in 1999. The sediment cores were obtained in very low flow areas near the former GDR-border. Gamma-spectrometric dating applied to a comparable sediment core at the location TK-BC (see Figure 1) indicated an age of $\geq$ 100 years at a depth of 95-100 cm and a sediment accumulation rate of up to 3-4 cm/year at the sediment top. These data will be applied for a rough estimation of the sedimentation time periods associated to the core samples investigated in the presented study.

Briefly, this study focuses on the vertical distribution of DDT and its metabolites in undisturbed sediment layers located in the same sampling area (location TKS, Figure 1). The correlation of the detected concentrations with the stratigraphical sediment profile reflects not only a geochronolgical increase or decrease of these compounds, but also the possibility to reflect the ban as well as the resulting restriction and termination of the DDT-production at this industrial area.

‡ Mainly adapted/reprinted from Ricking et al., 2004

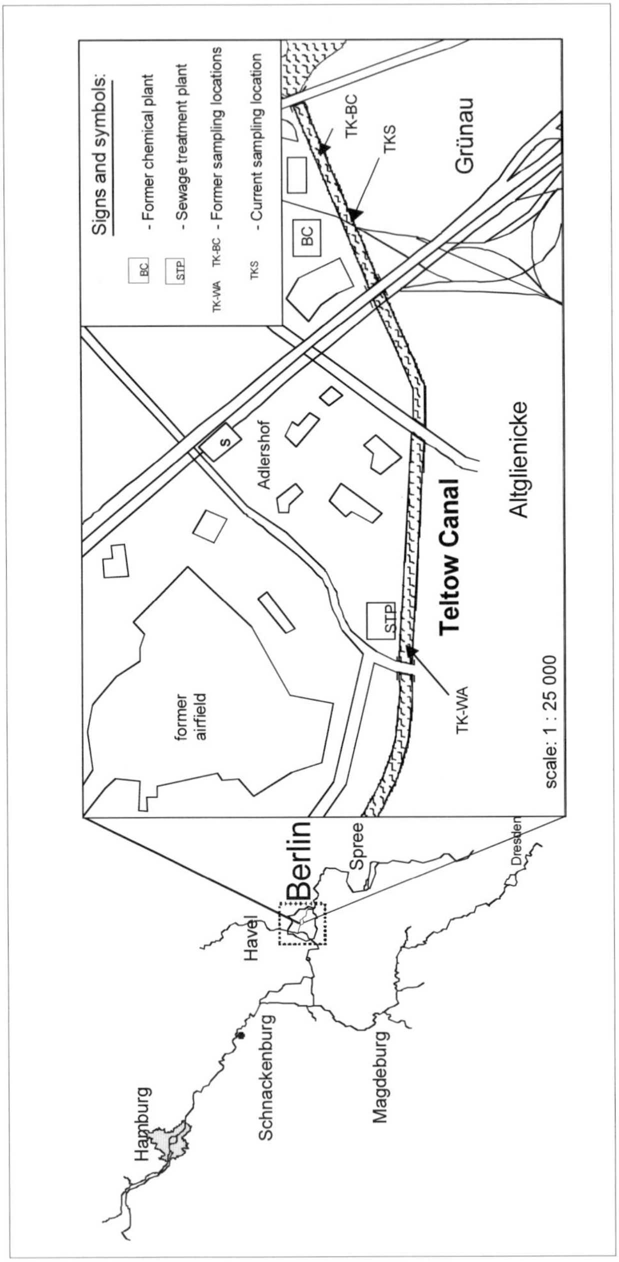


**Fig. 1:** Map of a part of the Teltow Canal, Berlin, Germany showing former and current sampling locations (from Schwarzbauer et al. 2003)

## Samples

Two 40 cm cores were taken by means of a tube coring system in May 1999 from a zodiac at location TKS (see Figure 1). The fresh sediment material was characterised as a black fine mud, without sand, with a high water content and an anoxic sediment-water interface. Considering former results of nearby located cores, the 40 cm of these cores represent an sedimentation period of approx. 40 to 50 years.

After sampling the sediment cores were frozen immediately and stored at –20°C before subdividing. As no substructures in the sediment material were indicated, the two cores were sliced into 2 cm layers and combined in one homogeneous sediment sample and freeze-dried under mild conditions (Ricking et al. 2003).

In addition, a fluffy layer sample of the 4 cm above the sediment/water boundary as well as a corresponding water sample representing the water phase directly above the sediments were obtained. Furthermore, the pore water of the first two sediment layers was separated and analysed.

## Results and discussion

Two combined sediment cores (TKS) of the Teltow Canal, Berlin, Germany (see Figure 1) were investigated by inorganic and organic geochemical analysis especially focused on the pesticides content. According to former investigations (Schwarzbauer et al. 2003a) elevated amounts of DDT and its metabolites were determined. Assuming undisturbed sediment layers the depth correlated identification and quantification of these compounds allowed a temporal integration of the historical input of DDT and its metabolites.

**Tab. 1:** Results of the inorganic geochemical analysis applied to pore water of a preliminary sample (TKWA), the first two sediment layers (TKS 0-2 and 2-4) of the current sediment cores and to a sample of the corresponding water column (TKS 4-0). (n.d.= below detection limit)

| Sample | **DOC** (mg/L) | **Cd** (mg/L) | **Cu** (mg/L) | **Pb** (mg/L) | **Zn** (mg/L) |
|---|---|---|---|---|---|
| TKWA | 17 | n.d. | 14 | n.d. | 0.10 |
| TKS 4-0 | 17 | n.d. | 1 | n.d. | 0.02 |
| TKS 0-2 | 60 | n.d. | 14 | n.d. | 0.03 |
| TKS 2-4 | 16 | n.d. | 9 | n.d. | 0.03 |

Prior to the organic analyses the samples were characterised by inorganic geochemical analyses of pore water of the first two upper sediment layers (see Table 1). In comparison to these data the values of a former investigated sample (TKWA) close to the location were also presented (Ricking, unpublished).

The pore water samples of the sediment layers close to the surface of both samples, TKWA and TKS, were characterised by DOC values between 16 and 60 mg/L. The maximum values were obtained in the top layer of the TKS sample. The results of the inorganic analyses revealed concentrations of copper and zinc within a range between 1 to 14 mg/L and 0,02 to 0,1 mg/L, respectively. For the TKS samples the highest values were determined in the top layer. The concentrations of lead and cadmium were below the detection limit in all samples investigated.

Organic analyses applied to the water and fluffy layer samples revealed also a huge contamination by DDT-related compounds as presented in Table 2. In the water samples the more hydrophilic compounds of the DDT-metabolites were detected. Further on, traces of DDOH, DDM and no HCH were analysed. Interestingly, four further pesticides (Chlorobenzilat, Etoxinol, Chloropropylate or Rospan, Kelthane) with an analogous structure to DDT were identified.

**Tab. 2:** Results of the organic geochemical analysis applied to fluffy-layer sample (FL), the first two sediment layers (TKS 0-2 and 2-4) of the current sediment cores and to a sample of the corresponding water column (TKS 4-0). (n.d.= below detection limit)

| **Compound** | **Samples** | | | **Compound** | |
|---|---|---|---|---|---|
| ng/L | TK 4-0 | TK 0-2 | TK 2-4 | ng/g | FL |
| 2,4'-DDA | 210 | 330 | 830 | 2,4'-DDA | 510 |
| 4,4'-DDA | 550 | 1500 | 3100 | 4,4'-DDA | 2000 |
| 4,4'-DDE | n.d. | n.d. | n.d. | 4,4'-DDE | 1100 |
| 2,4'-DDD | n.d. | n.d. | n.d. | 2,4'-DDMS | n.d. |
| 4,4'-DDD | n.d. | n.d. | n.d. | 4,4'-DDMS | 58 |
| 4,4'-DDMU | n.d. | n.d. | n.d. | 2,4'-DDD | 4100 |
| 2,4'-DBP | 270 | 210 | 59 | 4,4'-DDD | 8800 |
| 4,4'-DBP | 620 | 1600 | 350 | 4,4'-DDNU | 180 |
| 4,4'-DBPH | | 250 | | 4,4'-DDMU | 5600 |
| | | | | 2,4'-DBP | 450 |
| | | | | 4,4'-DBP | 3700 |

The fluffy-layer sample revealed the same compounds as in the sediment samples, the screening illustrated additionally all 4 HCHs (α-δ) and HCB, not detected in the water samples.

The organic geochemical analyses applied to all sediment samples of the core TKS reveals very high concentrations of DDT and its metabolites. These expected amounts were formerly attributed to a pesticides producing chemical plant (Heberer et al 1999). The quantitative analyses included the 4,4'-isomers of DDT, DDD, DDE, DDCN, DBP, DDMS, DDMU, DDEt and DDM. All corresponding 2,4′-isomers were also detected, but the recovery and calibration data were kept from the 4,4′-isomers. For quantification of DDMS and DDEt the calibration data and recovery of 4,4′-DDD were used due to commercial non available reference material. All quantitative data are presented in Table 3 and 4 as well as in Figure 2.

Following the occurrence of DDT and its metabolites will be discussed with respect to their vertical distribution in the sediment core investigated. Main contaminant was DDD, the main metabolite of the anaerobic degradation pathway, with maximum values of approx. 24000 ng/g dry weight in the deeper layers. The concentration detected in all sediment layers ranged between 2000 to 24000 ng/g.

A second group of metabolites including DDE, DDEt and DBP appeared with concentrations lower than DDD. Very low concentrations were determined for DDCN and DDM with values between 1 to 650 ng/g. The concentration of DDE, the metabolite predominately accumulated in the more aerobic environment, was in the range of 10% to 110 % of DDD. Regarding the maximum concentrations of DDD up to 24000 ng/g in comparison to the highest values of DDE (up to 900 ng/g) a predominantly anaerobic environment with an significant aerobic proportion can be stated.

DDT itself was determined with concentrations up to 2500 ng/g, reflecting a progressive degradation of the pesticide. Figure 2 illustrates the distribution of all 4,4′-isomers in relation to the sedimentation depth. The presented data are arranged according to their different concentration levels. For all compounds a sharp increase in concentration was observed relative to the top layer samples. In addition it is remarkable that the maximum concentration of 4,4′-DDD, 4,4′-DMU, 4,4′-DDT and 4,4′-DDMS peaked at a depth of 24 cm, whereas the maximum values of 4,4′-DBP, 4,4′-DDCN, 4,4′-DDEt and 4,4′-DDM, were located in a slightly deeper layer (28 and 30cm).

**Tab. 3:** Compound information for identification and quantification

| Compounds | Chemical structure |
|---|---|
| 4,4´-DDT<br>2,2-Bis(4-chlorophenyl)-<br>1,1,1-trichloroethane | 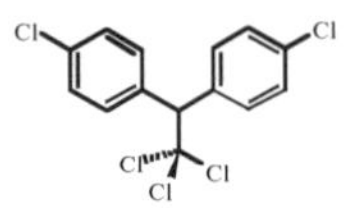<br>characteristic ion<br>(m/z): 235, 237<br>recovery: 64% |
| 4,4´-DDD<br>2,2-Bis(4-chlorophenyl)-<br>1,1-dichloroethane | 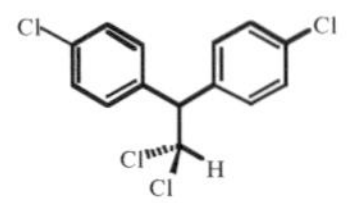<br>characteristic ion<br>(m/z): 235, 237<br>recovery: 70% |
| 4,4´-DDMS<br>2,2-Bis(4-chlorophenyl)-<br>1-chloroethane | 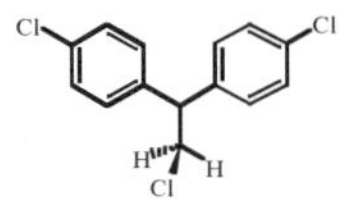<br>characteristic ion<br>(m/z): 235, 237<br>recovery: 70% |
| 4,4´-DDEt<br>1,1-Bis(4-chlorophenyl)ethane | 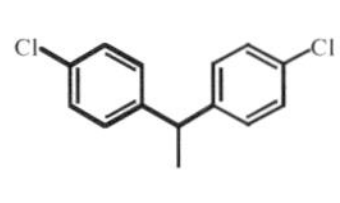<br>characteristic ion<br>(m/z): 235, 237<br>recovery: 70% |
| 4,4´-DDE<br>2,2-Bis(4-chlorophenyl)-<br>1,1-dichloroethene | 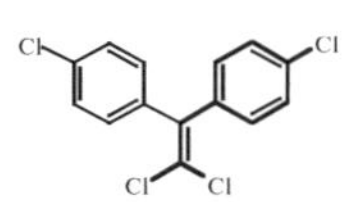<br>characteristic ion<br>(m/z): 246, 248<br>recovery: 89% |
| 4,4´-DDMU<br>2,2-Bis(4-chlorophenyl)-<br>1-chloroethene | 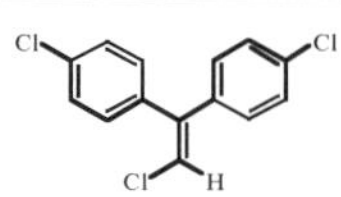<br>characteristic ion<br>(m/z): 212, 282<br>recovery: 75% |
| 4,4´-DDM<br>Bis(4-chlorophenyl)methane | 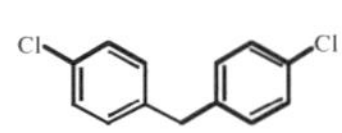<br>characteristic ion<br>(m/z): 236, 238<br>recovery: 68% |
| 4,4´-DDCN<br>Bis(4-chlorophenyl)-<br>acetonitrile | 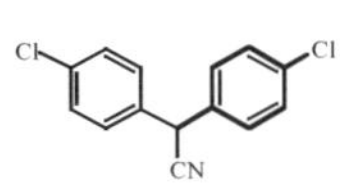<br>characteristic ion<br>(m/z): 226, 228<br>recovery: 90% |
| 4,4´-DBP<br>4,4-Dichlorobenzophenone | 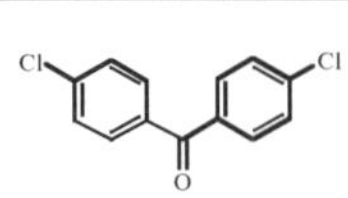<br>characteristic ion<br>(m/z): 139, 141<br>recovery: 77% |

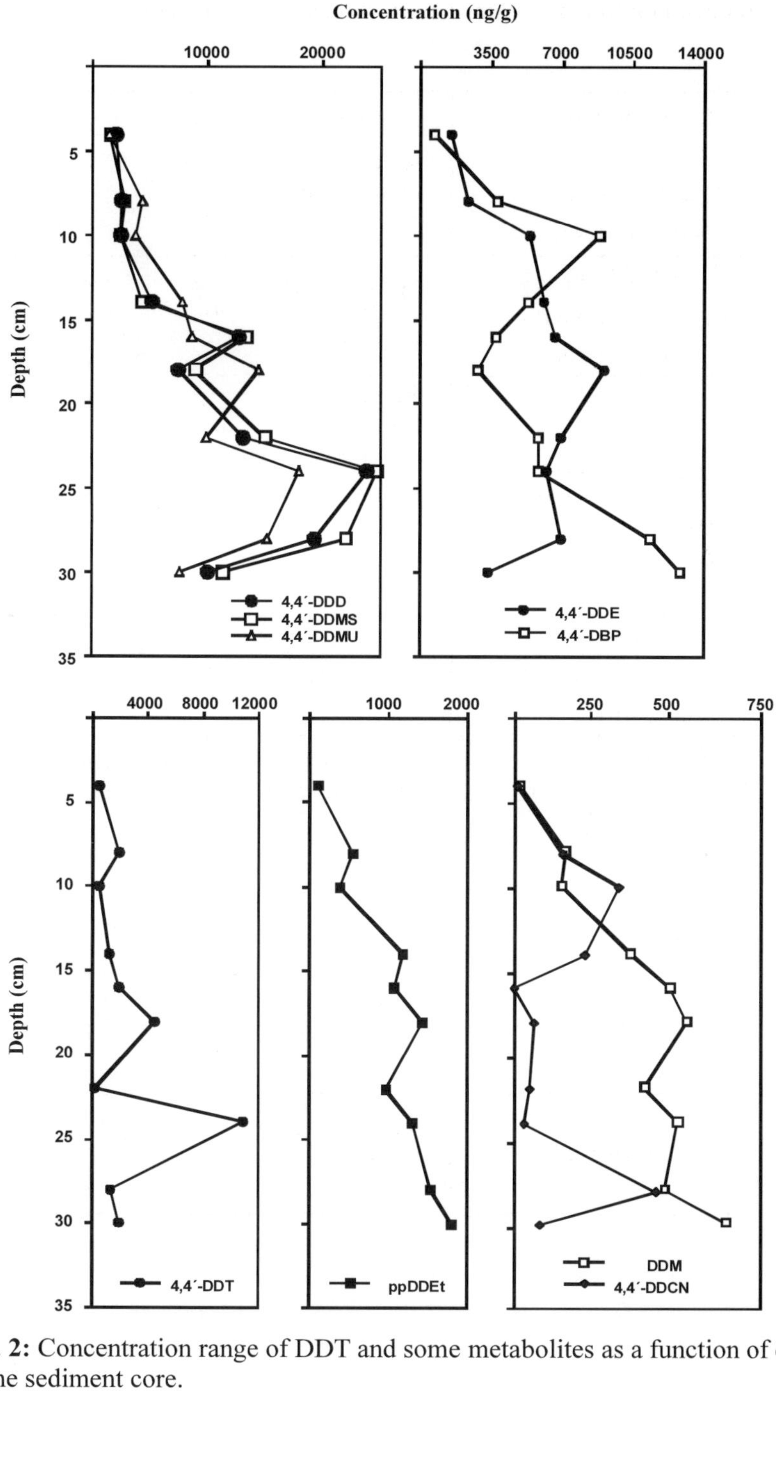


**Fig. 2:** Concentration range of DDT and some metabolites as a function of depth in the sediment core.

**Tab. 4:** Quantitative data of the 2,4´- and 4,4´-isomers of all DDT-related compounds detected in [ng/g] dry weight and the calculated 2,4´- to 4,4´- ratios.

| Depth | DDE | | | DDMU | | | DDCN | | | DBP | | |
|---|---|---|---|---|---|---|---|---|---|---|---|---|
| | 2,4'- | 4,4' | ratio | 2,4'- | 4,4' | ratio | 2,4'- | 4,4' | ratio | 2,4'- | 4,4' | ratio |
| [cm] | ng/g | ng/g | | ng/g | ng/g | | ng/g | ng/g | | ng/g | ng/g | |
| 2 | 120 | 1500 | 0.08 | n.d. | 1500 | n.c. | 1 | 19 | 0.05 | 130 | 720 | 0.18 |
| 8 | 260 | 2300 | 0.10 | n.d. | 4300 | n.c. | 4 | 160 | 0.02 | 850 | 3800 | 0.22 |
| 10 | 330 | 5400 | 0.06 | n.d. | 3700 | n.c. | 6 | 340 | 0.03 | 1600 | 8800 | 0.18 |
| 14 | 640 | 6100 | 0.10 | n.d. | 7800 | n.c. | 7 | 230 | 0.03 | 890 | 5300 | 0.17 |
| 16 | 650 | 6600 | 0.10 | n.d. | 8600 | n.c. | 1 | 14 | 0.07 | 780 | 3700 | 0.21 |
| 18 | 810 | 9000 | 0.09 | n.d. | 14000 | n.c. | 5 | 70 | 0.07 | 700 | 2800 | 0.25 |
| 22 | 610 | 6900 | 0.09 | n.d. | 10000 | n.c. | 7 | 60 | 0.12 | 1100 | 5800 | 0.19 |
| 24 | 720 | 6200 | 0.12 | n.d. | 18000 | n.c. | 3 | 40 | 0.07 | 1000 | 5800 | 0.17 |
| 28 | 800 | 6900 | 0.11 | n.d. | 15000 | n.c. | 14 | 460 | 0.03 | 1900 | 11000 | 0.17 |
| 30 | 650 | 3300 | 0.20 | n.d. | 7600 | n.c. | 7 | 91 | 0.08 | 1900 | 13000 | 0.15 |

| Depth | DDT | | | DDD | | | DDMS | | | DDEt | | | DDM |
|---|---|---|---|---|---|---|---|---|---|---|---|---|---|
| | 2,4'- | 4,4' | ratio | 2,4'- | 4,4' | ratio | 2,4'- | 4,4' | ratio | 2,4'- | 4,4' | ratio | 4,4' |
| [cm] | ng/g | ng/g | | ng/g | ng/g | | ng/g | ng/g | | ng/g | ng/g | | ng/g |
| 2 | 81 | 460 | 0.18 | 1600 | 2100 | 0.76 | 650 | 1400 | 0.46 | 28 | 110 | 0.25 | 30 |
| 8 | 93 | 1900 | 0.05 | 3000 | 2600 | 1.15 | 1500 | 2800 | 0.53 | 85 | 540 | 0.16 | 160 |
| 10 | 62 | 520 | 0.12 | 2600 | 2400 | 1.08 | 1200 | 2400 | 0.50 | 60 | 390 | 0.15 | 150 |
| 14 | 120 | 1200 | 0.10 | 4300 | 5200 | 0.83 | 1900 | 4300 | 0.44 | 230 | 1200 | 0.19 | 360 |
| 16 | 76 | 2000 | 0.04 | 11000 | 13000 | 0.85 | 5400 | 13000 | 0.41 | 210 | 1100 | 0.19 | 480 |
| 18 | 120 | 4500 | 0.03 | 8100 | 7500 | 1.08 | 4300 | 8800 | 0.49 | 260 | 1400 | 0.18 | 530 |
| 22 | 290 | 1700 | 0.17 | 9800 | 13000 | 0.75 | 7100 | 15000 | 0.47 | 190 | 980 | 0.19 | 400 |
| 24 | 1500 | 11000 | 0.10 | 18000 | 24000 | 0.75 | 12000 | 25000 | 0.48 | 260 | 1300 | 0.20 | 500 |
| 28 | 740 | 1300 | 0.57 | 14000 | 19000 | 0.74 | 11000 | 22000 | 0.50 | 300 | 1500 | 0.20 | 460 |
| 30 | 60 | 1900 | 0.03 | 7400 | 10000 | 0.79 | 5800 | 11000 | 0.53 | 280 | 1800 | 0.15 | 650 |

In comparison to the quantitative vertical distribution of all other metabolites the concentrations of 4,4′-DDCN showed a significant higher variation within the core. Higher concnetrations were observed in two sediment horizons, but noteworthy the highest concentration was also detected a depth of 28 cm.

As expected, the quantitative data of all 2,4′-Isomers detected indicate very similar trends in the concentration gradients as compared to the 4,4′-isomers. The concentration ranges vary approximately between 10 to 25 % as compared to 4,4′-isomers. Only the 2,4′-DDMS showed higher contributions of up to 50 % (see Table 4).Noteworthy the ratio of 2,4'-/4,4'-DDD differed significantly with values between 0-75 to 1.1. This observation is quite unusual and the reasons are not known so far.

With respect to the described results a significant contamination of sediment samples of the Teltow Canal, Berlin (Germany) with DDT and related compounds was pointed out for a longer time period. Considering the sedimentation rate and γ-spectrometric dating formerly applied to a comparable sediment core (unpublished results, M. Ricking) a geochronological description of the DDT related emissions into the sediments of Teltow Canal is enabled.

The decreasing concentrations from the 30 cm layer towards the top of the sediment core characterise a decline in emission. This can be attributed to the reduction of the industrial DDT-production since 1975. Following, the highest production rates can be attributed to a time period corresponding to the lowest sediment layers between 24 and 30 cm.

# 6 Summary and conclusions

Environmental studies were conducted dealing with the organic contamination of river and groundwater systems as the result of anthropogenic activities. The spatial distribution of the sampling areas is summarized in Fig. 9 of chapter 1. Furthermore, the individual studies can be attributed to different analytical tools or topics including qualitative and quantitative screening analyses, monitoring of transport and transformation processes, geochronological investigations, isotopic studies, application of the anthropogenic marker approach and characterization of bound residues. This attribution is visualized in Fig. 1. However, nearly all studies presented were based on very detailed GC/MS non-target screening analyses as the main important analytical tool for continuative environmental investigations on organic contaminations.

The main objectives, important data and essential conclusions of the individual studies are briefly summarized briefly below:

Groundwater contaminations (chapter 2):

The organic geochemical analyses applied to seepage water and leakage water samples of a waste deposit landfill revealed a wide variety of organic substances which were attributed to natural or xenobiotic waste components. Apart from plant material-derived compounds and degradation products of peptides, carbohydrates and lignin, numerous xenobiotic substances were identified and attributed to the groups of pharmaceuticals, plasticizers, pesticides or chlorinated aromatics. Not all of the substances identified in the seepage water samples were recovered in the leakage water sample due to degradation processes or dilution by uncontaminated water. Quantitative analyses discriminated contaminants which were affected by degradation processes and persistent compounds.

Several xenobiotics identified in both the seepage water and the leakage water on the same concentration level were not only persistent but also specific for the waste deposit-derived emission. Hence, these stable source indicators were quantitatively analysed in groundwater and drainage effluent samples. Using this organic marker compound approach the lateral and vertical distribution of the waste deposit derived contamination, the dilution factors as well as the long time emission behaviour within the corresponding aquifer were successfully determined and monitored. The applicability of the organic marker approach was confirmed by the continuously constant relative composition of the specific substances.

A second study characterizing a huge groundwater contamination caused by chemicals used for wood impregnation revealed mono- and dichlorinated naphthalenes, chlorobenzo(b)thiophene, acenaphthene and methylated naphthalenes as the main pollutants. Anaerobic microbial degradation of the main pollutant 1-chloronaphthalene was evident from identification of 1-chloro-4-naphthol and 1-chloronaphthoic acid as biotic transformation products. In contrast inorganic and compound specific stable carbon isotope analyses indicated only a minor degree of microbial transformation. Thus, sorption seemed to be the main attenuation process within the aquifer affected by the coal tar based contamination described above.

Besides the substances related to the impregnation activities, the pesticides atrazine and bromacil were detected and attributed to another contamination source.

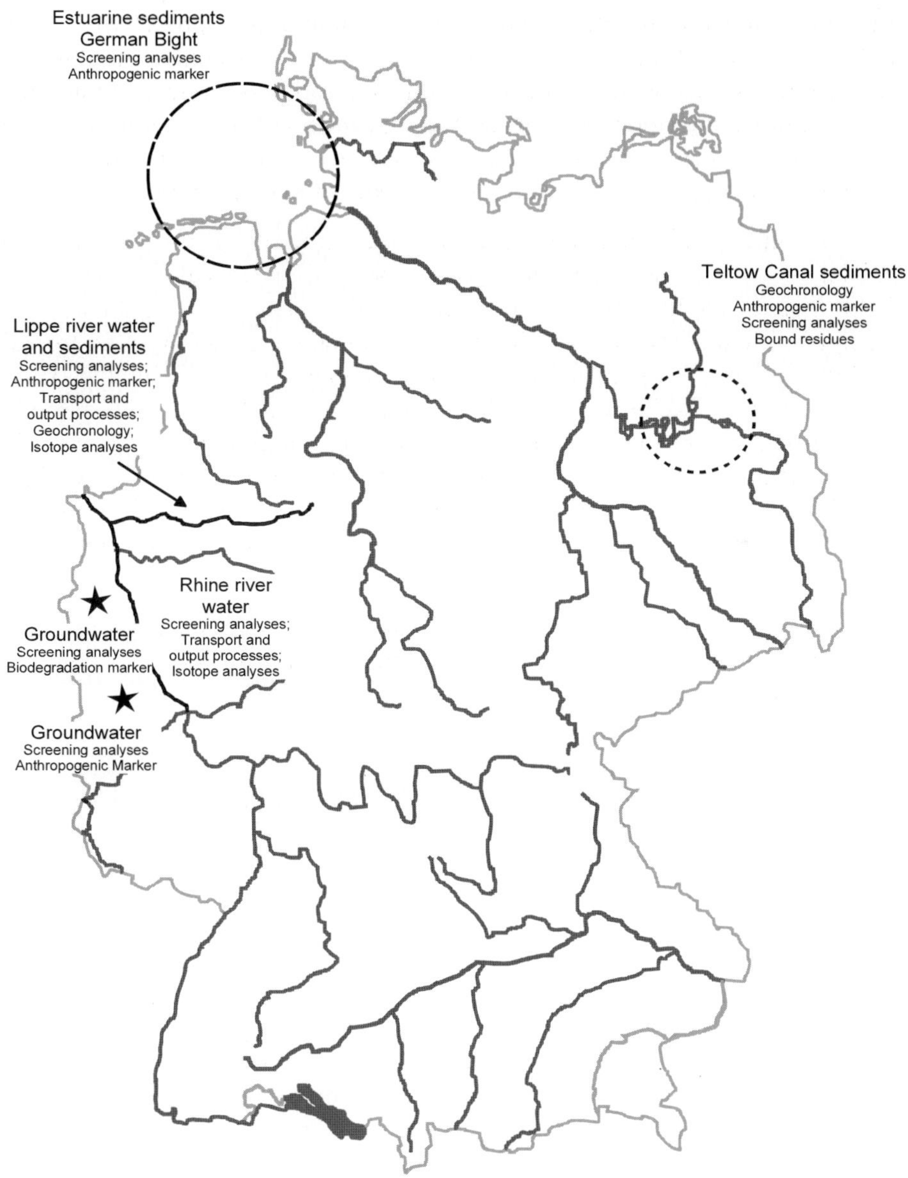


**Fig. 1:** Summary of different types of environmental investigations presented.

Screening analyses of various river systems (chapter 3):

*Lippe river*

GC/MS-screening analyses of water samples from the Lippe river, Germany, revealed the presence of a wide spectrum of low-molecular weight organic compounds ranging from nonpolar constituents like aliphatic hydrocarbons to polar constituents like n-carboxylic acids and phenols. Most of the identified compounds were attributed to anthropogenic input and are used as e.g. plasticizers, flame retardants, pharmaceutical drugs or fragrances. Some of them had rarely been noticed as organic pollutants of aquatic environments before. These are, among others, 9-methylacridine, the plasticizers 2,2,4-trimethyl-1,3-pentandioldi-*iso*-butyrate (TXIB) and triethyl citrate, the surfactant 2,4,7,9-tetramethyl-5-decyne-4,7-diol (TMDD), triphenylphosphinoxide and the flame retardant tris(chloropropyl) phosphate.

Input pathways of the detected compounds were traced back by sampling various input sources of organic matter like discharge of wastewater and Lippe river tributaries. The major proportion of contaminants were ubiquitous in Lippe river water and also occurred in effluent from a municipial sewage treatment plant and in samples from the tributaries characterizing a dominant contribution of typical sewage derived pollutants.

In a second part of this study the organic pollution was characterized on a quantitative basis. Dissolved organic loads of fourteen specific anthropogenic contaminants were calculated on the basis of their concentrations in water and river runoff on the day of sampling. The organic loads of each compound were compiled along the longitudinal section of the river in order to generate individual spatial pollution profiles. It was observed that distribution of organic loads along the river showed distinctive patterns, depending upon the input situation and physico-chemical properties of the compound. Three types of compounds were classified of which Type 1 compounds were of special interest due to their stability in the aqueous phase and potential application as anthropogenic markers.

Additionally, sediment samples of the Lippe river (Germany) were investigated by non-target screening analyses. Within the present study so far unknown or rarely detected contaminants as well as anthropogenic molecular marker compounds were selected for quantification. The obtained qualitative and quantitative analytical results were interpreted in order to characterize the anthropogenic contamination of the particulate matter of the Lippe river including spatial distribution, input effects and time depending occurrence. Anthropogenic molecular markers were identified discharged from (i) municipal sources characterized by diffusive effluents, (ii) point sources of three different industrial branches characterized by spatially limited appearance, (iii) diffusive agricultural sources. While municipal and agricultural contaminations were ubiquitous and diffusive, industrial emission sources were spatially isolated and assigned to mining activities, chlorochemical as well as organotin industry. Specific sources for selected nitrogen containing compounds comprising N-formylpiperidine, 3,6-dichlorocarbazole and bis(4-octylphenyl)amine, which are obviously of anthropogenic origin, could not be clarified

Beside a reproducible spatial distribution of the anthropogenic markers through all sampling campaigns, no preferred seasonal trends were observed.

*Rhine river*

The qualitative and quantitative GC/MS analyses constituted a comprehensive characterization of the low molecular weight organic contamination of Rhine river water. A minor proportion of the contaminants was attributed to biogenic sources. Within the group of anthropogenic contaminants only a few compounds were characterized as frequently detected or priority pollutants (phthalates, caffeine). Numerous compounds exhibiting physiological or ecotoxicological properties were identified which are only rarely reported or still unnoticed riverine contaminants (e.g. DEET, mephobarbital, DIPN).

Additionally, mono- and dibrominated (methoxyphenyl)propionic acids and hydroxymethylacetophenones were identified for the first time as environmental contaminants. Analytical data (MS, IR, GC) derived from their structure elucidating analyses are presented. The dominance of brominated compounds as compared to chlorinated substances in the Rhine water samples is an unusual state of riverine pollution with respect to halogenated compounds.

Quantitative analyses characterized up to five groups of pollutants with respect to their concentration profiles as the result of input and output processes. In particular the spatial distribution and the intensity of emission sources on the one hand and the environmental stability as well as the tendency to adsorb on the particulate matter on the other hand determined the quantitative occurrence of the individual compounds. Nevertheless, accumulation tendencies or elevated concentrations were observed for several contaminants with potential harmful effects (e.g. carbamazepine, brominated phenols, aromatic sulfones and phthalates).

*Havel and Spree rivers*

In sediment samples of the Havel and Spree river, which are tributaries to the Elbe river, specific molecular markers of anthropogenic activities were identified. Despite a wide variety of lipophilic organic compounds from diffuse anthropogenic contamination, a local emission of an industrial point source was reflected by specific markers including halogenated compounds and nitrogen containing substances (4-ethylnitrobenzene, formyl piperidine, acetyl piperidine). In addition to well-known anthropogenic markers various new molecular tracers were detected and discussed, namely plasticizers (alkylsulfonic acid aryl esters, tributyl and tricresyl phosphates), synthetic fragrances (galaxolide, tonalide, 4-oxoisophorone), additives of personal care products (4-methoxycinnamic acid 2-ethylhexyl ester, benzyl benzoate, dibenzyl ether, benzophenone), occurring due to sewage treatment plant input.

Fate and distribution of organic contaminants (chapter 4):

*Musk fragrances*

In a preliminary study the emission behaviour and source specifity of six polycyclic musk fragrances and the most prominent nitro musks were investigated in effluents derived from Swedish and Canadian waste water treatment plants by means of quantitative analyses. Generally, the Canadian samples were contaminated at a higher level than Swedish samples, by a maximum factor of approx. 10. Information on the variation within the emission pattern are provided in particular for galaxolide and tonalide expressed as HHCB/AHTN ratios. These ratios vary significantly for sewage effluents derived from different countries as well as compared to values analysed in natural samples of the aquatic environment. These variations seemed to be affected by a higher biodegradability of HHCB, source specific differences as the result of changing technical formulations as well as partition processes within the aquatic systems.

Therefore, the occurrence and distribution of polycyclic musks in the Lippe river system was investigated to observe the dynamic transport and partitioning of these compounds between aqueous and particulate phases after their discharge into the river by sewage effluents. HHCB, AHTN,

AHMI and ADBI were determined in 19 water and surface sediment samples which were taken from a longitudinal section. The load of dissolved HHCB and AHTN was calculated on the basis of compound concentrations in water and the corresponding river runoff data and ranged from 3 to 293 g/d and from 1 to 108 g/d, respectively. Increasing loads of HHCB and AHTN along the river reflect a high input of sewage effluents in the densely populated areas along the central part of the river. Decreasing loads at the lower reaches indicate that in the corresponding river sections the rate of removal of musks was higher than the rate of input. Degradation and/or adsorption to particulate matter are processes that might explain this phenomenon. Consequently, high concentrations of HHCB and AHTN were detected in surface sediments from the Lippe river (from 5 to 191 µg/kg and from 2 to 1399 µg/kg, respectively). HHCB/AHTN ratios in sediment samples were lower (average 1.2) than in water samples (average 2.9), suggesting the preferential adsorption of AHTN to particulate matter.

*Compound specific isotope analyses*

This study described the application of a common analytical procedure adapted for compound-specific stable carbon isotope analyses of riverine contaminants. To evaluate the sensitivity of the analytical method and the precision of the isotopic data obtained a set of numerous substances at different concentration levels were measured. For most of the anthropogenic contaminants investigated (including chlorinated aliphatics and aromatics, musk fragrances, phthalate based plasticizers and tetrabutyl tin) acceptable carbon isotope analyses could be obtained down to amounts of approx. 5 ng. These amounts correspond to concentrations in water samples at a natural abundance level of low to medium contaminated river systems. However, it has to be considered that the accuracy as well as the sensitivity of the analytical method depend partially on the chemical properties of the substances measured.

Five recovery experiments were conducted in order to assess changes of the carbon isotope ratios during sample preparation and measurement. The compounds selected for these experiments are known riverine contaminants. Isotopic shifts or higher variations of the isotope ratios as a result of the analytical procedures applied were observed only for a couple of contaminants.

Furthermore, compound specific carbon isotope analyses were performed on water extracts of the Rhine and Lippe rivers. By comparing the variation of the data of several individual compounds with the deviations obtained from the recovery experiments it was possible to differentiate contaminants with unaffected isotope ratios (e.g. squalene, bis(2-ethylhexyl)phthalate) and substances with significant alterations of the $\delta^{13}C$-values (e.g galaxolide, *iso*-propylpalmitate).

*Bound residues*

Anthropogenic activities induce significant alterations of the macromolecular organic matter in riverine systems mainly by emission of pollutants and their subsequent incorporation into geopolymers. The non-extractable residues of highly polluted riverine sediments (Spree river, Teltow Canal, Germany) were qualitatively analysed in order to monitor the occurrence, alteration and distribution of several organic xenobiotics in situ (e.g., plasticizers, pesticides and metabolites of brominated and chlorinated aromatics, fragrances, technical additives, nitro compounds). A comprehensive characterization combining different analytical techniques (pyrolytic analyses and chemical degradation techniques) provided information concerning the incorporation mechanism and the mode of binding of a variety of organic pollutants with different chemical properties.

In particular, the investigations focussed on the identification and quantitation of bound 2,2-bis(chlorophenyl)-1,1,1-trichlorethane (DDT) residues in order to obtain further information about the fate of DDT-derived compounds within the particulate matter of the aquatic environment.

Generally, the distribution of the bound DDT-related compounds was found to differ distinctly from the distribution within the extractable fraction. The main metabolite of the anaerobic degradation pathway (2,2-bis(chlorophenyl)-1,1-dichlorethane, DDD) is most abundant in the sediment extracts, but occurred only in insignificant concentrations in the degradation products of all procedures applied. The most abundant DDT-metabolites released after the degradation procedures were 4,4'-DBP, 4,4'-DDA and 4,4'-DDM. In addition, 4,4´-DDM was detected at rather high concentrations by pyrolytic analysis.

The results imply a weak association to the non-extractable particulate matter based on non-covalent inter-actions for the observed DDT-related contaminants. The release of these compounds was initiated by the modification and degradation of the organic macromolecular matrix as well as of the inorganic material.

Furthermore, numerous methoxychlor-related compounds were detected not only in the extracts but also in parts of the hydrolysis products.

Additionally, the bound fraction of numerous further anthropogenic contaminants were investigated by quantitation of the extractable and nonextractable matter. The selection of the contaminants (including chlorinated and brominated naphthalenes, 2,4,6-tribromoaniline, mono- and dibrominated phenols, phthalates, tri-*n*-butylphosphate, 2,4,4-trimethylpentane-1,3-dioldi-*iso*-butyrate, bisphenol A, butylated nitrophenols, 4-nitrobenzoic acid, galaxolide and tonalide) was based on the results of extended GC-MS-screening analyses applied to the extracts of the sediment samples as well to the extracts derived from selective chemical degradation procedures.

With respect to their distribution and concentration two groups of bound contaminants can be differentiated. A major portion of the contaminants appeared in the extractable as well as in the nonextractable fraction with concentrations generally higher in the latter. A second group occurred only in the nonextractable fraction with substantial concentrations but not in the extractable fraction.

Considering the different selectivities of the degradation methods a second differentiation was observed. Most contaminants appeared frequently and independently of the kind of degradation procedure applied. With respect to the molecular structures of the contaminants investigated a weaker noncovalent association to the nonextractable particulate matter has to be stated for these compounds.

Only few compounds were determined exclusively after application of one degradation method. For these contaminants (e.g. nitrophenols, brominated phenols) a more specific interaction with the macromolecular organic matter of the nonextractable fraction has to be assumed.

Particle associated output from riverine systems (chapter 5):

*German Bight*

A large proportion of pollution in the coastal zone derived from anthropogenic contamination is a result of riverine contributions. Accordingly, the pollution of North Sea estuaries and coastal areas by organic xenobiotics is dominantly characterized by land-derived contamination. GC/MS analyses applied to sediment samples of the German Bight documented the state of organic contamination and identified specific molecular markers that are appropriate to estimate the discharge of anthropogenic compounds derived from the Elbe river. The detailed screening analyses revealed a wide variety of organic lipophilic compounds of biogenic, petrogenic as well as anthropogenic origin.

Potential marker compounds indicating the contribution of the Elbe river are mainly chlorinated aromatic contaminants. Specifically, these markers include tetra- to hexachlorobenzenes, mono- to dichloronaphthalenes, hexachlorobutadiene, tetrabutyl tin, alkylsulfonic acid phenylesters, 1,2,3,6,7,8-hexahydro-1,1,6,6-tetramethyl-4-isopropyl-*as*-indacene and 4,4′-dichlorodiphenylsulfide.

For the evaluation of these proposed Elbe marker compounds the quantitative and spatial distribution was investigated. These data provide an important criterion to define their source specifity.

Two groups of Elbe-derived contaminants were differentiated. Substances of the first group include hexachlorobutadiene and alkylsulfonic acid phenyl esters and appear only at sampling locations directly influenced by the Elbe river. The second group consists of compounds, which occur also in minor concentrations at sites situated at greater distance from the coastal area, which are therefore less influenced by the Elbe river. However, their spatial and quantitative distribution characterizes them clearly as Elbe river derived components. Compounds of the second group are tetra- to hexachlorinated benzenes, mono- and disubstituted chloronaphthalenes, tetrabutyl tin, 4,4′-dichlorodiphenyl-sulfide and DDT related compounds.

In order to discern the discussed Elbe marker compounds from non-marker substances several nonspecific contaminants were analysed additionally. These contaminants include galaxolide, tonalide, tri-*n*-butylphosphate, 2,4,4-trimethylpentane-1,3-dioldi-*iso*-butyrate and 2,2,4,4,6,8,8-heptamethylnonane. No significant accumulation of these compounds was observed at sampling locations mainly influenced by either the Elbe river or the Ems and Weser rivers.

### *Geochronological studies*

Preliminary organic geochemical investigations were performed on layers of a formerly dated sediment core taken from a riparian wetland of the Lippe river (Germany). The samples were collected from two distinct depth intervals representing sedimentation periods of 1944-1952 and 1972-1984, respectively. The quantitative distribution of selected persistent lipophilic organic substances characteres municipal as well as industrial emissions. Considering both geochemical and geochronological data time-dependent information was deduced about industrial production and domestic effluents for the region investigated. In addition, usage and subsequent emission of characteristic pollutants into the riverine system were interpreted for both time periods monitored.

These investigations were extended by organic and inorganic geochemical and geochronological investigations performed on a second sediment core taken from the same sampling site representing an accumulation time period between 1935 and 1986.

In detail most important emission sources, information on technical production and usage as well as on the pathways of individual pollutants with appropriate environmental stability were considered for interpreting the sediment contamination during the past ~50 years and to determine the load history. In particular contaminants related to mining and industrial activities, as well as municipal sewage and agricultural effluents were recorded. Time depending concentration profiles are discussed in detail in order to gain a deeper insight into time depending contamination trends.

Two different pollution histories with different maximum emisson were discriminated. A common contamination as reflected by heavy metals, PACs, LABs, PCBs was detected in the time range between 1947 and 1955. In contrast more modern contaminants, e.g. selected organochlorine compounds, butyl tin compounds, methyltriclosan as well as tris(chloropropyl)phosphates (TCPP) appear only in the time period since 1970. Hence, these concentration profiles reflect in case of environmental stability the emission of the contaminants to the corresponding riverine system. In summary, the study revealed a comprehensive reconstruction of the pollution history of the Lippe river system.

A similar investigation was performed on two 40 cm sediment cores derived from the Teltow Canal in Berlin (Germany) representing a sedimentation period of approx. 40 to 50 years. This area was highly contaminated by DDT residues in the 80'ies. Therefore, this study focused in particular on different DDT-related metabolites. A significant contamination with the main metabolite DDD accompanied by lower concentrated metabolites including DDMS, DDEt, DDCN, DBP, DDM, DDMU and DDE was pointed out for a longer time period. The highest concentrations in the sediment layers between 32 and 40 cm can be related to the highest production rate in the 70'ies, whereas the decreasing concentrations from the most contaminated layers towards the top of the sediment core characterise a decline in emission in this area as a result of reduced industrial DDT-production and a ban of DDT including the relaxation of its envrionmental occurrence.

Besides the contamination-specific conclusions related to the individual studies as presented above, some superior and more comprehensive conclusions can be drawn taking into account all the results obtained and the analytical methods applied:

- riverine and groundwater systems exhibit individual and specific contamination profiles. They can only be reflected accurately by very detailed as well as comprehensive screening analyses. In contrast to conventional monitoring analyses regarding preselected pollutants or routinely analysed contaminants fail to notice a wide spectrum of riverine compounds, although ecotoxicological or harmful effects cannot be excluded. Consequently, non-target screening analyses as well as structure elucidating analyses represent an important skill within environmental investigations on aquatic systems. Consequently, during the analyses described numerous organic compounds were identified, which were unknown so far or formerly not described as environmental contaminants (e.g. bis(4-octylphenyl)amine, 4,4′-dichlorodiphenyl-sulfide, mephobarbital, N-formylpiperidine, di-*iso*-propylnaphthalenes, bromacil). Furthermore, various new pollutants were structurally elucidated and confirmed by synthetic reference material. Examples include 3,6-dichlorocarbazole, 1-chloronaphthoic acid, chlorobenzo(b)thiophene, mono- and dibrominated (methoxyphenyl)propionic acids as well as mono- and dibrominated hydroxymethylacetophenones.
- Numerous organic contaminants identified were important not only in terms of ecotoxicological risk assessment, but also in terms of monitoring environmental impairments with respect to different anthropogenic emission sources. It was demonstrated by various studies presented, that source specific and environmental stable organic substances are powerful indicators not only for landfill leachates or coal tar based contaminations affecting groundwater areas (chapters 2.1 and 2.2), but also for tracing anthropopgenic emissions and corresponding sources in riverine systems in terms of spatial distribution (chapter 3, chapter 5.1) as well as time depending occurrence (chapter 5.2). This anthropogenic marker approach has been inadequately considered in the environmental sciences so far. However, intense screening analyses combined with detailed information on the environmental behaviour are

essential requirements for a successful isolation and application of appropriate marker compounds.

- A comprehensive documentation of the state of pollution of complex natural systems (e.g. river systems) is provided solely by complex and complementary environmental investigations considering various aspects of discharge, distribution and fate of man made chemicals. This was demonstrated by the analyses of anthropogenic contaminants in water and particulate matter (chapters 3.1 and 4.1.2) combined with isotopic analyses (chapter 4.2) and geochronological studies (chapters 5.2.1 and 5.2.2), which allowed detailed insights into the complex anthropogenic pollution of the Lippe river system. Similarly, the huge DDT-related contamination of particulate matter of the Teltow Canal was detected comprehensively by screening analyses of the extractable fraction (chapter 3.3), investigations on the bound DDT-residues (chapter 4.3) as well as geochronological studies (chapter 5.2.3). In particular geochronological investigations allow to obtain information on former contamination levels and pollution histories reflecting for example the time depending efficiency of regulating sanctions and environmental protection measures. Nevertheless, this approach is not commonly used up to this time.
- For these purposes also complex analytical tools have to be further developed and to be applied. This includes for example the broad introduction of compound specific isotope analyses in environmental investigations. Although analytical limitations hindered an uncomplicated launching of isotope analyses in environmental studies so far (as described in chapter 4.2), the usefulness of this analytical tool requires an intensive development of appropriate analytical procedures for a successful implementation of this technique in environmental sciences. Similarly, also pyrolysis as well as chemical degradation techniques are not well introduced in environmental analyses compared to other organic geochemical studies.

# 7 Appendix

For identifying a broad spectrum of organic compounds, non-discriminating methods were applied for extraction and analyses of the sample material. The procedure included non-selective extraction of lipophilic, semi- to low-volatile organic compounds from the sample matrices and their concentration in organic solvents. Additionally, bound residues were investigated by chemical degradation techniques and pyrolysis. Beside qualitative and quantitative GC/MS-analysis also compound specific stable carbon isotope analysis were performed using a GC/irmMS-system

## General analytical procedures

### Chemicals and blank experiments

In order to minimize sample contamination only glass, metal and teflon equipment was used in the laboratory. The equipment was cleaned by ultrasonic agitation in detergent containing water (Extran, Merck, Germany) and rinsed with high-purity acetone and *n*-hexane prior to the extraction procedure. All solvents were purchased from Merck, Germany, and distilled over a 0.5 m packed column (reflux ratio approximately 1:25). The solvent purity was tested by gas chromatographic analyses. Anhydrous granulated sodium sulphate (Merck, Germany) and hydrochloric acid (Merck, Germany) which were needed for the analytic procedure, were cleaned with pure *n*-hexane. All reference substances were purchased from Promochem (Germany), Merck (Germany) and Aldrich (Germany).

Frequent execution of blank analyses confirmed continously the absence of laboratory derived contamination at concentration levels above the limit of detection (LOD) and quantification (LOQ)

## Determination of recovery rates

Recovery rates for water samples were determined by (i) spiking high-purity water (Lichrosolv, Merck, Germany) or precleaned sediments with respective reference compounds at concentration levels similar to the detected values and (ii) subsequent execution of the analytical procedure as described in the following chapters.

## Water sample preperation

### Sampling

Filled sample flasks were sealed free of air bubbles with glass stoppers. They were stored in the darkness at a temperature of approximately 4 °C.

### Extraction

Prior to extraction, water samples were filtered through precleaned Whatman GF/F filters (using pure acetone/*n*-hexane) in order to remove suspended particulate matter from the aqueous phase. A sequential liquid/liquid extraction procedure was applied to approximately 1000-mL aliquots of the water samples using the solvents *n*-pentane, dichloromethane and dichloromethane after acidification to pH 2. Each extraction step was carried out in a separating funnel with 50 mL of the solvent. The third extraction was applied to the pre-extracted water samples after addition of 2 mL of concentrated hydrochloric acid that was pre-cleaned by intense extraction with n-hexane. Subsequently the organic layers were separately dried by filtration over 1 g of anhydrous granulated sodium sulphate (Merck, Darmstadt, FRG) and 50 µL of an internal standard solution containing routinely $d_{34}$-*n*-hexadecane (6.0 ng/µL), $d_{10}$-anthracene (5.1 ng/µL) and $d_{12}$-chrysene (4.7 ng/µL) in *n*-hexane was added. Acidic compounds in the third extract were methylated by addition of a diazomethan solution and subsequent reconcentration. Prior to GC/MS analyses the extracts were reduced to a final volume of approximately 25 µL by rotary evaporation at room temperature.

## Sediment sample preperation

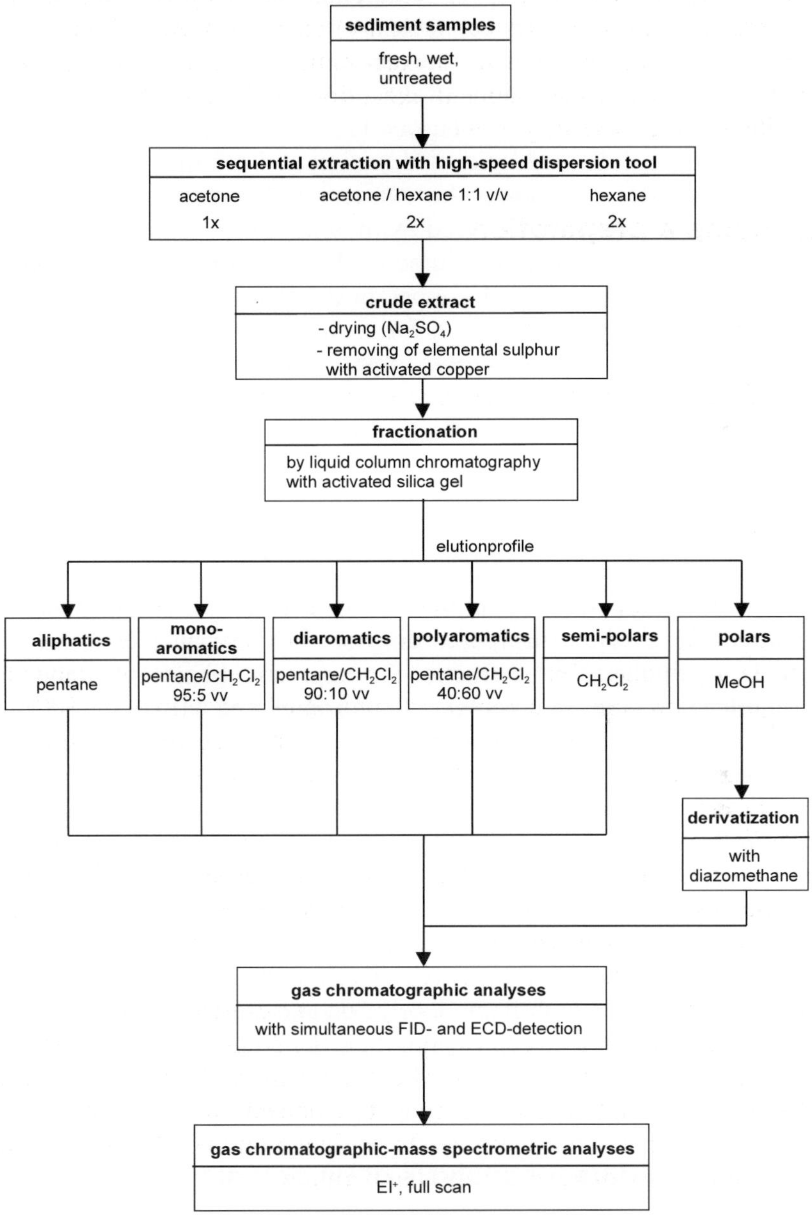


**Fig. 1:** Analytical procedure for non-target screening analyses of sediment samples.

## Extraction

Amounts of 10 to 400 g fresh wet sediments were extracted sequentially with 50 mL of acetone, twice with 50 mL *n*-hexane/acetone 50/50 (vol/vol) and twice with 50 mL *n*-hexane. Extraction was carried out by dispersing the samples in portions of 20 g for 5 min in the solvent using a high-speed dispersion tool (Ultra-Turrax T25, IKA, Stauffen, FRG ). Each extraction step was followed by centrifugation at 4000 rpm and separation of the solvent. After combining the extracts and separating the aqueous phase, the organic layer was dried with anhydrous granulated sodium sulphate and concentrated to a volume of 1 mL. Sulfur was removed by addition of 50 mg of activated copper powder and ultrasonic agitation. After 16 h, the extract was prepared for chromatographic fractionation by filtration over 1 g of anhydrous granulated sodium sulphate and concentration to 0.5 mL.

## Fractionation

Sediment extracts were separated into six fractions by column chromatography (Baker, silica gel 40 µm) using mixtures of *n*-pentane and dichloromethane as the eluent according to Schwarzbauer et al. (2000). Fraction 1: *n*-pentane (5 mL), fraction 2: *n*-pentane/dichloromethane 95/5 v/v (8.5 mL), fraction 3: *n*-pentane/dichloromethane 90/10 v/v (5 mL), fraction 4: *n*-pentane/dichloromethane 40/60 v/v (5 mL), fraction 5: dichloromethane (5 mL), fraction 6: methanol (5 mL). The acidic compounds of fraction 6 were methylated by addition of 0.5 mL of a methanolic diazomethane solution. Prior to analysis, 50 µL of an internal standard solution containing 6.0 ng/µL $d_{36}$-hexadecane, 5.1 ng/µL $d_{10}$-anthracene and 4.7 ng/µL $d_{12}$-chrysene in *n*-hexane were added to each fraction, and the volume was reduced to approximately 50 µL by rotary evaporation at room temperature. All fractions were analysed on a gas chromatograph equiped with flame ionization and electron capture detector (GC-FID/ECD) and on a gas chromatograph linked to a mass spectrometer (GC-MS).

## Extraction procedure for analysis of tin organic compounds

Since mono-, di- and trialkyltin compounds are not extracted by the procedure mentioned above, a second extractive derivatization was applied to the selected sediment samples. Dry pre-extracted sediment residues (1-2 g) were acidified by mixing with 1.5 mL of 10 % aqueous hydrochloric acid. After 0.5 h the pH of the slurry was adjusted to 4 with 10% aqueous sodium hydroxide and a sodium acetate/acetic acid buffer solution. One mL of a 10 % aqueous sodium tetraethylborate solution and 45 mL of hexane/acetone 67/33 (vol/vol) were added. Extraction and separation was performed with a high dispersion tool. The procedure was repeated three times and the combined extracts were prepared for chromatographic fractionation as described above.

### TOC and dry weight determination

Dry weights were determined by drying 3 to 4 g portions of sediments at 105°C to constant weight. The TOC content of dried particulate matter was measured using a CR 12 LECO carbon analyzer.

## Chemical degradation procedures

The chemical degradation steps were carried out in two different modes. In a first set aliquots of the preextracted samples were treated seperately with KOH/MeOH, $BBr_3$ and $RuO_4$. Following, in a second step the pre-extracted and saponified residues were treated once more with $BBr_3$ or $RuO_4$. In Fig. 2 the flow scheme of the analytical procedure is given.

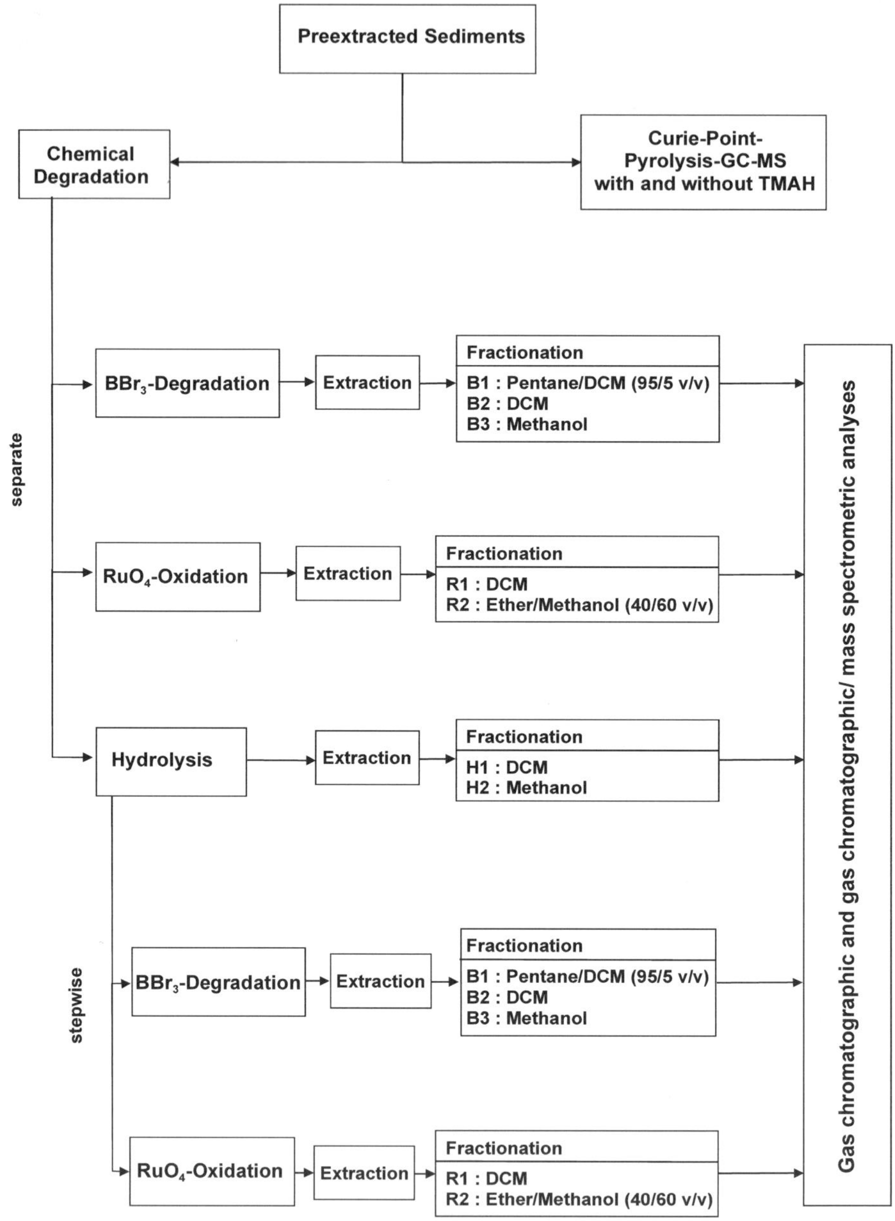


**Fig. 2:** Flow scheme of the analytical procedure for determining bound residues

## Pre-extraction

All sediment samples were pre-extracted first with 40 mL of methanol or butanol and subsequently with solutions of hexane/acetone (1:1 v/v) by means of shaking for 24 h in the dark and ultrasonication at 2 x 450 W (Badelin, Berlin, FRG) in a water bath for 30 minutes. Each extraction step was followed by centrifugation at 1800 x g and the combined extract was reduced in volume by rotary evaporation.

## Alkaline hydrolysis

Aliquots of 150 to 500 mg of the pre-extracted samples were placed in 8 mL vials and 120 mg of KOH dissolved in a mixture of 0.2 mL pre-extracted water and 8 mL of methanol were added. Following, the closed vials were heated for 24 h at 105° C. After cooling 3 mL of pre-extracted water was added and the solutions were filtered using Whatman glass fiber filters (0.7 µm pore diameter). The mixture was acidifed to pH 3 - 5 by addition of approx. 1 mL of a 10% hydrochloric acid solution. Subsequently the solution was extracted three times with 5 mL of dichloromethane. The combined organic layers were dried with anhydrous granulated sodium sulphate and concentrated to a volume of approx. 0.5 mL.

The crude extracts were separated into two fractions by column chromatography (Baker, 2g silica gel 40 µm) using dichloromethane (fraction 1) and methanol (fraction 2) as the eluent, respectively. Prior to analysis, 50 µL of an internal standard containing 90 ng/µL $d_{34}$-hexadecane in *n*-hexane were added to each fraction, and the volume was reduced to 200 µL by rotary evaporation at room temperature.

## $BBr_3$-treatment

To approx. 150 mg of pre-extracted or pre-extracted/saponified samples 5 mL of a 1.0 M boron tribromide solution in dichloromethane was added and the flasks were closed. Following, ultrasonication in a water bath for 2 h was performed. After 24 h of stirring at room temperature a second ultrasonic step was applied. Subsequently 2 mL of diethylether were added, the supernatant was decanted and filtered using a Whatman glass fiber filter (0.7 µm pore diameter). The solid residue was washed twice with diethylether and the combined organic solutions were added after filtration to the filtered reaction mixture. The combined organic layers were washed twice with 5 mL of pre-extracted water and dried with anhydrous granulated sodium sulphate. Prior to fractionation the solutions were concentrated to a volume of approx. 0.5 mL by rotary evaporation.

The crude extracts were separated into three fractions by column chromatography (Baker, 2 g silica gel 40 μm) using the following mixtures as the eluent: Fraction 1 – *n*-pentane/dichloromethane 95/5 v/v, fraction 2 - dichloromethane, fraction 3 - methanol. Prior to analysis, 50 μL of an internal standard containing 90 ng/μL $d_{34}$-hexadecane in *n*-hexane were added to each fraction, and the volume was reduced to 100 μL by rotary evaporation at room temperature.

## $RuO_4$-oxidation

A mixture of 8 mL tetrachloromethane, 8 mL acteone and 1 mL of pre-extracted water was added to aliquots of 20 to 170 mg of pre-extracted or pre-extracted/saponified samples. In addition, 1500 mg sodium perjodate and 10 mg of ruthenium(IV)oxide were added and the reaction mixture was stirred for 4 h in darkness. The reaction was stopped by addition of 50μL of methane and 2 drops of concentrated sulphuric acid. The liquid phase was separated by decantation and the residue was washed twice with tetrachloromethane. The combined organic layers were collected in a separatory funnel and washed with 5 mL of pre-extracted water as well as 1 mL of a sodium thiosulfate solution. All water layers were combined and re-extracted five times with 10 mL of diethylether. The diethylether solution was added to the organic solution, the extract was dried with anhydrous granulated sodium sulphate and concentrated to a volume of 0.5 mL prior to fractionation.

The crude extracts were separated into two fractions by column chromatography (Baker, 2 g silica gel 40 μm) using the following mixtures as the eluent: Fraction 1 – dichloromethane, fraction 2 – diethylether/methanol (40/60 v/v). Prior to analysis, 50 μL of an internal standard containing 90 ng/μL $d_{34}$-hexadecane in *n*-hexane was added to each fraction, and the volume was reduced to 100 μL by rotary evaporation at room temperature.

## Pyrolysis-gas chromatography-mass spectrometry

Pyrolysis-gas chromatography was performed using a Horizon Curie-Point Pyrolator with a pyrolsis temperature of 710°C held for 10 s. The interface was heated to 300°C and the capillary column was directly inserted into the pyrolysis chamber. Parameters for the gas chromatographic separation are described below. For Py-GC/MS analyses the same Pyrolysator device was linked to a GC/MS-system with chromatographic and mass spectrometric conditions as described below.

## Gas chromatography and gas chromatography/mass spectrometry

Principally, gas chromatographic analyses were carried out on a GC 8000 gas chromatograph (Fisons Instruments, Germany) equipped with a nonpolar fused silica capillary column (common dimensions: 25 m x 0.25 mm ID x 0.25 μm film). Flow velocity of the carrier gas (hydrogen) was approx. 40 cm/s. The GC oven was normally programmed from 60 to 300 °C at a rate of 3 °C/min after 3 min at the initial temperature, and was kept at 300 °C for 20 min. The injection was carried out on a split/splitless injector at 270 °C, splitless time was 60 s. The end of the capillary column was splitted to lead the eluate separately to a flame ionization detector (FID) and an electron capture detector (ECD) for a simultaneous detection of the analytes.

GC/MS analyses were carried out either with a double foccusing sector field mass spectrometer (Finnigan MAT 8222, Finnigan, Germany) linked to a HP 5890 gas chromatograph (Hewlett Packard, USA) or with a quadrupol mass spectrometer (Trace MS, Thermoquest, Egelsbach) linked to Mega Series 5140 gas chromatograph (Carlo Erba , Milano, Italy). The gas chromatographic conditions were the same as described above. Carrier gas velocity (helium) was approx. 35 cm/s.

The sector field mass spectrometer was operated in full-scan mode at a resolution of 1000 in $EI^+$-mode (70 eV), source temperature 200 °C, scanning from 35 to 700 amu with a rate of 1 s/decade and an inter-scan time of 0.1 s.

The quadrupol mass spectrometer was also operated in electron impact ionization mode ($EI^+$, 70 eV) with a source temperature of 200°C. The mass spectrometer was scanning from 35 to 700 amu at a rate of 0.5 s/decade with an inter-scan time of 0.1 s.

Principally, acidic compounds in the polar fractions were methylated prior to analysis by derivatisation with a diazomethane or TMSH solution.

## GC/irmMS analysis

Compound specific stable carbon isotope analyses were conducted on a Delta Plus XL mass spectrometer (ThermoFinnigan, Bremen, FRG) equipped with a GCC III combustion interface and linked to a gas chromatograph 6980A (Agilent, Waldbronn, FRG). Gas chromatographic separation was normally performed on a 60m x 0.25 mm i.d. x 0.25 µm film fused silica capillary column coated with a non polar film. Chromatographic conditions were: 1 µL split/splitless injection at 60 °C, splitless time 60s, 3 min hold, then programmed at 2 ° $min^{-1}$ to 300 °C, helium carrier gas velocity 35 cm/s. The oxidation of the eluting substances was carried out at 940 °C facilitated by a CuO/NiO/Pt-catalyst. All measurements were performed in triplicate. The carbon isotope ratio of the reference gas (carbon dioxide) was calibrated with a certified reference standard purchased from Chiron (Trondheim, NO) containing *n*-undecane (-26.11 ‰ vs PDB), *n*-pentadecane (-30.22 ‰ vs PDB) and *n*-eicosane (-33.06 ‰ vs PDB). All data presented are expressed relative to the PDB standard.

## Identification of organic compounds

Identification of individual compounds was based on comparison of $EI^+$-mass spectra with those of reference compounds, mass spectral data bases (NIST/EPA/NIH Mass Spectral Library NIST2000, Wiley/NBS Registry of Mass Spectral Data, 7th Ed., electronic versions) and gas chromatographic retention times, elution patterns or retention indices (e.g. Vassilaros *et al.*, 1982; Rostad and Pereira, 1986; Ballschmitter *et al.*; 1987, Bundt *et al.*, 1991; Paschke *et al.*, 1992; Peters and Moldowan, 1993; Wang *et al.*, 1994). For correction of injection time inaccuracies the retention times of the internal standard compounds were used.

## Quantitative analysis

Quantitative data of selected target compounds were obtained by integration of specific ion chromatograms extracted from the TIC. Injection volume and sample volume inaccuracies were corrected for by using internal standard compounds as a surrogate standard. An external four-point-calibration generated from reference compounds was used for quantification.

## Inorganic analysis

To investigate the content of heavy metals approximately 3 g of each investigated sample were prepared by chemical pulping with nitrohydrochloric acid in accordance to the DIN-norm 38414, part 7. The concentration of selected heavy metals (Cu, Cr, Cd, Ni, Pb, Zn) are determined by atom absorption spectrometry with ICP-OES, AAS-FIMS or HGA-AAS -system (Perkin Elmer, Optima 2000 DV).

# 8 References


Abalos M, Bayona JM, Compano R, Granados M, Leal C, Prat MD (1997) Analytical procedures for the determination of organotin compounds in sediment and biota: a critical review. *J Chromat* 788, 1-49.

Abke W, Korpien H, Post B (1993) Belastung des Grundwassers im Abstrom von Gleisanlagen durch Herbizide. *Vom Wasser* 81, 257-273.

Aboul-Kassim TAT, Simoneit BRT (1996) Aliphatic and aromatic hydrocarbons in particulate fallout of Alexandria, Egypt: sources and implications. *Environ Sci Technol* 29, 2473-2483.

Abramowicz DA, Brennan MJ, van Dort HM, Gallagher E (1993) Factors influencing the rate of polychlorinated biphenyl dechlorination in Hudson river sediments. *Environ Sci Technol* 27, 1125-1131.

Abu-Quare A, Abou-Donia M (2000) Increased 8-hydroxy-2'-deoxyguanosine, a biomarker of oxidative DNA damage in rat urine following a single dermal dose of DEET (*N,N*-diethyl-*m*-toluamide), and permethrin, alone and in combination. *Toxicol Lett* 117, 151-160.

Abu-Qare AW, Duncan HJ (2002) Photodegradation of the herbicide EPTC and the safener dichlormid, alone and in combination. *Chemosphere* 46, 1183-1189.

Adler P, Steger-Hartmann T, Kalbfus W ( 2001) Vorkommen natürlicher und synthetischer östrogener Steroide in Wässern des süd- und mitteldeutschen Raumes. *Acta Hydrochim Hydrobiol* 29, 4, 227-241.

Adolfsson-Erici M, Pettersson M, Parkkonen J, Sturve J (2002) Triclosan, a commonly used bactericide found in human milk and in the aquatic environment in Sweden. *Chemosphere* 46, 1485-1489.

Adriaens P, Grbic-Galic D (1994) Cometabolic transformation of mono- and dichlorobiphenyls and chlorohydroxy-biphenyls by methanotrophic groundwater isolates. *Environ Sci Technol* 28, 1325-1330.

Ahel M, Molnar E, Ibric S, Giger W (2000) Estrogenic metabolites of alkylphenol polyethoxylates in secondary sewage effluents and rivers. *Wat Sci Technol* 42, 15-22.

Akerblom N, Olsson K, Berg AH, Andresson PL, Tysklind M, Förlin L, Norrgren L (2000) Impact of Polychlorinated Naphthalenes (PCNs) in Juvenile Baltic Salmon, *Salmo salar:* Evaluation of Estrogenic Effects, Development, and CYPIA Induction. *Arch Environ Contam Toxicol* 38, 225-233.

Albaiges J, Casado F, Ventura F (1986) Organic indicators of groundwater pollution by a sanitary landfill. *Wat Res* 20, 1153-1159.

Alberts JJ, Takács M (1999) Importance of humic substances for carbon and nitrogen transport into southeastern United States estuaries. *Org Geochem* 30, 385-396.

Alcock RE, Jones KC (1999) New directions - 'New' organic compounds in the environment. *Atmos Environ* 33, 1645-1646.

Alder AC, Häggblom MM, Oppenheimer SR, Young LY (1993) Reductive dechlorination of polychlorinated biphenyls in anaerobic sediments. *Environ Sci Technol* 27, 530-538.

Alexander M (1995) How toxic are toxic chemicals in soil ? *Environ* Sci Technol 29, 2713-2717.

Allen JE, Forney FW, Markovetz AJ (1971) Microbial subterminal oxidation of alkanes and alk-1-enes. *Lipids* 6, 448-452.

Amaral OC, Otero R, Grimalt JO (1996) Volatile and semi-volatile organochlorine compounds in tap and riverine waters in the area influenced of a chlorinated organic solvent factory. *Wat Res* 30, 1876-1884.

Ansari AA, Singh IB, Tobschall HJ (1998) Organotin compounds in surface and pore waters of Ganga plain in the Kanpur-Unnao industrial region, India. *Sci Total Environ* 223, 157-166.

Annweiler E, Meckenstock RU, Michaelis W (2001) Anaerobic cometabolic conversion of benzothiophene by a sulfate-reducing enrichment culture and in a tar-oil-contaminated aquifer. *Appl Environ Microbiol* 67, 5077-5083.

Arpino PJ, Ignatiadis I, de Ryke G (1987) Sulphur-containing polynuclear aromatic hydrocarbons from petroleum. Examination of their possible statistical formation in sediments. *J Chromat* 390, 329-348.

Aschmann SM, Arey J, Atkinson R, Simonich SL (2001) Atmospheric Lifetimes and Fates of Selected Fragrance Materials and Volatile Model Compounds. *Environ Sci Technol* 35, 3595-3600.

Ash M, Ash I (1991) Concise Encyclopedia of Industrial Chemical Additives. Hodder and Stoughton.

Asperger A, Engewald W, Fabian G (1999) Analytical characterization of natural waxes employing pyrolysis-gas chromatography-mass-spectrometry. *J Anal Appl Pyrolysis* 50, 103-115.

Atwater JW, Jasper S, Mavinic DS, Koch FA (1983) Experiments using daphnia to measure landfill leachate toxicity. *Wat Res* 17, 1855–1861.

de Almeida Azevedo D, Lacorte S, Vinhas T, Viana P, Barcelo D (2000) Monitoring of priority pesticides and other organic pollutants in river water from Portugal by gas chromatography–mass spectrometry and liquid chromatography–atmospheric pressure chemical ionization mass spectrometry. *J Chromat* 879, 13-26.

Bailey RE (2001) Global hexachlorobenzene emissions. *Chemosphere* 43, 167-182.

Bajc S, Ambles A, Largeau C, Derenne S, Vitorovic D (2001) Precursor biostructures in kerogen matrix revealed by oxidative degradation: Oxidation of kerogen from Estonian kukersite. *Org Geochem* 32, 773-784.

Ballschmiter K, Zell M (1980) Analysis of polychlorinated biphenyls (PCB) by glass capillary gas chromatography. Composition of technical Aroclor- and Clophen-PCB mixtures. *Fresenius J Anal Chem* 302, 20-31.

Ballschmitter K, Schäfer W, Buchert H (1987) Isomer-specific identification of PCB congeners in technical mixtures and environmental samples by HRGC-ECD and HRGC-MSD. *Fresenius J Anal Chem* 326, 253-257.

Ballschmitter K (2003) Pattern and sources of naturally produced organohalogens in the marine environment: biogenic formation of organohalogens. *Chemosphere* 52, 313-324.

Balk F, Ford RA (1999a) Environmental risk assessment for polycyclic musks, AHTN and HHCB I. Fate and exposure assessment. *Toxicol Lett* 111, 57-79.

Balk F, Ford RA (1999b) Environmental risk assessment for polycyclic musks, AHTN and HHCB II. Effect assessment and risk characterization. *Toxicology Lett* 111, 81-94.

Barber L, Writer JH (1998) Impact of the 1993 Flood on the Distribution of Organic Contaminants in Bed Sediments of the Upper Mississippi River. *Environ Sci Techn* 32, 2077-2083.

Baronti C, Curini R, D'Ascenzo G, Di Corcia A, Gentili A, Samperi R (2000) Monitoring natural and synthetic estrogens at activated sludge sewage treatment plants and in a receiving river water. *Environ Sci Technol* 34, 5059-5066.

Barriuso E, Houot S (1996) Rapid mineralization of the s-trazine ring of atrazine in soils in relation to soil management. *Soil Biol Biochem* 28, 1341-1348.

Battaglin WA, Goolsby DA (1999) Are shifts in herbicide use reflected in concentration changes in Midwestern rivers? *Environ Sci Technol* 33, 2917-2925.

Bayona J, Cassellas M, Fernandez P, Solanas A, Albaiges J (1994) Sources and seasonal variability of mutagenic agents in the Barcelona City aerosol. *Chemosphere* 29, 441-450.

Beaulieu JC, Grimm CC (2001) Identification of volatile compounds in Cataloupe at various developmental stages using solid phase microextraction. *J Agric Res* 49, 1345-1352.

Belfroid AC, Van der Horst A, Vethaak AD, Schäfer AJ, Rijs GBJ, Wegener J, Cofino WP (1999) Analysis and occurrence of estrogenic hormones and glucuronides in surface water and wastewater in The Netherlands. *Sci Total Environ* 225, 101-108.

Bence AE, Kvenvolden KA, Kennicutt MC (1996) Organic geochemistry applied to environmental assessments of Prince William Sound, Alaska, after the Exxon Valdez oil spill-a review. *Org Geochem* 24, 7-42.

Benner BA, Wise SA, Currie LA (1995) Distinguishing the contributions of residential wood combustion and mobile source emissions using relative concentrations of dimethylphenanthrenes. *Environ Sci Technol* 29, 2382-2389.

Beringer FM, Falk RA, Karniol M, Lillien G, Masullo G, Mausner M, Sommer E (1959) Diaryliodonium Salts. IX. The Synthesis of Substituted Diphenyliodonium Salts. *J Am Chem Soc* 81, 342-351.

Bernerd F, Vioux C, Asselineau D (2000) Evaluation of the protective effect of sunscreens on *in vitro* reconstructed human skin exposed to UVB or UVA irradiation. *Photochem Photobiol* 71, 314-320.

Bester K, Hühnerfuss H, Lange W, Theobald N (1997). Results of non target screening of lipophilic organic pollutants in the German Bight I: Benzothiazoles. *Sci Total Environ* 207, 111-118.

Bester K, Hühnerfuss H, Lange W, Rimkus GG, Theobald N (1998a) Results of non-target-screening of lipophilic organic pollutants in the German Bight II: Polycyclic musk fragrances. *Wat Res* 32, 1857-1863.

Bester K, Biselli S, Gatermann R, Hühnerfuss H, Lange W, Theobald N (1998b) Results of non-target screening of lipophilic organic pollutants in the German Bight III: 2,5-dichloroaniline. *Chemosphere* 36, 1873-1883.

Bester K, Gatermann R, Hühnerfuss H, Lange W, Theobald N (1998c) Results of non-target screening of lipophilic organic pollutants in the Germann Bight IV: Identification and quantification of chloronitrobenzenes and dichloronitrobenzenes. *Environ Poll* 102, 163-169.

Bester K, Theobald N (2000) Results of non-target screening of lipophilic organic pollutants in the German Bight V: xanthen-9-one. *Wat Res* 34, 2277-2282.

Bethan B, Bester K, Hühnerfuss H, Rimkus G (1997) Bromocyclen contamination in surface water, waste water and fish from Northern Germany, and gas chromatographic chiral separation. *Chemosphere* 34, 2271-2280.

Bethan B, Dannecker W, Gerwig H, Hühnerfuss H, Schulz M (2001) Seasonal dependence of the chiral composition of □-HCH in coastal deposition at the North Sea. *Chemosphere* 44 , 591-597.

Böckle K, Stieber M, Werner P, Frimmel FH (1991) Degradation of polycyclical aromatic hydrocarbons in the subsurface. *Vom Wasser* 76, 309-319.

Birchenough AC, Barnes N, Evans SM, Hinz H, Krönke I, Moss C (2002) A review assessment of tributyltin contamination in the North Sea, based on surveys of butyltin tissue burdens and imposex/intersex in four species of neogastropods. *Mar Poll Bull* 44, 534-543.

Boelens M, de Valois PJ, Wobben HJ, van der Gen A (1971) Volatile Flavor Compounds from Onion. *J Agr Food Chem* 19, 984-991.

Bohlen H, Hicke K, Stöbel AO (1989) Die Belastung der Unterweser im bremischen Raum mit Halogenorganika und Phosphorsäureestern. Teil 1. *Vom Wasser* 72, 185-197.

Bohlen H, Hicke K, Stoebel AO, Thiemann W (1990) Die Belastung der Unterweser im bremischen Raum mit Halogenorganika und Phosphorsäureestern, Teil 2. *Vom Wasser* 74, 375-381.

Bollhöfer A, Mangini A, Lenhard L, Wessels M, Giovanoli F, Schwarz B (1994) High Resolution $^{210}$Pb dating of Lake Constance sediments: Stable lead in Lake Constance. *Environ Geol* 24, 267-274.

Bolz U, Hagenmaier H, Körner W (2001) Phenolic xenoestrogens in surface water, sediments, and sewage sludge from Baden- Württemberg, south-west Germany. *Environ Pollut* 115, 291-301.

Booker RS, Pavlostathis SG (2000) Microbial reductive dechlorination of hexachloro-1,3-butadien in a methanogenic enrichment culture, *Wat Res* 34 (18), 4437-4445.

Bopp RF, Simpson HJ, Olsen CR, Kostyk N (1981) Polychlorinated biphenyls in sediments of the tidal Hudson river, New York. *Environ Sci Technol* 15, 210-216.

Boul HL, Garnham MLG, Hucker D, Baird D, Alslable J (1994) Influence of agricultural practices on the levels of DDT and its residues in soil. *Environ Sci Technol* 28, 1397-1402.

Brand WA (1996) High precision isotope ratio monitoring techniques in mass spectrometry. *J Mass Spectrom* 31, 225-235.

Brauch HJ (1993) Pesticides in the River Rhine. *Acta Hydrochim Hydrobiol* 21, 137-144.

Breitung V (1992) Belastung der Saar mit Ugilec 141 und PCB. *Vom Wasser* 79, 185-197.

Brownawell BJ, Farrington JW (1986) Biogeochemistry of PCBs in interstitial waters of a coastal marine sediment. *Geochim Cosmochim Acta* 50, 157-169.

Bruijn J, Busser F, Seinen W (1989) Determination of octanol/water partition coefficients for hydrophobic organic chemicals with the 'slow stirring' method. *Environ Tox Chem* 8, 499-512.

Bruns WA, Mankiewicz PJ, Bence E, Page DS, Parker KR (1997) A principal-component and least-squares method for allocation polycyclic aromatic hydrocarbons in sediment to multiple sources. *Environ Toxicol Chem* 16, 1119-1131.

Brusseau GA, Tsien HC, Hanson RS, Wackett LP (1990) Optimization of trichloroethylene oxidation by methanotrophs and the sue of a colorimetric assay to detect soluble methane monooxygenase activity. *Biodegrad* 1, 19-29.

Bryant JG (1993) Chlorocarbons and Chlorohydrocarbons: Chlorinated benzenes. In Koschwitz JI, Howe-Grant M (Eds.) *Kirk-Othmer – Encyclopedia of Chemical Technology* 6, 4th Ed., John Wiley & Sons, New York, 87-101.

Buchert H, Bihler S, Ballschmiter K (1982) Untersuchungen zur globalen Grundbelastung mit Umweltchemikalien. VII Hochauflösende Gaschromatographie persistenter Chlorkohlenwasserstoffe (CKW) und Polyaromaten (AKW) in limnischen Sedimenten unterschiedlicher Belastung. *Fresenius J Anal Chem* 313, 1-20.

Buchert H, Ballschmiter K (1985) Mustererkennung von polychlorierten Biphenylen (PCB) in Umweltproben. *Fresenius J Anal Chem* 320, 709-710.

Buerge IJ, Poiger T, Müller MD (2003) Caffeine, an anthropogenic marker for waste water contamination of surface waters. *Environ Sci Technol* 37, 691-700.

Buffle J, Leppard GG (1995) Characterization of aquatic colloids and macromolecules: 1. Structure and behavior of colloidal material. *Environ Sci Technol* 29, 2169-2175.

Bulterman AJ, van Loon WMGM, Ghijsen RT, Brinkmann UAT, Huitema IM, de Groot B (1997) Highly selective determination of macromolecular chlorolignosulfonic acids in river and drinking water using curie-point pyrolysis-gas chromatography-tandem mass spectrometry. *Environ Sci Technol* 31, 1946-1952.

Bundt J, Herbel H, Steinhart H, Franke S, Francke W (1991) Structure-type separation of diesel fuels by solid phase extraction and identification of the two- and three-ring aromatics by capillary GC-mass spectrometry. *J High Resolut Chromat* 14, 91-98.

Bunge M, Ballerstedt H, Lechner U (2001) Regiospecific dechlorination of spiked tetra- and trichlorodibenzo-*p*-dioxins by anaerobic bacteria from PCDD/F-contaminated Spittelwasser sediments. *Chemosphere* 43, 675-681.

Burns IG, Hayes MHB, Stacey M (1973) Spectroscopic studies on the mechanisms of adsorption of paraquat by humic acid and model compounds. *Pestic Sci* 4, 201-209.

Burns W, Mankiewics P, Bence E, Page D, Parker K (1997) A principal-component and least squares methode for allocating polycyclic aromatic hydrocarbons in sediment to multiple sources. *Environ Tox Chem* 16, 1119-1131.

Buser HR, Zook DR, Rappe C (1992) Determination of methyl sulfone-substituted polychlorobiphenyls by mass spectrometric techniques with application to environmental samples. *Anal Chem* 64, 1176-1183.

Buser HR, Poiger T, Müller MD (1998a) Occurrence and Fate of the Pharmaceutical Drug Diclofenac in Surface Waters: Rapid Photodegradation in a Lake. *Environ Sci Technol* 32, 3449-3456.

Buser HR, Müller MD, Theobald N (1998b) Occurrence of the pharmaceutical drug clofibric acid and the herbicide mecoprop in various Swiss lakes and in the North Sea. *Environ Sci Technol* 32, 188-192.

Buser HR, Poiger T, Müller MD (1999) Occurrence and Environmental Behaviour of the Chiral Pharmaceutical Drug Ibuprofen in Surface Waters and in Wastewater. *Environ Sci Technol* 33, 2529-2535.

Cai Y, Rapsomanikis S, Andreae MO (1993) Determination of butyltin compounds in sediment using gas chromatography – atomic absorption spectrometry: comparison of sodium tetrahydroborate and sodium tetraethylborate derivatization methods. *Anal Chim Acta* 274, 243-251.

Cameron RD, Koch FA (1980) Toxicity of landfill leachates. *J Wat Pollut Control Fed* 52, 760–769.

Capriel P, Haisch A, Khan U (1985) Distribution and nature of bound (nonextractable) residues of atrazine in a mineral soil nine years after the herbicide application. *J Agric Food Chem* 33, 587-589.

Castillo M, Barceló D (1999) Determination of organic pollutants in industrial wastewater effluents. In: Sample Handling and Trace Analysis of Pollutants: Techniques, Applications and Quality Assurances, Barceló D (Ed.). Elsevier Science B.V., 537-583.

Catallo WJ, Schlenker M, Gambrell RP, Shane BS (1995) Toxic chemicals and trace metals from urban and rural Louisiana lakes: Recent historical profiles and toxicological significance. *Environ Sci Technol* 29, 1436-1445.

Chakhmakhchev S, Suzuki M, Takayama K (1997) Distribution of alkylated dibenzothiophenes in petroleum as a tool for maturity assessments. *Org Geochem* 26, 483-490.

Chalaux N, Takada H, Bayona JM (1995) Molecular markers in Tokyo Bay sediments: sources and distribution. *Marine Environ Res* 40, 77-92.

Chan KH, Lam MHW, Poon KF, Yeung HY, Chiu TKT (1998) Application of sedimentary fecal stanols and sterols in tracing sewage pollution in coastal waters. *Wat Res* 32, 225-235.

Chau YK, Wong PT (1991) Recent developments in speciation and determination of organometallic compounds in environmental samples. *Fresenius J Anal Chem* 339, 640-645.

Chen J, Wang M, Ho CT (1998) Volatile compounds generated from the thermal degradation of N-acetylglucosamine. *J Agric Food Chem* 46, 3207-3209.

Chian ESK, DeWalle FB (1977) Characterization of soluble organic matter in leachate. *Environ Sci Technol* 11, 158-163.

Chin YP, Gschwend PM (1992) Partitioning of polycyclic aromatic hydrocarbons to marine porewater organic colloids. *Environ Sci Technol* 26, 1621-1626.

Choudhary G (1995) Human health perspectives on environmental exposure to hexachlorobutadiene: A review. *Environmental carcinogenesis and ecotoxicology reviews* C13, 179-203.

Cikryt P (1991) Die Gefährdung des Menschen durch Dioxin und verwandte Verbindungen. *Nachr Chem Tech Lab* 39, 648-656.

Claus E, Heininger P, Bade M (1998) Mass spectrometric identification of polychlorinated dibenzothiophens (PCDBTs) in surface sediments of the River Elbe. *Fresenius J Anal Chem* 361, 54-58.

Colombo JC, Pelletier E, Brochu C (1989) Determination of hydrocarbon sources using *n*-alkane and polyaromatic hydrocarbon distribution indexes. Case study: Rio de la Plata estuary, Argentina. *Environ Sci Technol* 23, 888-894.

Coates JD, Chakraborty R, McInerney MJ (2002) Anaerobic benzene biodegradation - a new era. *Res Microbiol* 153, 621-628.

Colby J, Stirling DI, Dalton H (1977) The soluble methan monooxygenase of *Methylococcus capsulatus* (Bath). *Biochem J* 165, 395-402.

Connolly JP, Zahakos HA, Benaman J, Ziegler CK, Rhea JR, Russell K (2000) A model of PCB fate in the upper Hudson river. *Environ Sci Technol* 34, 4076-4087.

Corin NS, Backlund PH, Kulovaara MAM (2000) Photolysis of the Resin Acid Dehydroabietic Acid in Water. *Environ Sci Technol* 34, 2231-2236.

Countway RE, Dickhut RM, Canuel EA (2003) Polycyclic Aromatic Hydrocarbons (PAH) distributions and associations with organic matter in surface waters of the York river, VA Estuary. *Org Geochem* 34, 209-224.

Cozzarelli IM, Baedecker MJ, Eganhouse RP, Goerlitz DF (1994) The geochemical evolution of low-molecular-weight organic acids drived from the degradation of petroleum contaminants in groundwater. *Geochim Cosmochim Acta* 58, 863-877.

Cranwell PA (1981a) Diagenesis of free and bound lipids in terrestrial detritus deposited in a lacustrine sediment. *Org Geochem* 3, 79-89.

Cranwell PA (1981b) Alkyl esters in recent sediments of two productive lakes. In Bjoroy M *et al* (Eds.) *Advances in Org Geochem* 1981, John Wiley & Sons, Chichester, 299-308.

Cranwell PA, Eglinton G, Robinson N (1987) Lipids of aquatic organism as potential contributors to lacustrine sediments – II. *Org Geochem* 11, 513-527.

Dachs J, Bayona JM, Raoux C, Albaigés J (1997) Spatial, vertical distribution and budget of polycyclic aromatic hydrocarbons in the Western Mediterranian seawater. *Environ Sci Technol* 31, 682-688.

Dahlman O, Mörck R, Ljungquist P, Reimann A, Johansson C, Boren H, Grimvall A (1993) Chlorinated structural elements in high molecular weight organic matter from unpolluted waters and bleached-kraft mill effluents. *Environ Sci Technol* 27, 1616-1620.

Danish Environmental Protection Agency (2001) Environmental and Health Assessment of Alternatives to Phthalates and to flexible PVC. *Environ Project No. 590.*

Dannenberger D, Lerz A (1998) Chlorinated hydrocarbons in river sediments of Mecklenburg - West Pomerania - Patterns, levels and regional distribution. *Acta Hydrochim Hydrobiol* 26, 263-271.

Dannenberger D, Lerz A (1999) Occurrence and transport of organic micro-contaminants in sediments of the Odra river estuarine system. *Acta Hydrochim Hydrobiol* 27, 303-307.

Daughton CG, Ternes TA (1999) Pharmaceuticals and Personal Care Products in the Environment: Agents of subtile change?. *Environ Health Perspect* 107 Suppl 6, 907-944.

DeLorenzo ME, Scott GI, Ross PE (2001) Toxicity of pesticides to aquatic microorganism: A review. *Environ Tox Chem* 20, 84-98.

Dempster HS, Sherwood Lollar B, Feenstra S (1997) Tracing organic contaminants in groundwater: A new methodology using compound-specific isotopic analysis. *Environ Sci Technol* 31, 3193-3197.

Desbrow C, Routledge EJ, Brighty GC, Sumpter JP, Waldock M (1998) Identification of estrogenic chemicals in STW effluent. 1. Chemical fractionation and in vitro biological screening. *Environ Sci Technol* 32, 1549-1558.

De Wit CA (2002) An overview of brominated flame retardants in the environment. *Chemosphere* 46, 583-624.

Dewulf JP, van Langenhove LF (1998) Air/water exchange dynamics of 13 volatile chlorinated C1- and C2-hydrocarbons and monocyclic aromatic hydrocarbons in the southern North Sea and the Scheldt estuary. *Environ Sci Technol* 32, 903-911.

Di Corcia A, Capuani L, Casassa F, Marcomini A, Samperi R (1999) Fate of linear alkyl benzenesulfonates, coproducts, and their metabolites in sewage treatment plants and a receiving river water. *Environ Sci Technol* 33, 4119-4125.

Ding WH, Tsing SH (1998) Analysis of nonylphenol polyethoxylate and their degradation products in river water and sewage effluents by gas chromatography ion trap (tandem) mass spectrometry with electron impact and chemical ionization. *J Chromat* 824, 79-90.

Doll A, Püttmann W (1999) Natural Attenuation. Sanierung von Mineralölkontaminationen in Boden und Grundwasser durch natürliche Rückhalte- und Abbauprozesse in den USA. *Altlasten Spektrum*, 331-339.

Dominik A, Mangini A, Müller M (1981) Determination of recent deposition rates in Lake Constance with radioisotopic methods. *Sedimentol* 28, 653-677.

Doong RA, Sun YC, Liao PL (2002) Distribution and fate of organochlorine pesticide residues in sediments from selected rivers in Taiwan. *Chemosphere* 48, 237-246.

Dsikowitzky L (2002) PhD Thesis, Aachen University, 131 pp.

Dsikowitzky L, Schwarzbauer J, Littke R (2002) Distribution of polycyclic musks in water and particulate matter of the Lippe River. *Org Geochem* 33, 1747-1758.

Dsikowitzky L, Schwarzbauer J, Kronimus A, Littke R (2004a) The anthropogenic contribution to the organic load of the Lippe river (Germany). Part 1: Qualitative characterisation of low-molecular weight organic compounds. *Chemosphere* 57, 1275-1288 .

Dsikowitzky L, Schwarzbauer J, Kronimus A, Littke R (2004b) The anthropogenic contribution to the organic load of the Lippe river (Germany). Part 2: Quantification of specific organic contaminants. *Chemosphere* 57, 1289-1300.

Ducrotoy J-P, Elliott M, De Jonges VN (2000) The North Sea. *Mar Poll Bull* 41, 5-23.

Duffield P, Bourne D, Tan K, Garruto RM, Duncan MW (1994) Analysis of the neurotoxic plasticizer n-butylbenzenesulfonamide by gas chromatography combined with accurate mass selected ion monitoring. *J Anal Toxicol* 18, 361-368.

Duinker JC, Schulz DE, Petrick G (1988) Multidimensional gas chromatography with electron capture detection for the determination of toxic congeners in polychlorinated biphenyl mixtures. *Anal Chem* 60, 478-482.

Dunn WJ, Stalling DL, Schwartz JW (1984) Pattern recognition for classification and determination of polychlorinated biphenyls in environmental samples. *Anal Chem* 56, 1308-1313.

Eganhouse RP, Kaplan IR (1985) α-Tocopherol acetate as an indicator of municipal waste contamination in the environ-ment. *Environ Sci Technol* 19, 282-285.

Eganhouse RP (1997) Molecular markers and environmental organic geochemistry: an overview. In Eganhouse RP (Ed.) *Molecular markers in Environmental Geochemistry. ACS Symposium Series* 671, Washington D.C., 1-21.

Eganhouse RP, Sherblom PM (2001) Anthropogenic organic contaminants in the effluent of a combined sewer overflow: impact on Boston Harbour. *Mar Environ Res* 51, 51-74.

Ehmann J, Ballschmitter K (1989) Isomer-specific determination of tetrachlorobenzyltoluenes (TCBT) in the technical mixture Ugilec 141 by capillary gas chromatography. *Fresenius J Anal Chem* 332, 904-911.
Eisele M, Schorer M (1997) Schadstoffsorption an Biofilmen in Fließgewässern. *Vom Wasser 89, 139-151.*
Eisenreich SJ, Capel D, Robbins JA, Bourbonniere R (1989) Accumulation and diagenesis of chlorinated hydrocarbons in lacustrine sediments. *Environ Sci Technol* 23, 1116-1126.
Elizalde-Gonzalez MP, Hutfließ M, Hedden K (1996) Retention index system, adsorption characteristics and structure correlations of polycyclic aromatic hydrocarbons in fuels. *J High Res Chrom* 19, 345-352.
Engel MH, Macko SA (1993) *Organic Geochemistry – Principles and Applications*, Plenum, New York.
Eschke HD, Tarud J, Dibowski HJ (1994) Untersuchungen zum Vorkommen polycyclischer Moschus-Duftstoffe in verschiedenen Umweltkompartimenten (1. Mitteilung). *Umweltchem Ökotox* 6, 183-189.
Eschke HD, Tarud J, Dibowski HJ (1995a) Untersuchungen zum Vorkommen polycyclischer Moschus-Duftstoffe in verschiedenen Umweltkompartimenten (2. Mitteilung). *Umweltchem Ökotox* 7, 131-138.
Eschke HD, Dibowski HJ, Traud J (1995b) Nachweis und Quantifizierung von polycyclischen Moschus-Duftstoffen mittels Ion-Trap GC/MS/MS in Humanfett und Muttermilch. *Deutsche Lebensmittel-Rundschau* 12, 375-379.
Eschke HD (1996) Organische Spurenstoffe in der Ruhr. In: Ruhrwassergüte (1996), Ruhrverband, Essen, Germany, 86-90.
Espadaler I, Caixach J, Om J, Ventura F, Cortina M, Pauné F, Rivera J (1997) Identification of organic pollutants in Ter River and its system of reservoirs supplying water to Barcelona (Catalonia, Spain): A study by GC/MS and FAB/MS. *Wat Res* 31, 1996-2004.
Esteves VI, Duarte AC (2000) Differences between Humic Substances from Riverine, Estuarine, and Marine Environments Observed by Fluorescence Spectroscopy. *Acta Hydrochim Hydrobiol* 28,359-363.
Evers EHG, Re KCM, Olie K (1988) Spatial variations and correlations in the distribution of PCDDs, PCDFs and related compounds in sediments from the River Rhine - Western Europe. *Chemosphere* 17, 2271-2288.
Evers EH, Berghem JW, van Ollie K (1998) Exploratory data analysis of PCDD and PCDF measurements in sediments from industrialized areas. *Chemosphere* 39, 459-466.
Fabbri D, Trombini C, Vassura I (1998a) Analysis of polystyrene in polluted sediments by pyrolysis-gas chromatography-mass spectrometry. *J Chrom Sci* 36, 600-604.
Fabbri D, Mongardi M, Mintanari L, Galletti GC, Chiavari G, Scotti R (1998b) Comparison between CP/MAS $^{13}$C-NMR and pyrolysis-GC/MS in the structural characterization of humins and humic acids of soil and sediments. *Fresenius J Anal Chem* 362, 299-306.

Falandysz J, Strandberg L, Bergquvist P (1996) Polychlorinated naphthalenes in sediment and biota from the Gdansk Basin, Baltic Sea. *Environ Sci Technol* 30, 3266-3274.

Falandysz J (1998) Polychlorinated naphthalenes: an environmental update. *Environ Pollut* 101, 77-90.

Farre, M, Ferrer I, Ginebreda A, Figueras M, Olivella L, Tirapu Lluis, Vilanova M, Barcelo D (2001) Determination of drugs in surface water and wastewater samples by liquid chromatography – mass spectrometry: Methods and Preliminary results including toxicity studies with *Vibrio fischeri*. *J Chromat A* 938, 187-197.

Faure P, Landais P, Schlepp L, Michels R (2000) Evidence for Diffuse Contamination of River Sediments by Road Asphalt Particles. *Environ Sci Technol* 34, 1174-1181.

Fent K, Hunn J (1991) Phenyltins in water, sediment and biota of freshwater marines. *Environ Sci Technol* 25, 958-963.

Fent K (1996) Organotin in municipal wastewater and sewage sludge. In *Organotin – Environmental fate and effects,* Champ MA, Seligman PF, eds. Chapman and Hall, London, Chapter 28, 581-600.

Fernandes MB, Sicre MA, Boireau A, Tronczynski J (1997) Aquatic hydrocarbon distributions in the Seine estuary: biogenic polyaromatics and *n*-alkanes. *Estuaries* 20, 281-290.

Fernholz E (1938). On the constitution of α-tocopherol. *J Amer Chem Soc* 60, 700-705.

Fiehn O, Wegener G, Jochimsen J, Jekel M (1998) Analysis of the ozonation of 2-mercaptobenzothiazole in water and tannery wastewater using sum parameters, liquid- and gas chromatography and capillary electrophoresis. *Wat Res* 32, 1075-1084.

Flintrop C, Hohlmann B, Jasper T, Korte C, Podlaha OG, Scheele S, Veizer J (1996) Anatomy of pollution: Rivers of North Rhine – Westphalia, Germany. *Americ J Sci* 296, 58-98.

Focant J-F, Eppe G, Pirard C, De Pauw E (2001) Fast clean-up for polychlorinated dibenzo-*p*-dioxins, dibenzofurans and coplanar polychlorinated biphenyls analysis of high-fat-content biological symples. *J Chromat* 925, 207-221.

Ford RA (1998) The Human Safety of the Polycyclic Musks AHTN and HHCB in Fragrances – A Review. *Deutsche Lebensmittel-Rundschau* 8, 268-275.

Ford RA, Hawkins DR, Schwarzenbach R, Api AM (1999) The systemic exposure to the polycyclic musks, AHTN and HHCB, under conditions of use as fragrance ingredients: Evidence of lack of complete absorption from a skin reservoir. *Toxicol Lett* 111, 133-142.

Först C, Stieglitz L, Roth W, Kuhnmünch S (1989) Quantitative analysis of volatile organic compounds in landfill leachates. *Intern J Environ Anal Chem* 37, 287-293.

Foster GD, Lippa KA, Miller CV (2000) Seasonal concentrations of organic contaminants at the fall line of the Susquehanna river basin and estimated fluxes to the Northern Chesapeake Bay, USA. *Environ Tox Chem* 19, 992-1001.

Fox WM, Copplestone CD, Johnson MS, Leah RT (2001) The organochlorine contamination history of Mersey estuary, UK, revealed by analysis of sediment cores form salt marshes. *Marine Environ Res* 51, 213-227.

Francke W, Franke S, Hildebrandt S, Link M, Schwarzbauer J (1995) Spurenanalytik organischer Substanzen zur Charakterisierung der Belastung und Festlegung sanierungs-relevanter Leitparameter. In: *Die Belastung der Elbe. Teil I - Elbenebenflüsse*, Forschungszentrum Karlsruhe (Hrsg.), Bonn, 136-160.

Franke S, Hildebrandt H, Schwarzbauer J, Link M, Francke W (1995) Organic compounds as contaminants of the Elbe river and its tributaries. Part II: GC/MS screnning analyses for contaminants of the Elbe river water. *Fresen J Anal Chem* 353, 39-49.

Franke S, Schwarzbauer J, Link M, Francke W (1996) Identifizierung dimerer PAK und anderer ungewöhnlicher Aromaten in Umweltproben. In: *Neue Techniken der Bodenreinigung*, Stegmann, R. (Hrsg.), Economica Verlag, Bonn, 149-168.

Franke S, Schwarzbauer J, Francke W (1998a) Arylesters of alkylsulfonic acids in sediments. Part III of organic compounds as contaminants of the Elbe river and its tributaries.*Fresenius J Anal* Chem 360, 580-588.

Franke S, Hildebrandt S, Francke W, Bester K, Hühnerfuss H, Gatermann R (1998b) Chlorinated bis(propyl)ether and chlorinated bis(ethyl)formals in the German Bight of the North Sea. *Mar Poll Bull* 36, 546-551.

Franke S, Meyer C, Heinzel N, Gatermann R, Hühnerfuss H, Rimkus G, König WA, Francke W (1999) Enantiomeric Composition of the Polycyclic Musks HHCB and AHTN in Different Aquatic Species. *Chirality* 11, 795-801.

Friege H, Kirchner W, Leuchs W (1991) Pestizide in Oberflächengewässern. Teil 1: Herbizid-Belastungen in der Lippe und im westdeutschen Kanalnetz. *Vom Wasser* 76, 29-38.

Fries E, Püttmann W (2001) Occurrence of organophosphate esters in surface water and ground water in Germany. *Journal of Environmental Monitoring* 3, 621-626.

Frimmel FH, Grenz R, Kordik E, Dietz F (1989) Nitriloacetat (NTA) und Ethylendinitrilotetraacetat (EDTA) in Fließgewässern der Bundesrepublik Deutschland. *Vom Wasser* 72, 175-184.

Fromme H, Otto T, Pilz K, Neugebauer F (1999) Levels of synthetic musks: Bromocyclene and PCBs in eel (*Anguilla anguilla*) and PCBs in sediment samples from some waters of Berlin/Germany. *Chemosphere* 39, 1723-1735.

Fromme H, Otto T, Pilz K (2001) Polycyclic musk fragrances in different environmental compartments in Berlin (Germany). *Wat Res* 35, 121-128.

Fuoco R, Colombini MP, Samcova E (1993) Individual determination of ortho- and non-ortho substituted polychlorinated biphenyls (PCBs) in sediments by high performance liquid chromatographic pre-separation and gas chromatography - ECD detection. *Chromatographia* 36, 65-70.

Furlong ET, Kotels JC, Aiken GR (1997) Transport and degradation of semivolatile hydrocarbons in a petroleum-contaminated aquifer, Bemidji,

Minnesota. In: Eganhouse, R.P. (Ed.) *Molecular markers in Environmental Geochemistry, ACS Sympos Ser* 671, Washington D.C., 398-412.

Garrison AW, Pope JD, Allen FR (1979) In: *Identification and analysis of organic pollutants in water*. Ann Arbor Science Publ, Ann Arbor, MI, 517-523.

Gatermann R, Hühnerfuss H, Rimkus G, Wolf M, Franke S (1995) The distribution of nitrobenzene and other nitroaromatic compounds in the North Sea. *Marine Pollut Bull* 30, 221 - 227.

Gatermann R, Bester K, Franke S, Hühnerfuss H (1996) The distribution of O,O,O-trimethylthiophosphate and O,O,S-trimethyldithiophosphate in the North Sea. *Chemosphere* 32, 1907 - 1918.

Gatermann R, Hellou J, Hühnerfuss H, Rimkus G, Zitko V (1999) Polycyclic and nitro musks in the environment: A comparison between Canadian and European aquatic biota. *Chemosphere* 38, 3431-3441.

Gatermann R, Biselli R, Hühnerfuss H, Rimkus GG, Hecker M, Karbe L (2002) Synthetic musks in the environment. Part 1: species-dependent bioaccumulation of polycyclic and nitro musk fragrances in freshwater fish and mussels. *Archives Environ Contam Toxicol* 42, 437-446.

Gebauer H, Bouter T (1997) Moschus. *European Cosmetics* 1, 30-35.

Gelin F, Boogers I, Noordeloos AAM; Sinninghe Damste JS, Riegmann R, de Leeuw JW (1997) Resistant biomacromolecules in marine microalgae of the classes Eustigmatophyceae and Chlorophyceae: Geochemical implications. *Org Geochem* 26, 659-674.

Gevao B, Harner T, Jones KC (2000) Sedimentary record of poly-chlorinated naphthalene concentrations and deposition fluxes in a dated lake core. *Environ Sci Technol* 34, 33-38.

Gevao B, Mordaunt C, Semple KT, Piearce TG, Jones KC (2001) Bioavailability of nonextractable (bound) pesticide residues to earthworms. *Environ Sci Technol* 35, 501-507.

Geyer HJ, Rimkus G, Scheunert I, Kaune A, Schramm KW, Kettrup A, Zeeman M, Muir DCG, Hansen LG. Mackay D (2000) Bioaccumulation and occurrence of endocrine disrupting chemicals (EDCs), persistent organic pollutants (POPs), and other organic compounds in fish and other organisms including humans. In: Beek B (Eds.): *The handbook of environmental chemistry, Vol. 2, Part J: Bioaccumulation*, Springer, Berlin, Heidelberg.

Gintautas PA, Danial SR, Macalady DL (1992) Phenoxyalkanoic acid herbicides in municipal landfill leachates. *Environ Sci Techno* 26, 517-521.

Glinzer O, Severin D, Bedürftig C, Czogalla CD, Puttins U (1983) Mass spectrometry of sulphur compounds as model substances for crude oil analysis – bisanellated thiophenes. *Fresenius J Anal Chem* 315, 208-212.

Göbbels FJ, Püttmann W (1997) Structural investigation of isolated aquatic fulvic and humic acids in seepage water of waste deposits by pyrolysis-gas chromatography/mass spectrometry. *Wat Res* 31, 1609-1618.

Gomez-Ariza JL, Giraldez I, Morales E (2001a) Occurrence of organotin compounds in water, sediments and mollusca in estuarine systems in the southwest of Spain. *Wat Air Poll* 126, 253-270.

Gomez-Ariza JL, Morales E, Goraldez I, Sanchez-Rodas D, Velasco A (2001b) Sample treatment in chromatography-based speciation of organometallic pollutants. *J Chromat A* 938, 211-224.

Goßel W, Püttmann W (1991) Untersuchungen zum mikrobiologischen Abbau von Kohlenwasserstoffen im Grundwasserkontaktbereich. *GWF – Wasser, Abwasser* 132, 126-131.

Götz R (1985) Chlorinated dioxins (PCDD), chlorinated dibenzofurans (PCDF) and other toxic substances in leachate of the sanitary landfill Georgswerder/Hamburg. *Vom Wasser* 65, 215-228

Götz R, Friesel P, Roch K (1993) Polychlorinated-*p*-dioxins (PCDD), Dibenzofurans (PCDF) and other chlorinated compounds in the river Elbe: Results on bottom sediments and fresh sediments collected in sedimentation chambers. *Chemosphere* 27, 105-111.

Götz R, Enge P, Friesel P (1994a) Quantification of polychlorinated biphenyls (PCB) and hexachlorobenzene (HCB) in the picogram/liter range in water of the river Elbe. *Fresenius J Anal Chem* 348, 694-695.

Götz R, Friesel P, Roch K, Ebel C, Siebert J, Reincke H, Wolff S, Specht W, Weeren RD, Pahnke J, Rußbuelt P (1994b) Pesticides in the River Elbe, in tributaries of the River Elbe and in small surface waters of the River Elbe catchment area. *Vom Wasser* 83, 43-55.

Götz R, Steiner B, Friesel P (1995) Polychlorinated dibenzo-*p*-dioxins and dibenzofurans in suspended particulate matter of the River Elbe 1994. *Fresenius J Anal Chem* 353, 111-113.

Grasset L, Ambles A (1998) Structure of humin and humic acid from an acid soil as revealed by phase transfer catalyzed hydrolysis. *Org Geochem* 29, 881-891.

Green J, Dalton H (1989) Substrate specifity of soluble methane monooxygenase. *J Biol Chem* 264, 17698-17703.

Grimmer G, Jacob J, Naujack KW (1981a) Profile of the polycyclic aromatic hydrocarbons from lubricating oil – inventory by GC/MS. *Fresenius Z Analyt Chem* 306, 347-355.

Grimmer G, Jacob J, Naujack KW, Dettborn G (1981b) Profile of the polycyclic aromatic hydrocarbons from used engine oil – inventory by GC/MS. *Fresenius Z Analyt Chem* 309, 13-19.

Grohmann K, Gilbert E, Eberle SH (1998) Identifizierung von Stickstoffverbindungen niederer molarer Masse in Abläufen einer kommunalen Kläranlage mit biologischer Reinigungsstufe. *Acta Hydrochim Hydrobiol* 26, 20-30.

Guardiola J, Ventura F, Rivera J (1989) Occurrence of industrial organic pollution in a ground water supply: Screening, monitoring and evaluation of treatment processes. *Wat Supply* 7, 11-16.

Gupta LP, Subramanian V, Ittekkot V (1997) Biogeochemistry of particulate matter transported by the Godavaria River. *Biogeochem* 38, 103-128.

Gustafsson Ö, Axelman J, Broman D, Eriksson M, Dahlgaard H (2001) Process-diagnostid patterns of chlorobiphenyl congeners in two radiochronologically characterized sediment cores from the northern Buffin Bay. *Chemosphere* 45, 759-766.

Guthrie EA, Bortiatynsky JM, van Heemst JD, Richamn JE, Hardy KS, Kovach EM, Hatcher PG (1999) Determination of [$^{13}$C]pyrene sequestration in sediment microcosms using flash-pyrolysis-GC-MS and $^{13}$C-NMR. *Environ Sci Technol* 33, 119-125.

van Haelst AG, Bakboord J, Parsons JR, Govers HAJ (1995) Biodegradation of tetrachlorobenzyltoluenes and polychlorinated biphenyls in activated sludge and in cultures of alcaligenes sp. *Chemosphere* 31, 2799-2808.

Haderlein SB, Weissmacher KW, Schwarzenbach RP (1996) Specific adsorption of nitroaromatic explosives and pesticides to clay minerals. *Environ Sci Technol* 30, 612-622.

Haglund P, Jakobsson E, Asplund L, Athanasiadou M, Bergmann A (1993) Determination of polychlorinated naphthalenes in polychlorinated biphenyl products via capillary gas chromatography-mass spectrometry after separation by gel permeation chromatography. *J Chromat* 634, 79-86.

Halling-Sørensen B, Nors Nielsen S, Lanzky PF, Ingerslev F, Holten Lützhøft HC, Jørgensen SE (1998) Occurrence, Fate and Effects of Pharmaceutical Substances in the Environment – A Review. *Chemosphere* 36, 357-393.

Hansen PD (1986) Nitriloacetat in den Berliner Gewässern und in der Elbe. *Vom Wasser* 66, 167-176.

Hardy ML (2002) The toxicology of the three commercial polybrominated diphenyl oxide (ether) flame retardants. *Chemosphere* 46, 757-777.

Harino H, Fukushima M (1992) Simultaneous determination of butyltin and phenyltin compounds in the aquatic environment by gas chromatography. *Anal Chim Acta*, 264, 91-96.

Harmsen J (1993) Identification of organic compounds in leachate from a waste tipp. *Wat Res* 17, 699-705.

Harrington RR, Poulson SR, Drever JI, Colberg PJS, Kelly EF (1999) Carbon isotopic systematics of monoaromatic hydrocarbons: vaporization and absorption experimments. *Org Geochem* 30, 765-775.

Hart S (1996) In situ bioremediaton: defining the limits. new approaches to engineered and intrinsic bioremediation are being developed and field tested. *Environ Sci Technol* 30, 398A-401A.

Haslam J, Soppet W, Willis HA (1951) Analytical examination of plasticizers obtained from PVC compositions. *J Appl Chem* 1, 112-124.

Hatcher PG, Bortiatynsky JM, Minard RD, Dec J, Bolag J (1993) Use of high resolution $^{13}$C-NMR to examine the enzymatic covalent binding of $^{13}$C-labelled 2,4-dichlorphenol to humic substances. *Environ Sci Technol* 27, 2098-2103.

Hattori Y, Kobayashi A, Takemoto S (1984) Determination of trialkyltin, dialkyltin and triphenyltin compounds in environmental water and sediments. *J Chromat* 315, 341-349.

Hatzinger PB, Alexander M (1995) Effect of aging of chemicals in soil on their biodegradability and extractibilty. *Environ Sci Technol* 29, 537-545.

Headley J, Peru K, Friesen D, Neu T (2001) Liquid chromatography –tandem mass spectrometry determination of *N*-methylpyrrolidinone in riverine biofilms. *J Chromat* A 917, 159-165.

Heberer Th, Schmidt-Bäumler K, Stan H-J (1998) Occurrence and distribution of organic contaminants in the aquatic system in Berlin. Part I: Drug residues and other polar contaminants in Berlin surface and groundwater. *Acta Hydrochim Hydrobiol* 26, 272-278.

Heberer T, Dünnbier U (1999a) DDT metabolite bis(chlorophenyl)acetic acid: The neglected environmental contaminant. *Environ Sci Technol* 33, 2346-2351.

Heberer Th, Gramer S, Stan HJ (1999b) Occurrence and distribution of Organic contaminants in the aquatic system. Part III: Determination of synthetic musks in Berlin surface water applying solid phase microextraction (SPME) and gas chromatography-mass spectrometry in the full-scan mode. *Acta Hydrochim Hydrobiol* 27, 150-156.

Hedges JI, Parker PL, (1976) Land-derived matter in surface sediments from the gulf of mexico. *Geochim Cosmochim Acta* 40, 1019-1029.

Hedges JI, Cowie Richey JE, Quay PD, Brenner R, Strom M, Forsberg BR (1994) Origins and processing of organic matter in the Amazon river as indicated by carbohydrates and amino acids. *Limnol Oceanogr* 39, 743 – 761.

Hedges JI, Keil RG, Brenner R (1997) What happens to terrestrial organic matter in the ocean? *Org Geochem* 27, 195-212.

Heemken OP, Reincke H, Stachel B, Theobald N (2001) The occurrence of xenoestrogens in the Elbe river and the North Sea. *Chemosphere* 45, 245-259.

Heim S, Schwarzbauer J, Kronimus A, Littke R, Hembrock-Heger A (2003). Organic pollutants in riparian wetlands of the Lippe river (Germany). *Environ Chem Lett*, 1, 169-173

Heim S, Schwarzbauer J, Kronimus A, Littke R (2004) Geochronology of a load history: anthopogenic contaminants in riparian wetland sediments of the Lippe river (Germany). *Org. Geochem.* 35, 1409-1425.

Heinisch E, Klein S, Mohnke M, Sohr I, Schmidt B, Schmidt R, Franz P (1990) Stoffmusterverteilungen in Oberflächengewässern als Reflexion von Eintragsmustern – Schwewrflüchtige halogenierte Kohlenwasserstoffe. *Z Ges Hyg Grenzgeb* 36, 237-241.

Heinisch E (1992) Umweltbelastung in Ostdeutschland, Praxisbeiträge zur ökologischen Chemie von Chlorkohlenwasserstoffen. *Wissenschaftliche Buchgesellschaft, Darmstadt*, 283 p.Herbert BE, Bertsch PM, Novak JM (1993) Pyrene sorption by water-soluble organic Carbon. *Environ Sci Technol* 27, 398-403.

Hemmerich R (1980) Bestimmung des Pb-210 Flusses an Bodenprofilen. Staatsexamenarbeit, Institut für Umweltphysik, Universität Heidelberg, 34 pp.

Hendriks JA, Maas-Diepeveen JL, Noordsu A, van der Gaag MA (1994) Monitoring Response of XAD-concentrated water in the Rhine-Delta: A major part of the toxic compunds remains unidentified. *Wat Res* 28, 581-598.

Herbert BE, Bertsch PM, Novak JM (1993) Pyrene sorption by water-soluble organic Carbon. *Environ Sci Technol* 27, 398-403.

Herren D, Berset JD (2000) Nitro musks, nitro musk amino metabolites and polycyclic musks in sewage sludges. Quantitative determination by HRGC-ion-trap-MS/MS and mass spectral characterization of the amino metabolites. *Chemosphere* 40, 565-574.

Higgins IJ, Best DJ, Hammond RC (1980) New findings in methane-utilizing bacteria highlight their importance in the biosphere and their commercial potential. *Nature* 286, 561-564.

Hintelmann H, Wilken R-D (1994) Methylquecksilberverbindungen – Belastungsschwerpunkte im Längsschnitt der Elbe. *Vom Wasser* 82, 163173.

Hock B, Rothe S, Seifert M (2000) Hormone in der Umwelt- die Tests im Überblick. *Nachrichten aus der Chemie* 48, 918-924.

Hofbauer B, Jüttner F (1988) Occurrence of isopropylthio compounds in the aquatic ecosystem (Lake Neusiedl, Austria) as a chemical marker for *Mycrocystis flos-aquae. FEMS Microbiol Ecology* 53, 113-122.

Holm JH, Rügge K, Bjerg PL, ChristensenTH (1995) Occurrence and distribution of pharmaceutical organic compounds in the groundwater downgradient of a landfill (Grinsted, Denmark). *Environ Sci Technol* 29, 1415-1420.

Holoubek I, Caslavasky J, Helesic J, Vancura R, Kohoutek J, Kocan A, Petrik J, Chovancova J (1994) Project TOCOEN - the fate of selected organic pollutants in the environment. Part XXI - the contents of PAHs, PCBs, PCDD/Fs in sediments from the Danube river catchment area. *Tox Environ Chem* 43, 203-215.

Horsfield B, Disko U, Leistner F (1989) The micro-scale simulation of maturation: Outline of a new technique and its potential applications. *Geologische Rundschau* 78, 361-374.

Houot S, Benoit P, Charnay MP, Barriuso E (1997) Experimental techniques to study the fate of organic pollutants in soils in relation to their interactions with soil organic constituents. *Analysis Magazine* 25, 9-19.

Hsu TS, Bartha R (1976) Hydrolysable and nonhydrolysable 3,4-dichloroaniline-humus complexes and their respective rates of biodegradation. *J Agric Food Sci* 24, 118-222.

Huber SA, Scheunert I, Dörfler U, Frimmel FH (1992) Zum Einfluss des gelösten organischen Kohlenstoffs (DOC) auf das Mobilitätsverhalten einiger Pestizide. *Acta Hydrochim Hydrobiol* 20, 74-81.

Huhn W (1978) Über das Vorkommen von Phthalsäureestern im Oberflächenwasser. *Acta Hydrochim Hydrobiol* 6, 45-50.

Hühnerfuss H, Faller J, Kallenborn R (1993) Enantioselective and nonenantioselective degradation of organic pollutants in the marine ecosystem. *Chirality* 5, 393-399.

Hühnerfuss H, Bester K, Landgraff O, Pohlmann T, Selke K (1997) Annual balances of hexachlorocyclohexanes, polychlorinated biphenyls and triazines in the German Bight. *Mar Poll Bull* 34, 419-426.

Hunger K, Mischke P, Rieper W, Raue R (1998) Azo dyes. In: *Ullmanns Encyclopedia of industrial chemistry*, 6$^{th}$ ed, electronic release, Wiley VCH.

Huppert N, Würtele M, Hahn HH (1998) Determination of the plasticizer N-butylbenzenesulfonamide and the pharmaceutical Ibuprofen in waste water using solid phase microextraction (SPME). *Fresenius J Anal Chem* 362, 529-536.

Hyötyläinen J, Knuutinen J, Malkavaara P, Siltala J (1998a) Pyrolysis-GC-MS and CuO-Oxidation-HPLC in the characterization of HMMs from sediments and surface waters downstream of a pulp mill. *Chemosphere* 36, 291-314.

Hyötyläinen J, Knuutinen J, Malkavaara, P (1998b) Transport of high molecular mass lignin material in the receiving water system of a mechanical pulp mill. *Chemosphere* 36, 577-587.

ICPR (International Commission for the Protection of the Rhine), Aktionsprogramm Rhein, 1996. Vergleich der Gewässergüte des Rheins mit Zielvorgaben 1990-1993, Zwischenbilanz.

Ikonomou MG, Rayne S, Fischer M, Fernandez MP, Cretney W (2002) Occurrence and congener profiles of polybrominated diphenyl ethers (PBDEs) in environmental samples from coastal British Columbia, Canada. *Chemosphere* 46, 649-663.

Imagawa T, Lee CW (2001) Correlation of polychlorinated naphthalenes with polychlorinated dibenzofurans formed from waste incineration. *Chemosphere* 44, 1511-1520.

Ingerowski G, Friedle A, Thumalla J (2001) Chlorinated Ethyl and Isopropyl Phosphoric Acid Triesters in the Indoor Environment- An Inter-Laboratory Exposure Study. *Indoor Air* 11, 145-149.

Irmer U (2000) Die neue EG-Wasserrahmenrichtlinie: Bewertung der chemischen und ökologischen Qualität von Oberflächengewässern. *Acta Hydrochim Hydrobiol* 28, 7-14.

Ishikawa S, Taketomi M, Shinohara R (1985) Determination of trialkyl and triaryl phosphates in environmental samples. *Wat Res* 19, 119-125.

Ittekkot V, Arain R (1986) Nature of particulate organic matter in the river Indus, Pakistan. *Geochim Cosmochim Acta* 50, 1643-1653.

Ittekkot V (1988) Global trends in the nature of organic matter in river suspensions. *Nature* 332, 436-438.

Jackson A, Pardue J (1999) Quantifying the mineralization of contaminants using stable carbon isotope ratios. *Org Geochem* 30, 787.

Jacobson JL, Jacobson SW (1996) Intellectual impairment in children exposed to polychlorinated biphenyls in utero. *New Eng J Med* 335, 783-789.

Jakobsson E, Asplund L (2000) Polychlorinated naphthalenes (PCN's). In: *New Types of persistent halogenated compounds. The Handbook of Environmental Chemistry 3 K*, Paasivirta J (Ed.), Springer Verlag, Berlin – Heidelberg- New York, 97-126.

Jaffe R, Wolff GA, Cabrera AC, Chitty HC (1995) The biogeochemistry of lipids in river of the Orinoco Basin. *Geochim Cosmochim Acta* 59, 4507-4522.

Jaffe R, Elisme T, Cabrera AC (1996) Organic Geochemistry of seasonally flooded rain forest soils: molecular composition and early diagenesis of lipid components. *Org Geochem* 25, 9-17.

Jantzen E, Kuballa J, Wilken R-D (1993) Muster zinnorganischer Verbindungen in der Elbe. *Vom Wasser* 80, 245-252.

Jantzen E, Prange A (1995) Organometallic species of the elements tin, mercury and lead in sediments of the longitudinal profile of the River Elbe. *Fresenius J Anal Chem* 353, 28-33.

Jarman WM, Hilkert A, Bacon CE, Collister JW, Ballschmiter K, Risebrough RW (1998) Compound-specific carbon isotopic analysis of aroclors, clophens, kaneclors and phenoclors. *Environ Sci Technol* 32, 833-836.

Järnberg U, Asplund L, Wit C (1993) Polychlorinated biphenyls and polychlorinated naphthalenes in swedish sediment and biota: levels, patterns and time trends. *Environ Sci Technol* 27, 1364-1374.

Järnberg U, Asphund L, De Wit C, Crafström AK, Haglund P, Jansson B, Lexen K, Stradell M, Olsson M, Jonsson B (1993) Polychlorinated Biphenyls and *Polychlorinated* Naphthalenes in Swedish Sediment and Biota: Levels, Patterns, and Time Trends. *Environ Sci Technol* 27, 1364-1374.

Jensen-Korte U, Anderson C, Spitteller M (1987) Photodegradation of pesticides in the presence of humic substances. *Sci Total Environ* 62, 335-340.

Johnson WP, Amy GL (1995) Facilitated transport and enhanced desorption of polycyclic aromatic hydrocarbons by natural organic matter in aquifer sediments. *Environ Sci Technol* 29, 807-817.

Jones KC, de Voogt P (1999) Persistent organic pollutants (POPs): State of the science. *Environ Poll* 100, 209-221.

Jordan MJ, Margaria CA, Shaw PE, Goodner KL (2002) Aroma active components in aqueous Kiwi fruit essence and Kiwi fruit puree by GC-MS and multidimensional GC/GC-O. *J Agric Food Chem* 50, 5386-5390.

Jovancicevic B, Tasic L, Wehner H, Faber E (1997) Identification of oil-type pollution in recent sediments. *Fresenius Environ Bull* 6, 667-673.

Jovancicevic B, Tasic L, Wehner H, Markovic D, Polic P (1998) *n*-Alkane Distribution as a tool in the identification of organic type pollution in river sediments. *Fresenius Environ Bull* 7, 320-326.

Jovancicevic B, Polic P, Vrvic M, Sheeder G, Teschner T, Wehner H (2003) Transformations of *n*-alkanes from petroleum pollutants in alluvial ground waters. *Environ Chem Lett* 1, 73-81.

Käcker T, Haupt ETK, Garms C, Francke W, Steinhart H (2002) Structural characterization of humic acid-bound PAH residues in soil by $^{12}$C-CPMAS-NMR-spectroscopy: evidence of covalent bonds. *Chemosphere* 48, 117-131.

Kähkönen MA, Suominen KP, Manninen PKG, Salkinoja-Salonen MS (1998) 100 Years of sediment accumulation history of organic halogens and heavy metals in recipient and nonrecipient lakes of pulping industry in Finland. *Environ Sci Technol* 32, 1741-1746.

Kallenborn R, Gatermann R, Rimkus GG (1999) Synthetic musks in environmental samples: indicator compounds with relevant properties for environmental monitoring. *J Environ Monitoring* 1/4, 70N-74N.

Kallenborn R, Hühnerfuss H (2001) *Chiral Environmental Pollutants.* Trace Analysis and Ecotoxicology, Springer Verlag, Heidelberg, 209 pp.

Kaminsky R, Hites RA (1984) Octachlorostyrene in Lake Ontario: Sources and Fates. *Environ Sci Technol* 18, 275-279.

Kan AT, Fu G, Hunter MA, Tomson MB (1997) Irreversible adsorption of naphthalene and tetrachlorobiphenyl to Lula and surrogate sediments. *Environ Sci Technol* 31, 2176-2185.

Kaplan IR, Galperin Y, Lu ST, Lee RP (1997) Forensic environmental geochemistry: differentiation of fuel-types, their sources and release time. *Org Geochem* 27, 289-317.

Katase T, Kim Y (1999) Determination of adipic acid esters potentially migrated from plastic films for foods-wrapping by gas chromatography/ gas chromatography combined with mass spectrometry. *Bunseki Kagaku* 48, 649-655.

Kaufman DD, Still GG, Paulson GD, Bandal SK (Hrsg) (1976) Bound and conjugated pesticide residues. *ACS Symp Ser* 29.

Kawagoshi Y, Fukunaga I, Itoh H (1999) Distribution of organophosphoric acid triesters between water and sediment at a sea-based solid waste disposal site. *J Material Cycl Waste Management* 1, 53-61.

Keil RG, Mayer LM, Quay PD, Richey JE, Hedges JI (1997) Loss of organic matter from riverine particles in deltas. *Geochim Cosmochim Acta* 61, 1507-1511.

Keith LH, Telliard WA (1979) Priority pollutants: a perspective view. *Environ Sci Technol* 13, 416-423.

Kevekordes S, Mewrsch-Sundermann V, Diez M, Dunkelberg H (1997) In vitro genotoxicity of polycyclic musk fragrances in the micronucleus test. *Mutation Res* 395, 145-150.

Khan SU (1982) Bound pesticide residues in soil and plants. *Residue Reviews* 84, 1-25.

Killela DR, Aldstadt JH (2002) Identification of dimethylchloroarsine near a former herbicide factory by headspace solid-phase microextraction gas chromatography - mass spectrometry. *Chemosphere* 48, 1003-1008.

Kjeller LO, Rappe C (1995) Time trends in levels, patterns, and profiles for polychlorinated dibenzo-p-dioxins, dibenzofurans, and biphenyls in a sediment core from the Baltic proper. *Environ Sci Technol* 29, 346-355.

Klaus U, Mohamed S, Volk M, Spiteller M (1998a) Interaction of aquatic humic substances with anilazine and its derivatives: The nature of the bound residues. *Chemosphere* 37, 341-361.

Klaus U, Oesterreich T, Volk M, Spiteller M (1998b) Interaction of aquatic dissolved organic matter (DOM) with amitrole: The nature of bound residues. *Acta Hydrochim Hydrobiol* 26, 311-317.

Kleinschnitz M, Schreier P (1998) Identification and Semi-quantitative Determination of a Migration Contaminant from Beverage Carton Packages into Mineral Water by On-line Solid Phase Extraction Gas Chromatography-Mass Spectrometry (SPE-GC-MS). *Chromatographia* 48, 581-583.

Klös H, Schoch C (1992) Historische Entwicklung einer Sedimentbelastung: Gedächtnis einer Industrieregion. *Acta Hydrochim Hydrobiol* 21, 32-37.

Klös H, Schoch C (1993) Einfache Methoden zur Radiodatierung limnischer Sedimente. *Z Umweltchemie Ökotoxikol* 5, 2-6.

Knepper T, Pilz N, Seel P (1996) Das Insektenrepellent Diethyltoluamid (DEET) - Ein neuer Problemstoff für die Wasserwerke? *53. Jahresbericht der Arbeitsgemeinschaft Rhein-Wasserwerke e.V.*, 66-77.

Koch R (1995) Umweltchemikalien. VCH, Weinheim, 3. ed., 421 pp.

Kögel-Knabner I (2000) Analytical approaches for characterizing soil organic matter. *Org Geochem* 31, 609-626.

Koistinen J, Paasivirta J, Suonpera M (1995) Contamination of pike and sediment from Kymijoki river by PCDEs, PCDDs, and PCDFs: Contents and patterns compared to pike and sediments from Bothnian Bay and seals from Lake Saimaa. *Environ Sci Technol* 29, 2541-2547.

Kolpin DW, Thurman EM, Linhart SM (2001) Occurrence of cyanazine compounds in groundwater: Degradates more prevalent than the parent compound. *Environ Sci Technol* 35, 1217-1222.

Kolpin DW, Furlong ET, Meyer MT, Thurman EM, Zaugg SD, Barber LB, Buxton HT (2002) Pharmaceuticals, Hormones, and Other Organic Wastewater Contaminants in U.S. Streams, 1999-2000: A National Reconnaissance. *Environ Sci Technol* 36, 1202-1211.

König J, Balfanz E, Funcke W, Romanowski T (1983) Determination of oxygenated polycyclic aromatic hydrocarbons in airborne particulate matter by capillary gas chromatography and gas chromatography / mass spectrometry. *Anal Chem* 55, 599-603.

Kronimus A, Schwarzbauer J, Dsikowitzky L, Heim S, Littke R (2004) Anthropogenic contaminants in sediments of the Lippe River, Germany. *Wat Res* . 38, 3473-3484.

Krueger CJ, Barber LB, Metge DW (1998) fate and transport of linear alkylbenzensulfonates in a sewage-contaminated aquifer: A comparison of natural-gradient pulsed tracer tests. *Environ Sci Technol* 32, 1134-1142.

Kumke U, Zwiener C, Abbt-Braun G, Frimmel FH (1999) Spectroscopic Characterization of Fulvic Acid Fractions of a Contaminated Groundwater. *Acta Hydrochim Hydrobiol* 27, 409-415.

Kümmerer K (2001) Drugs in the environment : emission of drugs, diagnostic aids and disinfectants into wastewater by hospitals in relation to other sources – a review. *Chemosphere* 45, 957-969.

Kuruntachalam K, Takashi I (1998) Isomer-Specific Analysis and Toxic Evaluation of Polychlorinated Naphtalenes in Soil, Sediment and Biota Collected near the Site of a Formar Chlor-Alkali Plant. *Environ Sci Technol* 32, 2507-2514.

Lach G, Steffen D (1997) Orientierende Untersuchungen von Gewässersedimenten auf Nitro-/Polymoschusverbindungen und die Flammschutzmittel TCEP und TCPP. In: *Oberirdische Gewässer* 3/97, Ed. Niedersächsisches Landesamt für Ökologie, Germany.

Lahr J, Maas-Diepeveen JL, Stuijfzand SC, Leonards PEG, Drüke JM, Lücker S, Espeldoorn A, Kerkum LCM, van Stee LLP, Hendriks AJ (2003) Responses in sediments bioassays used in the Netherlands: can observed toxicity be explained by routinely monitored priority pollutants? *Wat Res* 37, 1691-1710.

Lang V (1992) Polychlorinated biphenyls in the environment, *J Chromat* 595, 1-43.

Lartiges SB, Garrigues PP (1995) Degradation kinetics of organophosphorus and organonitrogen pesticides in different waters under various environmental conditions. *Environ Sci Technol* 29, 1246-1254.

Later DW, Lee ML, Bartle KD, Kong RC, Vassilaros L (1981) Chemical class separation and characterization of organic compounds in synthetic fuels. *Anal Chem* 53, 1612-1620.

Latimer JS, Quinn JG (1996) Historical trends and current inputs of hydrophobic organic compounds in an urban estuary: the sediment record. *Environ Sci Technol* 30, 623-633.

Lee CL, Fang MD (1997) Sources and distribution of chlorobenzenes and hexachlorobutadiene in surficial sediments along the coast of southwestern Taiwan. *Chemosphere* 35, 2039-2050.

Lema JM, Mendez R, Blazques R (1988) Characteristics of landfill leachates and alternatives for their treatment: A review. *Water Air Soil Pollut* 40, 223-250.

Lenhart W, Lemm E (1993) Bestimmung und Vorkommen von Triarylphosphaten in verschiedenen Gewässern. *Vom Wasser* 81, 93-103.

Leppard GG, Flanniga DT, Mavrocordatos D, Marvin CH, Bryant DW, McCarry BE (1998) Binding of polycyclic aromatic hydrocarbons by size classes of particulate in Hamilton Harbor water. *Environ Sci Technol* 32, 3633-3639.

Li GC, Felbeck GT (1972) A study of the mechanism of atrazine adsorption by humic acid from muck soil. *Soil Sci* 113, 430-433.

Lichtfouse E (1997) Soil, a sponge for pollutants. *Analusis* 25, M16-M23.

Liechtenstein EP, Katan J, Anderegg BN (1977) Binding of "persistent" and "nonpersistent" $^{14}C$-labelled insecticides in an agricultural soil. *J Agrar Food Chem* 25, 43-47.

Liechtenstein EP (1980) Bound residues in soils and transfer of soil residues in crops. *Residue reviews* 76, 147-155.

Lindsey ME, Meyer M, Thurman EM (2001) Analysis of Trace Level of Sulfonamide and Tetracycline Antimicrobials in Groundwater and Surface Water Using Solid-Phase Extraction and Liquid Chromatography/Mass Spectrometry. *Anal Chem* 73, 4640-4646

Lindström A, Buerge IJ, Poiger T, Bergqvist PA, Müller MD, Buser HR (2002) Occurrence and Environmental Behavior of the Bactericide Triclosan and ist Methyl derivate in Surface Waters and in Waste Waters. *Environ Sci Technol* 36, 2322-2329.

Liu B, McConnell LL, Torrents A (2001) Hydrolysis of chlorpyrifos in natural waters of the Chesapeake Bay. *Chemosphere* 44, 1315-1323.

Loganathan BG, Kannan K, Watanabe I, Kawano M, Irvine K, Kumar S, Sikka HC (1995) Isomer-specific determination and toxic evaluation of polychlorinated biphenyls, polychlorinated/polybrominated dibenzo-*p*-dioxins and furans, polybrominated biphenylethers and extractable organic halogen in Carp from the Buffalo river, New York. *Environ Sci Technol* 29, 1832-1838.

Looser MO, Parriaux A, Bensimon M (1999) Landfill underground pollution detection and characterization using inorganic traces. *Wat Res* 17, 3609-3616.

Luthy RG, Aiken GR, Brusseau ML, Cunningham SD, Gschwend PM, Pignatello JJ, Reinhard M, Traina SJ, Weber WJ, Westall JC (1997) Sequestration of hydrophobic organic contaminants by geosorbents. *Environ Sci Technol* 31, 3341-3347.

Maguire RJ (1987) Environmental aspects of tribtuyltin. *Appl Organomet Chem* 1, 475-498.
Mai BX, Fu JM, Sheng GY, Kang YH, Lin Z, Zhang G, Min YS, Zeng EY (2002) Chlorinated and polycyclic aromatic hydrocarbons in riverine and estuarine sediments from Pearl River Delta, China. *Environ Poll* 117, 457-474.
Matthews DE, Hayes JM (1978) Isotope-ratio-monitoring gas chromatography-mass spectrometry. *Anal Chem* 50, 1465-1473.
Mattiasson B, Larsson PO, Mosbach K (1977) Polycyclic aromatic hydrocarbons in long-range transported aerosols. *Nature* 268, 518-520.
Mazeas, L, Budzinski H (2002) Stable carbon isotopic study ($^{12}C/^{13}C$) of the fate of petrogenic PAHs (methylphenanthrenes) during an in-situ oil spill simulation experiment. *Org Geochem* 33, 1253-1258.
Mazeas, L, Budzinski H, Raymond N (2002) Absence of stable carbon isotope fractionation of saturated and polycyclic aromatic hydrocarbons during aerobic bacterial biodegradation. *Org Geochem* 33, 1259-1272.
McCarthy JF, Zachara JM (1989) Subsurface transport of contaminants: Mobile colloids in the subsurface environment may alter the transport of contaminants. *Environ Sci Technol* 23, 496-502.
McCrody SE, Farrington JW, Gschwend PM (1996) Comparison of the in situ and desorption sediment-water partitioning of polycyclic aromatic hydrocarbons and polychlorinated biphenyls. *Environ Sci Technol* 30, 172-177.
McDonald TA (2002) A perspective on the potential health risks of PBDEs. *Chemosphere* 46, 745-755.
McGregor DB, Brown A, Cattanach P, Edwards I, McBride D, Riach C, Caspary WJ (1988) Responses of the L5178Y th+/tk- Mouse Lymphoma Cell Forward Mutation Assay. III. 72 Coded Chemicals. *Environ Mol Mutagen* 12, 85-154.
McNally DL, Mihelcic JR, Lueking DR (1998) Biodegradation of three and four-ring polycyclic aromatic hydrocarbons under aerobic and denitrifying conditions. *Environ Sci Technol* 32, 2633-2639.
McRae C, Sun CG, Snape C (1999) $\delta^{13}C$ values of coal-derived PAHs from different processes and their application to source apportionment. *Org Geochem* 30, 881-889.
Meckenstock RU, Anweiler E, Michaelis W, Richnow HH, Schink B (2000) Anaerobic naphthalene degradation by a sulfate-reducing enrichment culture. *Appl Environ Microbiol* 66, 2743-2747.
Meier-Augstein W (1999) Applied gas chromatography coupled to isotope ratio mass spectrometry. *J Chromat A* 842, 351-371.
Meijers AP, van der Leer RC (1976) The occurrence of organic micropollutants in the river Rhine and the river Maas in 1974. *Wat Res* 10, 597-604.
Menzie CA, Potocki BB, Santodonato J(1992) Exposure to cancerogenic PAHs in the environment. *Environ Sci Technol* 26, 1278-1284.
Mersch-Sundermanno V, Kevekordeso S, Jentero C (1998) Lack of mutagenicity of polycyclic musk fragrances in Salmonella typhimurium. *Toxicology In Vitro* 12, 389-393.

van Metre PC, Callender E, Fuller CC (1997) Historical trends in organochlorine compounds in river basins identified using sediment cores from reservoirs. *Environ Sci Technol* 31, 2339-2344.

Metzger JW, Möhle E (2001) Flammschutzmittel in Oberflächengewässern, Grundwässern und Abwässern - Eintragspfade und Gehalte. Forschungsbericht, Institut für Siedlungswasserbau, Wassergüte und Abfallwirtschaft, University of Stuttgart, 53 pp.

Michaelis W, Richnow HH, Seifert R (1995) Chemically bound chlorinated aromatics in humic substances. *Naturwissenschaften* 82, 139-142.

Miermans CJ, van der Velde LE, Frintrop PC (1999) Analysis of volatile organic compounds using the purge and trap injector coupled to a gas chromatograph/ion trap mass spectrometer: review of the results in Dutch surface water of the Rhine, Meuse, Northern Delta Area and Westerscheldt over the period 1992-1997. *Chemosphere* 40, 39-48.

Miikki V, Hänninen K, Knuutinen J, Hyötyläinen J (1999) Pyrolysis of humic acids from digested and composted sewage sludge. *Chemosphere* 38, 247-253.

Milliadis GE, Maltou PT (1997) Monitoring of the pesticide levels in natural waters of Greece. *Bull Environ Contam Toxicol* 59, 909-923.

Mongenot T, Derenne S, Largeau C, Tribovillard NP, Lallier-Verges E, Dessort D, Connan J (1999) Spectroscopic, kinetic and pyrolytic studies of kerogen from the dark parallel laminae facies of the sulphur-rich Orbangnoux deposit (Upper Kimmeridgian, Jura). *Org Geochem* 30, 39-56.

Möller K, Bretzke C, Hühnerfuss H, Kallenborn R, Kinkel JN, Kopf J, Rimkus G (1994) Durchlässigkeit der Blut-Hirn-Schranke von Seehunden für das Enantiomer (+)-alpha-1,2,3,4,5,6-Hexachlorcyclohexan und dessen absolute Konfiguration. *Angew Chem* 106, 911 - 912; *Angew Chem Int Ed Engl 33, 882 - 884.*

Moreira N, Mendes F, Pereira O, Guedes de Pinho P, Hogg T, Vasconcelos I (2002) Volatile sulphur compounds in wines related to yeast metabolism and nitrogen composition of grape musts. *Anal Chim Acta* 458, 157-167.

Moskalev NV, Sirotkina EE, Ogorodnikov VD (1985) A convenient method for obtaining 3-chloro- and 3,6-dichlorocarbazoles. *Chem Heterocycl Comp* 21, 542-543.

Mulders EJ (1973) Volatile components from the non-enzymic browning reaction of the cysteine/cystine-ribose system. *Z Lebensmittel Unters-Forsch* 152, 193-201.

Müller A, Heininger P, Wessels M, Pelzer J, Grünwald K, Pfitzner S, Berger M (2002) Contaminant levels and ecotoxicological effects in sediments of the River Odra. *Acta Hydrochim Hydrobiol* 30, 244-255.

Müller K, Deurer M, Hartmann H, Bach M, Spiteller M, Frede HG (2003) Hydrological characteristics of pesticide loads using hydrograph separation at different scales in a German catchment. *J Hydrol* 273, 1-17.

Müller S, Schmid P, Schlatter C (1996) Occurrence of nitro and non-nitro benzenoid musks in human adipose tissue. *Chemosphere* 33, 17-28.

Murphy EM, Zachara JM, Smith SC (1990) Influence of mineral-bound humic substances on the sorption of hydrophobic organic compounds. *Environ Sci Technol* 24, 1507-1516.

Murphy EM, Zachara JM, Smith SC (1994) Interaction of hydrophobic organic compounds with mineral-bound humic substances. *Environ Sci Technol* 28, 1291-1299.

Murray JP, Rouse JV, Carpenter AB (1981) Groundwater contamination by sanitary landfill leachate and domestic waste water in carbonate terrain: Principal source diagnosis, chemical transport characteristics and design implications. *Wat Res* 15, 745-757.

Nanny MA (1999) Deuterium NMR characterization of non-covalent interactions between monoaromatic compounds and fulvic acids. *Org Geochem* 30, 901-909.

Nerin C, Echarri I, Cacho J (1991) Determination of HCHs in sediments of the river Gallego. *Fresenius J Anal Chem* 339,
684-687.

Nerurkar VR, Wakayama I, Rowe T, Yanagihara R, Garruto RM (1993) Preliminary Observations on the in Vitro Toxicity of N-Butylbenzenesulfonamide: A Newly Discovered Neurotoxin. *Ann NY Acad Sc* 679, 280-287.

Nicholls CR, Allchin CR, Law RJ (2001) Levels of short and medium chain length polychlorinated *n*-alkanes in environ-mental samples from selected industrial areas in England and Wales. *Environ Poll* 114, 415-430.

Northcott GL, Jones KC (2000) Experimental approaches and analytical techniques for determining organic compounds bound residues in soil and sediment. *Environ Poll* 108, 19-43.

Northcott GL, Jones KC (2001) Partitioning, extractability and formation of nonextractable PAH residues in soil. 1. Compound differences in aging and sequestrian. *Environ* Sci Technol 35, 1103-1110.

Nystrom RF, Berger CRA (1958) Reduction of organic compounds by mixed hydrides. II Hydrogenolysis of ketones and alcohols. *J Amer Chem Soc* 80, 2896-2898.

Oesterreich T, Klaus U, Volk M, Neidhart B, Spiteller M (1999) Environmental fate of amitrole: influence of dissolved organic matter. *Chemosphere* 38, 379-392.

Öllers S, Singer HP, Fässler P, Müller SR (2001) Simultaneous quantification of neutral and acidic pharmaceuticals and pesticides at the low- ng/l level in surface and waste water. *J Chromat A* 911, 225-234.

Öman C, Hynning PA (1993) Identification of organic compounds in municipal landfill leachates. *Environ Pollut* 80, 265-271.

Orem WH, Colman SM, Lerch HE (1997) *Org Geochem* 27, 153.

Osemwengie LI, Steinberg S (2001) On-site solid-phase extraction and laboratory analysis of ultra-trace synthetic musks in municipal sewage effluent using gas-chromatography-mass spectrometry in the full-scan mode. *J Chromat* A 932, 107-118.

Oxynos K, Schramm KW, Marth P (1995) Chlorinated hydrocarbons- (CHC) and PCDD/F-levels in sediments and breams (Abramia brama) from the river Elbe (Contribution to the German Environmental Specimen Banking). *Fresenius J Anal Chem* 353, 98-100.

Paasivirta J (Ed.) (2000) New Types of persistent halogenated compounds. *The handbook of environmental chemistry 3 K*, Springer Verlag, Berlin - Heidelberg - New York, 366 p.

Pachur HJ, Ahrens M, Ricking M, Röper HP (1993) Heavy metal and chlorinated hydrocarbon contamination of selected aquatic sediments from the Berlin Area. *Land Degrad Rehabil* 4, 351-360.

Palm A, Cousins IT, Mackay D, Tysklind Mats, Metcalfe C, Alaee A (2002) Assessing the environmental fate of chemicals of emerging concern: a case study of the polybrominated diphenyl ethers. *Environ Poll* 117, 195-213.

Palm H, Lammi R (1995) Fate of pulp mill organochlorines in the Gulf of Bothnia sediments. *Environ Sci Technol* 29, 1722-1727.

Paneli M, Rosenberg E, Grasserbauer M, Ceulemans M, Adams F (1997) Assessment of Organolead Species in the Austrian Danube-basin Using GC-MIP-AED. *Fresenius J Anal Chem* 357, 756-762.

Papa AJ, Sherman PD (1981) Fragrances. In: *Kirk-Othmer-Encyclopedia of Chemical Technology* 13, Grayson M, Eckroth D (Eds.), 3rd Edition, John Wiley & Sons, New York, 894-941.

Parris GE (1980) Covalent binding of aromatic amines to humate.1. Reactions with carbonyls and quinones. *Environ Sci Technol* 14, 1099-1106.

Paschke A, Herbel W, Steinhart H, Franke S, Francke W (1992) Determination of mono- to tetracyclic aromatic hydrocarbons in lubricating oil. *J High Resolut Chromat* 15, 827-833.

Paxeus N (1996) Organic pollutants in the effluents of large wastewater treatment plants in Sweden. *Wat Res* 30, 1115-1122.

Paxeus N, Schröder HF (1996) Screening for non-regulated organic compounds in municipal wastewater in Göteborg, Sweden. *Wat Sci Technol* 33, 9-15.

Paxeus N (2001) Organic compounds in municipal landfill leachates. *Wat Sci Technol* 42, 323-333.

Pedersen SN, Lindholst C (1999) Quantification of the xenoestrogens 4- *tert.*-octylphenol and bisphenol A in water and in fish tissue based on microwave assisted extraction, solid-phase extraction and liquid chromatography-mass spectrometry. *J Chromat A* 864, 17-24.

Pellizzari ED, Moseley MA, Cooper SD (1985) Recent advances in the analysis of polychlorinated biphenyls in environmental and biological media.. *J Chromat* 334, 277-314.

Penalver A, Pocurull E, Borrull F, Marce RM (2001) Comparison of different fibers for the solids-phase microextraction of phthalate esters from water. *J Chromat A* 922, 377-384.

Peng P, Morales-Izquierdo A, Hogg A, Strausz OP (1997) Molecular structure of Athabasca asphaltene: Sulfide ether and ester linkages. *Energy Fuels* 11, 1171-1187.

Perdue EM, Wolfe NL (1982) Modification of pollutant hydrolysis kinetics in the presence of humic substances. *Environ Sci Technol* 16, 847-852.

Pereira WE, Rostad CE, Leiker TJ (1990) Determination of trace levels of herbicides and their degradation products in surface and ground waters by gas chromatography / ion-trap mass spectrometry. *Anal Chim Acta* 228, 69-75.

Pesonen M, Teivainen P, Lundström J, Jakobsson E, Norrgren L (2000) Biochemical Reponses of Fish Sac Fry and a Primary Cell Culture of Fish Hepatocytes Exposed to Polychlorinated Naphthalenes. *Arch Environ Contam Toxicol* 38, 52-58.

Petermann PH, Delfino JJ (1990) Identification of isopropyl-, alkyl-diphenylmethanes, diisopropylnaphtahalene, linear alkylbenzenes and other polycherloinated biphenyl replacement compounds in effluents, sediments and fish in the Fox river system, Wisconsin. *Biomed Environ Mass Spectromet* 19, 755-770.

Peters KE, Moldowan JM (1993) *The Biomarker Guide: Interpreting molecular fossils in petroleum and ancient sediments*, Prentice Hall, Englewood Cliffs, New Jersey.

Piccolo A, Celano G, De Simone C (1992) Interactions of atrazine with humic substances of different origins and their hydrolysed products. *Sci Total Environ* 117/118, 403-412.

Picer M, Kovac T, Britvic S, Picer N (2001) The chemical and biogenotoxic characterization of organic xenobiotics in aquatic sediment materials. 1. The application and comparison of chemically non-specific and biogenotoxic methods. *Chemosphere* 44, 1673-1683.

Pietsch J, Schmidt W, Sacher F, Fichtner S, Brauch HJ (1995) Pesticides and other organic micro pollutants in the River Elbe. *Fresenius J Anal Chem* 353, 75-82.

Pignatello JJ, Xing B (1996) Mechnanisms of slow sorption of organic chemicals to natural particles. *Environ Sci Technol* 30, 1-11.

Planas C, caixach J, Santos FJ, Rivera J (1997) Occurence of pesticides in spanish surface waters. Analysis by high resolution gas chromatography coupled to mass spectrometry. *Chemosphere* 34, 2393-2406.

Pommer EH (2001) Wood preservation. In: *Ullmann's Encyclopedia of industrial Chemistry*, 6$^{th}$ Ed., electronic release.

Poppe A, Alberti J, Friege H, Rönnefahrt B (1988) Chlorinated diphenylmethane (Ugilec 141) – a new dangerous impact on the environment. *Vom Wasser* 70, 33-42.

Poppe A, Alberti J, Bachhausen P (1991) Trends of river sediment contaminations with tetrachlorobenzyltoluenes and polychlorinated biphenyls in Northrhine-Westfalia. *Vom Wasser* 76, 191-198.

Postma JF, Tonkes M, Brils JM, de Poorter LRM, Botterweg J (1996) *Proceedings of Monitoring Tailor-made*, Nunspeet, The Netherlands, September 1996, RIZA 155-163.

Prahl FG, Pinto LA, (1987) A geochemical study of long-chain n-aldehydes in Washington coastal sediments. *Geochim Cosmochim Acta* 51, 1573-1582.

Prösch J, Puchert W (1998) Coffein: Vorkommen in Fließgewässern Mecklenburg-Vorpommerns. *Vom Wasser* 91, 207-214.

Puls W, Gerwinski W, Haarich M, Schirmacher M, Schmidt D (1997) Lead budget for the German Bight. *Mar Poll Bull* 34, 410-418.

Radach G, Heyer K (1997) A cadmium budget for the German Bight in the North Sea. *Mar Poll Bull* 34, 375-381.

Rapsomanikis S, Donard OF, Weber JH (1986) Speciation of lead and methyllead ions in water by chromatography/Atomic Absorption Spectrometry after ethylation with sodium tetraethylborate. *Anal Chem* 58, 35-38.

Reader S (1992) Identification and determination of butyltin compounds by gas chromatography - ion trap spectrometry. *Anal Chim Acta* 262, 307-314.

Reddersen K, Heberer T, Dünnbier U (2002) Identification and significance of phenazone drugs and their metabolites in ground- and drinking water. *Chemosphere* 49, 539-544.

Reemtsma T, Fiehn O, Kalnowski G, Jekel M (1995) Microbial Transformations and Biological Effects of Fungizide-Derived Benzothiazoles Determined in Industrial Wastewater. *Environ Sci Technol* 29, 478-485

Reinhard M, Goodman NL, Barker JF (1984) Occurrence and distribution of organic chemicals in two landfill leachate plumes. *Environ Sci Technol* 18, 953-961.

Reiser R, Toljander H, Albrecht A, Giger W (1997) Alkylbenzenesulfonates in recent lake sediments as molecular marker for the environmental behaviour of detergent-derived chemicals. In: Eagenhouse RP (Ed.) *Molekular markers in environmental geochemistry. ACS Symposium Series* 671, Washington D.C., 197-212.

Reiser R, Toljander H, Albrecht A, Giger W (1997) Alkylbenzenesulfonates in recent lake sediments as molecular marker for the environmental behaviour of detergent-derived chemicals. In: Eagenhouse R P (Ed.) Molekular markers in environmental geochemistry, *ACS Symposium Series* 671, Washington DC, 197-212.

Renner R (1998) Intrinsic remediation under the microscope. Concerns aboutr the use of "natural attenuation" have led to a major review of the method effectivenss. *Environ Sci Technol* 32, 180A-182A.

Rennie PJ (1988) Determination of 2-mercaptobenzothiazole in river water by high performance liquid chromatography. *Chromatographia* 26, 297-299.

Requejo AG, Brown J, Boehm PD (1985) Thermal degradation products of non-volatile organic matter as indicators of anthropogenic inputs to estuarine and coastal sediments. In: Sigleo AC, Hattori A (Ed.) *Marine and estuarine geochemistry*, Lewis Publishers, Chelsea (Michigan), 81-96.

Reupert R, Plöger E (1989) Bestimmung stickstoffhaltiger Pflanzenbehandlungsmittel in Trink-, Grund- und Oberflächenwasser. *Vom Wasser* 72, 211-232.

Richardson ML, Bowron JM (1985) The fate of pharmaceutical chemicals in the aquatic environment. *J Pharm Pharmacol* 37, 1-12.

Richnow HH, Seifert R, Hefter J, Kästner M, Mahro B, Michaelis W (1994) Metabolites of xenobiotica and mineral oil constituents linked to macromolecular matter in polluted environments. In: Advances in Org Geochem 1993, *Org Geochem* 22, 671-681.

Richnow HH, Seifert R, Hefter J, Link M, Francke W, Schaefer G, Michaelis W (1997) Organic pollutants associated with macromolecular soil organic matter – mode of binding. *Org Geochem* 26, 745-758.

Richnow HH, Eschenbach A, Mahro B, Seifert R, Wehrung P, Albrecht P, Michaelis W (1998) The use of $^{13}$C-labelled polycyclic aromatic hydrocarbons for the analysis of their transformation in soils. *Chemosphere* 36, 2211-2244.

Ricking M, Svenson A, Beckman EM (2002a) Polyaromatic compounds in sediment samples from the Stockholm area; *J Soils Sed*, 2 (3), 129-136.

Ricking M, Schwarzbauer J, Hellou J, Svenson A, Zitko V (2002b) Polycyclic aromatic musk compounds in Canada and Sweden – first results. *Marine Poll Bull* 46, 410-417.

Ricking M, Schwarzbauer J, Franke S (2003) Molecular markers of anthropogenic activities in sediments of the Havel and Spree rivers (Germany). *Wat Res* 37, 2607-2617.

Ricking M, Heim S, Schwarzbauer J (2004) Geochronology and specific anthropogenic markers in a sediment core from the Teltow Canal in Berlin, Germany. *Organohalogen compounds* 66, 1455-1462.

Riley G, Collier RJ, Jones DM, Eglinton G (1991) The biogeochemistry of Ellemsere Lake U.K. – I: source correlation of leaf wax inputs to the sedimentary lipid record. *Org Geochem* 17, 901-912.

Rimkus G, Brunn H (1996) Synthetische Moschusduftstoffe – Anwendung, Anreicherung in der Umwelt und Toxikologie, Teil 1: Herstellung, Vorkommen in Lebensmitteln, Aufnahme durch den Menschen. *Ernährungs-Umschau* 43, 442-449.

Rimkus GG, Butte W, Geyer H (1997) Critical considerations on the analysis and bioaccumulation of musk xylene and other synthetic nitro musks in fish. *Chemosphere* 35, 1497-1507.

Rimkus G (1999) Polycyclic musk fragrances in the aquatic environment. *Toxicol Lett* 111, 37-56.

Robertson BK, Alexander, M (1998) Sequestration of DDT and Dieldrin in soil: Disappearance of acute toxicity but not the compounds. *Environ Tox Chem* 17, 1034-1038.

Roempp (1995) Römpp's Chemie Lexikon, electronic relaease, Georg Thieme Verlag, Stuttgart

Roose P, Brinkmann UAT (2000) Volatile compounds in various marine organisms from the southern North Sea. *Mar Poll Bull* 12, 1167-1177

Rostad CE (1997) From the 1988 drought to the 1993 flood: Transport of halogenated organic compounds with the Mississippi river suspended sediment at Thebes, Illinois. *Environ Sci Technol* 31, 1308-13012.

Rostad CE, Pereira J (1986) Kovats and Lee retention indices determined by gas chromatography/mass spectrometry for organic compounds of environmental interest. *J High Res Chrom* 9, 328-334.

Rügge K, Bjerg PL, Christensen TH (1995) Distribution of organic compounds from municipal solid waste in the groundwater downgradient of a landfill. *Environ Sci Technol* 29, 1395-1400.

Ruzo L, Jones D, Safe S, Hutzinger O (1976) Metabolism of chlorinated naphthalenes. *J Agric Food Chem* 24, 581-583.

Sacher F, Lange FT, Brauch HJ, Blankenhorn I ( 2001) Pharmaceuticals in groundwaters: Analytical methods and results of a monitoring program in Baden- Württemberg, Germany. *J Chromat A* 938, 199-210.

Sacher F, Lochow E, Brauch HJ (1998) Synthetische organische Komplexbildner - Analytik und Vorkommen in Ober-flächengewässern. *Vom Wasser* 90, 31-41.

Saeger VW, Hicks O, Kaley RG, Michael PR, Mieure JP, Tucker ES (1979) Environmental fate of selected phosphate esters. *Environ Sci Technol* 13, 840-844.

Safe SH, Gaido K (1998) Phytoestrogens and anthropogenic estrogenic compounds. *Environ Tox Chem* 17, 119-126.

Sakkas VA, Konstantinou IK, Albanis TA (2002) Aquatic phototransformation study of the antifouling Sea-Nine 211: identification of byproducts and the reaction pathway by gas chromatography - mass spectrometry. *J Chromat* 959, 215-227.

Saliot A, Tronczynsk, J, Scribe P, Letolle R (1988) The application of isotopic and biogeochemical markers to the study of organic matter in a macrotidal estuary, the Loire, France. *Estuarine Coastal Shelf Sci* 27, 645-699.

Salvito DT, Senna RJ, Federle TW (2002) A framework for prioritising fragrance materials for aquatic risk assessment. *Environmental Toxicology and Chemistry* 21, 1301-1308.

Sanders G, Jones KC, Hamilton-Taylor J (1992) Historical input of polychlorinated biphenyls and other organochlorines to a dated lacustrine sediment core in rural England. *Environ Sci Technol* 26, 1815-1821.

Sanders G, Jones KC, Hamilton-Taylor J (1993) Concentrations and depositon fluxes of polynuclear aromatic hydrocarbons and heavy metals in the dated sediments of a rural english lake. *Environ Toxicol Chem* 12, 1567-1581.

Sano M, Yotsui Y, Abe H, Sasaki S (1976) A new technique for the detection of metabolites labelled by the isotope $^{13}C$ using mass fragmentometry. *Biomed Mass Spectrom* 3, 1-3.

Santos EBH, Duarte AC (1998) The influence of pulp and paper mill effluents on the composition of the humic fraction of aquatic organic matter. *Wat Res* 32, 597-608.

Scheringer M (1999) *Persistenz und Reichweite von Umweltchemikalien,* Wiley-VCH, Weinheim-New York, 203 p.

Schink B, Philipp B, Müller J (2000) Anaerobic degradation of phenolic compounds. *Naturwissenschaften* 87, 12-23.

Schlautmann MA, Morgan JJ (1993) Binding of fluorescent hydrophobic organic probe by dissolved humic substances and organically-coated aluminium oxide surfaces. *Environ Sci Technol* 27, 2523-2532.

Schlör H (1970) Chemie der Fungizide. In Wegler R. (Ed.) *Chemie der Pflanzenschutz- und Schädlingsbekämpfungsmittel* 2, Springer Verlag, Berlin, 45-156.

Schmegel C (1995) Struktur-Wirkungsuntersuchungen in der prospektiven Analyse von Umweltgefährdungen am Beispiel der Substanzklasse der Benzothiazole. PhD Thesis, University of Bremen.

Schmidt S, Hintner JP, Fortnagel P, Franke S, Francke W ISEB 4 (4th International Symposium on Environmental Biotechnology), 10-12. April 2000, Noordwijkerhout, The Netherlands, Poster Abstracts.

Schöberl P, Spilker R (1996) Alkylbenzosulfonat (LAS)-Konzentrationen im Lippe Sediment eines Rhein-Altarms. *Tenside Surf Det* 33, 400-403.

Schuetzle D, Lee FSC, Prater TJ (1981) The identification of polynuclear aromatic hydrocarbon (PAH) derivatives in mutagenic fractions of diesel particulate extracts. *Intern J Environ Anal Chem* 9, 93-144.

Schullerer S, Frimmel F (1993) Characterization of organic sulphur compounds in surface water by ion-pair adsorption under different conditions. *Anal Chim Acta* 283, 251-257.

Schulte E, Malisch R (1983) Berechnung der wahren PCB-Gehalte in Umweltproben. *Fresenius J Anal Chem* 314, 545-551.

Schulten HR, Leinweber P (1996) Characterisation of humic and soil particles by analytical pyrolysis and computer modelling. *J Anal Appl Pyrolysis* 38, 1-53.

Schultz B, Kjeldsen P (1986) Screening of organic matter in leachates from sanitary landfills using gas chromatography combined with mass spectrometry. *Wat Res* 20, 965-970.

Schulz HM, Püttmann W, Riedel F (1991) Analytik von polycyclischen Aromatischen Kohlenwasserstoffen in Schwebstäuben eines Stadtgebietes (Aachen). *Wissenschaft und Umwelt* 3-4, 127-136.

Schulze T, Wetterauer B, Schwarzbauer J, Hollert H, Braunbeck T, Ricking M (2003) DDT und Metaboliten in Sedimenten Berliner Gewässer. *UWSF - Z Umweltchem Ökotox* 15, 71-77.

Schwarzbauer J (1997) PhD-Thesis, Department of Chemistry, University Hamburg, 1-282.

Schwarzbauer J, Littke R, Weigelt V (2000) Identification of specific organic contaminants for estimating the contribution of the Elbe river to the pollution of the German Bight. *Org Geochem* 31, 1713-1731.

Schwarzbauer J, Ricking M, Franke S, Francke W (2001) Halogenated organic contaminants in sediments of the Havel and Spree rivers (Germany). Part 5 of Organic compounds as contaminants of the Elbe river and its tributaries. *Environ Sci Technol* 35, 4015-4025.

Schwarzbauer J, Heim S, Brinker S, Littke R (2002) Occurrence and alteration of organic contaminants in seepage and leakage water from a waste deposit landfill. *Wat Res* 36, 2275-2287

Schwarzbauer J, Ricking M, Littke R (2003a) DDT-related compounds bound to the nonextractable particulate matter in sediments of the Teltow Canal, Germany. *Environ Sci Technol* 37, 488-495.

Schwarzbauer J, Franke S (2003b) Identification and quantitation of dinaphthyl sulfones as contaminants in particulate matter of the Elbe river, Germany.Part VI of organic compounds as contaminants of the Elbe river and its tributaries. *Chemosphere* 51, 973-981.

Schwarzbauer J, Ricking M, Littke R (2003c) Quantitation of nonextractable anthropogenic contaminants released from Teltow Canal sediments after chemical degradation. *Acta Hydrochim Hydrobiol* 31, 469-481.

Schwarzbauer J, Littke R (2004) Quantitation of selected organic contaminants in sediment samples of the German Bight. *J Soils Sed* 4, 177-183.

Schwarzbauer J, Ricking M, Keller R, Gieren B, Littke R (2005a) Anthropogenic organic contaminants incorporated into the non-extractable organic matter of riverine sediments from the Teltow Canal (Berlin). In *Environmental Chemistry*, Lichtfouse E, Schwarzbauer J, Didier R (Eds.), Springer Verlag, Weinheim, 329-352.

Schwarzbauer J, Dsikowitzky L, Heim S, Littke R (2005b) Determination of $^{13}C/^{12}C$-ratios of anthropogenic organic contaminants in river water samples by GC-irmMS. *Intern J Environ Anal Chem* 85, 349-364.

Schwarzbauer J, Heim S (2005c) Lipophilic organic contaminants in the Rhine river (Germany). *Water Res* 39, 4735-4748.

Schwarzenbach RP, Gschwend PM, Imboden DM (1993) *Environmental Organic Chemistry*. John Wiley and Sons, New York, 681 p.

Scrimshaw MD, Bubb JM, Lester JN (1996) Organochlorine contamination of UK Essex coast salt-marsh sediments. *J Coast Res* 12, 246-255.

Seinen W, Lemmen JG, Pieters RH, Berbruggen EM, van den Burg B (1999) AHTN and HHCB show weak estrogenic – but no uterotrophic activity. *Toxicology Lett* 111, 161-168.

Sellström U, Kierkegaard A, De Wit C, Jansson B (1998) Polybrominated Diphenyl Ethers and Hexabromocyclodadecane in Sediment and Fish from a Swedish River. *Environ Tox Chem* 17, 1065-1072.

Senesi N, Testini C, Miano TM (1987) Interaction mechanism between HAs of different origin and nature and electron-donor herbicides – a comparative IR and electron-spin-resonance study. *Org Geochem* 11, 25-30.

Senesi N (1992) Binding mechanisms of pesticides to soil humic substances. *Sci Total Environ* 123/124, 63-76.

Senthilkumar K, Duda CA, Villeneuve DL, Kannan K, Falandysz J, Giesy, JP (1999) Butyltin compounds in sediment and fish from the Polish coast of the Baltic Sea. *Environ Sci Poll Res* 6, 200-206.

Shawky S, Emons H (1998) Distribution Pattern of Organotin Compounds at Different Trophic Levels of Aquatic Ecosystems. *Chemosphere* 3, 523-535

Sheldon LS, Hites RA (1979) Sources and movement of organic chemicals in the Delaware river. *Environ Sci Technol* 13, 574-579.

Sherwood-Lollar B, Slater GF, Ahad J, Sleep B, Spivack J, Brennan M, Mackenzie P (1999) Contrasting carbon isotope fractionation during biodegradation of trichloroethylene and toluene: Implications for intrinsic remediation. *Org Geochem* 30, 813-820.

Simonich SL, Begley WM, Debaere G, Eckhoff WS (2000) Trace analysis of fragrance materials in wastewater and treated wastewater. *Environ Sci Technol* 34, 959-965.

Simonich SL, Federle TW, Eckhoff WS, Rottiers A, Webb S, Sabaliunas D, de Wolf W (2002) Removal of fragrance materials during U.S. and European wastewater treatment. *Environ Sci Technol* 36, 2839-2847.

Skark C, Zullei-Seibert N (1994) Auftreten von Pflanzenschutzmitteln in Grundwässern unter Berücksichtigung diffuser Stoffeinträge von Totalherbiziden. *Vom Wasser* 82, 91-105.

Slater GF, Sherwood-Lollar B, Allen King R, O'Hannesin S (2002) Isotopic fractionation during reductive dechlorination of trichloroethene by zero-valent iron: Influence of surface treatment. *Chemosphere* 49, 587-596.

Somasundaram L, Coats JR, Racke KD, Stahr HM (1990) *Bull Environ Contamin Toxicol* 44, 254-259.

Sonnenberg LB, Johnson JD, Christman, RF (1989) Aquatic Humic Substances: Influence on Fate and Treatment of Pollutants. American Chemical Society, Washington DC, 2-23.

Spies RB, Andresen BD, Rice DW (1987) Benzothiazoles in estuarine sediments as indicators of street runoff. *Nature* 327, 697-699.

Stan HJ, Linkerhäger M (1994) Vorkommen von Clofibrinsäure im aquatischen System - Führt die therapeutische Anwendung zu einer Belastung von Oberflächen-, Grund- und Trinkwasser? *Vom Wasser* 83, 57-68.

Standley LJ, Kaplan LA, Smith D (2000) Molecular Tracers of Organic Matter Sources to Surface Wat Resources. *Environ Sci Technol* 34, 3124-3130.

Stankiewicz BA, van Bergen PF, Smith MB, Carter JF, Briggs DEG, Evershed RP (1998) Comparison of the analytical performance of filament and curie-point pyrolysis devices. *J Anal Appl Pyrolysis* 45, 133-151.

Stark A, Abrajano jr TA, Hellou J, Metcalf-Smith JL (2003) Molecular and isotopic characterization of polycyclic aromatic hydrocarbon distribution and sources at the international segment of the St. Lawrence River. *Org Geochem* 34, 225-237.

Steffen D, Kuballa J, Jantzen E, Hempel M (1995) Zinnorganische Verbindungen in den Kornfraktionen von Gewässersedimenten der Elbe, Weser, Ems und Aller. *Vom Wasser* 39, 200-206.

Steinberg CEW, Sturm A, Kelbel J, Kyu Lee S, Hertjorn N, Freitag D, Kettrup AA (1992) Changes of acute toxicity of organic chemicals to daphnia magna in the presence of dissolved humic material (DHM). *Acta Hydrochim Hydrobiol* 20, 326-332.

Steinberg P, Fischer T, Arand M, Park E, Elmadfa I, Rimkus G, Brunn H, an Dienes HP (1999) Acute hepatotoxicity of polycyclic musk 7-acetyl-1,1,3,4,4,6-hexamethyl-1,2,3,4-tetrahydronaphthaline (AHTN). *Toxicol Lett* 111, 151-160.

Stephanou E (1989) Long-chain n-aldehydes. *Naturwissenschaften* 76, 464-467.

Stupp HD, Püttmann W (2001) Migrationsverhalten von PAK in Grundwasserleitern. Hinweise für die Anwendung von MNA bei sanierugnsbedürftigen Grundwasserverunreinigungen durch PAK. *Altlasten Spektrum* 3/2001.

Stuer-Lauridsen F, Mikkelsen S, Havelund S, Birkved M, Hansen LP (2001) Environmental and Health Assessment of Alternatives to Phthalates and to flexible PVC. *Environ Project* 590, *COWI Consulting Engineers and Planners AS for Danish EPA* 1-274.

Stumpf M, Ternes TA, Haberer K, Baumann W (1996a) Nachweis von natürlichen und synthetischen Östrogenen in Kläranlagen und Fließgewässern. *Vom Wasser* 87, 251-261

Stumpf M, Ternes TA, Haberer K, Seel P, Baumann W (1996b) Nachweis von Arzneimittelrückständen in Kläranlagen und Fließgewässern. *Vom Wasser* 86, 291-303.

Sturchio NC, Clausen JL, Heraty LJ, Huang L, Holt BD, Abrajano, TA (1998) Chlorine isotope investigation on natural attenuation of trichloroethylene in an aerobic aquifer. *Environ Sci Technol* 32, 3037-3042.

Sun YC, Doong RA, Liao PL (2002) Distribution and fate of organochlorine pesticide residues in sediments from the selected rivers in Taiwan. *Chemopshere* 48, 237-246.

Sündermann J, Radach G (1997) Fluxes and budgets of contaminants in the German Bight. *Mar Poll Bull* 34, 395-397.

Suzuki T, Kondo H, Yaguchi K (1998) Estimate the leachability and persistence of pesticides at golf courses from point-source monitoring and model to predict pesticide leaching to groundwater. *Environ Sci Technol* 32, 920-929.

Svenson A, Allard AS, Viktor T, Örn S, Parkkonen J, Olsson PE, Förlin L, Norrgren L (accepted) Estrogenicity of domestic and industrial effluents in Sweden. *Aquatic Ecosystem Management Health.*

Swedish Society for Nature Conservation (2000) Fragrances 2000. http://www.leffingwell.com/rap-bmv-fragrances2000.pdf, 1-49.

Szostek B, Aldstadt JH (1998) Determination of organoarsencials in the environment by solid-phase microextraction - gas chromatography - mass spectrometry. *J Chromat* 807, 253-263.

Takada H, Farrington JW, Bothner MH, Johnson CG, Tripp BW (1994) Transport of sludge-derived organic pollutants in deep-sea sediments at Deep Water Dumpe Site 106. *Environ Sci Technol* 28, 1062-1072.

Takada H, Futoshi S, Bothner MH, Tripp BW, Johnson CG, Farrington JW (1997) Anthropogenic molecular markers: Tools to identify the sources and transport pathways of pollutants. In: *Molecular markers in environmental geochemistry, ACS Symposium Series* 671, Washington DC, pp. 178-195.

Takada H, Eganhouse RP (1998) Molecular markers of anthropogenic waste. In: *Encyclopedia of Environmental Analyses and Remediation*, Meyers RA (Ed.), John Wiley & Sons, 2883-2940.

Takada H, Ishiwatari R (1990) Biodegradation experiments of linear alkylbenzenes (LABs): isomeric composition of $C_{12}$ LABs as an indicator of the degree of LAB degradation in the aquatic environment. *Environ Sci Technol* 24, 86-91.

Takada H, Satoh F, Bothner H, Tripp BW, Johnson CG, Farrington JW (1997) Anthropogenic molecular markers: tools to identify the sources and transport pathways of pollutants. In Eganhouse RP (Ed.) *Molecular markers in Environmental Geochemistry. ACS Symposium* Series 671, Washington D.C., 178-195.

Takizawa Y (2000) Unterstandig minamata disease and strategies to prevent further environmental contamination by methylmercurry. *Wat Sci Technol* 42, 139-146.

Tan GH, Vijayaletchumy K (1994) Determination of organochlorine pesticide residue in river sediments by soxhlet extraction with hexane-acetone. *Pesticide Sci* 40, 121-126.

Tan GH (1995) Residue levels of Phthalate Esters in Water and Sediment samples from the Klang River Basin. *Bull Environ Chem Tox* 54, 171-176.

Ternes TA (1998) Occurrence and Fate of Drugs in German Sewage Treatment Plants and Rivers. *Wat Res* 32, 3245-3260.

Ternes T, Hirsch R (2000) Occurrence and behavior of X-ray contrast media in sewage facilities and the aquatic environment. *Env Sci Technol* 34, 2741-2748.

Ternes T, Bonerz M, Schmidt T (2001) Determination of neutral pharmaceuticals in Wastewater and rivers by liquid chromatography- electrospray tandem mass spectrometry. *J Chromat A* 938, 175- 185.

Terytze K (1995) Low volatile chlorinated hydrocarbons and heavy metals in aquatic sediments in Berlin and environs. *Intern J Environ Anal Chem* 59, 197-211.

Thake DC, Mays DC, Leber AP, Dacre JC (1993) p-Chlorophenyl methyl sulfide, p-chlorophenyl methyl sulfoxide and p-chlorophenyl methyl sulfone. II . Subchronic effects in rodents and rhesus monkeys. *J Americ College Toxicol* 12, 377-395.

Theobald N, Lange W, Gählert W, Renner F (1995) Mass spectrometric investigations of water extracts of the river Elbe for the determination of potential inputs of pollutants into the North Sea. *Fresenius J Anal Chem* 353, 50-56.

Thurman EM, Malcolm RL (1983) Structural study of humic substances--New approaches and methods. I*n:* Christman RF, Gjessing ET (eds.) Aquatic and terrestrial humic materials: Ann Arbor, Michigan., Ann Arbor Science, p. 1-24.

Tissot BP und Welte DH (1984) *Petroleum Formation and Occurence*, Springer, Berlin.

Tomkins BA, Sega GA, Macnaughton SJ (1998) The quantitation of sulfur mustard by-products, sulfur containing herbicides and organophosphonates in soil and concrete. *Anal Lett* 31, 1603-1622.

Tong HY, Karasek FW (1984) Quantification of polycyclic aromatic hydrocarbons in diesel exhaust particulate matter by high-performance liquid chromatography fractionation and high-resolution gas chromatography. *Anal Chem* 56, 2129-2134.

Trapp S, Brüggemann R, Kalbfus W, Frey S (1992) Organische und Anorganische Stoffe im Main. *Wasser Abwasser* 133, 495-504.

Tran K, Yu CC, Zeng EY (1997) Organic pollutants in the coastal environment of San Diego, California. 2. Petrogenic and biogenic sources of aliphatic hydrocarbons. *Environ Toxicol Chem* 16, 189-195.

Ullmann (2001) Ullmann's Encyclopedia of Industrial Chemistry; 6th Edition, 2001 electronic release.

Van Metre PC, Callender E, Fuller CC (1997) Historical trends in organochlorine compounds in river basins identified using sediment cores from reservoirs. *Environ Sci Technol* 31, 2339-2344.

van Urk G, Kerkum F, van Leeuwen CJ (1993) Insects and insecticides in the Lower Rhine. *Wat Res* 27, 205-213.

Vassilaros DL, Kong RC, Later DW, Lee ML (1982) Linear retention index system for polycyclic aromatic compounds. Critical evaluation and additional indices. *J Chrom* 252, 1-20.

Venkatesan MI, de Leon RP, van Geen A, Luoma SN (1999) Chlorinated hydrocarbon pesticides and polychlorinated biphenyls in sediment cores from San Francisco Bay. *Marine Chem* 64, 85-97.

Venkatesan MI, Ruth E, Steinberg S, Kaplan I (1987) Organic chemistry of sediments from the continental margin of Southern New England, USA – Part II. Lipids. *Marine Chem* 21, 267-299.

Verbruggen EMJ, van Loon WMGM, Tonkes M, van Duijn P, Seinen W, Hermens JLM (2000) Biomimic extraction as a tool to identify chemicals with high bioconcentration potential: An illustration by two fragrances in sewage treatment plant effluents and surface waters. *Environ Sci Technol* 33, 801-808.

Verschueren K (1996) *Handbook of Environmental Data on Organic Chemicals.* International Thompson Publishing Company, New York – Albany, 2064 pp.

Vilomet JD, Veron A, Ambrost JP, Moustier S, Bottero JY, Chatelet-Snidaro L (in press) Isotopic tracing of landfill leachates and pollutant lead mobility in soil and groundwater. *Environ Sci Technol* :doi 10.1021/es010245r

Vinzelberg G, Schwarzbauer J, Littke R (2005) Groundwater contamination by chlorinated naphthalenes and related substances caused by activities of a former military base. *Chemosphere* 61, 770-782.

Volkman JK, Revill AT, Murray AP (1997) Applications of biomarkers for identifying sources of natural and pollutant hydrocarbons in aquatic environments. In: *Molecular markers in Environmental Geochemistry, ACS Symp.Series* 671, 110-132.

de Voogt P, Häggberg L (1993) Isolation and determination of methylsulfonated-PCB's in biological samples by sulfuric acid partitioning, adsorption chromatography and GC-ECD. *Chemosphere* 27, 271-278.

Waggot A (1981) Trace organic substances in the River Lee (Great Britain). *Chem Wat Reuse*, Cooper WJ (ed), Vol 2, 55-99. Ann Arbor Sci , Ann arbor, MI.

Wakeham SG, Schaffner C, Giger W (1980) Polycyclic aromatic hydrocarbons in recent lake sediments-I. Compounds having anthropogenic origins. *Geochim Cosmochim Acta* 44, 403-413.

Wang Z, Fingas M, Li K (1994) Fractionation of a light crude oil and identification and quantification of aliphatic, aromatic and biomarker compounds by GC-FID and GC-MS, Part 2. *J Chromat Sci* 32, 367-382.

Wang Z, Fingas M (1995) Use of methyldibenzothiophenes as markers for differentiation and source identification of crude and weathered oils. *Environ Sci Technol* 29, 2842-2849.

Wang Z, Fingas M, Sergy G (1995) Chemical characterization of crude oil residues from an Arctic beach by GC/MS and GC/FID. *Environ Sci Technol* 29, 2622-2631.

Warren N, Allan IJ, Carter JE, House WA, Parker A (2003) Pesticides and other micro-organic contaminants in freshwater sedimentary environments – a review. *Appl Geochem* 18, 159-194.

Watanabe I, Kashimoto T, Tatsukawa R (1986) Hexabromobenzene and its debrominated compounds in river and estuary sediments in Japan. *Bull Environ Contam Toxicol* 36, 778- 784.

Weber WJ, Lebouef EJ, Young TM, Huang W (2001) Contaminant interactions with geosorbent organic matter: insights drawn from polymer sciences. *Wat Res* 35, 853-868.

Weigel S, Kuhlmann J, Hühnerfuss H (2002) Drugs and personal care products as ubiquitous pollutants: occurrence and distribution of clofibric acid, caffeine and DEET in the North Sea. *Sci Total Environ* 295, 131-141.

Wheeler WB, Stratton GD, Twilley RR, Ou LT, Carlson DA, Davidson JM (1979) Trifluralin degradation and binding in soil. *J Agr Food Chem* 27, 702-706.

Wenzel-Hartung R, Brune H, Grimmer G (1990) Evaluation of the carcinogenic potency of four environmental polycyclic aromatic compounds following intrapulmonary application in rats. *Exp Pathol* 40, 221-227.

West WR, Smith PA, Booth GM (1986) Determination and genotoxicity of nitrogen heterocycles in a sediment from the Black river. *Environ Tox Chem* 5, 511-519.

Whiticar MJ (1999) Carbon and hydrogen isotope systematics of bacterial formation and oxidation of methane. Chem Geol 161, 291-314.

Wilken RD, Kuballa J, Jantzen E (1994) Organotins: their analysis and assessment in the Elbe river system, Northern Germany. *Fresenius J Anal Chem* 350, 77-84.

Wilken RD, Wirth H (1986) The adsorption of hexachlorobenzene on naturally occurring adsorbents in water. In: *International Agency for Research on Cancer*, Lyon, 75-81.

Winkels HJ, Kroonenberg SB, Lychagin MY et al (1998) Geochronology of priority pollutants in sedimentation zones of the Volga and Danube delta in comparison with the Rhine delta: *Appl Geochem* 13, 581-591.

Winkler M, Kopf G, Hauptvogel C, Neu T (1998) Fate of artificial musk fragrances associated with suspended particulate matter (SPM) from the river Elbe (Germany) in comparison to other organic contaminants. *Chemosphere* 37, 1139-1156.

Wong KC, Ong CH (1993) Volatile components of the fruits of bachang (Mangifera foetida Lour.) and kuini (Mangifera odorata Griff.). *Flavour Fragrances J* 8, 147-151.

Yamamoto T, Yasuhara A, Shiraishi H, Nakasugi O (2001) Bisphenol A in hazardous waste landfill leachates. *Chemosphere* 42, 415-418.

Yan CT, Jen JF (1992) Determination of volatile fatty acids in landfill leachates by gas chromatography with distillation pretreatment. *Anal Chim Acta* 259, 259-264.

Yanik PJ, O'Donnell TH, Macko SA (2003a) Source apportionment of polychlorinated biphenyls using compound specific isotope analysis. *Org Geochem* 34, 239-251.

Yanik PJ, O'Donnell TH, Macko SA, Qian Y, Kennicutt II MC (2003b) The isotopic compositions of selected crude oil PAHs during biodegradation. *Org Geochem* 34, 291-304.

Yuan SY, Chang JS, Yen JH, Chang BV (2001) Biodegradation of phenanthrene in river sediment. *Chemosphere* 43, 273-278.

Yunker MB, MacDonald RW, Fowler BR, Cretney WJ, Dallimore SR, McLaughlin FA (1991) Geochemistry and fluxes of hydrocarbons to the Beaufort Sea shelf: A multivariate comparison of fluvial inputs and coastal erosion of peat using principal components analysis. *Geochim Cosmochim Acta* 55, 255-273.

Yunker MB, MacDonald RW, Cretney WJ, Fowler BR, McLaughlin FA (1993) Alkane, terpene and polycyclic aromatic hydrocarbon geochemistry of the Mackenzie River and Mackenzie shelf: Riverine contributions to Beaufort Sea coastal sediment. *Geochim Cosmochim* Acta 57, 3041-3061.

Zhang G, Min YS, Min BX, Sheng GY, Fu JM, Wang ZS (1999) Time trend of BHCs and DDTs in a sedimentary core in Macao estuary, Southern China. *Marine Pollut Bull* 39, 326-330.

Zhang X, Young LY (1997) Carboxylation as an initial reaction in the anaerobic metabolism of naphthalene and phenanthrene by suldiogenic consortia. *Appl Environ Microbiol* 63, 4759-4764.

Zhou JL, Fileman TW, House WA, Long JLA, Mantoura RFC, Meharg AA, Osborn D, Wright J (1998) Fluxes of organic containants from the river catchment into, through and out of the Humber Estuary, UK. *Marine Poll Bull* 37, 330-342.

Zegough Y, Derenne S, Largeau C, Saliot A, (1996) Organic matter sources and early diagenetic alterations in Arctic surface sediments of Lena River delta and Laptev Sea, Eastern Siberia., II. Analysis of the carboxylic acids released via sequential treatments. *Org Geochem* 24, 841-857.

Zegough Y, Derenne S, Largeau C, Bardoux G, Mariotti A (1998) Organic matter sources and early diagenetic alterations in Arctic surface sediments , Lena River delta and Laptev Sea, Eastern Siberia., II. Molecular and isotopic studies of hydrocarbons. *Org Geochem* 28, 571-583.

Ziechmann W (1972) Über die Elekronen-Donator und Acceptor-Eigenschaften von Huminstoffen. *Geoderma* 8, 111-113.

Zwiener C, Kumke MU, Abbt-Braun G, Frimmel FH (1999) Absorbed and bound residues in Fulvic Acid fractions of a contaminated groundwater - Isolation, chromatographic and spectroscopic characterization. *Acta Hydrochim Hydrobiol* 27, 208-213.

# Index

Printing: Krips bv, Meppel
Binding: Stürtz, Würzburg